Volume II

# Elements of Digital Satellite Communication

Channel Coding and
Integrated Services
Digital Satellite
Networks

# ELECTRICAL ENGINEERING
# COMMUNICATIONS AND SIGNAL PROCESSING

**Raymond L. Pickholtz, Series Editor**

Anton Meijer and Paul Peeters
*Computer Network Architectures*

Marvin K. Simon, Jim K. Omura, Robert A. Scholtz, and Barry K. Levitt
*Spread Spectrum Communications, Volume I*

Marvin K. Simon, Jim K. Omura, Robert A. Scholtz, and Barry K. Levitt
*Spread Spectrum Communications, Volume II*

Marvin K. Simon, Jim K. Omura, Robert A. Scholtz, and Barry K. Levitt
*Spread Spectrum Communications, Volume III*

William W Wu
*Elements of Digital Satellite Communication: System Alternatives, Analyses, and Optimization, Volume I*

William W Wu
*Elements of Digital Satellite Communication: Channel Coding and Integrated Services Digital Satellite Networks, Volume II*

**Also of interest:**

Victor B. Lawrence, Joseph L. Lo Cicero, and Laurence B. Milstein, editors
*IEEE Communication Society's Tutorials in Modern Communications*

Wushow Chou, Editor-in-Chief
*Journal of Telecommunication Networks*

Volume II

# Elements of Digital Satellite Communication

## Channel Coding and Integrated Services Digital Satellite Networks

William W Wu

*International Telecommunications Satellite Organization (INTELSAT)*

# COMPUTER SCIENCE PRESS

*Computer Science Press*
*1803 Research Boulevard*
*Rockville, Maryland 20850*
1 2 3 4 5 6  Printing                             Year  89 88 87 86 85

**Library of Congress Cataloging in Publication Data**

Wu, William W
  Elements of digital satellite communications.

  Bibliography: v. 1, p.:                                    v. 2, p.,:
  Includes index.
  Contents: v. 1. System alternatives, analyses, and optimization—v. 2.
Channel coding and integrated services digital satellite  networks.
    1. Artificial satellites in telecommunication.
2. Digital communications. I. Title.
TK5104.W8      1985            621.38′0422                      85-5243
ISBN 0-91489-439-0 (v. 1)
ISBN 0-88175-000-X (v. 2)

# CONTENTS

# PREFACE

This is the second volume of the two-volume work *Elements of Digital Satellite Communication*. The chapters of each volume are listed below.

Volume I: SYSTEM ALTERNATIVES, ANALYSES, AND OPTIMIZATION

1. Overview
2. Transmission Link, Channel Characterization, and INTELSAT VI
3. Multiple Access
4. Digital Modems and Receivers
5. Synchronization
6. Satellite Switching and Onboard Processing
7. System Optimization

Volume II: CHANNEL CODING AND INTEGRATED SERVICES DIGITAL SATELLITE NETWORKS

1. Overview
2. Theoretical Foundations
3. Applications of Combinatorial Sets in Satellite Communication
4. Cryptology and Message Security
5. Channel Coding: Performance and Variations
6. Error Codec Implementation
7. Integrated Services Digital Satellite Networks and Protocols

Volume I of *Elements of Digital Satellite Communication* explores system alternatives including the techniques, analyses, and tradeoffs of multiple access, modulation, syncronization, satellite switching, and onboard processing. System modeling is presented in terms of satellite link channels in cascade. System optimization is addressed through a set of sample problems that can be solved not only by linear and nonlinear programming but by dynamic, integer, probabilistic, and/or combinatorial programming methods.

In Volume II we continue to identify, to explore, and to analyze selected subjects that are pertinent to either present or future digital satellite communication. The objectives of Volume II are the same as those of Volume I. The emphases on methodology, unified analyses, and the applications of alien theories that characterize Volume I are carried over to Volume II.

To cover any subject in depth, it must be at the expense of not covering other topics. Each chapter subject in this volume can be a book by itself. In fact, for error-correcting codes and cryptographic methods there are good books already written. What makes this book different is that it is written with satellite applications in mind. As elements of a system, the subjects can no longer be treated separately. In addition, it is not a simple task to ascertain the capabilities and limitations of each scheme when it is integrated into a particular system. However, such a task will be simpler if we know the fundamentals, alternatives, interface problems, and a satellite system environment. Chapters 4, 5, and 6 address this need.

The subjects discussed in Chapters 3, 7, and parts of Chapter 2 have not previously appeared in any text or any other literature. To make the material more useful, particularly Chapter 3, Chapter 2 on theoretical foundations has been provided. Further, in addition to the theoretical derivations, useful source data are included in the appendices.

The level of treatment in this volume is more in depth than in Volume I. This depth is reflected in the mathematical maturity, as in Chapter 3, and in detail, as found in Chapter 7. If the reader is interested only in the mechanics of the procedures without requiring the knowledge of why and how they are formulated, step-by-step algorithms and examples are provided separately. Otherwise, the material in Chapter 2 and other derivations cannot be omitted. One exception is the TDMA network protocol presented in Chapter 7.

Some parts of Chapters 2, 3, 5, and 6 were written and copyrighted (U.S. Library of Congress Registration TXU 28-031) as lecture notes. Originally, Volume I was concerned with the overall system aspect of satellite communication and Volume II provided the details of the subsystem aspect. The end result is almost as planned, with the exception that in Chapter 7 of the present Volume II, the issues of Integrated Services Digital Satellite Networks (ISDSN) turn out to be more extensive than any element presented in Volume I.

It should be apparent that some unique characteristics of a satellite network in general cannot be found in any terrestrial cable network. These characteristics include:

- *Broadcasting Capability*
  With a satellite, a signal from one station can be received simultaneously at all stations in the coverage area of the satellite. Within a satellite coverage area, additional stations can be installed without affecting the satellite system.
- *Distance Capability*
  With a satellite, message transmission is independent of distance and terrain.
- *Processing Capability*
  With a satellite, onboard processing can provide signal enhancement, error rate reduction, efficiency improvement, and network switching flexibility.
- *Transmission Capability*
  Through multiple access techniques multi-destinational high-speed multi-way transmission is not only feasible but effective.
- *Implementation Advantage*
  It costs less and takes less time to launch a satellite, and to build a group of earth stations, than to utilize long-distance cables or fiber optics.

For the above reasons we can see that a mere extension of the terrestrial-based ISDN (Integrated Services Digital Network) experience limits the range and scope of a truly integrated services digital network. Therefore, we define and discuss in Chapter 7 the criteria and issues concerning ISDSN and suggest that ISDN be included for an effective digital communication network including satellites. In pursuit of this objective, we discuss the INTELSAT TDMA network as a possible cornerstone for such an integrated network.

For comments and corrections on this second volume, I wish to thank Drs. Elwyn Berlekamp, Willis Gore, C. L. (David) Liu, and P. Tong for Chapters 2, 3, and 5; Dr. James Massey for Chapter 4; and Dr. Leonard S. Golding and Keiichiro Koga for Chapter 7. However, none of them has seen the final version. Thus any additional error, or omissions, are my responsibility.

In Chapter 4 on cryptology and message security, Sections 4.4.1, 4.4.3, 4.6.1–4.6.5, parts of 4.7 and 4.8 are the work of Dr. J. Massey. I have only added a few illustrations and a number of examples. Therefore, I am most grateful for Dr. Massey's generosity in permitting the use of his lecture notes, a fact which not only renders that chapter outstanding but has served to improve the overall quality of this book.

I have also benefited from the use of the original derivations or the answers of my questions about their previous work, and express my thanks to Drs. R. C. Bose, David Chase, David G. Leeper, James Morakis, Neil J. A. Sloane, and Andrew Viterbi.

The computer program of the INTELSAT TDMA Network has been jointly developed by KDD Laboratories and Mitsubishi Software Company. The competence and dedication of K. Koga, H. Shinonaga, and H. Maruo are reflected in the end results, as described in Chapter 7. I wish to thank for their support J. Dicks, G. Forcina, S. Kahng, J. Jankowski, G. Paine, E. Podraczky, and O. Shimbo from INTELSAT; H. Kaji and T. Muratani from KDD Laboratories.

I wish also to thank L. Blue, B. Brienza, A. Gorelick, S. Kaplan, B. Kogut, R. Liu, and R. Ramminger for their efforts on the second volume.

---

The views and judgments expressed in this book are solely those of the author, and do not necessarily reflect the official policies of INTELSAT, an organization owned and participated in by 109 countries.

# Chapter 1

# OVERVIEW

In the first volume of this two-volume set on elements of digital satellite communication, we discussed essential elements such as satellite transmission links, channel characterization, multiple access schemes with queues, and digital modulations including amplitude, frequency, phase, and their combinations. Some modulation performance analyses included the effects of interferences, nonlinearities, and additive white Gaussian noise. Channel equalization and interference cancellation techniques were presented. As a practical maximum likelihood sequence detector, the results of combining equalization and sequence estimation were illustrated in terms of complexity and performance.

In Volume I we considered almost all the relevant subjects related to frame synchronization in digital satellite communication from theory to practice, and included useful results and guidelines, from the past to the most recent; we provided construction procedures for synchronization sequence generation and gave actual sync sequences with their characteristics for a number of satellite systems. We know the tradeoffs of onboard processing along with onboard switching, computational complexity, and traffic algorithms. Through mathematical programming methods, the last chapter of Volume I demonstrates the usefulness and the power of optimization in a variety of problems within the sphere of satellite communication.

In this second volume we continue to identify, to explore, and to analyze essential elements in digital satellite communication. We begin by presenting in Chapter 2 the theoretical foundations to be used in the later chapters. These theoretical foundations include number systems, congruences, finite field theory, and combinatorial sets. The reason we include this material is not merely for the sake of convenience, but we feel it is absolutely essential to the full appreciation of Chapters 3 through 6. Furthermore, the presentation on the combinatorial Euclidean sets is not available elsewhere.

Building on the results of Chapter 2, Chapter 3 demonstrates the usefulness of these combinatorial sets in satellite communication for a wide variety of areas. These areas include the generation of signature sequences for random multiple access systems with prescribed characteristics, sidelobe control in phase array antennas, control aliasing in signal processing, determination of transponder frequency assignment, collision control in satellite packet transmission, minimization of beam interference, and the generation of error-correcting codes for either error detection, retransmission, or forward error correction. In all cases

1

the sets provide the optimal solutions. Once the reader is familiar with the basic properties of sets, a single set can often serve for a large number of unrelated applications in satellite communication, as we shall demonstrate in this chapter.

The advantage of the multidestination transmission capability of a satellite system can turn into a disadvantage if sensitive information is received by the wrong people. Information transmitted through a satellite is also vulnerable to message tampering and identity falsification. As the nature of the information transmitted through a satellite network becomes increasingly complex, message security is no longer limited to the domain of military, diplomatic, or governmental operations. Chapter 4 relates cryptology to digital satellite communication. The needs, justifications, and examples of cryptosystems using satellites are brought forth. The characteristic distinctions between communication systems with and without encryption/decryption are observed. Cryptology is concerned with the degree of unrecognizability of a message; cryptosystems make transmitted messages unrecognizable to all but the intended receivers. All significant cryptographic algorithms and cryptosystems are discussed in this chapter.

The problems of studying cryptology are also mentioned. These problems include the difficulty of evaluating a cryptosystem in terms of its cryptographic strength, and the unavailability of some of the ''secrets'' on a cryptographic algorithm for nontechnical reasons. Because of the importance of secret sharing systems to satellite communication, we devote one section of Chapter 4 to the first published survey of all the work done in this area, and another section to presenting—also for the first time—a unified construction method for secret sharing systems.

Chapter 5 provides the justifications, variations, classifications, performance measures, and limitations of basic error coding applications for satellite communication channels. Justifications are discussed through tradeoffs of power, bandwidth, and efficiency. Variations are illustrated by the diversity of applicable codes and decoding techniques. Such diversity can be unified into a simple classification. For practical purposes, critical distinctions and applicabilities among the codes are examined. Potentially useful codes and coding techniques are suggested, and their coding gain expressions are provided. Code performance measures are compared on a sample of applicable codes for channels under consideration.

Code and decoding selection criteria are put forward for the satellite transmission environment. A number of implementable codes are listed in the Appendix.

Coding limitations are sized up through performance bounds. New formulations are derived for forward error correction, error detection, and their combination. We also include both analytical and experimental results of soft decision and code concatenations.

Because encryption and decryption devices are in general simpler to implement than error-correcting codecs, we discuss the theoretical foundations of the cryptological algorithms in Chapter 4 without considering implementation. In Chapter 6 we are concerned with the complexity and design of error-correcting codecs, whose performance is described in Chapter 5. The designs for some decoders are illustrated at the element logical level in Chapter 6.

In Chapter 7, we embark on a number of topics that relate to networks, satellites, and protocols. When we consider issues related to integrated digital satellite network services, these issues are no longer elements of satellite communication, but, instead, satellite communication becomes an element of such network services. We are interested in this area because it is important and it is challenging. In this chapter, some criteria have been outlined toward the establishment of such a global service network. The international organization responsible for the development and standardization of such a task is introduced. In some areas its work is criticized, while in other areas its work is highly praised.

A large portion of Chapter 7 is devoted to the INTELSAT TDMA network protocol operation, which includes network architecture, network elements, and detailed protocol procedures. The objectives are twofold: the first is to reveal the inner workings of this first international high-speed commercial digital satellite network and the second is to explore some of its merits, which may be introduced as international standards.

Among its theoretical aspects, Chapter 7 includes the connection between network protocol information and rate-distortion theory, and the equivalence between an open network of series queues (as in a satellite link) and a closed network with feedback queues. Since the main purpose of satellite communication is to transmit and receive messages, or to carry traffic, and as the volume of traffic increases congestion will result, Chapter 7 concludes with a brief discussion on congestion control.

# Chapter 2

# THEORETICAL FOUNDATIONS

## 2.1 FUNDAMENTALS

From a practical standpoint, the material in this chapter could be relegated to an appendix. We choose to include it here, however, because we feel it is essential to the thorough appreciation of the rest of this volume in the following ways:

1. Most other chapters of this volume are based on material in this chapter, which includes the fundamental theory on which different aspects of digital satellite communication are based. A thorough understanding of the basics will not only help the reader to appreciate the methods described in subsequent chapters, but will also encourage the reader to develop and extend these concepts.
2. Some of the material described in this chapter is not available anywhere else. Without a clear exposition of the fundamental subjects, the applications appear limited and weak.
3. Some of the background material in this chapter is too old to be conveniently looked up. Once located, it may not be easily comprehensible.
4. Some simple subjects may be so far removed from the daily experiences of a practical designer that a few definitions and illustrative examples may be necessary to bring back previously well-understood subjects and bring the rest of the chapters of this volume into good harmony and continuity.

Therefore, the materials in this chapter are not merely mathematical exercises, but concepts that apply directly to the different elements of satellite communication to be described.

## 2.2 NUMBER SYSTEMS

Among the fundamental subjects in number systems are the concepts of prime, composite numbers, finite fields, and congruence relations. In this section we shall define and outline the essential number theories as a basis for subsequent applications and discussions.

Number systems concern the classification of numbers. A simple classification, shown in Figure 2.1, begins with complex numbers that consist of real and imaginary numbers. Under the real number system we have rational numbers and integers. Mathematically, 0 is considered to be an integer. Negative integers are, of course, integers. Positive integers are also called natural numbers. Under positive integers we have prime numbers. The purpose of this introduction is to see how prime numbers, which are the concern of this subsection, fit into the number system.

### 2.2.1 Prime Numbers

An integer $p > 1$ is defined as a prime number (or a prime) if it is divisible only by 1 and itself. That is to say, prime numbers refuse to be divided into an integer by any integers other than themselves and 1. For small integers, primes can be obtained by striking out all the composite numbers from the list of integers. For example, let us determine the prime numbers from 1 to 12 as shown on the second row of Table 2.1. The left column of the table indicates the possible nonunity divisors 2 to 12. In the columns, we locate the number that can only be divided by itself and mark it with x. These numbers are 2, 3, 5, 7, and 11 and they are all primes.

It is a difficult task to determine whether an integer is a prime if the integer is a very large number. If an integer is not a prime, it is a composite number, which can be expressed as a product of primes. The standard notation $a|b$ means $a$ divides $b$, or $b$ is divisible by $a$; $(a, b) = 1$ denotes $a$ and $b$ are relatively prime, i.e., they have no common divisor other than 1.

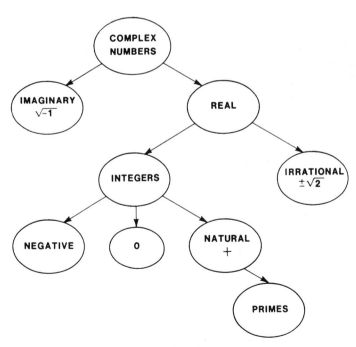

**Figure 2.1** Number systems.

**Table 2.1**
Prime number determination.

| Primes \ Integers \ Divisors | 1 | 2 | 3 | 4 | 5 | 6 | 7 | 8 | 9 | 10 | 11 | 12 |
|---|---|---|---|---|---|---|---|---|---|---|---|---|
| (Primes) | | x | | x | | x | | | | x | | |
| 2 | | 1 | | 2 | | 3 | | 4 | | 5 | | 6 |
| 3 | | | 1 | | | 2 | | | 3 | | | 4 |
| 4 | | | | 1 | | | | 2 | | | | 3 |
| 5 | | | | | 1 | | | | | 2 | | |
| 6 | | | | | | 1 | | | | | | 2 |
| 7 | | | | | | | 1 | | | | | |
| 8 | | | | | | | | 1 | | | | |
| 9 | | | | | | | | | 1 | | | |
| 10 | | | | | | | | | | 1 | | |
| 11 | | | | | | | | | | | 1 | |
| 12 | | | | | | | | | | | | 1 |

## 2.3  CONGRUENCES

A common example of a congruence relation is the analogue clock, which has the numbers from 1 to 12 and the time, although indefinitely continuous, always divided by 12, i.e., modulo 12. For 12-hour counting, the difference of any hours cannot be divided by 12, but for 24-hour counting the difference of certain hours can be divided by 12. This leads to the following definition of congruences: If an integer $m \neq 0$ divides the difference $a - b$, we say that $a$ is congruent to $b$ modulo $m$ and the congruence or congruence relation is written as $a \equiv b$ (mod $m$); otherwise $a \not\equiv b$ (mod $m$). Immediately it is not difficult to verify that if $a, b, c, d, x, y$ are all integers we have from [1, 2]

1.     $a \equiv b$ (mod $m$)

   $b \equiv a$ (mod $m$)

 $a - b \equiv 0$ (mod $m$)

   are equivalent.

2. If $a \equiv b$ (mod $m$) and $b \equiv c$ (mod $m$), then $a \equiv c$ (mod $m$).
3. If $a \equiv b$ (mod $m$) and $c \equiv d$ (mod $m$), then $ax + cy \equiv bx + dy$ (mod $m$).
4. If $a \equiv b$ (mod $m$) and $c \equiv d$ (mod $m$), then $ac \equiv bd$ (mod $m$).
5. If $a \equiv b$ (mod $m$) and $d$ divides $m$, $d > 0$, then $a \equiv b$ (mod $d$).
6. If $x \equiv y$ (mod $m$), then $(x,m) = (y,m)$, where $y$ is called a residue of $x$ modulo $m$.

### 2.3.1 First Order Congruences

The congruence relations discussed so far for integers can be extended to polynomials. Solving congruence relations with polynomials is like solving algebraic equations. Let

$$f(x) = a_0 + a_1x + \ldots + a_{n-1}x^{n-1} + a_nx^n. \tag{2.1}$$

If $a_n \not\equiv 0 \pmod{m}$, then the degree of the congruence $f(x) \equiv 0 \pmod{m}$ is $n$. If $\alpha$ is an integer such that $f(\alpha) \equiv 0 \pmod{m}$, then $\alpha$ is a solution to the congruence relation $f(x) \equiv 0 \pmod{m}$, or $x \equiv \alpha \pmod{m}$ is a solution to $f(x) \equiv 0 \pmod{m}$.

When $n = 1$ in $f(x)$, we have the first order congruences which have the form $(a_0 + a_1x) \equiv 0 \pmod{m}$. If $(a_1, m) = 1$, then the first order congruence has exactly one solution: $x \equiv x_1 \pmod{m}$. We are interested in this first order congruence solution because of the Chinese Remainder Theorem, which will be used in Chapter 4 for the construction of secret sharing systems.

THEOREM 2.1 (*Chinese Remainder Theorem*): Let $m_1, m_2, \ldots, m_r$ denote $r$ positive integers which are relatively prime in pairs, and let $a_1, a_2, \ldots, a_r$ be any $r$ integers. Then the first order congruences $x \equiv a_i \pmod{m_i}$, $i = 1, 2, \ldots, r$, have common solutions. In addition, any two solutions are congruent modulo $(m_1 m_2 \ldots m_r)$.

*Proof.* Let $m = m_1 m_2 \ldots m_r$, because the $m$'s are pairwise prime, $m/m_j$ is relatively prime to $m_j$ for $j = 1, 2, \ldots, r$. There exist integers $b_j$ such that

$$\left(\frac{m}{m_j}\right) b_j \equiv 1 \pmod{m_j} \tag{2.2}$$

and

$$\left(\frac{m}{m_j}\right) b_j \equiv 0 \ (m_i) \tag{2.3}$$

for $i \neq j$. If $i = j$,

$$x_0 = \sum_{j=1}^{r} \frac{m}{m_j} b_j a_j \pmod{m_j}$$

$$= \sum_{\substack{j=1 \\ i \neq j}}^{r} \frac{m}{m_j} b_j a_j \pmod{m_i}$$

$$+ \frac{m}{m_i} b_i a_i \pmod{m_i}$$

$$= a_i \pmod{m_i}. \tag{2.4}$$

Thus $x_0$ is a common solution to the first order congruences, and completes the proof.

### 2.3.2  Higher Order Congruences

When the order $n$ of $f(x)$ in a congruence relation is larger than 1, the congruences are said to be higher order. There is no general technique for solving higher order congruences; the general approach has been to reduce higher order to the first order and then use the Chinese Remainder Theorem to obtain the common solution. A special case of the higher order congruences is of the form $x^n \equiv a$ mod $(m)$, which can be effectively used in public key cryptography.

## 2.4  EULER'S FUNCTION

Euler's $\phi$-function is a number denoted as $\phi(m)$, which is the number of positive integers less than or equal to $m$ that are relatively prime to $m$, i.e., having no common factor with $m$.

*Example 2.1:*  Let $m = 31$, since every integer less than 31 is relatively prime to 31, thus $\phi(31) = 30$.

Calculation of $\phi(m)$ depends on whether $m$ is a prime. If $m$ is factorable, then

$$\phi(m) \simeq m\pi \qquad (2.5)$$

for large prime factors of $m$ and $\pi < 1$. A computer program to calculate prime factors from a given integer $m$ is described in the following example.

*Example 2.2:*  Any integer $m$ greater than 1 can be uniquely represented in the canonical form as

$$m = p_1^{\alpha_1} p_2^{\alpha_2} p_3^{\alpha_3} \cdots p_r^{\alpha_r} \qquad (2.6)$$

where $p_1, p_2, p_3, \ldots, p_r$ are distinct primes, and $\alpha_1, \alpha_2, \alpha_3, \ldots, \alpha_r$ are positive integers. Combining (2.6) and the definition of the prime number, the steps to test an integer $m$ to see if it is a prime number are:

Step 1.  Find a list of prime numbers $p_1, p_2, \ldots, p_k$, where $p_k < m/2 < p_{k+1}$, $p_i > p_j$ if $i > j$.

Step 2.  Test the number $m$ to see if there are any prime factors; i.e., once $m$ can be divided by any $p_i$ $(i = 1, \ldots, k)$, then $m$ is not a prime number. Otherwise, $m$ is a prime number.

Euler's $\phi$-function will be used in Chapter 4 in connection with cryptology. In addition to other applications, the function has been useful in calculating the number of maximum length sequences. For a $k$-stage shift register generator which is capable of generating a binary sequence length of $2^k - 1$ with proper feedback connections, by Zieler's theorem [3] the number of such sequences is $\phi(2^k - 1)/k$.

*Example 2.3:*  For $k = 5$ we have $2^5 - 1 = 31$, from Example 2.1 $\phi(31) = 30$, thus the number of maximum length sequences with $k = 5$ is $\phi(31)/5 = 6$.

Other interesting properties concerning Euler's ϕ-function are listed below:

*Property 1.*   If $m$ and $n$ denote any two positive, relatively prime integers, then $\phi(mn) = \phi(m)\phi(n)$.

*Property 2.*   If $m > 1$, then

$$\phi(m) = m \prod_{p|m}\left(1 - \frac{1}{p}\right) \tag{2.7}$$

for all primes $p$ that divides $m$.

*Example 2.4:*   If $m = p_1^{e_1} p_2^{e_2} \ldots p_r^{e_r}$, then

$$\prod_{p|m}\left(1 - \frac{1}{p}\right) = \prod_{j=1}^{r}\left(1 - \frac{1}{p_i}\right). \tag{2.8}$$

*Property 3* (Euler).   If $p_1, p_2, \ldots, p_n$ are distinct prime factors of $m$, $m \geqslant 2$, and the integers $a$ and $m$ are relatively prime, then

$$a^{\phi(m)} \equiv 1 \pmod{m}. \tag{2.9}$$

*Property 4.*   If $m = p^e$ with $p$ being a prime, and any positive integer $e$, then

$$m = \sum_{d|m}\phi(d). \tag{2.10}$$

To see that Property 4 is true, we extend the summation

$$\sum_{d|m}\phi(d) = \phi(1) + \phi(p) + \phi(p^2) + \ldots + \phi(p^e)$$

$$= 1 + (p - 1) + p(p - 1) + \ldots + p^{e-1}(p - 1)$$

$$= p^e = m.$$

Other verifications can also be obtained from Niven and Zuckerman [1].

## 2.5   FERMAT'S THEOREM

THEOREM 2.2:   If any integer $a$ is not divisible by a prime $p$, then

$$a^{p-1} \equiv 1 \pmod{p} \tag{2.11}$$

or

$$a^p \equiv a \pmod{p}.$$

*Proof.*   If $p$ does not divide $a$, then $p$ and $a$ are relatively prime. Since $\phi(p)$ is the number of positive integers less than or equal to $p$ that are relatively prime to $p$, we have from Property 3 equation (2.9),

$$a^{\phi(p)} \equiv 1 \pmod{p}.$$

Because $p$ is a prime itself, $\phi(p) = p - 1$, the theorem follows immediately. Fermat's Theorem is also used in Chapter 4 for the public key cryptosystems.

## 2.6  IRREDUCIBLE POLYNOMIALS

A polynomial, $f(x)$, is said to be irreducible if it cannot be factored into two or more polynomials of degrees less than $f(x)$ over the same field of consideration. There are several ways of determining irreducible polynomials. A brute-force approach is through a computer search beginning with lowest degree irreducible polynomials.

A systematic method for determining irreducible polynomials is the sieve method, which proceeds, in the binary field, as follows. An irreducible polynomial, $f(x)$, must not be divisible by $x$; hence, we need only consider polynomials of the form $1 + k_1 x + \ldots , k_n x^m$. The polynomial $x + 1$ is irreducible, but no polynomial of degree $m \geq 2$ is irreducible if it is divisible by $x + 1$. Nevertheless $(x + 1)|f(x)$ if and only if $f(1) = 0$, and $f(1) = 0$ if and only if the number of nonzero terms in $f(x)$ is even. Hence, we need only consider polynomials $f(x)$ with an odd number of nonzero terms when the degree of $f(x)$ is larger than 1; i.e., among polynomials of degree 2, one needs only consider $x^2 + x + 1$, which is irreducible.

Among polynomials of degree 3, we need only consider $x^3 + x^2 + 1$, $x^3 + x + 1$. Since the only possible irreducible factors of these polynomials must be $(x + 1)^3$ or $(x + 1) (x^2 + x + 1)$, and since we have already eliminated $(x + 1)$ as a factor, these polynomials are irreducible.

To determine the fourth-degree irreducible polynomials, we consider the set of fourth-degree polynomials with an odd number of nonzero terms. From these we eliminate the polynomial $(x^2 + x + 1)^2$. In this way, we continue by writing all polynomials of degree 5 which are not divisible by $(x + 1)$ and by eliminating from this set the collection of fifth-degree polynomials which are obtainable by multiplying an irreducible polynomial of degree 2 by an irreducible polynomial of degree 3. Although this method is tedious, it can determine all irreducible polynomials.

## 2.7  FINITE FIELD AND ARITHMETIC INVERSES

A finite field $GF(p^m)$ can be generated from an irreducible polynomial of degree $m$. If the coefficients of the irreducible polynomial are binary numbers 0 or 1, then the finite field is $GF(2^m)$, which is convenient for describing algebraic error-correcting codes. Finite fields have been defined and applied to optimal correlation sequence generation in Chapter 5 of Volume I for synchronization.

For the purpose of constructing sets for practical usage in Section 2.11 and 2.12, an understanding of field theory is essential. The following known results are stated without proofs; the proofs may be found in [4].

THEOREM 2.3:   For every prime power $n = p^m$, there exists a finite field $GF(p^m)$, which contains $p^m$ elements. $GF(p^m)$ can be obtained as the residue classes of polynomials modulo an irreducible polynomial of degree $m$ over $GF(p)$.

THEOREM 2.4: The $p^m - 1$ nonzero elements of $GF(p^m)$ form a cyclic group under multiplication. A generator of this multiplication group is called a primitive element of $GF(p^m)$.

For a prime power $p^m$ the field $GF(p^{m(t+1)})$ is called an extension field of $GF(p^m)$. As in Theorem 2.4, $GF(p^{m(t+1)})$ is obtained as the set of residue classes of polynomials modulo an irreducible polynomial of degree $t + 1$ and irreducible over $GF(p^m)$.

THEOREM 2.5: The subfields of $GF(p^{m(t+1)})$ are the fields $GF(p^s)$ where $s$ divides $m(t + 1)$. For each $s$ dividing $m(t + 1)$, $GF(p^{m(t+1)})$ has a unique subfield $GF(p^s)$.

THEOREM 2.6: A primitive element $x$ of $GF(p^{m(t+1)})$ is a root of a polynomial $g(x)$ of degree $m(t + 1)/s$ and is irreducible over $GF(p^s)$.

There are two types of inverses, multiplicative and additive. An additive inverse is simply an integer $x$ which produces 0 when added to a given positive integer $b$; i.e., $b + x = 0$. Multiplicative inverses are defined in an integer domain over a finite field, and they are usually computed with moduli. A multiplicative inverse is an integer $S$ which produces 1 when multiplied by a given integer $t$; i.e., $S \cdot t = S \cdot S^{-1} = 1$. For example, in $2 \times 3 = 1 \pmod 5$, 3 is the multiplicative inverse of 2 under modulo 5 arithmetic. Inverses are commonly used in both channel coding and cryptology.

## 2.8 THE EUCLIDEAN ALGORITHM

The Euclidean algorithm is a well-known procedure in fundamental number theory to find the greatest common divisor (gcd) of any two positive numbers. For those readers who have been away from the fundamentals for too long, we include the subject here for ready reference. The algorithm will be applied in Chapter 4.

Let $a$ and $b$ be two positive numbers and assume that $a \geq b$, then there exist unique integers $q_1$ and $r_1$ such that

$$a = bq_1 + r_1,$$

where $q_1$ is the quotient and $r_1$ $(0 \leq r_1 < a)$ is the remainder following the first step in the division process; i.e., $a$ divided by $b$. If we divide $b$ by $r_1$ we always have

$$b = r_1 q_2 + r_2$$

with $0 \leq r_2 < r_1$. Continuing this process, we have

$$r_1 = r_2 q_3 + r_3, \quad 0 \leq r_3 < r_2$$
$$r_2 = r_3 q_4 + r_4, \quad 0 \leq r_4 < r_3$$

$$\vdots$$

$$r_{k-1} = r_k q_{k+1}, \qquad r_{k+1} = 0.$$

From the last expression where the remainder is zero, $r_{k-1}$ is divisible by $r_k$. If $a$ divides $b$, and $b$ divides $c$, then $a$ divides $c$. This chain reaction traces the divisibility all the way from $r_k$ to $r_1$, from $r_1$ to $b$, from $b$ to $a$, with $r_k < r_{k-1} \ldots < r_2 < r_1$, and all $r$'s are nonnegative integers. Thus $r_k$, the gcd of $a$ and $b$, is obtained by the iterative procedure.

*Example 2.5:*  If we let $a = 21$ and $b = 372$, the gcd is calculated by first reversing the roles of $a$ and $b$ because $a < b$, and then proceeding as

$$372 = 21 \times 17 + 15$$
$$21 = 15 \times 1 + 6$$
$$15 = 6 \times 2 + 3$$
$$6 = 3 \times 2 + 0.$$

Thus the gcd of 21 and 372 is 3.

## 2.9   INFORMATION MEASURES

In Volume I we used freely (with proper references) the channel capacity and the transmission aspects of information theory, which was almost a prerequisite to the subjects under discussion. For the purpose of Chapter 4 in this volume, however, when we discuss Shannon's ideal secrecy system, the following simple relations shall suffice.

A measure of information is expressed as a probabilistic phenomenon. There are two types of information, i.e., self-information and mutual information. Self-information deals with either marginal or conditional probabilistic events of random variables, while mutual information is concerned with the joint events of more than one random variable. If we denote $\square$ as a random variable, and let $\Pr(\square)$ be the probability of the event that $\square$ occurs, then the self-information is defined as

$$I(\square) = \log \frac{1}{\Pr(\square)}, \tag{2.12}$$

and the self-information that corresponds to the conditional probability $\Pr(\square \mid \triangle)$ is

$$I(\square \mid \triangle) = \log \frac{1}{\Pr(\square \mid \triangle)}. \tag{2.13}$$

Similarly, the joint events of $\square$ and $\triangle$, $\square$ and $\triangle \mid \bigcirc$ define the mutual information between these random variables as

$$I(\square, \triangle) = \log \frac{\Pr(\square, \triangle)}{\Pr(\square)} \tag{2.14}$$

$$I(\square, \triangle | \bigcirc) = \log \frac{\Pr(\square | \triangle, \bigcirc)}{\Pr(\square | \bigcirc)} . \tag{2.15}$$

If $\overline{I(\square)}$ represents the average value of the corresponding self-information $I(\square)$, then the term entropy is defined as

$$H(\square) = \overline{I(\square)} \tag{2.16}$$

$$H(\square | \triangle) = \overline{I(\square | \triangle)}. \tag{2.17}$$

However, there is no specific term for the average value of mutual information, but we denote the average mutual information $I(\square, \triangle)$ as $\overline{I(\square, \triangle)}$. By simple probabilistic identity we have the entropy relations

$$\overline{I(\square, \triangle)} = H(\square) - H(\square | \triangle)$$
$$= H(\triangle) - H(\triangle | \square)$$
$$= H(\square) + H(\triangle) - H(\square, \triangle) \tag{2.18}$$

$$H(\square | \triangle) = H(\square, \triangle) - H(\triangle). \tag{2.19}$$

The best reference on information theory is the book by Gallager [5].

## 2.10 COMBINATORIAL SETS

Combinatorics or combinatorial theory is a branch of applied mathematics that deals with the combination, permutation, arrangement, or enumeration of discrete physical objects or situations to achieve the optimal solution. In Chapter 7 of Volume I we discussed combinatorial programming as part of the technique for system optimization, and demonstrated its usefulness for satellite communication. In the next two sections we shall explore in greater depth a specific area of combinatorial theory which is concerned with a collection of elements of a set with specific useful properties. We refer to these classes of sets as combinatorial sets. We are interested in combinatorial sets because of their potential powerful impact on satellite communication system designs. (For readers who are only interested in the applications, these sections may be omitted.) When one tries to use the technique and apply it to a different instance for which an existing set is not available, then one needs to come back to the proper section for the purpose of generating a particular combinatorial set. Combinatorial sets consist of two types of sets, the additive sets and the difference sets. From a practical standpoint we shall only address the latter.

### 2.10.1 Difference Sets

A general $(v, k, \lambda)$ combinatorial difference set

$$\{D\} = \{d_0, d_1, \dots, d_{k-1}\} \tag{2.20}$$

was defined in Chapter 7 of Volume 1 as a collection of $k$ residues (or elements) modulo $v$, such that for any positive integer $\delta \not\equiv 0 \pmod{v}$, which is not necessarily an element of $\{D\}$, the congruence

$$d_i - d_j \equiv \delta \ (\mathrm{mod}\ v) \tag{2.21}$$

has $\lambda$ solution pairs $(d_i, d_j)$ with $d_i$ and $d_j$ in $\{D\}$. $v$, $k$, and $\lambda$ are referred to as the parameters of $\{D\}$. There exists more than one way to construct $\{D\}$ that satisfies (2.21). Except for the sets to be described in Section 2.11, a comprehensive treatment of $\{D\}$ can be found in [6].

A set with the above property can be derived in a number of ways, which we can classify into three basic categories: finite geometries, residue class, and number theory.

The most useful and relevant to satellite communication are the sets derivable from finite geometries, to which we shall devote most of this chapter. The emphasis will be on their derivation and their characteristics.

The sets derivable from the residue classes include quadratic residues, biquadratic residues, octic residues, and sextic residues. Because the structures are too rigid and the number of sets is smaller we are not interested in these sets from a practical standpoint. Among the sets which can be derived from number theory we provide the following summary:

THEOREM 2.7:   Let $\alpha$ be a primitive element of both $p$ and $p + 2$, where $p$ and $p + 2$ are both prime, then the numbers $\alpha^0$, $\alpha$, $\alpha^2$, . . . , $\alpha^{(p^2-3)/2}$; 0, $p + 2$, $2(p + 2)$, . . . , $(p - 1)(p + 2)$ form a difference set with parameters $v = p(p + 2)$, $k = (v - 1)/2$, and $\lambda = (v - 3)/4$.

THEOREM 2.8 (Whiteman):   Let $p$ and $q$ be distinct primes such that $(p - 1, q - 1) = 4$ and let $d = (p - 1)(q - 1)/4$. Let $\alpha$, $\alpha'$ be distinct common primitive roots of $p$, $q$ with $\alpha' \not\equiv \alpha^r \ (\mathrm{mod}\ v)$ for any $r$. Then one (but not both) of the sets

$$\boxed{\alpha^0, \alpha, \alpha^2, \ldots, \alpha^{d-1};\ \ 0, q, 2q, \ldots, (p-1)q}$$

$$\boxed{1, \alpha', \alpha'^2, \ldots, \alpha'^{d-1};\ \ 0, q, 2q, \ldots, (p-1)q}$$

is a difference set with parameters $v = pq$, $k = (v - 1)/4$, $\lambda = (v - 5)/16$ if and only if $q = 3p + 2$ and $k$ is an odd square.

The proofs of the two theorems are lengthy and can be found in the lecture notes by Baumert [6]. We shall simply present Example 2.6 as the result. From the above theorems we also can sense the rigidness of the structure.

*Example 2.6:*   A difference set with parameters $v = 901$, $k = 225$, and $\lambda = 56$ is derived by Baumert from the number theoretical approach. The elements of the set are:

| 0 | 1 | 5 | 9 | 12 | 13 | 14 | 16 | 22 | 25 | 41 | 43 | 45 | 47 | 53 | 59 | 60 | 65 | 69 | 70 |
|---|---|---|---|----|----|----|----|----|----|----|----|----|----|----|----|----|----|----|----|
| 71 | 79 | 80 | 81 | 89 | 92 | 93 | 106 | 108 | 109 | 110 | 114 | 117 | 124 | 125 | 126 | 133 | 139 | 144 | 147 |
| 152 | 156 | 159 | 167 | 168 | 169 | 173 | 174 | 182 | 183 | 192 | 194 | 196 | 198 | 202 | 203 | 205 | 208 | 209 | 212 |
| 214 | 215 | 219 | 222 | 223 | 224 | 225 | 226 | 229 | 231 | 232 | 233 | 235 | 244 | 254 | 256 | 259 | 264 | 265 | 274 |
| 277 | 286 | 292 | 293 | 295 | 296 | 300 | 307 | 308 | 313 | 318 | 319 | 325 | 326 | 345 | 350 | 352 | 355 | 363 | 369 |
| 371 | 379 | 382 | 387 | 394 | 395 | 397 | 400 | 401 | 402 | 405 | 407 | 419 | 422 | 423 | 424 | 433 | 445 | 447 | 460 |
| 461 | 465 | 467 | 469 | 477 | 484 | 492 | 498 | 502 | 503 | 516 | 523 | 526 | 529 | 530 | 531 | 533 | 536 | 540 | 543 |
| 545 | 550 | 559 | 564 | 570 | 571 | 574 | 577 | 579 | 581 | 583 | 585 | 587 | 596 | 599 | 602 | 611 | 617 | 618 | 620 |
| 621 | 622 | 625 | 630 | 634 | 636 | 639 | 641 | 656 | 658 | 661 | 664 | 665 | 688 | 689 | 691 | 694 | 695 | 706 | 708 |
| 711 | 713 | 720 | 721 | 724 | 729 | 735 | 737 | 742 | 746 | 752 | 760 | 766 | 767 | 772 | 778 | 780 | 786 | 795 | 801 |
| 813 | 824 | 826 | 827 | 828 | 835 | 837 | 840 | 843 | 845 | 848 | 849 | 852 | 853 | 859 | 862 | 863 | 865 | 870 | 874 |
| 878 | 881 | 886 | 897 | 898 | | | | | | | | | | | | | | | |

## 2.11   PROJECTIVE GEOMETRY SETS

All the applications cited in Chapter 3 of this volume depend on the existence of a difference set, which is defined in Section 2.10.1. As stated, from a practical viewpoint only two special classes of sets are particularly useful: projective geometry sets and Euclidean geometry sets. These two classes of sets have a common characteristic that can be described in terms of finite geometries. An understanding of the fundamentals is necessary not only to apply them, but also, possibly, to generate new sets to provide more freedom in system design.

In this section we shall provide illustrations, tabulations, and construction algorithms of projective geometry difference sets.

### 2.11.1   Geometries over Finite Fields

In the studies of finite geometry there exist two fundamental geometries, Euclidean and projective [7]. These geometries are represented by their dimension $t$ over a finite field of $n$ elements. $t$ or $t + 1$ is the size of the ordered collection of elements that belong to the field. An ordered set of $t$ or $t + 1$ elements is called a point in the geometry, and any linear combination of two or more points is called a line in the geometry. Euclidean and projective geometries are denoted by $\mathrm{EG}(t, n)$ and $\mathrm{PG}(t, n)$, respectively. The sets derived from $\mathrm{PG}(t, n)$ and $\mathrm{EG}(t, n)$ are denoted as $\{D_p\}$ and $\{D_E\}$.

### 2.11.2   Planar Projective Geometry Sets $\{D_p\}$

Let $X_0, X_1, \ldots, X_t$ be the ordered $t + 1$ elements and $\alpha_0, \alpha_1, \ldots, \alpha_t$ be the $t + 1$ field elements from $\mathrm{GF}(n)$, which has $n - 1$ nonzero elements. A point in $t$-dimension $\mathrm{PG}(t, n)$ is a set of such $t + 1$ elements. Thus there are $n^{t+1} - 1$ nonzero points in $\mathrm{PG}(t, n)$. Because the points $p_0 = (\alpha_0, \alpha_1, \ldots, \alpha_t)$ and $p_0 = (\xi\alpha_0, \xi\alpha_1, \ldots, \xi\alpha_t)$ are the same for every field element $\xi \neq 0$, the total number of points in $\mathrm{PG}(t, n)$ may be divided into groups of $n - 1$ nonzero field elements with the points in the same group representing the same point. Hence $\mathrm{PG}(t, n)$ has $(n^{t+1} - 1)/(n - 1)$ distinct points. For any two distinct points $p_0 = (\alpha_0, \alpha_1, \ldots, \alpha_t), p_1 = (\beta_0, \beta_1, \ldots, \beta_t)$, we can define the line joining these two points as the set of points of the form

$$\lambda_0 p_0 + \lambda_1 P_1 = (\lambda_0 \alpha_0 + \lambda_1 \beta_0, \ldots, \lambda_0 \alpha_t + \lambda_1 \beta_t) \qquad (2.22)$$

with $\lambda_0, \lambda_1$ in GF($n$). The points of a line form a one-dimensional geometry and the line has $(n^{1+1} - 1)/(n - 1) = n + 1$ points.

In PG($t, n$) the points that satisfy a set of $t - r$ linear independent equations

$$\alpha_{i0} X_0 + \alpha_{i1} X_1 + \ldots + \alpha_{it} X_t = 0 \qquad (2.23)$$

for $i = 1, 2, \ldots, t - r$ are said to form an $r$-dimensional subspace (or flat) PG($r, n$). Let $p_0, p_1, \ldots, p_r$ be $r + 1$ linear independent points; i.e.,

$$\xi_0 p_0 + \xi_1 p_1 + \ldots + \xi_r p_r = (0, \ldots, 0). \qquad (2.24)$$

This implies $\xi_0 = \ldots = \xi_r = 0$ for nonzero points in PG($r, n$). Consider all points of the form $\xi_0 p_0 + \xi_1 p_1 + \ldots + \xi_r p_r$. For $r \geq 1$ the subspace contains the line joining any two points in the subspace. Hence, every PG($r, n$) has $(n^{r+1} - 1)/(n - 1)$ points.

The number of PG($r, n$) for $r < t$ can be obtained as follows. Since every PG($r, n$) is determined by $r + 1$ independent points, the first point may be chosen in $(n^{t+1} - 1)/(n - 1)$ ways. With the first point deleted, the second point may be chosen in $[(n^{t+1} - 1)/(n - 1)] - 1$ ways. The third point may be chosen from any point that is not on the line through the previous two points. Since a line contains $(n^2 - 1)/(n - 1)$ points, the third points can be chosen in $[(n^{t+1} - 1)/(n - 1)] - [(n^2 - 1)/(n - 1)]$ ways. Similarly the $r + 1$ point may be chosen in

$$\frac{n^{t+1} - 1}{n - 1} - \frac{n^i - 1}{n - 1} = \frac{n^i(n^{t-i+1} - 1)}{n - 1} \qquad (2.25)$$

ways. Hence, there exist $\prod_i^r n^i \, (n^{t-i+1})/(n - 1)$ ordered sets of $(r + 1)$ independent points in PG($t, n$). By the same argument there are

$$\prod_i^r n^i \, (n^{r-i+1} - 1)/(n - 1)$$

ordered sets of $(r + 1)$ points in each PG($r, n$). The number of $r$-dimensional subspaces in PG($t, n$) is the ratio of these two products. That is,

$$
\begin{aligned}
R(t, r, n) &= \frac{\displaystyle\prod_{i=0}^{r} (n^{t-i+1} - 1)}{\displaystyle\prod_{i=0}^{r} (n^{r-i+1} - 1)} \\[2mm]
&= \frac{(n^{t+1} - 1)(n^t - 1) \ldots (n^{t-r+1} - 1)}{(n^{r+1} - 1)(n^r - 1) \ldots (n - 1)} \\[2mm]
&= \frac{(1 + \ldots + n^t)(n + \ldots + n^t) \ldots (n^r + \ldots + n^t)}{(1 + \ldots + n^r) \ldots (n^{r-1} + n^r)n^r}.
\end{aligned} \qquad (2.26)
$$

The last expression of (2.26) is obtained by converting the ratios into series form, i.e., $(n^{t+1} - 1)/(n - 1) = 1 + n + \ldots + n^t$.

Next, for $s > r$, we want to find the number of PG($s, n$) in PG($t, n$) that contain a PG($r, n$). First we choose a point $p_{r+1}$ not contained in the given PG($r, n$). $p_{r+1}$ may be chosen out of $n^{r+1} + \ldots + n^t$ points. We then choose $p_{r+2}$ out of the $n^{r+2} + \ldots + n^t$ points not contained in PG($r+1, n$), which contains $p_{r+1}$ and the given PG($r, n$). Continuing in this manner we can obtain a PG($s, n$) containing the given PG($r, n$) in $(n^{r+1} + \ldots + n^t) \ldots (n^s + \ldots + n^t)$ ways. Letting $t = s$ implies every PG($s, n$) is obtained in $(n^{r+1} + \ldots + n^s) \ldots (n^{s-1} + n^s)n^s$ ways. Hence

$$
\begin{bmatrix}
\text{The number of different} \\
\text{PG}(s, n) \text{ in PG}(t, n) \text{ which} \\
\text{contain a given PG}(r, n)
\end{bmatrix}
$$

$$
= \frac{(n^{r+1} + \ldots + n^t) \ldots (n^s + \ldots + n^t)}{(n^{r+1} + \ldots + n^s) \ldots (n^{s-1} + n^s)n^s}. \qquad (2.27)
$$

As a consequence of the above discussions we have:

- Every PG($t, n$) contains exactly $1 + n + \ldots + n^t$ points.
- Every PG($t, n$) contains exactly

$$
\frac{(1 + n + \ldots + n^t) \ldots (n^r + \ldots + n^t)}{(1 + n + \ldots + n^r) \ldots (n^{r-1} + n^r)n^r}
$$

  number of PG($r, n$)'s.

- Every PG($r, n$) in PG($t, n$) is contained in

$$
\frac{(n^{r+1} + \ldots + n^t) \ldots (n^s + \ldots + n^t)}{(n^{r+1} + \ldots + n^s) \ldots (n^{s-1} + n^s)n^s}
$$

  PG($s, n$)'s.

- For $r = 0$, every point is contained in

$$
\gamma = \frac{(n + \ldots + n^t) \ldots (n^s + \ldots + n^t)}{(n + \ldots + n^s) \ldots (n^{s-1} + n^s)n^s} \qquad (2.28)
$$

  PG($s, n$) of a PG($t, n$) for $t \geq s > 0$.

- For $r = 1$, every line is contained in

$$
\lambda = \frac{(n^2 + \ldots + n^t) \ldots (n^s + \ldots + n^t)}{(n^2 + \ldots + n^s) \ldots (n^s)} \qquad (2.29)
$$

  PG($s, n$) for $t \geq s > 1$.

Note that every PG($s, n$) contains the whole line joining every pair of points. Thus, every pair of points is contained in $\lambda$ different PG($s, n$). If we identify the number of points as $v$, and the number of subspaces PG($s, n$) as $b$, then the

PG($s, n$) contained in a PG($t, n$) form a balanced incomplete block (BIB) design
with

$$b = \frac{(1 + n + \ldots + n^t) \ldots (n^s + \ldots + n^t)}{(1 + \ldots + n^s) \ldots (n^{s-1} + n^s)n^s} \qquad (2.30)$$

$$v = 1 + n + \ldots + n^t \qquad (2.31)$$

$$k = 1 + n + \ldots + n^s \qquad (2.32)$$

$$\gamma = \frac{(n + \ldots + n^t) \ldots (n^s + \ldots + n^t)}{(n + \ldots + n^s) \ldots (n^{s-1} + n^s)n^s} \qquad (2.33)$$

$$\lambda = \begin{cases} 1, & \text{if } s = 1 \\[2mm] \dfrac{(n^2 + \ldots + n^t) \ldots (n^s + \ldots + n^t)}{(n^2 + \ldots + n^s) \ldots (n^{s-1} + s^n)s^n}, & \text{if } s > 1. \end{cases} \qquad (2.34)$$

*Example 2.7:*   For PG(3,2), let $s = 1$.

$$b = \frac{(1 + 2 + 2^2 + 2^3)(2 + 2^2 + 2^3)}{(1 + 2)2} = 35 \text{ lines}$$

$$v = 1 + 2 + 2^2 + 2^3 = 15 \text{ points}$$

$$k = 1 + n = 3 = \text{Every line of PG(3,2) contains 3 points.}$$

$$\gamma = 7$$

$$\lambda = 1$$

Since $n = 2$, GF(2) contains only 0,1 elements. The points are

$$\begin{array}{llll} p_0 = 1000 & p_5 = 1010 & p_{10} = 1110 \\ p_1 = 0100 & p_6 = 1001 & p_{11} = 1101 \\ p_2 = 0010 & p_7 = 0110 & p_{12} = 1011 \\ p_3 = 0001 & p_8 = 0101 & p_{13} = 0111 \\ p_4 = 1100 & p_9 = 0011 & p_{14} = 1111. \end{array}$$

The lines of PG(3,2) can be obtained by taking pairs of points as

$$\lambda_0 p_0 + \lambda_1 p_1 \text{ for } (\lambda_0, \lambda_1) = (0, 1)$$
$$(1, 0)$$
$$(1, 1).$$

*Example 2.8:*   The points in the $p_0, p_1$ are

$$p_0, p_1, \text{ and } p_0 + p_1 = 1000 + 0100$$
$$= 1100 = p_4 .$$

The lines through $p_0$, $p_4$ and $p_1, p_4$ need not be constructed if the line through
$p_0, p_1$ has already been counted. The 35 lines are

| | | | | |
|---|---|---|---|---|
| $p_0p_1p_4$ | $p_1p_2p_7$ | $p_2p_4p_{10}$ | $p_3p_{10}p_{14}$ | $p_5p_{11}p_{13}$ |
| $p_0p_2p_5$ | $p_1p_3p_8$ | $p_2p_6p_{12}$ | $p_4p_5p_7$ | $p_6p_7p_{14}$ |
| $p_0p_3p_6$ | $p_1p_5p_{10}$ | $p_2p_8p_{13}$ | $p_4p_6p_8$ | $p_6p_{10}p_{13}$ |
| $p_0p_7p_{10}$ | $p_1p_6p_{11}$ | $p_2p_{11}p_{14}$ | $p_4p_9p_{14}$ | $p_7p_8p_9$ |
| $p_0p_8p_{11}$ | $p_1p_9p_{13}$ | $p_3p_4p_{11}$ | $p_4p_{12}p_{13}$ | $p_7p_{11}p_{12}$ |
| $p_0p_9p_{12}$ | $p_1p_{12}p_{14}$ | $p_3p_5p_{12}$ | $p_5p_6p_9$ | $p_8p_{10}p_{12}$ |
| $p_0p_{13}p_{14}$ | $p_2p_3p_9$ | $p_3p_7p_{13}$ | $p_5p_8p_{14}$ | $p_9p_{10}p_{11}.$ |

The lines are formed with the constraints. For example: $p_0p_1p_4 = (1000) + (0100) + (1100) = 0$ as described by Mann [8].

From BIB designs with parameters $b$, $v$, $k$, $\gamma$, $\lambda$, if we let $t = 2$, $s = 1$, then we have designs with parameters $b = v = 1 + n + n^2$, $k = \gamma = 1 + n$, and $\lambda = 1$. This reduces to the well-known result of Singer's Theorem for difference sets [9]. The theorem provides a constructive procedure for obtaining the elements of the sets. As it turns out, this class of sets, denoted $[D_p]$, is most fundamental and useful.

We have demonstrated the origin of planar projective geometry sets and their relationship to projective geometry. For the Singer sets, the parameters are only a function of $n$, the number of elements in a finite field. In general, finite geometries can be represented by a finite field, which, in turn, can be generated from the corresponding irreducible polynomial. These two subjects are discussed in Sections 2.6 and 2.7.

## 2.11.3   The Generation of Planar Projective Geometry Sets

Now that we have defined and described planar projective geometry sets, we shall describe their construction algorithmically in steps. As we shall show, the procedure is much simpler when $n$ is a prime, but not a prime power; that is, $m \equiv 1$. In general, however, we need the following:

Step 1:  Given any nonzero positive integer $m$ and a prime $p$, let $n = p^m$. We identify the set parameters as

$$v = p^{2m} + p^m + 1$$
$$k = p^m + 1$$
$$\lambda = 1.$$

An extension field $GF(n^{t+1})$ can be generated from an irreducible primitive polynomial $f_o(x)$ of degree $m(t + 1)$ over $GF(p)$, the ground field. For planar geometry sets, $t = 2$.

$$f_o(x) = p_{m(t+1)}x^{m(t+1)} + \ldots + p_1x + p_0$$
$$= P_{3m}x^{3m} + \ldots + p_1x + p_0. \tag{2.35}$$

For $p = 2$, the coefficients of $f_o(x)$ are either 0 or 1. Such polynomials are listed in the Appendix of Reference [10].

Step 2: From subfield theory, GF($n$) can be taken as the field over GF($p$) for $n = p^m$, identifying the elements of GF($p$) and its corresponding generator polynomial $f(w)$. Start with a primitive element $\beta$ of GF($n^3$) satisfying the irreducible equation

$$f(\beta) = \beta^3 + w\beta^2 + w\beta + w = 0 \tag{2.36}$$

over GF($n$).

Step 3: Calculate all the powers of $\beta$ from $\beta_0 = 1, \ldots, \beta^{\nu-1}$ modulo $f(w)$ obtained from Step 2, identifying the zero coefficients of the $\beta^2$ term. The powers of $\beta$ corresponding to the zero coefficients are the residues of the sets.

*Example 2.9:* Let $p = m = 2, n = 2^2$.

Step 1: The parameters of $\{D_p\}$ are

$$v = 2^4 + 2^2 + 1 = 21$$
$$k = 2^2 + 1 = 5$$
$$\lambda = 1.$$

An extension field GF($n^{t+1}$) = GF($p^{m(t+1)}$) = GF($2^6$) can be generated by

$$f_o(x) = x^6 + x^5 + x^3 + x^2 + 1,$$

which is irreducible over GF(2).

Step 2: GF($2^6$) can be taken as the field over GF(2). Let $f(w) = w^2 + w + 1 = 0$ with elements 0, 1, $w$, $1 + w$. GF($n^3$) satisfies

$$\beta^3 = w\beta^2 + w\beta + w.$$

Step 3: Calculate the powers of $\beta$ as follows:
Start with

$$\beta^0 = 1$$
$$\beta^1 = \beta$$
$$\beta^2 = \beta^2$$
$$\beta^3 = w\beta^2 + w\beta + w$$
$$\beta^4 = w\beta^3 + w\beta^2 + w\beta$$
$$\quad = w(w\beta^2 + w\beta + w) + w\beta^2 + w\beta$$
$$\quad = w^2\beta^2 + w^2\beta + w^2 + w\beta^2 + w\beta.$$

Using the relation $f(w) = 0$, or $w^2 = 1 + w$, we have

$$\beta^4 = (1 + w) + \beta + \beta^2.$$

All the powers of $\beta$ are listed below:

|          | $\beta^0$ | $\beta^1$ | $\beta^2$ |
|----------|-----------|-----------|-----------|
| $\beta^0$  | 1     | 0     | 0     |
| $\beta^1$  | 0     | 1     | 0     |
| $\beta^2$  | 0     | 0     | 1     |
| $\beta^3$  | $w$   | $w$   | $w$   |
| $\beta^4$  | $1 + w$ | 1   | 1     |
| $\beta^5$  | $w$   | 1     | $1 + w$ |
| $\beta^6$  | 1     | $1 + w$ | 0   |
| $\beta^7$  | 0     | 0     | $1 + w$ |
| $\beta^8$  | 1     | 1     | 0     |
| $\beta^9$  | 0     | 1     | 1     |
| $\beta^{10}$ | $w$ | $w$   | $1 + w$ |
| $\beta^{11}$ | 1   | $1 + w$ | $1 + w$ |
| $\beta^{12}$ | 1   | 0     | $w$   |
| $\beta^{13}$ | $1 + w$ | $w$ | $1 + w$ |
| $\beta^{14}$ | 1   | $w$   | $1 + w$ |
| $\beta^{15}$ | 1   | 0     | $1 + w$ |
| $\beta^{16}$ | 1   | 0     | 1     |
| $\beta^{17}$ | $w$ | $1 + w$ | $w$ |
| $\beta^{18}$ | $1 + w$ | 1   | 0     |
| $\beta^{19}$ | 0   | $1 + w$ | 1     |
| $\beta^{20}$ | $w$ | $w$   | 1     |

For the zero coefficients of the $\beta^2$ terms, the corresponding powers of $\beta$ are 0, 1, 6, 8, and 18. These $k = 5$ integers are the residues of the set $\{D_p\}$. This example, presented in a different context, was taken from [4].

*Example 2.10:*   The $\{D\}$ set construction procedure is considerably simpler if $n$ is a prime (i.e., $m = 1$). In this case we only need to generate $GF(n^3)$ from a primitive polynomial.

With $m \equiv 1$, $p = 2$, $GF(2^3)$ can be generated by

$$f_o(x) = 1 + x + x^3.$$

The parameters of the set are

$$v = 2^2 + 2 + 1 = 7$$
$$k = 2 + 1 = 3$$
$$\lambda = 1.$$

The field table of $f_o(\beta)$ is:

|          | $\beta^0$ | $\beta^1$ | $\beta^2$ |
|----------|-----------|-----------|-----------|
| $\beta^0$ | 1        | 0         | 0         |
| $\beta^1$ | 0        | 1         | 0         |
| $\beta^2$ | 0        | 0         | 1         |
| $\beta^3$ | 1        | 1         | 0         |
| $\beta^4$ | 0        | 1         | 1         |
| $\beta^5$ | 1        | 1         | 1         |
| $\beta^6$ | 1        | 0         | 1.        |

The powers of $\beta$ corresponding to $\beta^2 = 0$ are 0, 1, 3. These are the residues of the set with $k = 3$ integers.

To verify that $\lambda = 1$, the differences can be easily obtained as $4 - 2 \equiv 2 \pmod 7$, $2 - 1 \equiv 1 \pmod 7$, $4 - 1 \equiv 3 \pmod 7$, $2 - 4 \equiv 5 \pmod 7$, $1 - 4 \equiv 4 \pmod 7$, and $1 - 2 \equiv 6 \pmod 7$. The differences of the set of congruence relations occur distinctly only once. That is, $\lambda = 1$. We have already seen this example used for two-level correlation sequence generation in Chapter 5 and for pairwise network designs in Chapter 6 of Volume I. We shall use this example again in the next chapter to demonstrate additional applications in satellite communication.

### 2.11.4  Properties of Planar Projective Geometry Sets

A planar projective geometry difference set $\{D_p\}$ with parameters $v$, $k$, $\lambda$ satisfies

$$k(k - 1) = \lambda(v - 1). \qquad (2.37)$$

This relationship comes from the definition of $\{D\}$ as described in Section 2.10.1; that is, there are $k(k - 1)$ pairs among the $k$ residues of a difference set. This number is exactly the same as the number of pair solutions among the $v - 1$ integers.

In addition to (2.37), the following characteristics are common to all $\{D_p\}$ [6]:

LEMMA 2.1   If $\{D_p\} = \{d_0, d_i, \ldots, d_{k-1}\}$ with parameters $v$, $k$, $\lambda$, then

$$\{D_i\} = \{d_0 + i, \ d_i + i, \ \ldots, \ d_{k-1} + i\} \bmod v \qquad (2.38)$$

with $i = 0, 1, \ldots, v - 1$, is also a difference set with the same parameters as $\{D\}$.

LEMMA 2.2   Given a $\{D_p\}$, then

$$c\{D_p\} = \{cd_0, \ cd_i, \ \ldots, \ cd_{k-1}\} \bmod v$$

is also a difference set with the same parameters as $\{D_p\}$ for any positive integer $c$ relatively prime to $v$.

LEMMA 2.3   Given $\{D_p\}$ of $(v, k, \lambda)$, then $c\{D_p\} + i \pmod v$ is also a difference set, where $c$ is relatively prime to $v$.

*Definition:*   An equivalent difference set is obtained by adding the positive integers $1, 2, \ldots, v$ to its residues of a given difference set modulo $v$. A cyclic difference set contains $v$ equivalent difference sets. Equivalence refers to the properties described in Lemmas 2.1 to 2.3.

THEOREM 2.9:   For a $\{D_p\}$ of parameters $v, k, \lambda$ and the incidence matrix $Q$, constructed from its complete equivalent sets,

$$QQ^T = (k - \lambda)I + \lambda J \tag{2.39}$$

where $I$ is the $v \times v$ identity matrix, $J$ is the $v \times v$ matrix in which all the entries are 1s, $Q^T$ is the transpose of $Q$.

*Proof.*   Let $t_{ij}$ denote the entry in the $i$th row and the $j$th column of $QQ^T$, which is again a $v \times v$ matrix. $t_{ij}$ is the value of the inner product of the $i$th and the $j$th row of $Q$. That is,

$$t_{ij} = \sum_{l=1}^{v} q_{il} q_{jl} \tag{2.40}$$

where $q_{il}$ is the entry in the $i$th row and the $l$th column of $Q$. For $i = j$, $t_{ii}$ is the number of sets that the $i$th symbol is in; that is, $t_{ii} = k$ for $i = 0, 1, \ldots, v - 1$. For $i \neq j$, $t_{ij}$ is the number of sets in which both the $i$th and the $j$th symbol appear; that is, $t_{ij} = \lambda$ for $i = 0, 1, \ldots, v - 1; j = 0, 1, \ldots, v - 1;$ and $i \neq j$. This completes the proof.

THEOREM 2.10:   Between any two equivalent difference sets of a given projective geometry difference set $\{D_p\}$, there are exactly $\lambda$ integers in common.

*Proof.*   The incidence matrix $Q$ of a difference set is a $v \times v$ square matrix, which is nonsingular and whose inverse $Q^{-1}$ exists [11]. In terms of $Q^{-1}$ and the relationships of Theorem 2.9, we have:

$$\begin{aligned} Q^T Q &= (Q^{-1}Q)Q^T Q \\ &= Q^{-1}(QQ^T)Q \\ &= Q^{-1}[(k - \lambda)I + \lambda J]Q \\ &= (k - \lambda)Q^{-1}IQ + \lambda Q^{-1}JQ \\ &= (k - \lambda)I + \lambda Q^{-1}JQ. \end{aligned} \tag{2.41}$$

To simplify (2.41), we need to show that

$$Q^{-1}JQ = J \quad \text{or} \quad JQ = QJ.$$

Because there are $k$ 1s in each row of $Q$, we can write

$$QJ = kJ$$

or

$$\frac{J}{k} = Q^{-1}J. \tag{2.42}$$

Now,

$$QQ^TJ = [(k - \lambda)I + \lambda J]J$$
$$= [(k - \lambda) + \lambda v]J \tag{2.43}$$

because

$$IJ = J \quad \text{and} \quad JJ = vJ.$$

(2.43) can be rewritten as

$$Q^TJ = Q^{-1}[(k - \lambda) + \lambda v]J$$
$$= [(k - \lambda) + \lambda v]Q^{-1}J$$
$$= \frac{1}{k}(k - \lambda + \lambda v)J \tag{2.44}$$

by (2.42). Transposing both sides of (2.44), we obtain

$$JQ = \frac{1}{n}(k - \lambda + \lambda v)J. \tag{2.45}$$

Since $J^T = J$, it follows that

$$JQJ = \frac{1}{n}(k - \lambda + \lambda v)JJ$$

$$= \frac{1}{n}(k - \lambda + \lambda v)vJ. \tag{2.46}$$

Since $J(QJ) = J(kJ) = kJJ = kvJ$, (2.46) becomes

$$kvJ = \frac{1}{k}(k - \lambda + \lambda v)vJ;$$

that is,

$$kv = \frac{1}{k}(k - \lambda + \lambda v)v$$

or

$$k^2 = k - \lambda + \lambda v.$$

Therefore, (2.45) becomes

$$JQ = \frac{1}{k}(k^2 J) = kJ. \tag{2.47}$$

Combining (2.42) and (2.47), we have

$$QJ = JQ. \tag{2.48}$$

Thus, (2.41) becomes

$$Q^T Q = (k - \lambda)I + \lambda J$$

$$= \begin{vmatrix} k & \lambda & \lambda & \cdot & \cdot & \cdot & \lambda \\ \lambda & k & \lambda & \cdot & \cdot & \cdot & \lambda \\ \cdot & \cdot & \cdot & & & & \cdot \\ \cdot & \cdot & & \cdot & & & \cdot \\ \cdot & \cdot & & & \cdot & & \cdot \\ \lambda & \lambda & \lambda & \cdot & \cdot & \cdot & k \end{vmatrix}. \tag{2.49}$$

This means that the inner product of every two columns (or rows) in $Q$ is equal to $\lambda$, and the theorem follows immediately.

*Example 2.11:*   From Example 2.10, the residues of the $\{D\}$ with $v = 7$, $k = 3$, and $\lambda = 1$, are $\{0, 1, 3\}$. By Lemma 2.1, the equivalent difference sets are $\{1, 2, 4\}$, $\{2, 3, 5\}$, $\{3, 4, 6\}$, $\{0, 4, 5\}$, $\{1, 5, 6\}$, and $\{0, 2, 6\}$, from which the incidence matrix is

$$Q = \begin{bmatrix} 1 & 1 & 0 & 1 & 0 & 0 & 0 \\ 0 & 1 & 1 & 0 & 1 & 0 & 0 \\ 0 & 0 & 1 & 1 & 0 & 1 & 0 \\ 0 & 0 & 0 & 1 & 1 & 0 & 1 \\ 1 & 0 & 0 & 0 & 1 & 1 & 0 \\ 0 & 1 & 0 & 0 & 0 & 1 & 1 \\ 1 & 0 & 1 & 0 & 0 & 0 & 1 \end{bmatrix}$$

$$QQ^T = \begin{bmatrix} 3 & 1 & 1 & 1 & 1 & 1 & 1 \\ 1 & 3 & 1 & 1 & 1 & 1 & 1 \\ 1 & 1 & 3 & 1 & 1 & 1 & 1 \\ 1 & 1 & 1 & 3 & 1 & 1 & 1 \\ 1 & 1 & 1 & 1 & 3 & 1 & 1 \\ 1 & 1 & 1 & 1 & 1 & 3 & 1 \\ 1 & 1 & 1 & 1 & 1 & 1 & 3 \end{bmatrix},$$

which shows that $k = 3$ and $\lambda = 1$.

## 2.12   EUCLIDEAN GEOMETRY SETS

In our discussion of projective geometry we have demonstrated its unique representation by finite fields, which also form the foundation for sets derivable from Euclidean geometries. For practical Euclidean geometry sets $\{D_E\}$, we will be interested only in the geometries EG($t, n$) with $t = 2$, $F_1 = \text{GF}(p^m\ n)$, and $F_2 = \text{GF}(p^{2m} = n^2)$. In practice, very few $\{D_P\}$ are available, so we shall call on $\{D_E\}$ to augment the number of usable difference sets. An additional advantage of $\{D_E\}$ is the set parameters, which may in practice be constrained and are not available from $\{D_P\}$. However, one of the prime reasons for our interest in $\{D_E\}$ is its capability of producing random multiple access sequences, which cannot be derived from $\{D_p\}$. In the discussion to follow, the parameters of $\{D_E\}$ are denoted as $v$, $n$, and $\lambda$.

A primitive element $\sigma_1$ of $F_1$ is a root of a primitive polynomial $g(x)$, of degree $b = m/s$, irreducible over $\text{GF}(p^s)$ for any integer $s$ that divides $m$. Further, every element of $F_1$ can be written as a polynomial of degree $b - 1$ in $\sigma_1$ over $\text{GF}(p^s)$.

Let $\sigma_2$ be a primitive element of $F_2$. Thus, $\sigma_2$ does not belong to $F_1$. The nonzero elements of $F_2$ are

$$\sigma_2^0 = 1, \sigma_2^1, \sigma_2^2, \ldots, \sigma_2^{p^{2m}-2}. \tag{2.50}$$

If $p^m = n$, then

$$p^{2m} - 2 = n^2 - 2 = (n - 1)(n + 1) - 1 \tag{2.51}$$

which will be useful later.

Because both $\sigma_1$ and $\sigma_2$ are primitive elements,

$$\sigma_1^{p^m-1} = 1, \sigma_2^{p^{2m}-1} = 1 \tag{2.52}$$

or

$$\sigma_1^{p^m-1} = \sigma_2^{p^{2m}-1} \tag{2.53}$$

$$\sigma_1^{n-1} = \sigma_2^{n^2-1}$$

$$(\sigma_1 - \sigma_2^{n+1})^{n-1} = 0$$

$$\sigma_1 = \sigma_2^{p^m+1} = \sigma_2^{n+1} \tag{2.54}$$

where $\sigma_1$ belongs to $F_1$ and $F_2$ and $\sigma_2 = \sigma$ belongs to $F_2$ only. (2.54) can be expanded as:

$$
\begin{array}{cccc}
\sigma^0 & \sigma^1 & \ldots, & \sigma^n \\
\sigma^{n+1} & \sigma^{n+2} & \ldots, & \sigma^{2n+1} \\
\sigma^{2(n+1)} & \sigma^{2(n+1)+1} & \ldots, & \sigma^{3n+2} \\
\sigma^{3(n+1)} & \sigma^{3(n+1)+1} & \ldots, & \sigma^{4n+3} \\
\cdot & \cdot & \cdot & \cdot \\
\cdot & \cdot & \cdot & \cdot \\
\cdot & \cdot & \cdot & \cdot \\
\sigma^{(n-2)(n+1)}, & \sigma^{(n-2)(n+1)+1}, & \ldots, & \sigma^{(n-2)(n+1)+(n+1)}.
\end{array} \tag{2.55}
$$

(2.55) displays all the nonzero elements of $F_2$. Among these $p^{2m} - 1 = n^2 - 1$ elements, those with exponents divisible by $n + 1$ are the elements belonging to $F_1$; this follows from (2.54). These elements of $F_1$ correspond to the elements in the 0th column in (2.55) as indicated by the dotted lines. Since there are $p^m - 1$ nonzero elements in $F_1$, these are all of the nonzero elements of $F_1$. Thus, $\sigma_1^0 = \sigma_2^0 = 1$, $\sigma_1 = \sigma_2^{n+1}$, $\sigma_1^2 = \sigma_2^{2(n+1)}$, . . . , $\sigma_1^{n-2} = \sigma_2^{(n-2)(n+1)}$ are the elements of $F_1$ expressed in terms of the elements of $F_2$.

Since $F_2$ is the second extension of $F_1$ the element $\sigma_2$ must be the root of an irreducible quadratic equation whose coefficients belong to $F_1$. Such an equation is of the form

$$\sigma_2^2 = E_0\sigma_2 + E_1 \tag{2.56}$$

where $E_0$, $E_1$ are elements of $F_1$. Since every element in the $\varepsilon$th extension of GF($q$) can be written as a polynomial of degree $\varepsilon - 1$ over GF($q$), we have, 1 over GF($q$), we have, for any element $\rho$ of $F_2$

$$\rho = x_0\sigma_2 + x_1 \tag{2.57}$$

where $x_0$, $x_1$ are elements in $F_1$.

## 2.12.1   Euclidean Geometries

In Euclidean space EG($t$, $n$), each $t$-tuple $(x_1, x_2, \ldots, x_t)$ is called a point, where the elements $x_1, x_2, \ldots, x_t$ belong to GF($n$). Two points, $\xi_b = (b_1, b_2, \ldots, b_t)$ and $\xi_c = (c_1, c_2, \ldots, c_t)$, represent the same point if and only if $b_i = c_i$, $i = 1, 2, \ldots, t$.

If $\xi_0, \ldots, \xi_r$ are $r + 1$ linearly independent points in EG($t$, $n$), or if $\xi_0 = 0$ and $\xi_1, \ldots, \xi_r$ are $r$ linearly independent points, then all linear combinations of these points can be expressed as

$$P = \xi_1 + a_1(\xi_1 - \xi_0) + \ldots + a_r(\xi_r - \xi_0), \tag{2.58}$$

which defines a subspace in EG($t$, $n$) called an $r$-flat that passes through the points $\xi_0, \xi_1, \ldots, \xi_r$. ($a_1, a_2, \ldots a_r$ are elements of GF($n$).) Let $P_i = \xi_i - \xi_0$ for $i = 1, 2, \ldots, r$, and $P_0 = \xi_0$; then (2.58) can be rewritten as

$$P = P_0 + a_1P_1 + a_2P_2 \ldots + a_rP_r. \tag{2.59}$$

From Theorem 2.3, the $n = p^m$ elements of GF($n$) can be represented by the set of $m$-tuples corresponding to the residue class of polynomials over GF($p$) modulo an irreducible polynomial of degree $m$. Such a set of $m$-tuples can be represented by the powers of a primitive element of GF($n$). When $t = m$, the set of $m$-tuples are the points in EG($t$, $n$). If $\sigma$ is a primitive element of GF($n^t$), then the points of EG($t$, $n$) may be represented by the first $v = n^t - 1$ powers of $\sigma$ and 0, i.e., by

$$0, \sigma^0, \sigma^1, \sigma^2, \ldots, \sigma^{v-1}.$$

Let $P(\sigma^i)$ denote the point corresponding to $\sigma^i$ and let $P(0)$ denote the element 0. For arbitrary $k$ we have

$$\sigma^k = b_t\, \sigma^{t-1} + b_{t-1}\, \sigma^{t-2} + \ldots + b_1\sigma^0, \tag{2.60}$$

which provides the correspondence

$$\sigma^k \leftrightarrow (b_1, b_2, \ldots, b_t).$$

For construction of Euclidean geometry difference sets in the next section, 1-flats or lines are of particular interest. From now on we will be concerned only with geometries for which $t = 2$ and $n = p^m$; with $p$ being a prime. In EG(2, $p^m$) and $r = 1$, (2.58) and (2.59) then reduce to

$$P = \xi_0 + a_1(\xi_1 - \xi_0)$$

$$= P_0 + a_1 P_1, \tag{2.61}$$

where $a_1$ is an element of $F_1$ and $P_0$ and $P_1$ are elements of $F_2$. For example, the $p^m$ points in EG(2, $p^m$) on the line joining the points $P_0 = P(0)$ and $P_1 = P(1)$ can be written as

$$P = P_0 + a_1 P_1 = P(0) + a_1 P(1). \tag{2.62}$$

Since $a_1$ can be represented by the powers of a primitive element $\sigma_1$ of $F_1$, we thus represent the points on a line joining the points $P(0)$, $P(1)$ as

$$P(0), P(\sigma_1^0 = 1), P(\sigma_1), P(\sigma_1^2), \ldots, P(\sigma_1^{n-2}). \tag{2.63}$$

From (2.54), $\sigma_1 = \sigma_2^{n+1}$ and, letting $\sigma_2 = \sigma$ from now on, then (2.63) becomes

$$P(0), P(1), P(\sigma^{n+1}), P(\sigma^{2(n+1)}), \ldots, P(\sigma^{(n-2)(n+1)}). \tag{2.64}$$

EG(2, $p^m$) will be referred to as the finite Euclidean plane (FEP). A point in FEP is denoted by the 2-tuple $(x_0, x_1)$ where the coordinates $x_0$, $x_1$ are elements of $F_1$. If $(x_0, x_1)$, $(y_0, y_1)$ are two different points in FEP, then the points on the line that contains the points $(x_0, x_1)$ and $(y_0, y_1)$ are

$$(z_0, z_1) = (x_0, x_1) + a_1(y_0 - x_0, y_1 - x_1) \tag{2.65}$$

with $z_0 = x_0 + a_1(y_0 - x_0)$ and $z_1 = x_1 + a_1(y_1 - x_1)$.

### 2.12.2  Linear Transformations of Vector Spaces over Finite Fields

Let the vector space of GF($n^t$) over GF($n$) be denoted as $V_t(n)$, and let a $k$-dimensional subspace of $V_t(n)$ be denoted as $V_t^k(n)$. The number of ways of choosing a basis for $V_t(n)$ is

$$(n^t - 1)(n^t - n)(n^t - n^2) \ldots (n^t - n^{t-1})$$

$$= n^{[(t-1)t]/2} \prod_{i=1}^{t} (n^i - 1), \tag{2.66}$$

where the basis is ordered and there are $n^t - 1$ ways of choosing the first basis element, $n^t - n$ ways of choosing the second, since all linear combinations of the first are deleted, and so forth.

To find the number of distinct subspaces $V_t^k(n)$, we note that there are $(n^t - 1)(n^t - n) \ldots (n^t - n^{k-1})$ ways of choosing $k$ linearly independent vectors.

Each such set of vectors generates a $k$-dimensional subspace $V_t^k(n)$ and each subspace can be generated in $(n^k - 1)(n^k - n) \ldots (n^k - n^{k-1})$ ways. If we denote the number of distinct $V_t^k(n)$ by $\begin{bmatrix} t \\ k \end{bmatrix}$, then

$$\begin{bmatrix} t \\ k \end{bmatrix} = \frac{(n^t - 1)(n^t - n) \ldots (n^t - n^{k-1})}{(n^k - 1)(n^k - n) \ldots (n^k - n^{k-1})}$$

$$= \prod_{i=0}^{k-1} \frac{(n^{t-i} - 1)}{(n^{k-i} - 1)}. \tag{2.67}$$

The quantities $\begin{bmatrix} t \\ k \end{bmatrix}$ are called the Gaussian coefficients and satisfy the recursion relations

$$\begin{bmatrix} t \\ k \end{bmatrix} = \begin{bmatrix} t-1 \\ k-1 \end{bmatrix} + n^k \begin{bmatrix} t-1 \\ k \end{bmatrix} \tag{2.68}$$

and

$$\begin{bmatrix} t \\ k \end{bmatrix} = \frac{n^t - 1}{n^k - 1} \begin{bmatrix} t-1 \\ k-1 \end{bmatrix}. \tag{2.69}$$

When $n = 1$, the Gaussian coefficients degenerate to the well-known binomial coefficients, and we have the familiar combinatorial identities

$$\binom{t}{k} = \binom{t-1}{k-1} + \binom{t-1}{k} \tag{2.70}$$

and

$$\binom{t}{k} = \frac{t}{k} \binom{t-1}{k-1}. \tag{2.71}$$

A concise description of linear transformations of vector spaces over finite fields can be found in [12].

Let $T$ be a $t \times t$ nonsingular matrix over GF($n$), and let $x$, $y$, and $\sigma$ be vectors of $t$ components over GF($n$). Then the set of all transformations of the form

$$y = Tx + \sigma \tag{2.72}$$

forms a group, where $\sigma$ is a primitive root in some extension field of GF($n$). The number of nonsingular $t \times t$ matrices over GF($n$) is the number of ways that a basis of $V_t(n)$ can be mapped into another basis. This is simply the number of bases of $V_t(n)$, which is (2.66).

For the special case of $t = 2$ and $n = p^m$, a linear transformation $T$ can be defined by a $2 \times 2$ nonsingular matrix with elements $E_{11}$, $E_{12}$, $E_{21}$, $E_{22}$ over $F_1$ as

$$T = \begin{bmatrix} E_{11} & E_{12} \\ E_{21} & E_{22} \end{bmatrix}, \tag{2.73}$$

and for any vector $V$ in the vector space of $F_2$ over $F_1$

$$V' = T(V) \tag{2.74}$$

where $V'$ is the linear transformation of $V$. The elements in $T$ are elements of $F_1$.

Since a point $(x_0, x_1)$ in FEP is simply a vector with two components, we then have

$$P' = (x'_0, x'_1) = T(x_0, x_1) = T(P). \tag{2.75}$$

This is a one-to-one transformation sending points into points, with $x'_0$ and $x'_1$ being the transforms of $x_0$ and $x_1$.

If $\xi_0$ and $\xi_1$ are two distinct points in FEP, then a line through these two points as shown in (2.61) is

$$L(\xi_0, \xi_1) = \{\xi_1 a_1 + (1 - a_1)\, \xi_0 | a_1 \varepsilon F_1\}$$

$$= P. \tag{2.76}$$

If $P \varepsilon L(P_0, P_1)$, then the linear transformation of the line vector is

$$TL(P_0, P_1) = \{T(P)|P\varepsilon L(P_0, P_1)\}. \tag{2.77}$$

Since                         $$T(P) = T(\alpha P_0 + \beta P_1), \tag{2.78}$$

$$TL(P_0, P_1) = \alpha T(P_0) + \beta T(P_1)$$

$$= L\,[T(P_0),\, T(P_1)]$$

$$= L(P'_0, P'_1). \tag{2.79}$$

Thus $T$ transforms lines to lines in FEP.

THEOREM 2.11: If $\rho \in F_2$, and for some $k \in \{0, 1, \ldots, n^2 - 2\}$, $\rho = \sigma^k(\sigma = \sigma_2)$, there exists a unique one-to-one transformation

$$\rho = \sigma^k = \alpha_k \sigma + \beta_k \tag{2.80}$$

and every element pair in $F_2$ is of this form. With $\alpha_k, \beta_k \in F_1$.

*Proof:* By induction for $k = 1$, this implies $\alpha_1 = 1$, $\beta_1 = 0$ in (2.80). For $k + 1$ and letting $\sigma = \sigma_2$ in (2.56), then

$$\sigma^{k+1} = \sigma \cdot \sigma^k = \alpha_k \sigma^2 + \beta_k \sigma$$

$$= \alpha_k(E_0 \sigma + E_1) + \beta_k \sigma$$

$$= (\alpha_k E_0 + \beta_k)\sigma + \alpha_k E_1$$

$$= \alpha'_k \sigma + \beta'_k, \tag{2.81}$$

which provides the point transformation of the point $(\alpha_k, \beta_k)$; i.e.,

$$(\alpha'_k, \beta'_k) = (\alpha_{k+1}, \beta_{k+1}) = T(\alpha_k, \beta_k) \tag{2.82}$$

where

$$T = \begin{bmatrix} E_0 & 0_1 \\ E_1 & 1_1 \end{bmatrix}. \tag{2.83}$$

$0_1$, $1_1$ represent the additive and multiplicative identities of $F_1$. The uniqueness is illustrated as follows. From (2.57),

$$\rho = x_0 \sigma_2 + x_1 = 0_2$$

where $0_2$ is the additive identity of $F_2$. If $x_0 = 0_1$, then we must have $x_1 = 0_1$; if $x_0 \neq 0_1$, then $\sigma_2 = -(x_1/x_0) \in F_1$, which is an impossibility. Thus

$$x_0' \sigma_2 + x_1' = x_0'' \sigma_2 + x_1'',$$

so we have

$$x_0' = x_0'', \ x_1' = x_1''. \qquad \text{Q.E.D.}$$

From the above discussions, $T$ transforms the points, except the origin $P(0)$, according to the cycle

$$P(\sigma^0) \to P(\sigma^1) \to P(\sigma^2) \to \ldots \to P(\sigma^{p^{2m}-2}) \tag{2.84}$$
$$= P(\sigma^{(n-1)(n+1)-1}) \to P(\sigma^{p^{2m}-1}) = P(\sigma^0).$$

We have

$$T\sigma^t = \sigma^{t+1}$$
$$T^i \sigma^t = \sigma^{t+i} \tag{2.85}$$

and

$$T^{p^{2m}-1} = T^{(n-1)(n+1)} \tag{2.86}$$

being the identity transformation. The origin $P(0)$ remains unchanged by the transformation.

Let $T$ act on the sets of points containing $P(0)$. Then $T$ will transform these sets of points into sets of points containing $P(0)$.

From (2.54), a line through the origin and $\sigma^0 = 1$ is expressed in terms of the points on this line as $(0)$, $(\sigma^0)$, $(\sigma^{n+1})$, . . . , $(\sigma^{(n-2)(n+1)})$. Applying the successive transformations $T^0$, $T^1$, $T^2$, . . . , $T^n$ to the points on this line, we obtain $n+1$ lines through the origin, as in (2.87). We omit the notation $P$ in front and represent the point by $(\cdot)$ in (2.87).

$$\ell_0 = (0), (\sigma^0), (\sigma^{n+1}), (\sigma^{2(n+1)}) \ldots ,(\sigma^{(n-2)(n+1)})$$
$$\ell_1 = (0), (\sigma^1), (\sigma^{n+2}), (\sigma^{2(n+1)+1}), \ldots ,(\sigma^{(n-2)(n+1)+1}):$$
$$\vdots$$
$$\ell_n = (0), (\sigma^n), (\sigma^{2(n+1)-1}),$$
$$(\sigma^{3(n+1)-1}), \ldots , (\sigma^{(n-1)(n+1)-1}). \tag{2.87}$$

Since the elements of $F_2$ are all distinct, the lines of (2.87) are distinct.

The line joining the points $P(\sigma^0)$, and $P(\sigma^1)$ does not pass through the origin, for, if it did, it would be coincident with the first line of (2.87). This is impossible, since $P(\sigma^1)$ is not a point on that line.

Let the points on the line joining $P(\sigma^0)$ and $P(\sigma^1)$ be expressed as elements of $F_2$. The elements of $F_2$ are again expressed in terms of the powers of a primitive element $\sigma$. If we let the powers of $\sigma$ be denoted as $d_0, d_1, \ldots, d_{n-1}$, then the points on the line joining $P(\sigma^0)$ and $P(\sigma^1)$ are

$$P(\sigma^{d_0}), P(\sigma^{d_1}), \ldots, P(\sigma^{d_{n-1}}) \tag{2.88}$$

where $d_0 = 0$ and $d_1 = 1$, to correspond to the points $P(\sigma^0)$ and $P(\sigma^1)$.

If $v$ is the smallest integer for which $T^v(P_i) = P_i$, we call $v$ the period of $T$ with respect to the point $P_i$. Since the period of $T$ with respect to a point $P_i \neq 0$ is $p^{2m} - 1 = n^2 - 1$, after applying the transformation $T^0, T^1, T^2, \ldots T^{n^2-1}$ in succession to (2.88), we get $n^2 - 1$ lines as (again with $P$ omitted)

$$L_0 = (\sigma^{d_0}), (\sigma^{d_1}), \ldots, (\sigma^{d_{n-1}})$$
$$L_1 = (\sigma^{d_0+1}), (\sigma^{d_1+1}), \ldots, (\sigma^{d_{n-1}+1})$$
$$\vdots \quad \vdots \qquad\qquad \vdots$$
$$L_{n^2-2} = (\sigma^{d_0+n^2-2}), \ldots, \sigma(\sigma^{d_{n-1}+n^2-2}). \tag{2.89}$$

In (2.89), the exponents of $\sigma$ can be reduced $\mod(n^2 - 1)$, and it is easily verified that none of the lines goes through the origin.

We now investigate the total number of lines in a FEP. From (2.83), for $T$ to be nonsingular, $E_0$ cannot be $0_1$, thus $E_0$ can assume one of $n - 1$ values and $E_1$, which does not have this restriction, can assume one of $n = p^m$ values. There can be $n(n - 1)$ pairs of $E_0$ and $E_1$ such that $T$ is nonsingular. If we make $\ell_1$ of $T$ variable in $F_1$ we then can have $n(n - 1)^2$ triplets. Since the set of linear transformations of $t$-dimensional Euclidean space forms a group, and the set of linear transformations of 2-dimensional Euclidean space forms an Abelian group, we can use the relation between a group and its subgroup to obtain the number of lines in FEP.

Let $G$ be the general linear group of all linear nondegenerative transformations. Let $L(0)$ denote the lines passing through the origin and $L(1)$ denote those not passing through the origin. Then $G$ operates on both $L(0)$ and $L(1)$, and $G$ is transitive on both $L(0)$ and $L(1)$.

The order of the group $G$ is the number of nonsingular $2 \times 2$ $T$ matrices over $F_1$. This is the number of ways that a basis of the vector space of $F_2$ over $F_1$ can be transformed into another basis. This is in turn simply the number of bases of $V_2(n)$ from (2.66), i.e.,

$$\begin{bmatrix} \text{The order of} \\ \text{the group } G \end{bmatrix} = (n^2 - 1)(n^2 - n) = N_G. \tag{2.90}$$

Let $N_{G,0}$ denote the number of triplet combinations described above, $N_{G,0} = (n - 1)^2 n$. It turns out that the triplets are the cosets. If we let $L(0)$ be the subgroup of $G$, then from group theory we have

$$\begin{bmatrix} \text{The order of} \\ \text{the subgroup } L(0) \end{bmatrix} = \frac{[\text{The order of the group } G]}{\begin{bmatrix} \text{The number of coset of } G \\ \text{with respect to the sub-} \\ \text{group } L(0) \end{bmatrix}}$$

$$= \frac{N_G}{N_{G,0}} = \frac{(n^2 - 1)n(n - 1)}{(n - 1)^2 n}$$

$$= n + 1. \tag{2.91}$$

Now considering $L(1)$ as a subgroup, the corresponding number of cosets $N_{G,L}$ is $n(n - 1)$. Then the number of lines not going through origin is

$$\begin{bmatrix} \text{The order of the} \\ \text{subgroup } L(1) \end{bmatrix} = \frac{N_G}{N_{G,L}} = \frac{(n^2 - 1)n(n - 1)}{n(n - 1)}$$

$$= n^2 - 1. \tag{2.92}$$

The total number of lines in FEP is the sum of the orders of the subgroups of $L(0)$ and $L(1)$,

$$N_{L_0 + L_1}; N_{L,1} = n^2 - 1 + n + 1 = n^2 + n. \tag{2.93}$$

### 2.12.3  Properties of Transformed Lines in a Finite Euclidean Plane

THEOREM 2.12: The lines of (2.89), which do not go through $P(0)$, are all distinct.

*Proof.* From (2.89) it can be seen that none of the $n^2 - 1$ lines contains the point $P(0)$. Thus, the lines do not go through $P(0)$. First, the lines $L_0$ and $L_1$ must be distinct. If this were not true, multiplication by $\sigma$ would merely permute the elements

$$\sigma^{d_0}, \sigma^{d_1}, \ldots, \sigma^{d_{n-1}}$$

among themselves. This means that every line of (2.89) consists of the same set of points but not in the same order. Hence there would occur in (2.89) only $n$ distinct elements of $F_2$, whereas in every column all the $n^2 - 1$ nonzero elements occur. This leads to $n^2 - 1 \leqslant n$, which is impossible since $n = p^m \geqslant 2$.

The first two lines of (2.89) have a point in common; i.e., $P(\sigma^{d_1}) = P(\sigma^{d_0 + 1})$, since $d_0 = 0$ and $d_1 = 1$. Since the $d$'s are all distinct, these two lines of (2.89) cannot have more than one point in common. Thus, $d_i \neq d_j + 1 \bmod (n^2 - 1)$ for any $i$ and $j$ but $i = 0, j = 1$; or $d_i - d_j \neq 1 \bmod (n^2 - 1)$ for any $i$ and $j$ but $i = 0, j = 1$; or there is only one pair of exponents in $L_0$ such that $d_v - d_u = 1 \bmod (n^2 - 1)$.

We can now prove that no two lines in (2.89) are identical. Suppose that $L_i$ and $L_j$, $i \neq j$, are identical. Then for $d_0, d_1$ in the $i$th row and some $d_u, d_v$ in the $j$th row, in order for the points $P(\sigma^{d_0+i}) = P(\sigma^{d_u+j})$ and $P(\sigma^{d_1+i}) = P(\sigma^{d_v+j})$, we have

$$d_0 + i = d_u + j \qquad \mathrm{mod}(n^2 - 1) \qquad\qquad (2.94)$$

$$d_1 + i = d_v + j \qquad \mathrm{mod}(n^2 - 1) \qquad\qquad (2.95)$$

Furthermore, $d_v \neq 1 \bmod (n^2 - 1)$, since $i \neq j$.
This gives

$$d_v - d_u = d_1 - d_0 = 1, \; d_v \neq 1 \bmod(n^2 - 1), \qquad (2.96)$$

which is impossible because $\lambda = 1$. This shows the validity of Theorem 2.12.

We have produced $n^2 - 1$ lines (2.89) that do not pass through $P(0)$, and $n + 1$ lines (2.87) that do pass through $P(0)$. Together, this makes $n(n + 1)$ lines. Recall from the last part of Section 2.12.2 that the number of lines EG(2, $n$) is exactly $n(n + 1)$. Thus the two sets of lines described above make up all the lines in EG(2, $n$).

### 2.12.4 Euclidean Geometry Difference Sets

As in $\{D_p\}$, $v$, $n$, and $\lambda$, are the parameters of Euclidean geometry difference sets $\{D_E\}$. Inequality (2.97) holds among the three parameters:

$$n(n - 1) \leq \lambda \, (v - 1) \qquad\qquad (2.97)$$

since there are $n(n - 1)$ pairs among the $n$ integers of a $\{D_E\}$. This number is not greater than the number of allowable solutions among the $v - 1$ nonzero integers, which leads to (2.97). The purpose of this section is to construct the set of $d$'s of $\{D_E\}$ based upon the background information provided thus far.

The following theorem provides some of the important properties of a Euclidean geometry difference set. In particular it proves that all the pairwise differences among the set of residues of $\{D_E\}$ are distinct and not divisible by $n + 1$. From the definition of a difference set, the number of solution pairs in $d_i - d_j \equiv \delta \pmod{v}$ is at most $\lambda$; thus the proof of the distinctiveness of the difference pairs directly shows that $\lambda = 1$ [13].

THEOREM 2.13:   Given an integer $n = p^m \geq 2$, we can find $n$ integers $d_0$, $d_1$, . . . , $d_{n-1}$ such that among the $n(n - 1)$ differences $d_i - d_j$, $(i, j = 0, 1, 2, . . . , n - 1$ and $i \neq j$) reduced modulo $n^2 - 1$, all the positive integers less than $n^2 - 1$ and not divisible by $n + 1$ occur exactly once [14].

*Proof.*   Among the $n^2 - 1$ lines in (2.89) there are $n$ lines of the form

$$(\sigma^{d_0 - d_j}), \; (\sigma^{d_1 - d_j}), \; . . . , \; (\sigma^{d_{n-1} - d_j}) \qquad\qquad (2.98)$$

$j = 0, 1, 2, . . . , n - 1$ going through the point $P(\sigma^0)$.

The points on the $n$ lines of the form of (2.98) contain the $n(n - 1)$ differences $d_i - d_j$, $i \neq j$ as well as the $n$ differences $d_i - d_i = 0$ as exponents. It was shown in (2.87) that one of the lines through the origin, $\ell_0$, contains $\sigma^0$, 0 and all the points whose exponents are multiples of $(n + 1)$. Therefore, since none of (2.89) goes through the origin, and each contains $\sigma^0$, none of the other points can have exponents that are multiples of $(n + 1)$. Thus

$$d_i - d_j \not\equiv 0 \bmod(n + 1), \quad i \neq j. \tag{2.99}$$

Since the lines are distinct and each line contains the same point, i.e., $\sigma^0 = \sigma^{d_j - d_j}$, it follows that each of the other points on the lines of (2.89) occurs just once. Since there are $(n - 1)$ multiples of $(n + 1)$ in $n^2 - 1$, there are $n^2 - 1 - (n - 1) = n(n - 1)$ integers less than $n^2$ not divisible by $(n + 1)$. Thus the $n(n - 1)$ distinct differences $d_i - d_j \bmod(n^2 - 1)$, $i \neq j$, are not divisible by $(n + 1)$. Q.E.D.

THEOREM 2.14: If $\sigma$ is a nonzero element in $F_2$, and $d_{k+1}$ is defined by

$$1 + \sigma^{z + k(n + 1)} = \sigma^{d_{k+1}} \tag{2.100}$$

for $k = 0, 1, \ldots, n - 2$, with a fixed value of $z = 1, 2, \ldots, n$, then

$$\{D\} = \{d_0 = 0, d_1, d_2, \ldots, d_{n-1}\} = \{D_E\} \tag{2.101}$$

*Proof.* Since the integers $z = 1, 2, \ldots, n$ are not divisible by $n + 1$, $\sigma^z$ is not an element of $F_1$, because the nonzero elements of $F_1$ are $0, 1, \sigma^{n+1}, \sigma^{2(n+1)}$, $\sigma^{(n-2)(n+1)}$. Hence $1 + \sigma^z$ is not an element of $F_1$. Let $1 + \sigma^z = \sigma^t$; then, for the same reason, $t$ is not divisible by $n + 1$. Then the line in FEP joining the points $P(\sigma^0)$ and $P(\sigma^t)$ does not pass through the origin, for, if it did, $P(\sigma^t)$ would be on the line determined by the points $P(0)$ and $P(\sigma^0)$ and $t$ would be divisible by $n + 1$.

The points on the line that join $P(\sigma^0)$ and $P(\sigma^t)$ are $x\sigma^0 + (1 - x)\sigma^t$. Since $1 - x$ is an element of $F_1$, it is equivalent to $\sigma^{k(n+1)}$. For $k = 0, 1, \ldots,$ $n - 2$, the $n - 1$ points on the line joining $P(\sigma^0)$ and $P(\sigma^t)$ are

$$\begin{aligned}
x\sigma^0 + (1 - x)\sigma^t &= (1 - \sigma^{k(n+1)})\sigma^0 + (\sigma^{k(n+1)})\sigma^t \\
&= 1 - \sigma^{k(n+1)}(1 - \sigma^t) \\
&= 1 - \sigma^{k(n+1)}(-\sigma^z) \\
&= 1 + \sigma^{k(n+1)} + z \\
&= \sigma^{d_{k+1}}. \tag{2.102}
\end{aligned}$$

Thus if $z$ is fixed and $d_{k+1}$ is defined by

$$1 + \sigma^{k(n+1)+z} = \sigma^{d_{k+1}}, \tag{2.103}$$

then the points $\sigma^{d_{k+1}}$ for $k = 0, 1, \ldots, n - 2$ plus $\sigma^{d_0} = \sigma^0$ are the points on a line that does not go through $P(0)$. Thus $P(\sigma^0), P(\sigma^{d_1}), \ldots, P(\sigma^{d_{n-1}})$ is the line $L_0$ of (2.89) with $d_0 = 0$. The properties of the lines of (2.89) and the exponents of the points on the lines have been treated in detail in Theorems 2.12 and 2.13. Thus the integers $d_0, d_1, d_2, \ldots, d_{n-1}$ are the residues of a Euclidean geometry difference set with properties described in Theorems 2.12 and 2.13. Q.E.D.

### 2.12.5   The Set-Generation Algorithm

Based upon the analyses in Sections 2.12 to 2.12.4, the construction procedure for a $\{D_E\}$ is presented in algorithmic steps as follows.

Step 1:   Given a prime or prime power $n = p^m$, $p^{2m} = n^2$. Identify the nonzero elements of the corresponding finite field $F_2$ in terms of the primitive element $\sigma$ satisfying a primitive polynomial $F_o(\sigma)$ of degree $2m$. Let $f_i \in GF(p^m)$, $i = 0, 1, \ldots, 2m - 1$;

$$F_o(\sigma) = \sigma^{2m} + f_{2m-1}\sigma^{2m-1} + \ldots + f_1\sigma + f_0 = 0, \quad (2.104)$$

then

$$\sigma^{2m} = -(f_{2m-1}\sigma^{2m-1} + \ldots + f_1\sigma + f_0). \quad (2.105)$$

Step 2:   With the selected integer $n$, form two sets of integers $z = 1, 2, \ldots,$ $n$; and $k = 0, 1, \ldots, n - 2$. Each value of $z$ provides a $\{D_E\}$. Choose a value of $z$ and run through the $k$ values to obtain

$$d(z,k) = z + k(n + 1). \quad (2.106)$$

That is,

$$d(z,0) = z + 0(n + 1)$$
$$d(z,1) = z + 1(n + 1)$$

$$\vdots$$

$$d(z, n - 2) = z + (n - 2)(n + 1).$$

The set of exponents $d_{k+1}$ can be obtained from $d(z,k)$ through the element $\sigma$ as in (2.107):

$$\sigma^{d_{k+1}} = 1 + \sigma^{d(z,k)} \quad (2.107)$$

for

$$k = 0, 1, \ldots, n - 2.$$

Step 3:   With $d_0 = 0$, the $\{D_E\}$ is

$$\{D\} = \{0, d_1, d_2, \ldots, d_{n-1}\}.$$

*Example 2.12:*   If we let $p = 3$, $m = 1$, $n = p^m = 3$, then in terms of the primitive root $\sigma$ the nonzero elements of $GF(3^2)$ can be obtained by finding a primitive polynomial of degree 2 which is irreducible over $GF(3)$.

$$F_o(x) = x^2 - 2x - 1.$$

Step 1:   Since $\sigma$ is a root of $F_o(x)$ we have $F_o(\sigma) = 0$, or

$$\sigma^2 = 2\sigma + 1.$$

Then the other elements of $GF(3^2)$ are:

$$\sigma^0 = 1$$
$$\sigma^1$$
$$\sigma^2 = 2\sigma + 1$$
$$\sigma^3 = \sigma \cdot \sigma^2 = \sigma(2\sigma + 1) = 2\sigma^2 + \sigma$$
$$= 2(2\sigma + 1) + \sigma = 2(\sigma + 1)$$
$$\sigma^4 = 2$$
$$\sigma^5 = 2\sigma$$
$$\sigma^6 = \sigma + 2$$
$$\sigma^7 = \sigma + 1.$$

Step 2: Choosing $z = 2$ and $k = 0$, we have

$$d(2,0) = 2 + 0(3 + 1) = 2$$
$$\sigma^{d_1} = 1 + \sigma^{d(2,0)} = 1 + \sigma^2 = 1 + 2\sigma + 1$$
$$= 2\sigma + 2 = \sigma^3.$$

Therefore $d_1 = 3$. Next, with $k = 1$ and the same value $z$, we have

$$d(2,1) = 2 + (3 + 1) = 6$$
$$\sigma^{d_2} = 1 + \sigma^6 = 1 + \sigma + 2 = \sigma.$$

Therefore, $d_2 = 1$.

Step 3: With $d_0 = 0$, the $\{D_E\}$ is

$$\{D\} = \{0, 3, 1\}.$$

### 2.12.6   Table of Primitive Polynomials over GF($p \neq 2$)

From Section 2.12.5 we observed that the starting point of the $\{D_E\}$ construction algorithm reduces to the selection of a primitive polynomial $F_o(\sigma)$ of degree $2m$. In Appendix 1 a table of such primitive polynomials is provided for $p$ from 3 to 31; $m$ varies only from 1 to 3, and in most cases $m = 1$. The table is an abstraction of the computational results from Alanen and Knuth's study of finite fields [15].
From (2.104) in general:

$$F_o(x) = x^{2m} + f_{2m-1}x^{2m-1} + \ldots + f_1 x + f_0. \tag{2.108}$$

Let

$$f_{2m-i} = a_i, \tag{2.109}$$

then,

$$F_o(x) = x^{2m} + a_1 X^{2m-1} + \ldots a_{2m}. \tag{2.110}$$

The primitive polynomials are represented by the coefficients $a_1, a_2, \ldots, a_{2m}$ of $F_o(x)$, as shown in Appendix 1.

In the table for each prime $p$, the values of $2m$, $n^2 = p^{2m}$, the factorization of $n^2 - 1$, and the number of polynomials $\phi(n^2 - 1)/2m$ are also indicated, where $\phi(n^2 - 1)$ is the number of integers less than $n^2 - 1$ that have no common factors with $n^2 - 1$ as described in Section 2.3. For example, there exist two primitive polynomials of degree 2 (for $m = 1$), $p = 3$ with $n^2 - 1 = 8$. From the list of values of $a_1$ and $a_2$, the two polynomials are $x^2 + x + 2$ and $x^2 + 2x + 2$.

## 2.13  LATIN SQUARES

A Latin square of order $n$ is a square matrix of size $n \times n$ such that each row and column is a permutation of its row or column elements, which are represented by positive integers $0, 1, 2, \ldots, n - 1$.

*Example 2.13:*  For $n = 3$, we have

$$\begin{bmatrix} 0 & 1 & 2 \\ 1 & 2 & 0 \\ 2 & 0 & 1 \end{bmatrix}.$$

For $n = 4$ we have

$$\begin{bmatrix} 0 & 1 & 2 & 3 \\ 1 & 2 & 3 & 0 \\ 2 & 3 & 0 & 1 \\ 3 & 0 & 1 & 2 \end{bmatrix}$$

where the second row is the cyclic permutation of the first row, and the third row is the cyclic permutation of the second row. In general, the $i$th row of a Latin square is obtained by the cyclic permutation of the $(i - 1)$th row elements.

Let a Latin square of order $n$ be denoted as

$$L = \begin{bmatrix} L_{(0,0)} & L_{(0,1)} & \cdots & L_{(0,n-1)} \\ \vdots & \vdots & & \vdots \\ L_{(n-1,0)} & L_{(n-1,1)} & \cdots & L_{(n-1,n-1)} \end{bmatrix} \tag{2.111}$$

The first general result concerning Latin squares is

THEOREM 2.15:   A Latin square exists for any positive integer $n$.

*Proof.* The proof is provided by the congruence relation from number theory. For an arbitrary entry $L(i,j)$ in $L$ $(i,j = 0, 1, \ldots, n - 1)$, we observe the congruence relation

$$L(i,j) \equiv i + j \pmod{n} \tag{2.112}$$

for the case of $0 \leqslant L(i,j) \leqslant n - 1$. If $L(i,j) = L(i,k)$ for some $k \neq j$, this implies

$$i + j \equiv i + k \pmod{n}$$
$$j \equiv k \pmod{n}, \tag{2.113}$$

which means $j = k$, a contradiction. Thus, for $L(i,j) = L(i,k)$, $j$ must equal $k$, which says the entries are all distinct for the $i$th row of $L$. This holds true for $i = 0, 1, \ldots, n - 1$; i.e., all the rows in $L$. The same argument holds true for all the columns of $L$; hence we have proved the theorem.

An interesting property of a Latin square is the fact that interchanging columns in a Latin square yields a Latin square. This property will be useful when we present the unified construction method for the secret sharing systems to be discussed in Chapter 4.

Let $L_1$ and $L_2$ be two Latin squares of order $n$. Let the elements in $L_1$ and $L_2$ be denoted as $L_1(i,j)$ and $L_2(i,j)$ respectively. $L_1$ and $L_2$ are said to be orthogonal if the $n^2$ ordered pairs $L_1(i,j), L_2(i,j)$ for $i,j = 0, 1, \ldots, n - 1$ are all distinct.

*Example 2.14:* For $n = 3$ from Example 2.13, we have

$$L_1 = \begin{bmatrix} 0 & 1 & 2 \\ 1 & 2 & 0 \\ 2 & 0 & 1 \end{bmatrix}.$$

After permuting the last two rows, we have

$$L_2 = \begin{bmatrix} 0 & 1 & 2 \\ 2 & 0 & 1 \\ 1 & 2 & 0 \end{bmatrix}.$$

Superimposing the entries in $L_1$ and $L_2$, we obtain the ordered pairs

$$\begin{bmatrix} (0,0) & (1,1) & (2,2) \\ (1,2) & (2,0) & (0,1) \\ (2,1) & (0,2) & (1,0) \end{bmatrix}.$$

As can be easily seen, the pairs are all distinct.

Let $L_1, L_2, L_3, \ldots, L_n$ be a set of Latin squares of order $n$. They are said to be a set of orthogonal Latin squares if and only if every pair of squares in the set is orthogonal. There is only one trivial Latin square of order 1. There are two Latin squares of order 2, but they are not orthogonal. Thus orthogonal Latin squares exist only for $n \geqslant 3$.

*Example 2.15:*  A pair of Latin squares of order 5 may be represented by

| α | β | γ | δ | ξ |
|---|---|---|---|---|
| δ | ξ | α | β | γ |
| β | γ | δ | ξ | α |
| ξ | α | β | γ | δ |
| γ | δ | ξ | α | β |

$$(2.114)$$

and

| A | B | C | D | E |
|---|---|---|---|---|
| C | D | E | A | B |
| E | A | B | C | D |
| B | C | D | E | A |
| D | E | A | B | C |

$$(2.115)$$

Superimposing the two Latin squares of (2.114) and (2.115), we have another Latin square of order 5:

| αA | βB | γC | δD | ξE |
|----|----|----|----|----|
| δC | ξD | αE | βA | γB |
| βE | γA | δB | ξC | αD |
| ξB | αC | βD | γE | δA |
| γD | δE | ξA | αB | βC |

$$(2.116)$$

where all the entries are distinct. This implies that the pair of Latin squares is mutually orthogonal.

In general, the number of orthogonal Latin squares of order $n$ depends on the value of $n$. If $n$ is a prime or a prime power, then there exist $n - 1$ such squares; if $n$ is not a power of any prime, then $n$ can be factored into prime powers; i.e., $n = p_1^{e_1}, p_2^{e_2}, \ldots, p_s^{e_s}$ for $p_i \neq p_j$. Let $g_{i_1}^{(1)}, g_{i_2}^{(2)}, \ldots, g_{i_s}^{(s)}$ denote the elements of $GF(p_1^{e_1})$, $GF(p_2^{e_2})$, $\ldots$, $GF(p_s^{e_s})$ respectively, where $g_0^{(i)}$ is the 0 element and $g_1^{(i)}$ is the unit element of $GF(p_i^{e_i})$. Form the points (as shown in Mann [8]),

$$\gamma = (g_{i_1}^{(1)}, g_{i_2}^{(2)}, \ldots, g_{i_s}^{(s)}),$$

which are multiplied and added by multiplying and adding their coordinates. Further, let

$$\gamma_j = (g_j^{(1)}, \ldots, g_j^{(s)}) \qquad (2.117)$$

$$0 < j \leqslant r = \min(p_i^{e_i} - 1) \qquad (2.118)$$

and number the remaining points in any arbitrary way from $r + 1$ to $n = p_1^{e_1}$ $\ldots p_s^{e_s}$ in such a way that $\gamma_m = 0 = (g_0^{(1)} \ldots g_0^{(s)})$. Then the arrays

$$
L_j = \begin{bmatrix}
0 & 1 & \cdots & \gamma_{n-1} \\
\gamma_j & 1 + \gamma_j & & \gamma_{n-1} + \gamma_j \\
\gamma_j \gamma_2 & 1 + \gamma_j \gamma_2 & & \gamma_{n-1} + \gamma_j \gamma_2 \\
\vdots & \vdots & & \vdots \\
\gamma_j \gamma_{n-1} & 1 + \gamma_j \gamma_{n-1} & & \gamma_{n-1} + \gamma_j \gamma_{n-1}
\end{bmatrix} \tag{2.119}
$$

form a set of $r$ orthogonal Latin squares. From (2.118) we can see that the number of Latin squares of nonprime order is no more than $r = \min(p_i^{e_i} - 1)$.

## 2.14 CONCLUDING REMARKS

With Chapter 2 as a foundation, we can understand the validity and optimality of the procedures to be presented in this volume and the reasons for the conclusions that will be drawn.

Some of the derivations and examples in this chapter are original and appear here for the first time. All subjects covered in Chapter 2 have either direct application or potential usefulness in the areas of digital satellite communication to be discussed throughout the rest of this volume. Appendix 1 provides a table of primitive polynomials over $GF(p \pm 2)$; these polynomials are the starting point for the derivation of Euclidean geometry difference sets $\{D_E\}$. Appendix 2 lists known sets which are complementary to $\{D_E\}$.

## 2.15 REFERENCES

1. Niven, I. and Zuckerman, H. *An Introduction to the Theory of Numbers.* New York: John Wiley & Sons, 1967.
2. Leveque, W. *Topics in Number Theory.* Reading, Mass.: Addison-Wesley, 1965.
3. Zieler, N. "Linear Recurring Sequence." *Siam Journal of Applied Mathematics* I (1959).
4. Hall, M. *Combinatorial Theory.* Waltham, Mass.: Blaisdell Publishing Co., 1967.
5. Gallager, R. *Information Theory and Reliable Communications.* New York: John Wiley & Sons, 1968.
6. Baumert, L.D. *Cyclic Difference Sets: Lecture Notes in Mathematics No. 182.* Pasadena, California: California Institute of Technology, 1971.
7. Dembowski, P. *Finite Geometries.* New York: Springer-Verlag, 1968.
8. Mann, H.B. *Addition Theorems.* New York: John Wiley & Sons, 1965.
9. Singer, J. "A Theorem in Finite Projective Geometry and Some Applications to Number Theory." *Transactions of the American Mathematical Society* 43 (1938): 377–85.
10. Peterson, W.W. and Weldon, E., Jr. *Error Correcting Codes.* Cambridge, Mass.: M.I.T. Press, 1972, Appendix C.
11. Liu, C. *Introduction to Combinatorial Mathematics.* New York: McGraw-Hill, 1968.

**12.** Mullin, R.C. and Blake, I.F. *The Mathematical Theory of Coding*. New York: Academic Press, 1975.

**13.** Bose, R.C. "An Affine Analogue of Singer's Theorem." *Journal of the Indian Mathematical Society* 6 (1942).

**14.** Srivastava, J., Harary, F., Rao, C., Rota, G., and Shrikhande, S., editors. *A Survey of Combinatorial Theory*. New York: North-Holland, 1973.

**15.** Alanen, J. and Knuth, D. "Table of Finite Fields." *Sankhya*, Series A, 26, 4 (December 1964): 305–28.

## 2.16 PROBLEMS

1. Prove that every integer larger than 6 can be represented as a sum of two integers greater than 1.

2. For any positive integer $m$, is $2^m - 1$ factorable into a product of primes?

3. If $2^m + 1$ is an odd prime, prove that $m$ is a power of 2.

4. If $2^m - 1$ is a prime, prove that $m$ must be a prime.

5. Solve the set of congruences

$$x \equiv 2 \ (\text{mod } 3)$$
$$x \equiv 3 \ (\text{mod } 5)$$
$$x \equiv 5 \ (\text{mod } 2).$$

6. Solve the congruence $x^2 + x - 1 \equiv 0 \ (\text{mod } 13)$.

7. Find solutions $x$ for $\phi(x) = 63$.

8. By Fermat's Theorem and without enumeration, show that $11 \times 31 \times 61$ divides $20^{15} - 1$.

9. Show that the number of divisions in the Euclidean algorithm is less than $2 \ln b / \ln 2$ if each nonzero remainder $r_i$ ($i \geq 2$) is less than $r_{i-2}/2$, with $b$ being a positive integer.

10. For any two random variables $\square$ and $\triangle$, prove that

$$H(\square | \triangle) = H(\square, \triangle) - H(\triangle).$$

11. For any three random variables $\square$, $\triangle$, and $\bigcirc$, is

$$H(\square, \triangle, \bigcirc) - H(\square, \triangle) = H(\square, \bigcirc) - H(\square)$$

   valid?

12. A condition for the existence of difference sets is the existence of symmetric block designs. Referring to the incidence matrix $Q$ (Theorem 2.9), when (2.37) is valid, then (2.39) exists if and only if (1) for $v$ even, $k - \lambda$ is a square; (2) for $v$ odd, the equation

$$z^2 = (k - \lambda)x^2 + (-1)^{(v-1)/2}\lambda y^2$$

has a solution in integers $x$, $y$, $z$ not all zero. This is known as the Bruck-Ryser-Chowla theorem. Verify this result.

13. With

$$\frac{n^{t+1} - 1}{n - 1} = 1 + n + \ldots + n^t,$$

verify in (2.26) that

$$\frac{(n^{t+1} - 1)(n^t - 1) \ldots (n^{t-r+1} - 1)}{(n^{r+1} - 1)(n^r - 1) \ldots (n - 1)}$$
$$= \frac{(1 + \ldots + n^t)(n + \ldots + n^t) \ldots (n^r + \ldots + n^t)}{(1 + \ldots + n^r) \ldots (n^{r-1} + n^r)n^r}.$$

14. In the development of Euclidean geometry sets, the lines in a Euclidean geometry plane have been classified as those lines going through the point of origin $P(0)$ and those lines not going through $P(0)$. Why do the successive transformations $T^i$ apply $n + 1$ times to the points on a line through $P(0)$ and $n^2 - 1$ times to the points on a line not going through $P(0)$?

15. Verify that in the proof of Theorem 2.13 there are $n$ lines of the form of (2.98) going through the point $P(\sigma^0)$ among the $n^2 - 1$ lines of (2.89).

16. One of the restrictions of Euclidean geometry sets is the fact that $\lambda = 1$, as indicated by Theorem 2.12. Investigate the existence of sets for $\lambda \neq 1$.

17. $F_o(x) = 1 + x + x^4$ is primitive over GF(2) in the finite field GF($2^4$). Generate the residues of the corresponding projective geometry set.

18. $F_o(x) = x^2 + x + 2$ is irreducible over GF(3); i.e., $m = 1, p = 3, n^2 - 1 = 8$. Obtain the corresponding Euclidean geometry set residues.

19. Derive the residues for the projective geometry sets with the following parameters $(v, k, \lambda)$

   a. $(21, 5, 1)$ from PG(2, 4)
   b. $(13, 4, 1)$ from PG(2, 3).

# Chapter 3

# APPLICATIONS OF COMBINATORIAL SETS IN SATELLITE COMMUNICATION

### 3.1 THE POWER OF COMBINATORIAL SETS

Among classical and modern communication theories, few simple analytical techniques other than the Fourier transform and Shannon's theorems have more impact in their diverse applications than the combinatorial sets described in Chapter 2.

Fourier transforms often assist only in computation, with the transformation itself not contributing to a better solution. Shannon's capacity theorem provides a design limit, but does not lead to a specific technique for approaching or reaching the limit. We are interested in combinatorial sets because we can use them not only to model a physical problem, but also to derive—or immediately obtain—the solution effortlessly. One set can almost always provide the optimum solution deterministically for many different applications. The key to a successful application is to match the structure of the set to the physical parameters of the problem in question, or at least as closely as possible. The sets can provide an instant optimal guideline, where other methods fail.

Following is a list of applications of the combinatorial sets to signal processing, detection, and transmission. Some of these subjects are unique to digital satellite communication, while others pertain to various other satellite-related applications.

1. Elimination of intermodulation products in FDMA systems through frequency selection.
2. Generation of synchronizable sequences with prescribed correlation properties in TDMA systems.
3. Minimization of message collision in superpacket transmission.
4. Optimal arrangement in pairwise network connections for onboard switching.
5. Sub-Nyquist temporal sampling for whitening of aliasing effects.
6. Sidelobe control in antenna arrays.
7. Mechanization of random multiple access systems.
8. Minimization of multiple beam interference.

9. Mechanization of transmission ciphering for encryption and decryption.
10. Generation of error-correcting code for coded systems.

Item 2 is described in Chapter 5 of Volume I and Item 4 is discussed in Chapter 6 of Volume I. Using the sets discussed in Chapter 2, the other items (i.e., 1, 3, 5, 6, 7, 8) will be treated in this chapter. Item 9 will be included in Chapter 4, and Item 10 will be elaborated upon in Chapters 5 and 6.

Through error-coding applications, combinatorial sets are capable of generating:

1. One-step majority logic decodable random error-correction block codes
2. Threshold decodable convolutional self-orthogonal random error-correction codes
3. Multiple burst error-correction codes.

We shall provide the derivation of convolutional self-orthogonal codes from combinatorial sets as a sample. We shall then provide references for the other code generations.

The purposes of this chapter are to demonstrate the usefulness of combinatorial sets and to present new algorithms for satellite communication applications based on the sets.

## 3.2 RANDOM MULTIPLE ACCESS SYSTEMS

In Chapter 3 (Sections 3.5.0–3.5.3 and Section 3.6.5) of Volume I we discussed the advantages and configuration of random multiple access (RMA) systems and derived an error probability expression for them. However, in Volume I we could not present the foundation that mechanizes such systems because we lacked the background given in Chapter 2 of Volume II. Now, after the treatment of Section 2.12, we are in a position to see how a random multiple access system can be efficiently implemented.

Recall that the basic transmission scheme of an RMA system is the transformation of binary bits into multiple-level time-frequency (TF) symbols, and the receiver of the system performs the inverse transformation from a sequence of multiple level symbols back to binary digits. The error performance and system efficiency are mostly affected by the construction of the multiple level symbol sequences for any given number of access stations. In this section, therefore, we shall present a systematic procedure for constructing RMA sequences. This procedure is optimum in the sense that no other sets of sequences of the same length can provide a better sequence distance property, which will be defined in the next section.

Before we plunge into the new algorithm a few words on previous related work are in order. Bluestein and Greenspan investigated large-distance codes for efficient approximation of orthogonal waveforms [1]. Their work is concerned with determining the number of ordered $\ell$-tuples $(\tau_1, \tau_2, \ldots, \tau_\ell)$ that can be formed from an alphabet of $A$ symbols under the constraint that no two $\ell$-tuples overlap in more than $K$ places. They considered $M$ $\ell$-tuples or signals. Each one of $M$ transmitted waveforms consists of $\ell$ elementary waveforms, and each elementary waveform is one of a set of $A$ mutually orthogonal waveforms. Under the restrictions that $A$ is a power of a prime $p$, $\ell \leq A + 1$, and $A$ forms a field of order $p^m$ for a positive integer $m$, they demonstrated that not more than $M = A^2$ signals can be generated with one or fewer overlaps. They also provide a procedure to obtain a code of length $A + 1$ over GF($A$). The maximum number of code words is $A^2$ with minimum distance $A$, which is a power of a prime. Their set of sequences, as it turns out, is an extended Reed-Solomon code, as can be seen by appending two complementary binary symbols to the first two columns of a Reed-Solomon code generator matrix (see Sections 4.12.3, 5.5.2.1).

RMA problems were also studied by Schuleter [2]. He applied time-frequency orthogonal biphase coding with emphasis on the effect of interference problems due to multiple users. Solomon considered optimal frequency-hopping sequences as a subset of a Reed-Solomon code over a field that is isomorphic to the number of frequencies [3]. He treated the problem as one-dimensional along the scale of frequency. Kasami and Lin discussed coding and decoding algorithms for a multiple access channel [4]. However, they considered only nonrandom multiple access.

Mersereau and Seay extended earlier work and provided frequency-hopping pattern construction procedure by dividing Reed-Solomon codewords into disjoint sets, and then selecting one member from each set to form the desired sequences [5]. For available frequencies allowing $k$ frequency overlaps, their procedure produces $f^{k-1}$ hopping sequences. Although the scheme was primarily developed for radar signaling, it may apply to satellite communication with limitations: signaling needs to be periodic, single frequency overlapping is not possible, the number of sequence patterns is small in comparison to the scheme to be presented, and their result has not been shown to be optimum. For these reasons, the following solution to the random multiple access problem is needed.

The nonbinary symbols of an RMA sequence come from the entry representation of a time-frequency matrix as described in Section 3.5.2 of Chapter 3, Volume I. The number of time divisions is denoted as $n$ and the number of frequency divisions is denoted as $f$; the total number of symbols (entries) is $m = nf$. The objective is to construct a sequence of $m$-ary $nf$ symbols so that a pair of sequences represents binary signaling from a particular station. Complete orthogonality among the set of sequences provides the maximum cross-correlation property and thus minimizes the probability of false detection error. However, the number of usable sequences in such a signaling scheme is very limited. Hence, one overlapping symbol is considered in order to increase the number of such sequences. Before we proceed with the sequence construction algorithm, let us outline the procedure and results as follows:

The algorithm begins by establishing two matrices $[W]_0$ and $[W]_1$ with size $(n^2-1) \times n$ and $(n+1) \times n$ respectively. The rows of $[W]_0$ consist of the elements of a Euclidean geometry difference set and its equivalent sets. The elements in the rows of $[W]_1$ are so constructed that the first element in its 0th row cannot be any difference modulo $(n^2 - 1)$ among the set of residues. A row matrix $R_j$ is then defined in terms of the row representation of time-frequency matrix $M$, whose elements are the code symbols. If the set of rows of $M$ is substituted into $R_j$, we obtain a set of matrices $B_j$. Next, a set of matrices $A_k$ is constructed and each $A_k$ is used as an ordering transformation on the $B_j$'s to obtain a set of matrices $U(j,k)$. The rows of $B_j$, $U(j,k)$, and the $f$ rows of $M$ constitute the set of desired sequences.

In terms of a sequence length $n$, the algorithm can generate a set of sequences with the following properties:

- Number of time divisions $= n$
- Number of symbols $= n^3$
- Sequence minimum distance $= n - 1$
- Number of frequency divisions $= n^2$
- Number of sequences $= n^2 (n^2 + n + 1)$

These properties will be proved in Section 3.4.

## 3.3 THE ALGORITHM FOR RMA SEQUENCE GENERATION

Step 1: Let $[W]_0$ be a matrix of $n^2 - 1$ rows. Each row consists of a set of $n$ integers. The integers of the nonzero row, i.e., $w_{i \neq 0}$, are determined from the set of integers in the 0th row, i.e., $w_0$.

$$[W]_0 = \begin{bmatrix} w_0 \\ w_1 \\ \vdots \\ w_{n^2 - 2} \end{bmatrix}. \tag{3.1}$$

The integers in $w_0$ come from the residues of a set $\{D_E\}$ (Section 2.12). Let $k = n$ and the residues of a $\{D\}$ be $d_0, d_1, \ldots, d_{n-1}$. Then

$$w_0 = [d_0, d_1, \ldots, d_{n-1}]. \tag{3.2}$$

The $i$th row $(i = 1, 2, \ldots, n^2 - 2)$ of $[W]_0$ is

$$w_i = [d_0 + i, d_1 + i, \ldots, d_{n-1} + i] \tag{3.3}$$

Next, we construct a matrix $[W]_1$ with $n + 1$ rows and $n$ columns. The rows are denoted as $W_{n^2-1}, W_{n^2}, \ldots, W_{n^2+n-1}$.

$$[W]_1 = \begin{bmatrix} W_{n^2-1} \\ W_{n^2} \\ \vdots \\ W_{n^2+n-1} \end{bmatrix}. \tag{3.4}$$

The 0th row of $[W]_1$ is

$$w_{n^2-1} = [0(n+1), \; 1(n+1), \; \ldots,$$

$$(n-2)(n+1), \; (n-1)(n+1)]. \tag{3.5}$$

The $(i')$th row of $[W]_1$ is obtained by the addition of $i'$ to each of the elements in the row $w_{n^2-1}$, except the last element, which always remains as $n^2-1$.

$$w_{n^2-1+i'} = [0(n+1)+i', \; 1(n+1)+i', \; \ldots,$$

$$(n-2)(n+1)+i', \; n^2-1]$$

$$i' = 1, 2, \ldots, n. \tag{3.6}$$

Step 2:  We define a matrix

$$R_j = [\bar{r}_{w_{j,0}} \; \bar{r}_{w_{j,1}} \ldots \bar{r}_\xi \ldots \bar{r}_{w_{j,n-1}}] \tag{3.7}$$

where

$$w_j = [w_{j,0} \; w_{j,1} \ldots w_{j,n-1}]$$

$$j = 0, 1, 2, \ldots n^2-2, n^2-1, \ldots, n^2+n-1 \tag{3.8}$$

($w_j$ is a row of $[W]_0$ or $[W]_1$)
and where in (3.7)

$$\bar{r}_\xi = \begin{bmatrix} n\xi \\ n\xi+1 \\ \vdots \\ n\xi+n-1 \end{bmatrix} \tag{3.9}$$

is a column matrix transposed from the $(\xi+1)$th row of a frequency-time matrix $M$ with size $f \times n$ containing the code symbols. The $M$ matrix is shown in (3.10). The algorithm is designed for $f=n^2$ only.

$$M = \begin{bmatrix} 0 & 1 & \ldots & n-1 \\ n & n+1 & \ldots & 2n-1 \\ \vdots & \vdots & & \\ n\xi & n\xi+1 & \ldots & (\xi+1)n-1 \\ \vdots & \vdots & & \\ (f-1)n & (f-1)n+1 & \ldots & fn-1 \end{bmatrix} = \begin{bmatrix} \bar{r}_0^T \\ \bar{r}_1^T \\ \cdot \\ \bar{r}_\xi^T \\ \cdot \\ \bar{r}_{f-1}^T \end{bmatrix} \tag{3.10}$$

$$B_j = \begin{bmatrix} nw_{j,0} & nw_{j,1} & \ldots & nw_{j,n-1} \\ nw_{j,0}+1 & nw_{j,1}+1 & \ldots & nw_{j,n-1}+1 \\ \vdots & \vdots & & \vdots \\ nw_{j,0}+n-1 & nw_{j,1}+n-1 & \ldots & nw_{j,n-1}+n-1 \end{bmatrix} \tag{3.11}$$

with

$$0 \leq j \leq n^2 + n - 1$$

$$A_k = \begin{bmatrix} \alpha_0 & \alpha_{\Delta\oplus2} & \alpha_{\Delta\oplus3} & \cdots & \alpha_{\Delta\oplus n} \\ \alpha_{\Delta\oplus n} & \alpha_0 & \alpha_{\Delta\oplus2} & \cdots & \alpha_{\Delta\oplus(n-1)} \\ \alpha_{\Delta\oplus(n-1)} & \alpha_{\Delta\oplus n} & \alpha_0 & \cdots & \alpha_{\Delta\oplus(n-2)} \\ \vdots & \vdots & \vdots & & \vdots \\ \alpha_{\Delta\oplus2} & \alpha_{\Delta\oplus3} & \alpha_{\Delta\oplus4} & \cdots & \alpha_0 \end{bmatrix} \tag{3.12}$$

where

$$a\oplus b = \begin{cases} (a+b) \bmod n & \text{if } a+b < n \\ (a+b+1) \bmod n & \text{if } a+b \geq n \end{cases}$$

and:   $\Delta = n - k - 2$

After substituting $\bar{r}_\xi$ into $R_j$ we have a square matrix $B_j$ of size $n \times n$ and with elements from the code symbols. The rows of $M$ and $B_j$ constitute the first part of the set of code words. $B_j$ is shown in (3.11).

Step 3:   Let us derive a set of permutation matrices $A_k$ from the set of integers $\{\alpha_0, \alpha_1, \ldots, \alpha_{(n-1)}\}$. The matrices are constructed as follows: The 0th row of $A_0$ consists of the complete set of the ordered integers. The 0th row of $A_1$ consists of the same set of integers with $i$ cyclic shifts to the right of all elements except the element in the 0th location. The $\ell$th row of the $A_k$ is obtained by $\ell$ cyclic shifts of all elements to the right. $A_k$ is expressed in (3.12) as a result of the above shifts.

Step 4:   Based on each $B_j$ of Step 2 and each $A_k$ of Step 3, we derive a $U(j,k)$ from an ordering transformation as in (3.13).

$$U(j,k) = F(A_k)\,[B_j]$$
$$j = 0, 1, \ldots, n^2 + n - 1$$
$$k = 0, 1, \ldots, n-2. \tag{3.13}$$

$F(A_k)\,[B_j]$ signifies that the column elements in each $B_j$ are rearranged so that their order of magnitude corresponds to the ordering of the elements in the corresponding column of $A_k$. This procedure applies to all columns in $B_j$.

The complete set of code words consists of the rows of $B_j$ (before the ordering transformation), and $f$ rows of $M$, and all the rows of $U(j,k)$ for all $j$ and $k$.

*Example 3.1:*    $\{D\} = \{d_0 = 1,\ d_1 = 6,\ d_2 = 7\}$
$n = 3,\ \bmod (n^2 - 1) = \bmod 8,\ f = 9$

$$[W]_0 = \begin{bmatrix} 1 & 6 & 7 \\ 2 & 7 & 0 \\ 3 & 0 & 1 \\ 4 & 1 & 2 \\ 5 & 2 & 3 \\ 6 & 3 & 4 \\ 7 & 4 & 5 \\ 0 & 5 & 6 \end{bmatrix} = \begin{bmatrix} w_0 \\ w_1 \\ w_2 \\ w_3 \\ w_4 \\ w_5 \\ w_6 \\ w_7 \end{bmatrix},\quad [W]_1 = \begin{bmatrix} 0 & 4 & 8 \\ 1 & 5 & 8 \\ 2 & 6 & 8 \\ 3 & 7 & 8 \end{bmatrix} = \begin{bmatrix} w_8 \\ w_9 \\ w_{10} \\ w_{11} \end{bmatrix}$$

Let

$$M = \begin{bmatrix} 0 & 1 & 2 \\ 3 & 4 & 5 \\ 6 & 7 & 8 \\ 9 & 10 & 11 \\ 12 & 13 & 14 \\ 15 & 16 & 17 \\ 18 & 19 & 20 \\ 21 & 22 & 23 \\ 24 & 25 & 26 \end{bmatrix} = \begin{bmatrix} \bar{r}_0^T \\ \bar{r}_1^T \\ \cdot \\ \cdot \\ \cdot \\ \cdot \\ \cdot \\ \cdot \\ \bar{r}_8^T \end{bmatrix}$$

representing nine frequency divisions and three time divisions. Each row of $M$ is denoted by $\bar{r}_\xi^T$. Then, from Step 2, the set of $B_j$'s are

$$B_0 = [\, \bar{r}_1 \quad \bar{r}_6 \quad \bar{r}_7\,]$$
$$= \begin{bmatrix} 3 & 18 & 21 \\ 4 & 19 & 22 \\ 5 & 20 & 23 \end{bmatrix}$$

$$B_1 = [\, \bar{r}_2 \quad \bar{r}_7 \quad \bar{r}_0\,]$$
$$= \begin{bmatrix} 6 & 21 & 0 \\ 7 & 22 & 1 \\ 8 & 23 & 2 \end{bmatrix}$$

$$B_2 = [\, \bar{r}_3 \quad \bar{r}_0 \quad \bar{r}_1\,]$$
$$= \begin{bmatrix} 9 & 0 & 3 \\ 10 & 1 & 4 \\ 11 & 2 & 5 \end{bmatrix}$$

$$B_3 = [\, \bar{r}_4 \quad \bar{r}_1 \quad \bar{r}_2\,]$$
$$= \begin{bmatrix} 12 & 3 & 6 \\ 13 & 4 & 7 \\ 14 & 5 & 8 \end{bmatrix}$$

$$B_4 = [\, \bar{r}_5 \quad \bar{r}_2 \quad \bar{r}_3\,]$$
$$= \begin{bmatrix} 15 & 6 & 9 \\ 16 & 7 & 10 \\ 17 & 8 & 11 \end{bmatrix}$$

$$B_5 = [\, \bar{r}_6 \quad \bar{r}_3 \quad \bar{r}_4\,]$$
$$= \begin{bmatrix} 18 & 9 & 12 \\ 19 & 10 & 13 \\ 20 & 11 & 14 \end{bmatrix}$$

$$B_6 = [\ \bar{r}_7 \quad \bar{r}_4 \quad \bar{r}_5]$$

$$= \begin{bmatrix} 21 & 12 & 15 \\ 22 & 13 & 16 \\ 23 & 14 & 17 \end{bmatrix}$$

$$B_7 = [\ \bar{r}_0 \quad \bar{r}_5 \quad \bar{r}_6]$$

$$= \begin{bmatrix} 0 & 15 & 18 \\ 1 & 16 & 19 \\ 2 & 17 & 20 \end{bmatrix}$$

$$B_8 = [\ \bar{r}_0 \quad \bar{r}_4 \quad \bar{r}_8]$$

$$= \begin{bmatrix} 0 & 12 & 24 \\ 1 & 13 & 25 \\ 2 & 14 & 26 \end{bmatrix}$$

$$B_9 = [\ \bar{r}_1 \quad \bar{r}_5 \quad \bar{r}_8]$$

$$= \begin{bmatrix} 3 & 15 & 24 \\ 4 & 16 & 25 \\ 5 & 17 & 26 \end{bmatrix}$$

$$B_{10} = [\ \bar{r}_2 \quad \bar{r}_6 \quad \bar{r}_8]$$

$$= \begin{bmatrix} 6 & 18 & 24 \\ 7 & 19 & 25 \\ 8 & 20 & 26 \end{bmatrix}$$

$$B_{11} = [\ \bar{r}_3 \quad \bar{r}_7 \quad \bar{r}_8]$$

$$= \begin{bmatrix} 9 & 21 & 24 \\ 10 & 22 & 25 \\ 11 & 23 & 26 \end{bmatrix}$$

By Step 3, $U(0,0) = F(A_0)\ [B_0]$

$$B_0 = \begin{bmatrix} 3 & 18 & 21 \\ 4 & 19 & 22 \\ 5 & 20 & 23 \end{bmatrix} \qquad A_0 = \begin{bmatrix} 0 & 1 & 2 \\ 2 & 0 & 1 \\ 1 & 2 & 0 \end{bmatrix}$$

$$U(0,0) = \begin{bmatrix} 3 & 19 & 23 \\ 5 & 18 & 22 \\ 4 & 20 & 21 \end{bmatrix}$$

$$U(4,0) = \begin{bmatrix} 15 & 7 & 11 \\ 17 & 6 & 10 \\ 16 & 8 & 9 \end{bmatrix}$$

$$\vdots$$

$$U(11,0) = \begin{bmatrix} 9 & 22 & 26 \\ 11 & 21 & 25 \\ 10 & 23 & 24 \end{bmatrix}$$

$$U(0,1) = F(A_1) [B_0]$$

$$B_0 = \begin{bmatrix} 3 & 18 & 21 \\ 4 & 19 & 22 \\ 5 & 20 & 23 \end{bmatrix} \qquad A_1 = \begin{bmatrix} 0 & 2 & 1 \\ 1 & 0 & 2 \\ 2 & 1 & 0 \end{bmatrix}$$

$$U(0,1) = \begin{bmatrix} 3 & 20 & 22 \\ 4 & 18 & 23 \\ 5 & 19 & 21 \end{bmatrix}$$

$$\vdots$$

$$U(11,1) = F(A_1) [B_{11}]$$

$$= \begin{bmatrix} 9 & 23 & 25 \\ 10 & 21 & 26 \\ 11 & 22 & 24 \end{bmatrix}.$$

The rows of $B_j$ ($j = 0, 1, \ldots, 11$); the rows of $M$, and all the rows of $U(j,k)$ for $k = 0, 1$ constitute the set of code words.

*Example 3.2:*   Using the algorithm described in this section, a computer program can be easily developed. From such a program, the $n = 4$ with $\{D_E\} = \{1,3,4,12\}$ RMA sequences can be generated with an $M$ matrix of size $16 \times 4$. The sequences generated are shown in Appendix 2.

## 3.4   CHARACTERISTICS OF THE RMA SEQUENCES

This section is designed to help the reader understand the algorithm presented in Section 3.3, and also to demonstrate the validity of the sequence construction algorithm. The results shown here will establish the foundations for practical applications.

The first theorem in this section is concerned with the distance measure of this class of RMA sequences. A minimum distance is defined and obtained in terms of the sequence length.

To obtain the overall minimum distance of the sequence, we first analyze the minimum distance of the rows of $[W]_0$. Then we use the results from Chapter 2 to derive the minimum distance of the rows of $[W]_1$. We can then obtain the minimum distance of the set of RMA sequences, based upon the relationships of the rows of $[W]_0$ and $[W]_1$. Finally, we obtain the number of sequences, also in terms of the sequence length.

The assumptions made in this section are those described in the construction procedure; i.e., the number of symbols is $fn$. We consider the case here only when $f = n^2$, and $n = k$, which is a parameter of $\{D_E\}$ as described in Chapter 2. The distance measure for this class of sequence is defined as follows:

*Definition:* The minimum distance ($d_{min}$) of the RMA sequence is defined as the number of symbol disagreements between any pair of different sequences in the set with the largest number of symbol agreements. Symbols are considered to be in agreement if they appear in any position of any two sequences.

The above definition differs from either Hamming or Lee distance measures. The RMA sequence definition refers to the number of common symbols between a pair of sequences, regardless of the positions of the common symbol. Thus, the minimum distance defined here is more restricted than the commonly used definitions in the coding literature. This is one of the reasons why nonbinary BCH codes and Reed-Solomon codes are not suitable for RMA applications.

We shall now show that the minimum distance of the RMA sequences is $n-1$. The minimum distance derivation is based on the definition and the assumptions stated above.

From Step 1 of the RMA sequence construction procedure we recognize that $w_i$ is an equivalent $\{D\}$ for each $i = 1, 2, \ldots, n^2 - 2$ from the initial $w_0$.

$$
\begin{aligned}
w_0 &= \{d_0 &, d_1 &, d_{n-1} &\} \\
w_1 &= \{d_0+1 &, d_1+1 &, d_{n-1}+1 &\} \\
w_2 &= \{d_0+2 &, d_1+2 &, d_{n-1}+2 &\} \\
&\vdots & & & \\
w_{n^2-2} &= \{d_0+n^2-2, d_1+n^2-2, d_{n-1}+n^2-2\}
\end{aligned}
\tag{3.14}
$$

(3.14) and the exponents of $\sigma$'s in (2.89) of Chapter 2 are the same. From Theorem 2.13, the lines of (2.89) are all distinct. The points on the line $L_i$ are

$$
P(\sigma^{d_0+i}), P(\sigma^{d_1+i}), \ldots, P(\sigma^{d_{n-1}+i})
\tag{3.15}
$$

for $i = 0, 1, \ldots, n^2 - 2$. If these lines are distinct, no two lines can have more than one point in common. Thus we have proved

LEMMA 3.1: The minimum distance of the rows of $[W]_0$ is

$$
d_{min}[W]_0 = n-1.
\tag{3.16}
$$

For a given $\{D\}$ no multiple of $(n+1)$ from 0 to $(n-1)$ can be a difference modulo $n^2 - 1$ among the set of residues, according to Theorem 2.14 (Chapter 2). This result says that the pairwise differences do not contain $n + 1$. Let us consider two arbitrary residues $d_j$ and $d_{j'}$ of $\{D\}, j \neq j'$. If $d_j \neq 0$ we can always subtract $d_j$ to make its value equal to 0; i.e., $d_j = d_j \equiv 0 \bmod (n^2 - 1)$. Similarly, $d_j - d_{j'} \equiv k \bmod (n^2-1)$ for some integer $k$. Theorem 2.14 says that among all the differences of $\{D\}$ $n+1 \neq k$. Or

$$
(d_j+i') - (d_{j'}+i') \not\equiv N \bmod (n^2-1)
\tag{3.17}
$$

where $N$ is a multiple of $(n+1)$. From the construction procedure of (3.6) for $[W_1]$:

$$
\begin{aligned}
w_{n^2-1+i'} &= [0(n+1)+i', 1(n+1)+i', \ldots, (n^2-1)] \\
i' &= 0, 1, \ldots, n.
\end{aligned}
\tag{3.18}
$$

Expanding the right-hand side of (3.18) for all $i'$, we have

$$
\begin{array}{lllll}
0 & (n+1) & 2(n+1) & \cdots & (n^2-1) \\
1 & (n+1)+1 & 2(n+1)+1 & \cdots & (n^2-1) \\
2 & (n+1)+2 & 2(n+1)+2 & \cdots & (n^2-1) \\
\vdots & \vdots & \vdots & & \vdots \\
n & (n+1)+n & 2(n+1)+n & \cdots & (n^2-1).
\end{array}
\tag{3.19}
$$

If we exclude the last entry $n^2-1$ in (3.19) which is common to every row of $[W]_1$, no two entries of the form $\ell(n+1)+i'$ are the same; hence we have just proved

LEMMA 3.2:   The minimum distance of the rows of $[W]_1$ is $n-1$.

$$
d_{\min}[W]_1 = n-1. \tag{3.20}
$$

Since from Step 2 of the code construction procedure, $w_j$'s are selected from the rows of $[W]_0$ and $[W]_1$, we need to show:

LEMMA 3.3:   The minimum distance between any row from $[W]_0$ and any row from $[W]_1$ is $n-1$.

*Proof:*   Since

$$
w_i = [d_0+i, \ldots, d_\gamma+i, \ldots, d_\beta+i, \ldots, d_{n-1}+i]
$$

$$
0 \le i \le n^2-2
$$

$$
w_{n-1+\phi}^2 = [\phi, (n+1)+\phi, \ldots, \gamma'(n+1)+\phi, \ldots,
$$

$$
\beta'(n+1)+\phi \ldots, n^2-1].
$$

If these two rows have one element in common:

$$
d_\gamma+i = \gamma'(n+1) + \phi, \tag{3.21}
$$

then it is impossible to have another element in common. Let us assume the countrary; i.e.,

$$
d_\beta+i = \beta'(n+1) + \phi.
$$

Then

$$
(d_\beta+i) - (d_\gamma+i) = (\beta'(n+1)+\phi) - (\gamma'(n+1)+\phi) \tag{3.22}
$$

or

$$
d_\beta - d_\gamma = (\beta'-\gamma')(n+1). \tag{3.23}
$$

But, from Theorem 2.13, no multiple of $(n+1)$ is a difference of the elements in a $\{D_E\}$. Thus we have proved Lemma 3.3.

Step 4 of the sequence generation algorithm requires a transformation by $A_k$. The following lemma examines the distance property resulting from this transformation.

LEMMA 3.4:   The minimum distance property is invariant under the transformation $F$ due to $A_k$.

*Proof:*   Since the code symbols in each $B_j$ matrix are distinct, it is evident that the column ordering transformation from each $A_k$ does not alter the distance property of the subset of code words produced by $A_k$. Next, we will show the distance property of any pair of code words generated from two different $A_k$.

The minimum distance of the set of rows of $W$ was established to be $n-1$ where

$$W = \begin{cases} [W]_0 \\ [W]_1 \end{cases}. \tag{3.24}$$

This distance property reflects the rows of $B_j$ as follows: for each $j$ the rows of $B_j$ are distinct, i.e., $d_{min} = n$. If two rows from two different $B_j$'s have a code symbol in common they must have the same symbol in the corresponding rows of $w$, i.e., the same $\bar{r}_\xi$. As it has been shown before, only one $\bar{r}_\xi$ can exist between any pair of rows in $W$. When this happens, there exists a common column $\bar{r}_\xi$ between two different $B_j$'s. It looks like:

$$R_j = [\ \bar{r}_a \quad \bar{r}_b \quad \ldots \quad \ldots \quad \bar{r}_\xi \quad \ldots \quad \bar{r}_z\ ] \tag{3.25}$$

$$B_j = \begin{bmatrix} a_0 & b_0 & \ldots & \ldots & \xi_0 & \ldots & z_0 \\ a_1 & b_1 & & & \xi_1 & \ldots & z_1 \\ \vdots & \vdots & & & \vdots & & \vdots \\ a_{n-1} & b_{n-1} & \ldots & \ldots & \xi_{n-1} & \ldots & z_{n-1} \end{bmatrix} \tag{3.26}$$

$$R_j' = [\ \bar{r}_a' \quad \bar{r}_b' \quad \ldots \quad \bar{r}_\xi \quad \ldots \quad \ldots \quad \bar{r}_z'\ ] \tag{3.27}$$

$$B_{j'} = \begin{bmatrix} a_0' & b_0' & \ldots & \xi_0 & \ldots & \ldots & z_0' \\ a_1' & b_1' & \ldots & \xi_1 & \ldots & \ldots & z_1' \\ \vdots & \vdots & & \vdots & & & \vdots \\ a_{n-1}' & b_{n-f}' & \ldots & \xi_{n-1} & \ldots & \ldots & z_{n-1}' \end{bmatrix} \tag{3.28}$$

where $\bar{r}_\xi$ can appear in different positions in $R_{j'}$ or $R_j$.

Since only $\bar{r}_\xi$ can be in common between any $R_j$ and $R_{j'}$ ($j' \neq j$), it is not only $\bar{r}_a \neq \bar{r}_a', \bar{r}_b \neq \bar{r}_b', \ldots$, but $\bar{r}_a \neq \bar{r}_b', \ldots \neq \bar{r}_z'$, also. In other words, no column symbols are in common either within a $B_j$, or between two different $B_j$'s. Since all the $\bar{r}$'s are different except $\bar{r}_\xi$ between two $B_j$'s, all the sequence symbols are different except $\xi_0, \xi_1, \ldots, \xi_{n-1}$. Different $\bar{r}_\xi$'s lead to distinctive symbols. Any nonrepeatable column permutations of the distinctive symbols do not change the minimum distance of the set of sequences. What remains to be shown is that the construction of $A_k$ is systematic; i.e., the orders in its columns are nonrepeatable. Actually, $A_k$ is a set of circulant matrices with elements $\bar{\alpha} = \alpha_0, \alpha_1, \ldots, \alpha_{n-1}$. A circulant matrix is completely determined by the elements in its 0th row. By permutations of the elements in its 0th row, the elements $\alpha_1, \alpha_2, \ldots, \alpha_{n-1}$ are cyclically shifted and the complete set of circulant matrices $A_k$ is then obtained.

The distinctiveness of the order of the $\alpha$ elements in the columns of $A_k$ can be shown in Figure 3.1, where the diagonal lines represent the $\alpha$ elements. The dotted region, which is the size of $A_k$, is cut with vertical lines representing the columns. It is evident that no vertical line intersects two identical $\bar{\alpha}$ lines twice in the region. In addition, it can be seen that each $\bar{\alpha}$ line intersects the vertical lines at distinct positions. As $k$ increases, the parallel $\bar{\alpha}$ lines (except $\alpha_0$) rotate upward diagonally as shown. For each rotation, the set of vertical lines never intersects the same $\bar{\alpha}$ line twice. Thus, the order of all columns $A_k$ is distinct.

As a consequence, the code distance property is thus invariant under the transformation $F(A_k)$ due to $A_K$; thus Lemma 3.3 is proved; and such transformation only increases the number of sequences.

From Lemma 3.1, the minimum distance of the rows of $[W]_0$ is equal to $n-1$. This property of the row distances reflects immediately on the subset of RMA sequences derived from the rows of $[W]_0$. From Lemma 3.2 the minimum distance of the rows of $[W]_1$ is $n-1$. This is also the minimum distance of the RMA sequences generated from the rows of $[W]_1$. Lemma 3.3 implies that the minimum distance between any pair of sequences, one from $[W]_0$ and the other from $[W]_1$, is $n-1$. Lemma 3.4 ensures that the distance property is unchanged by the transformation of $A_k$ regardless of whether the sequences before transformation come from $[W]_0$ or $[W]_1$. Since the $A_k$ are circulant, the column order of $A_k$ is distinct and the distance property is also $n-1$ for different $A_k$. As a result, we have proved the following:

THEOREM 3.1:   The class of RMA sequences, with length $n$, generated by the algorithm of Section 3.3 has minimum distances

$$d_{\min} = n - 1. \tag{3.29}$$

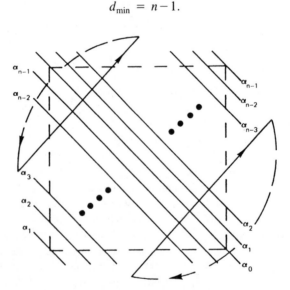

**Figure 3.1** The $\alpha$ lines of $A_k$.

THEOREM 3.2:  The number of the RMA sequences from the algorithm is $n^2(n^2+n+1)$.

*Proof:*  There are $n^2-1$ rows $w_i$ in $[W]_0$, and $n+1$ rows $w_i$ in $[W]_1$. Each of these $n^2+n$ rows produces a $B_j$ matrix of $n$ rows. The rows of $B_j$ are subsets of the sequences. There are $n(n^2+n)$ sequences in this subset; i.e., before transformation. Since there can be $n-1$ number of $A_k$ matrices, and each $A_k$ exhausts all the $B_j$ matrices for the ordering transformation, the number of sequences after $F(A_k)$ is $(n-1)\,n(n^2+n)\,+\,n(n^2+n)+n^2$.

As consequences of the four lemmas and two theorems presented in this section, we have established the results that for a sequence length $n$ with $n$ time divisions and $n^2$ frequency divisions, there exist $n^2(n^2+n+1)$ RMA sequences with pairwise distances of $n-1$. Table 3.1 lists the number of symbols and the number of sequences for $n \le 31$. Sections 3.2 and 3.3 are part of the results originally derived by Wu [6].

## 3.5  PERFORMANCE DERIVATIONS OF RMA SYSTEMS

Based on the discussions in the previous sections we can now proceed to establish some references of measure and the derivations of some criteria in order to estimate the performaces of coded RMA systems under most general conditions. We first provide a simple derivation of a known result in combinatorics, which otherwise would be difficult to justify. This result is shown to be useful in

**Table 3.1**

Parameters for RMA sequences up to length 31.

| $n$ | $p^m$ | Number of Symbols $n^3$ | Number of Sequences $n^2(n^2+n+1)$ |
|---|---|---|---|
| 1 | $1^1$ | 1 | 3 |
| 2 | $2^1$ | 8 | 28 |
| 3 | $3^1$ | 27 | 117 |
| 4 | $2^2$ | 64 | 336 |
| 5 | $5^1$ | 125 | 775 |
| 7 | $7^1$ | 343 | 2793 |
| 8 | $2^3$ | 512 | 4672 |
| 9 | $3^2$ | 729 | 7371 |
| 11 | $11^1$ | 1331 | 16093 |
| 13 | $13^1$ | 2197 | 30927 |
| 16 | $2^4$ | 4096 | 69888 |
| 17 | $17^1$ | 4913 | 88723 |
| 19 | $19^1$ | 6859 | 137541 |
| 23 | $23^1$ | 12167 | 292537 |
| 27 | $3^3$ | 19683 | 551853 |
| 29 | $29^1$ | 24389 | 732511 |
| 31 | $31^1$ | 29791 | 954273 |

describing both the access probability and the false detection probability. Next the miss-detection probability is formulated in terms of a set of code symbol transition probabilities. Finally, in this section we relate the channel utilization factor to the number of simultaneous active users in the system, the time and frequency division units, and the length of an RMA code word.

### 3.5.1.  System Access Probability

The objective here is to find the probability of the total number of active users simultaneously (as opposed to the larger number of potential users designed into the system) accessing the satellite. We shall begin with elementary algebra.

We recall that the binomial coefficients in the expansion of $(x+y)^n$ for $n = 1,2,\ldots$ can be obtained by Pascal's triangle, which can be easily constructed as follows:

$$
\begin{array}{ccccccccccccc}
 & & & & & & 1 & & & & & & \\
 & & & & & 1 & & 2 & & 1 & & & \\
 & & & & 1 & & 3 & & 3 & & 1 & & \\
 & & & 1 & & 4 & & 6 & & 4 & & 1 & \\
 & & 1 & & 5 & & 10 & & 10 & & 5 & & 1 \\
 & 1 & & 6 & & 15 & & 20 & & 15 & & 6 & & 1 \\
 & & & & & & \vdots & & & & & &
\end{array}
$$

Next we put alternating signs on each row of the triangle and, summing up each row except the last entry of each row, the triangle becomes:

$$
\begin{array}{cccccccc}
 & & & & 1 & & & \\
 & & & 1 & -2 & & = -1 & \\
 & & 1 & -3 & & 3 & & = 1 \\
 & 1 & -4 & & 6 & & -4 & = -1 \\
1 & -5 & & 10 & & -10 & & 5 & = 1 \\
1 & -6 & 15 & & -20 & & 15 & & -6 & = -1 \quad (3.30)\\
 & & & & \vdots & & &
\end{array}
$$

We notice that the row sums of the triangle become $1, -1, 1, -1, \ldots$ The elements in the Pascal triangle can now be expressed in combinatorial form as

$$1 = 1$$

$$1 - \binom{2}{1} = -1$$

$$1 - \binom{3}{2} + \binom{3}{1} = 1$$

$$1 - \binom{4}{3} + \binom{4}{2} - \binom{4}{1} = -1$$

$$1 - \binom{5}{4} + \binom{5}{3} - \binom{5}{2} + \binom{5}{1} = 1$$

$$\vdots \qquad\qquad \vdots$$

$$1 - \binom{n}{n-1} + \binom{n}{n-2} - \ldots \pm \binom{n}{1} = \pm 1. \tag{3.31}$$

Multiplying the first row equation of (3.31) by $S_1$, the second row equation by $S_2$, the third row equation by $S_3$, . . ., and the $n$th row equation by $S_n$, we get

$$S_1 = S_1$$

$$\left[1 - \binom{2}{1}\right] S_2 = -S_2$$

$$\left[1 - \binom{3}{2} + \binom{3}{1}\right] S_3 = S_3$$

$$\left[1 - \binom{4}{3} + \binom{4}{2} - \binom{4}{1}\right] S_4 = -S_4$$

$$\left[1 - \binom{5}{4} + \binom{5}{3} - \binom{5}{2} + \binom{5}{1}\right] S_5 = S_5$$

$$\vdots \qquad \vdots$$

$$\left[1 - \binom{n}{n-1} + \binom{n}{n-2} - \ldots \pm \binom{n}{1}\right] S_n = \pm S_n. \tag{3.32}$$

Adding all the equations in (3.32), we see that the righthand side sum is $P_1$ of (3.3) in Section 3.3.3 of Chapter 3, Volume I. The lefthand sides of (3.32) become

$$P_1 = S_1 + \left[1 - \binom{2}{1}\right] S_2 + \left[1 - \binom{3}{2} + \binom{3}{1}\right] S_3 +$$

$$\ldots + \left[1 - \binom{n}{n-1} + \ldots \pm \binom{n}{1}\right] S_n$$

$$= S_1 - \binom{2}{1} S_2 + \binom{3}{1} S_3 - \binom{4}{1} S_4 + \ldots$$

$$+ S_2 - \binom{3}{2} S_3 + \binom{4}{2} S_4 - \binom{5}{2} S_5 + \ldots$$

$$+ S_3 - \binom{4}{3} S_4 + \binom{5}{3} S_5 - \binom{6}{3} S_6 + \ldots$$

$$\vdots$$

$$+ S_{n-2} - \binom{n-1}{n-2} S_{n-1} + \binom{n}{n-2} S_n$$

$$+ S_{n-1} - \binom{n}{n-1} S_n$$

$+ S_n.$

$$= P(1) + P(2) + P(3) + \ldots + P(n-2)$$
$$+ P(n-1) + P(n)$$

$$= \sum_{z=1}^{n} P(z) \tag{3.33}$$

where

$$P(1) = S_1 - \binom{2}{1} S_2 + \binom{3}{1} S_3 - \binom{4}{1} S_4 + \ldots$$

$$P(2) = S_2 - \binom{3}{2} S_3 + \binom{4}{2} S_4 - \binom{5}{2} S_5 + \ldots$$

$$\vdots \qquad\qquad\qquad \vdots$$

$$P(n-1) = S_{n-1} - \binom{n}{n-1} S_n$$

$$P(n) = S_n. \tag{3.34}$$

Therefore, for any positive integer $z = 1, 2, \ldots, n$,

$$P(z) = S_z - \binom{z+1}{z} S_{z+1} + \binom{z+2}{z} S_{z+2} - + \ldots$$

$$\ldots \pm \binom{n}{z} S_n. \tag{3.35}$$

In an RMA system let us assume there are $K$ potential users. We define as $E_j$ the event that user $j$ desires access to communicate through the satellite. Since the needs of access are independent among the set of users, the events $E_1$, $E_2$, $\ldots$, $E_K$ are statistically independent, with probabilities $P(E_1)$, $P(E_2)$, $\ldots$, $P(E_K)$. For any number of users $k$ ($K > k \geq 1$), the probability of exactly $k$ users desiring access simultaneously is as follows. Let $z=k$ in (3.35):

$$P(k) = S_k - \binom{k+1}{k} S_{k+1} + \binom{k+2}{k} S_{k+2} + \ldots.$$

$$\pm \binom{K}{k} S_K, \tag{3.36}$$

where $S_k$ is the sum of the probabilities of $k$ events; i.e.,

$$S_k = \underbrace{\sum_{a,b,\ldots}}_{k} P \, [E_a \underbrace{E_b \ldots .}_{k}] \tag{3.37}$$

$$S_{k+1} = \sum_{\underbrace{a,b,\ldots}_{k+1}} P[\underbrace{E_a E_b \ldots \ldots}_{k+1}] \tag{3.38}$$

$$\vdots \tag{3.39}$$

$$S_K = \sum_{\underbrace{a,b,\ldots}_{K}} P (\underbrace{E_a E_b \ldots}_{K}).$$

Through (3.35) we have related the individual user access probabilities to system access probability.

Since the access events from the users are independent, we can write

$$P(E_a E_b \ldots \ldots E_x) = P(E_a) P(E_b) \ldots \ldots P(E_x). \tag{3.40}$$

If each user is accessing with Poisson point process messages with different message arrival rates, then for message source $a$ in an interval $t$

$$P(E_a) = e^{-\lambda_a t} \frac{(\lambda_a t)^{k_a}}{k_a!}, \tag{3.41}$$

where $k_a$ is the number of messages of user $a$ in time $t$. Similarly, we have

$$P(E_b) = e^{-\lambda_b t} \frac{(\lambda_b t)^{k_b}}{k_b!} \tag{3.42}$$

for source $b$, and so on. As a result, $P(k)$ can be obtained by substituting the $P(E_a)$, $P(E_b)$, . . ., for $S_k$, $S_{k+1}$, . . ., $S_K$.

*Example 3.3:*   Let there be $k = 5$ potential users. Each has a Poisson arrival rate of $\lambda_a = .1$, $\lambda_b = .2$, $\lambda_c = .3$, $\lambda_d = .4$, $\lambda_e = .5$ messages per microsecond. The corresponding numbers of messages per $t = 1.0$ second are $k_a = 1.0$, $k_b = 2.0$, $k_c = 3.0$, $k_d = 4.0$, and $k_e = 5.0$. The probability of exactly $k = 3$ users simultaneously accessing the satellite can be calculated as follows: $S_3 = 3! P(E_a E_b E_c) = 6 P(E_a) P(E_b) P(E_c) = 3 \times 10^{-5}$; $S_4 = 7.7 \times 10^{-8}$; $S_5 \approx 6 \times 10^{-11}$. We substitute $S_3$, $S_4$, and $S_5$ in (3.36) and get

$$P(3) = 3 \times 10^{-5} - \binom{4}{3} 7.7 \times 10^{-8} + \binom{5}{3} 6 \times 10^{-11}.$$

$$\approx 3 \times 10^{-5}.$$

### 3.5.2   Error Probability Derivations

In this section both false alarm and miss-detection probabilities are developed for nonperiodic RMA signals. The false alarm probability is expressed in terms of a set of combinatorial quantities. The miss-detection probability is developed through multinominal argument.

### 3.5.2.1   *False Detection Probability*

False detection error probability is based on the conditional probability that an RMA sequence is considered as detection but in fact the sequence was not sent. The error probability derivation is based on the detection process, in which two RMA sequences of $n$ positions each are matched against each other. An overlap occurs if identically matched symbols occupy the same position. If we let $z$ be the number of overlaps, then $z$, a random variable, takes on the values of 0, 1, . . ., $n$. Let $P(z)$ be the probability that there are exactly $z$ overlaps in $n$ symbols. Then we can use the expression (3.35) to obtain $P(z)$ in terms of the $S_z$'s.

To formulate an error probability expression we need to find the probability of overlapping code symbols beyond the designed tolerance of the sequence. Since we consider $m = n^3$ symbols of length $n$, the sample space under consideration is $\binom{m}{n}$. Let $P_i$ be the probability of an overlap at the $i$th digit. This implies that the remaining $n - 1$ symbols may be in any arbitrary arrangement in a sequence. After the $i$th symbol is fixed, there exist $\binom{m-1}{n-1}$ combinations of $m - 1$ other symbols to choose from. This gives

$$P_i = \frac{\binom{m-1}{n-1}}{\binom{m}{n}} = \frac{n}{m} = n^{-2}. \tag{3.43}$$

For two overlaps occurring at $i, j$ the probability of such an event is

$$P_{i,j} = \frac{\binom{m-2}{n-2}}{\binom{m}{n}}. \tag{3.44}$$

Accordingly, the probability of $z$ overlaps

$$P_{i,j, \ldots, z} = \frac{\binom{m-z}{n-z}}{\binom{m}{n}}. \tag{3.45}$$

If we assume the overlapped $z$ symbols are equally likely in $n$ positions, then we have

$$S_z = \frac{\binom{n}{z}\binom{m-z}{n-z}}{\binom{m}{n}}. \tag{3.46}$$

System error occurs when $\varepsilon + 1$ or more symbol overlaps can occur, where $\varepsilon$ is the number of predetermined tolerable overlaps. Therefore, the false detection probability is from (3.35).

$$f(m, n, \varepsilon) = \sum_{z=\varepsilon+1}^{n} P(z) = \sum_{z=\varepsilon+1}^{n} (-1)^{z-1} S_z. \tag{3.47}$$

If we substitute $S_z$ of (3.46) into (3.47), we have

$$f(m, n, \varepsilon) = \sum_{z=\varepsilon+1}^{n} (-1)^{z-1} \frac{\binom{n}{z}\binom{m-z}{n-z}}{\binom{m}{n}}. \tag{3.48}$$

If we substitute $m = n^3$ into (3.48), after canceling terms we have the general expression for the false detection probability of an RMA system as

$$f(n, \varepsilon) = \sum_{z=\varepsilon+1}^{n} (-1)^{z-1} \left[ \frac{n!}{(n-z)!} \right]^2 \frac{(n^3-z)!}{n^3!\, z!}. \tag{3.49}$$

### 3.5.2.2   Miss-Detection Probability

The miss-detection error probability $P_m(\lambda_Q)$, due to $\lambda_Q$ repeated symbols with lengths $n$ and symbol transition probability $p$ $(I/Q)$, is derived as follows: if we assume the sequence $W_0$ was transmitted, then,

$$P_m (\lambda_Q) = P_r \text{ (reject } W_0/W_0 \text{ was sent)}$$

$$= P_r \text{ (reject } W_0 \text{ due to error caused}$$

$$\text{by repeated symbols).}$$

We set out to derive the probability of $\lambda_Q$ repeated code symbols existing in an $n$-digit sequence based upon a set of given symbol transition probabilities.

Let the probability of realization of the RMA symbol $Q(q = 0, 1, \ldots, Q, \ldots, m-1)$ appearing at the $\ell$th position of a sequence be $P_\ell$ $(I/Q)$, where $\ell = 0, 1, \ldots, n-1$, and $I = 0, 1, \ldots, m-1$.

$$P_\ell(I|q) = \begin{bmatrix} P_\ell(0/0) & \cdots & P_\ell[(m-1)/0] \\ P_\ell(0/1) & & P_\ell[(m-1)/1] \\ \vdots & & \\ P_\ell[0/(m-1)] & \cdots & P_\ell[(m-1)/(m-1)] \end{bmatrix} \tag{3.50}$$

The reason this symbol transition probability exists is that the frequencies of adjacent symbols are more likely to be converted than the nonadjacent ones. Obviously,

$$\sum_I P_\ell (I|Q) = 1 \tag{3.51}$$

and a nonzero matrix element of (3.50) for each received sequence position. Next, denote the possible outcomes of each symbol by $e_0, e_1, \ldots e_m - 1$. The probability that an $n$-digit sequence $e_0$ occurs $\lambda_0$ times, $e_1$ occurs $\lambda_1$ times, . . ., $e_{m-1}$ occurs $\lambda_{m-1}$ times is the multinomial distribution:

$$P_m(\lambda_0, \lambda_1, \ldots, \lambda_{m-1}) = \frac{n!}{\lambda_0!\lambda_1! \ldots \lambda_{m-1}!} P_\ell^{\lambda_0} (e_0)$$
$$\cdot P_\ell^{\lambda_1} (e_1) \ldots, P_\ell^{\lambda_{m-1}} (e_{m-1}), \quad (3.52)$$

with

$$\lambda_1 + \lambda_2 + \ldots \lambda_{m-1} = m,$$
$$\lambda_1, \lambda_2, \ldots \lambda_{m-1} \geq 0. \quad (3.53)$$

For a miss-detection word containing $\lambda_Q$ repeats of the symbol $Q$, there are $n - \lambda_Q$ nonrepeated symbols; i.e., $n - \lambda_Q$ $\lambda$'s $= 1$. The rest of the $m - n$ $\lambda$'s are zero, which means there are $m - n$ probabilities in the product expression of (3.52) equal to unity. The remaining $n$ probabilities are less than or equal to unity for each of the $\lambda \geq 0$. Thus (3.52) becomes

$$P_m(\lambda_Q) = \frac{n!}{\lambda_Q!} P_\ell^{\lambda_Q} (e_Q) \prod_{q \neq Q}^{n-\lambda_Q} P_\ell^{\lambda_\sigma} (I|q), \quad (3.54)$$

with $\lambda_Q \geq 1$ and $\lambda_\sigma = 1$. We obtain the miss-detection probability

$$P_m(\lambda_Q) = \sum_{\lambda_Q=1}^{n} \frac{n!}{\lambda_Q!} P_\ell^{\lambda_Q} (e_Q) \prod_{\substack{\ell \neq Q \\ \ell=0}}^{n-\lambda_Q} P (e_\ell). \quad (3.55)$$

The evaluation of $P_m(\lambda_Q)$ depends on the symbol transition probabilities $P_\ell (e_\ell)$. For large numbers of code symbols the evaluation of $P_m(\lambda_Q)$ can be tedious.

### 3.5.3 Channel Utilization Factor of RMA Systems

The channel utilization factor $\mu_t$ of any communication system is defined as the ratio of average total information rate to the channel bandwidth. There are no exceptions for the RMA systems under consideration. Let the total average information rate of an RMA system be denoted as $\overline{R}$, then

$$\overline{R} = \frac{\begin{pmatrix} \text{Average number of} \\ \text{active message users} \end{pmatrix} \begin{pmatrix} \text{Information conveyed} \\ \text{by one user} \end{pmatrix}}{\text{Duration of the TF matrix}}. \quad (3.56)$$

If we specify a TF matrix, then the number of time divisions is its duration and equal to $nt$ where $t$ is the time unit after the division. If we assume a pair of RMA sequences of length $n$ represents the binary addressed signaling, the maximum number of message users in the system is $n^2(n^2+n+1)/2$. The average number of active users is a fraction of this maximum number of users, or

$Fn^2(n^2+n+1)/2$, for $F<1$. The binary information conveyed by one user is just one bit. Thus

$$\overline{R} = \frac{[F \cdot n^2(n^2+n+1)/2]}{nt}$$

$$= Fn(n^2+n+1)/2t, \tag{3.57}$$

where $F$ is a fraction in percentage.

If the system bandwidth $W$ contains $n^2$ frequency divisions each having bandwidth $f$, then

$$\mu_t = \frac{\overline{R}}{W} = \frac{Fn(n^2+n+1)/2t}{n^2f}$$

$$= \frac{F(n^2+n+1)}{2nft}. \tag{3.58}$$

Thus we have related the RMA channel utilization factor to the time division unit $t$, the frequency division unit $f$, the RMA sequence length $n$, and the fraction $F$ of the total number of users. If the number of simultaneous active users is given as $k$, then (3.58) becomes

$$\mu_t = \left[\frac{Fn^2(n^2+n+1)}{2}\right]\left(\frac{1}{n^3ft}\right)$$

$$= \frac{k}{n^3ft}, \tag{3.59}$$

which says that for fixed sequence length and TF units the channel utilization factor is proportional to the number of active simultaneous users. On the other hand, for a given average number of active users the channel utilization factor is inversely proportional to the time and frequency division units. Because the length of the RMA sequence $n$ is essentially a spreading factor in the time domain, we would expect that $\mu_t$ decreases with $n$. But (3.59) tells us how $\mu_t$ exactly depends on $n$.

*Example 3.4:*   For simplicity let $P(e_\ell)$ be equal for all $\ell$; then (3.55) reduces to

$$P_m(\lambda_Q) = \sum_{\lambda_Q=1}^{n} \frac{n!}{\lambda_Q!} [P_\ell(e_Q)]^{\lambda_Q}$$

$$\cdot [P_\ell(e_\ell)]^{n-\lambda_Q}. \tag{3.60}$$

A sample evaluation of the miss-detection error rates is shown in Figures 3.2 to 3.4, in which $n$ denotes the RMA sequence length and the pair probability combinations are designated as shown.

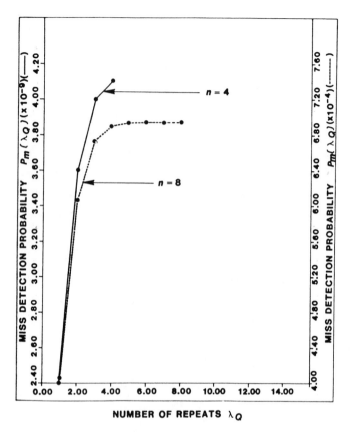

**Figure 3.2** Miss-detection error rate vs. the number of repeats of an RMA symbol for $P_\ell(e_Q) = 10^{-1}$, $P_\ell(e_\ell) = 10^{-1}$.

## 3.6   ANTENNAS

Normally the subject of antennas is beyond the scope of a book on the elements of digital satellite communication, but multiple beam, scanning beam, and discrete phase array antennas are more closely involved with communication than are spacecraft hardware designs. However, the purpose of this section and the next is not to provide an introduction to or a survey of satellite communication antennas, which can be found in references such as [7, 8, 9]; rather, the emphasis is on how to space the phase array antennas to systematically minimize sidelobe levels. The method is based once again on difference sets.

Combining the advantages of the high gain of a spot beam antenna and the transmission efficiency of TDMA, Reudink and Yeh proposed a movable scanning spot-beam antenna system for satellite communication [10]. The result is either to reduce the satellite transmitter power, to increase transmission capacity, or to decrease the size of earth station antennas. Two approaches for forming

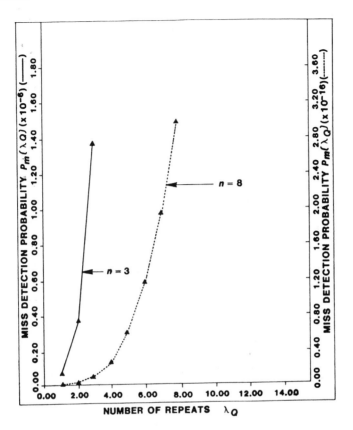

**Figure 3.3** Miss-detection error rate vs. the number of repeats of an RMA symbol for $P_\ell(e_Q) = 10^{-2}$, $P_\ell(e_\ell) = 10^{-3}$.

rapidly scanning spot-beams were proposed. The first approach uses a single reflector with a multiple feed-horn array; the other employs a phased array. The latter can generally provide better directivity with an equal size aperture than can multiple beam structures. We are most interested in phase array antennas, however, for their flexibility in sidelobe controls. We believe that phase array antennas will be better utilized in future satellite communication when the present technological problems such as weight and matching characteristics of array elements are resolved. For this reason we shall discuss phase array antennas and their sidelobe control problems based on the above assumptions in the following section.

### 3.6.1   Phase Array Antennas

Most antennas presently used in satellite communication for both earth stations and spacecraft are the reflecting type. In contrast with the reflectors, phase array

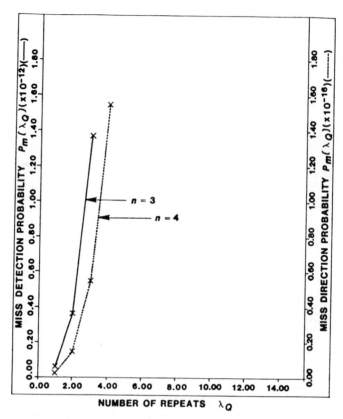

**Figure 3.4** Miss-detection error rate vs. the number of repeats of an RMA symbol for $P_\ell(e_Q) = 10^{-4}$, $P_\ell(e_\ell) = 10^{-5}$.

antennas do not rely on microwave reflection; instead, this class of antennas consists of (a) array elements—radiating dipoles or open-ended waveguides—which interface with free space; (b) phase shifters, which feed each element and are used to form and direct the beams; and (c) the feed system, which energizes and distributes power to the set of phase shifters. With the discrete and multilevel power distribution of the feed system, phase array antennas enjoy a large degree of flexibility as microwave radiators. Some of this flexibility is manifest in the controlling of the sidelobe levels.

An array is an aperture which is active (either radiating or receiving) only at designated discrete points. The points are either the radiators or the receivers. In practice, there are two types of arrays, linear and planar, each one with the set of points properly spaced. The spacings or distances can be all equal or different. When the antenna is directional and only the phase of the carrier is controlled for reception or transmission in an array, we have a phase array antenna. In this section we shall consider only linear phase array antennas, the planar phase array antennas being extensions of the linear case.

The beam pattern of a linear array with simple point sources can be expressed as

$$b(\theta) = \sum_{\ell=0}^{v-1} a_\ell \exp\{j(\ell+1)[d(\theta)-\phi]\}, \qquad (3.61)$$

where a minus sign inside exp { } is arbitrary, and

$$d(\theta) = \left(\frac{2\pi}{\lambda}\right) d \sin \theta$$

$\lambda$ = wavelength (meters)

$\theta$ = angle between incident ray and the normal to the array axis (rad)

$\phi$ = phase shift difference between two adjacent elements (rad)

$d$ = spacing between elements (assuming equal distance, meters)

$v$ = total number of elements

$a_\ell$ = amplitude of the $\ell$th element

$j = \sqrt{-1}$

If we let $x_0, x_1, x_2, \ldots, x_{v-1}$ denote the locations of $v$ possible equally spaced array elements, the distance between $x_{v-1}$ and $x_0$ is called the aperture, and the path length between $x_0$ and $x_1$ is $(x_1 - x_0) \sin \theta$, as shown in Figure 3.5. This represents a phase difference of $2\pi (x_1 - x_0) \sin \theta/\lambda$ radians if the $x_i$'s are measured in meters. But it is sometimes more convenient to express the distances of the elements in wavelength. When the peak of the mainlobe occurs at some angle $\theta_0$, the phase shift difference becomes

$$\phi = d(\theta) = \frac{2\pi d \sin \theta_0}{\lambda}. \qquad (3.62)$$

This is because $b(\theta)$ is maximum when the exponent, i.e., $[d(\theta)-\phi]$ in (3.61), is zero. If we let $\theta_0 = 0$ as a reference, then $\phi = 0$. As a consequence, (3.61) reduces to

$$b(\theta) = \sum_{\ell=0}^{v-1} a_\ell \exp\{j(\ell+1)d(\theta)\}. \qquad (3.63)$$

For equal amplitude excitation $a_i$ can be normalized to unity,

$$b(\theta) = \sum_{\ell=0}^{v-1} \exp\{j(\ell+1)d(\theta)\}. \qquad (3.64)$$

Comprehensive treatment of phase array antennas can be found in references [11, 12, 13]. Next we shall present an actual design of a phase array antenna for space communication before we turn to the problem of sidelobe control.

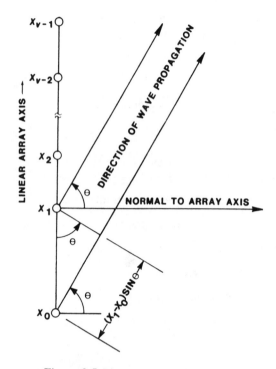

**Figure 3.5** Linear array configuration.

Although this design example is a two-dimensional array, the principle of side-lobe control to be discussed for linear array can be extended to two dimensions.

*Example 3.5:*   Under the sponsorship of NASA, the M.I.T. Center For Space Research proposed and designed an experimental satellite system called Sunblazer [14]. The purpose of the satellite was to measure electron density and other features of the solar corona by means of a two-frequency transmission. By measuring the relative time delay between transmitted solid-state pulse-powered signals of 225 MHz and 75 MHz, the electron density distribution was obtained. The relatively slow motion of the solar satellite is shown in Figure 3.6. Because of the weak signals from this satellite, a phased array antenna was designed and constructed near El Campo, Texas. This phasing of the array elements is accomplished through phase shifting of the local oscillator carriers and mixing these phase shifted carriers with the incoming radio frequency (RF) signal at each dipole location so that the resulting intermediate frequency (IF) energy is in phase. The overall geometric pattern of the array elements is shown in Figures 3.7 and 3.8.

Located at each dipole element is an image rejection (IR) filter and RF amplifiers (RFA), as shown in Figure 3.9, where the contents of Box A and Box B are illustrated. The amplified RF signal from each dipole of a 16-dipole group is mixed with a preshifted local oscillator carrier and then summed. This phasing

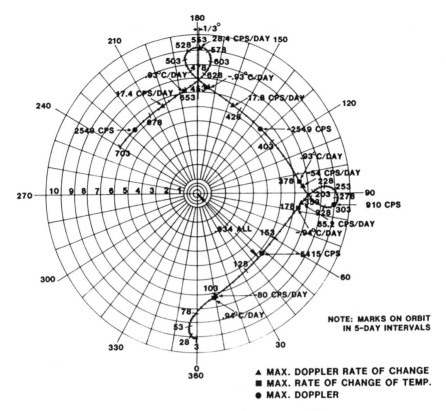

**Figure 3.6** Relative motion of solar satellite.

and summing function is accomplished in Box B, which is physically located in the geometrical center of each 16-dipole group. The notations used are:

$f_{\ell o} \lfloor \phi^+$ —local oscillator frequency with phase angle of plus $\phi$ degrees (lag) with respect to a reference.

$f_{\ell o} \lfloor \phi^-$ —local oscillator frequency with phase angle of $(360 - \phi^+)$ degrees (lead) with respect to a reference.

$f_{\ell o} \lfloor \phi_1^+$ —local oscillator frequency with phase angle $\phi_1$.

$f_{\ell o} \lfloor \phi_2^+$ —local oscillator frequency with phase angle $\phi_2$.

The Box B connections are shown in Figure 3.10, where the superscripts $H$ and $V$ represent horizontal and vertical polarization, $M$ is the mixer, and the outputs of Box B (Figure 3.10) are the vertical and horizontal components of the IF signal from 16 dipoles, which are in turn phased and summed with other 16-dipole groups.

The electronics of Box C, as indicated in Figure 3.11, functionally adds the IF signal energy from the four sets of 16-element groups. The signals of the

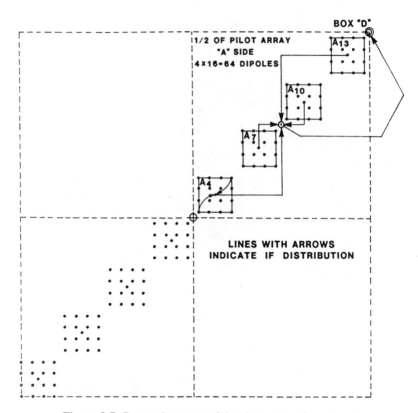

**Figure 3.7** Geometric pattern of the phase array elements—A.

four sets must, however, be first shifted in phase before they may be summed together. This phasing may be accomplished by mixing the IF signal with another shifted carrier ($f_{\ell_o} \underline{|\phi}$), as indicated in Figure 3.11.

In Figure 3.11, both the vertical and horizontal components of the incoming IF signals are (mixed) with 2.8 MHz carriers, with phase relationships $\phi_1^+$, $\phi_1^-$, and $\phi_2^+$, $\phi_2^-$. The resulting 3 MHz IF signals are then in phase and may be summed directly.

The IF signal outputs from each half of the pilot array (64 dipoles) are then summed in Box D, as indicated in Figure 3.12. Figure 3.13 gives a block diagram of the Box D electronics subsystem. As indicated in Box D, the horizontal and vertical polarized IF signals are phased by delaying the signal with a phase shifter. The amount of phase shift is controlled through a set of control lines from the central terminal.

## 3.6.2   Sidelobe Controls

Sidelobe problems exist in practical antenna and filter designs, and phase array antennas are no exception. The objective is to minimize the level of the sidelobe

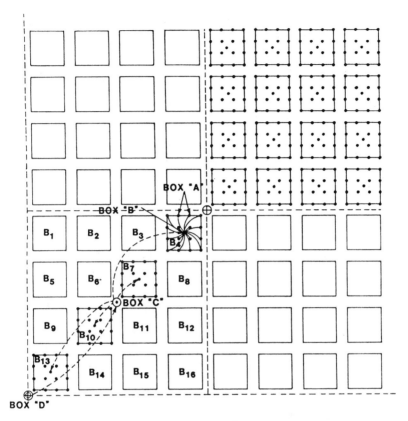

**Figure 3.8** Geometric pattern of the phase array elements—B.

for any given main beam pattern. This is a classic problem. For linear arrays of point sources with nonuniform amplitude distribution, the most significant result for sidelobe control is described by Dolph [15]. He optimally relates beam width and sidelobe level by using the Chebyshev polynomials. This method provides the minimum sidelobe level when the beam width between the first nulls is specified, or, conversely, the method provides the minimum beam width when the sidelobe level is specified. The Dolph-Chebyshev technique is powerful for linear arrays when all elements are used and with nonuniform amplitude distribution; the optimization procedure applies effectively only when the array elements are spaced more than half a wavelength apart. However, in the case of uniform amplitude and not so dense elements, the technique is no longer effective. The reason for sparsely spaced array elements arises in applications where a tradeoff antenna spacing and sidelobe is warranted [16]. In such situations, most solutions have been empirical. Systematic methods have been limited to the application of dynamic programming by Skolnik et al., and the statistical (or random) approaches, for example, by Lo and Steinberg [17–19]. But for the

BASIC DIPOLE CONNECTION    BASIC 16 DIPOLE GROUP

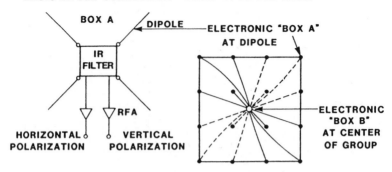

**Figure 3.9** Basic dipole group connection.

dynamic programming approach (Chapter 7, Volume I), the amount of computation can be huge, and the statistical approach lacks the predictability of actual engineering design.

In 1977, Leeper, in his doctoral dissertation for the University of Pennsylvania, applied the difference sets to the sidelobe control problem of phase array antennas [20]. We shall show how cleverly he establishes the connection between the array parameters and the difference set parameters, and then renders the results practical [21].

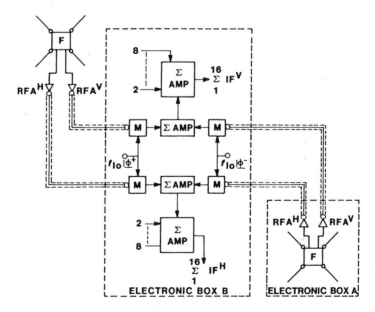

**Figure 3.10** Box B connections.

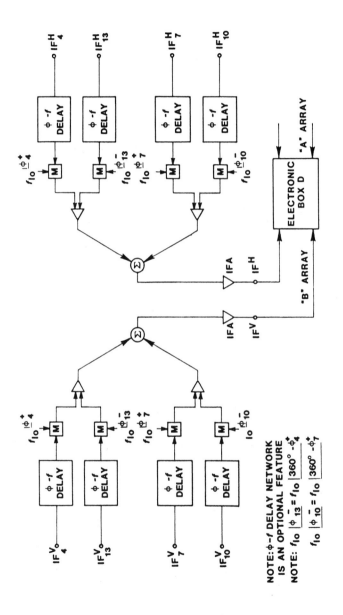

**Figure 3.11** Box C connections.

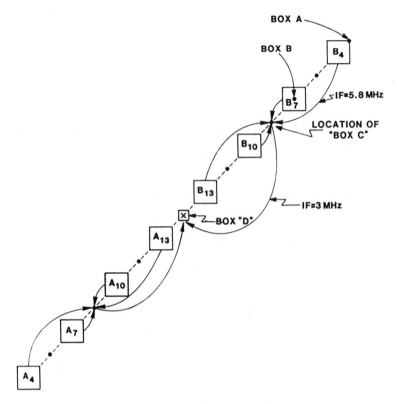

**Figure 3.12** IF return system—128-dipole array.

### 3.6.3  Difference Set Arrays

From Chapter 2, we know that a general $(v, k, \lambda)$ combinatorial difference set $\{D\} = \{d_0, d_1, \ldots, d_{k-1}\}$ is defined as a collection of $k$ residues (or elements) modulo $v$, such that for any positive integer $\delta \not\equiv 0 \pmod{v}$, which is not necessarily an element of $\{D\}$, the congruence

$$d_i - d_j \equiv \delta \pmod{v}$$

has $\lambda$ solution pairs $(d_i, d_j)$, with $d_i$ and $d_j$ in $\{D\}$. $v$, $k$, and $\lambda$ are referred to as the parameters of $\{D\}$. For phase array antenna application, the $\{D\}$'s are $\{D_p\}$.

A linear difference set array (DSA) is formed by first letting

$$x_\ell = d_\ell \tag{3.65}$$

for $\ell = 0, 1, 2, \ldots, k-1$, and in (3.63)

$$a_\ell = \begin{cases} 1, & \text{if } d_\ell \neq 0 \\ 0 & \text{otherwise.} \end{cases} \tag{3.66}$$

(3.65) simply says that a difference set linear array is formed by placing an array

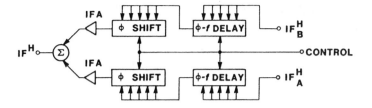

**Figure 3.13** Electronic box D.

element $x_\ell$ at the location corresponding to the set element $d_\ell$ for $\ell = 0, 1,$
$\ldots, k-1$. The array so constructed has the following characteristics: the aperture
size is $v$, the actual number of array elements is $k$, and it has a uniform element
amplitude.

*Example 3.6:* $\{D\} = \{d_0 = 0, d_1 = 1, d_2 = 3\}$ is a difference set with parameters
$v = 7$, $k = 3$, and $\lambda = 1$. The placement of the array elements is shown in Figure 3.14
with $x_0 = d_0$, $x_1 = d_1$, and $x_2 = d_2$.

In the next section we shall explore the sidelobe characteristics and sidelobe
controls of such difference set arrays, and demonstrate their superior sidelobe
behavior in comparison with random arrays.

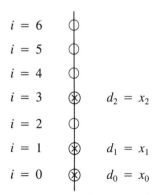

**Figure 3.14** Difference set phase array element location.

## 3.7   SIDELOBE CHARACTERISTICS OF DIFFERENCE SET ARRAYS

From (3.64), if we write the beam pattern in terms of the array element locations $x_0, x_1, \ldots, x_{v-1}$, we have

$$b(u) = \sum_{\ell=0}^{v-1} \exp(j2\pi x_\ell u) \tag{3.67}$$

without the minus sign inside exp(   ). Where

$$x_\ell = \frac{(\ell+1)d}{\lambda} \tag{3.68}$$

expressed in wavelength and $u = \sin\theta$. Using the sampling theorem, $b(u)$ can be constructed from $v$ samples taken at intervals $u = 1/L$, where $L$ is expressed in wavelength. In terms of its Fourier transformed samples, we have

$$b(u) = \sum_{\ell=-\infty}^{\infty} b\left(\frac{\ell}{L}\right) \frac{\sin[\pi L(u-\ell/L)]}{\pi L(u-\ell/L)}. \tag{3.69}$$

For an array with a spacing $d$ and aperture $(v-1)d$, if we let $vd = L$, and note that $b(u)$ has period $1/d$, (3.69) can be rewritten as

$$b(u) = \sum_{\ell=0}^{v-1} b\left(\frac{\ell}{vd}\right) \sum_{m=-\infty}^{\infty} \frac{\sin\left[\pi vd\left(u - \frac{\ell}{vd} - \frac{m}{d}\right)\right]}{\left[\pi vd\left(u - \frac{\ell}{vd} - \frac{m}{d}\right)\right]}. \tag{3.70}$$

The summation over $m$, to be denoted as $S$, may be rewritten

$$S = \sum_{m=-\infty}^{\infty} \frac{\sin[\pi v(ud - \ell/v - m)]}{[\pi v(ud - \ell/v - m)]}. \tag{3.71}$$

Letting $a = ud - \ell/v$, the numerator in (3.71) may be written

$$\sin[\pi v(a-m)] = \sin(\pi va)\cos(\pi vm) - \cos(\pi va)\sin(\pi vm)$$

$$= \sin(\pi va)\cos(\pi vm). \tag{3.72}$$

Note that if $v$ is even, the cosine term is always $+1$, and if $v$ is odd, the cosine term is $(-1)^m = \cos m\pi$. Thus

$$S = \begin{cases} \dfrac{\sin\pi va}{\pi v} \displaystyle\sum_{m=-\infty}^{\infty} \dfrac{1}{a-m} & , v \text{ even} \\[3em] \dfrac{\sin\pi va}{\pi v} \displaystyle\sum_{m=-\infty}^{\infty} \dfrac{\cos m\pi}{a-m} & , v \text{ odd.} \end{cases} \tag{3.73}$$

Applying the identity

$$\sum_{r=-\infty}^{\infty} \frac{\cos r\theta}{g-r} = \frac{\pi \cos g(\pi - \theta)}{\sin g\pi} \tag{3.74}$$

results in

$$S = \begin{cases} \dfrac{\sin \pi va}{v \tan \pi a}, & v \text{ even} \\[4mm] \dfrac{\sin \pi va}{v \sin \pi a}, & v \text{ odd}. \end{cases} \tag{3.75}$$

Substituting (3.75) back into (3.70), we have

$$b(u) = \sum_{\ell=0}^{v-1} b(\ell/vd) \frac{\sin [\pi vd (u - \ell/vd)]}{v \tan [\pi d (u - \ell/vd)]}, \quad v \text{ even}$$

$$\sum_{\ell=0}^{v-1} b(\ell/vd) \frac{\sin [\pi vd (u - \ell/vd)]}{v \sin [\pi d (u - \ell/vd)]}, \quad v \text{ odd}. \tag{3.76}$$

For $d = 1/2$ wavelength, the interval becomes $u = 1/L = 1/vd = 2/v$. In general $b(u)$ is completely specified by the $v$ samples of $b(u)$ at intervals of $u = 1/vd$ and at the locations $m/vd$ for $m = 0,1,2; \ldots, v-1$. For the locations halfway between the samples, the points may be represented as $(m+0.5)/vd$. For these nonsampled points of $u = (m+0.5)/vd$, the corresponding $b(u)$ from (13.76) when $v$ is odd is

$$b(m) = \sum_{\ell=0}^{v-1} b(\ell/vd) \frac{\sin \pi (m+0.5-\ell)}{v \sin[\pi(m+0.5-\ell)/v]}. \tag{3.77}$$

We shall pause here and come back later to (3.77) after we define and derive a quantity called the mainlobe-to-sidelobe ratio.

From (3.67) the array power of the beam pattern is

$$b\,b^*(u) = b(u)\,b^*(u) \tag{3.78}$$

$$= \sum_{m=0}^{v-1} \sum_{\ell=0}^{v-1} \exp[j2\pi (x_m - x_\ell)u]$$

The autocorrelation of the antenna element locations is just the transform of $b\,b^*(u)$, i.e., ($a_m = 1$, or 0 denotes whether an element is present)

$$B(\tau d) = \sum_{m=0}^{v-1} a_m \sum_{\ell=0}^{v-1} a_\ell \, \delta \, [\tau d - (m-\ell)d]. \tag{3.79}$$

From (3.67) the Fourier transform of $b(u)$ is

$$B(x) = \int_{-\infty}^{\infty} b(u) \exp(-j2\pi\, xu)du \qquad (3.80)$$

or

$$B(x) = \sum_{\ell=0}^{v-1} a_\ell\, \delta\, (x-x_\ell) \qquad (3.81)$$

where $a_\ell = 0$ or $1$ depending on whether $\delta(\cdot)$ exists. (3.80) represents all the antenna element locations as a series of impulses in $x$. For equal element spacing, (3.81) becomes

$$B(x) = \sum_{\ell=0}^{v-1} \delta(x-\ell d). \qquad (3.82)$$

The autocorrelation of $B(x)$ with the difference set of parameters $v$, $k$, and $\lambda$ is given with $\delta_{vd}(x)$, $\delta_d(x)$ denoting unit impulse trains

$$C(\tau d) = (k-\lambda)\, \delta_{vd}(x) + \lambda\delta_d(x). \qquad (3.83)$$

Recognizing that the Fourier transform of the autocorrelation function is the power pattern of (3.78), we then have

$$bb^*(u) = F[C(\tau d)]$$

$$= (k-\lambda)\, F[\delta_{vd}(x)] + \lambda F[\delta_d(x)]. \qquad (3.84)$$

To find $F[\delta_{vd}(x)]$ and $F[\delta_d(x)]$ in (3.84), we need the following:
From (3.82) we may approximate the Fourier transform of the beam pattern as

$$B(x) = \sum_{\ell=0}^{v-1} \delta(x-\ell d) \simeq \delta_d(x), \qquad (3.85)$$

which is a sequence of identical impulses with $d$ spacings. The Fourier series representation of $\delta_d(x)$ is

$$F\,[\delta_d(x) \simeq B(x)] = \sum_{\ell=-\infty}^{\infty} C_\ell \exp(j2\pi\ell x/vd), \qquad (3.86)$$

where

$$C_\ell = \frac{1}{vd} \int_{0-}^{vd-} B(x) \exp(-j2\pi\ell x/d)dx. \qquad (3.87)$$

The $(-)$ symbols on the integration limits indicate that the interval includes the impulse at $x = 0$ but not the one at $vd$. When $v = 1$, (3.87) can be reduced to

$$C_\ell = \frac{1}{d}. \qquad (3.88)$$

Substituting (3.88) into (3.86), we have

$$\delta_d(x) = \sum_{\ell=-\infty}^{\infty} \frac{1}{d} \exp(j2\pi\ell x/d). \tag{3.89}$$

The Fourier transform of (3.89) is

$$F[\delta_d(x)] = \frac{1}{d} \sum_{\ell=-\infty}^{\infty} \delta(u - \frac{\ell}{d})$$

$$= \frac{1}{d} \delta_{1/d}(u). \tag{3.90}$$

Similarly, we have

$$F[\delta_{vd}(x)] = \frac{1}{vd} \delta_{1/vd}(u). \tag{3.91}$$

Substituting (3.90) and (3.91) back into (3.84), we get

$$bb^*(u) = (k-\lambda)\left[\frac{\delta_{1/vd}(u)}{vd}\right] + \left[\frac{\lambda\,\delta_{1/d}(u)}{d}\right]. \tag{3.92}$$

Now if we let $B = (k-\lambda)/vd$, and $A = B+(\lambda/d)$ in (3.92), we have the array power pattern as shown in Figure 3.15 in terms of the difference set parameters, where $A$ and $B$ indicate the mainlobe and sidelobe respectively. It can be observed that when the array element spacing $d$ and the difference set are chosen, the magnitudes of both the mainlobe and sidelobe are constant, and the fact that the sidelobes all have the same amplitude is responsible for the improved sidelobe behavior.

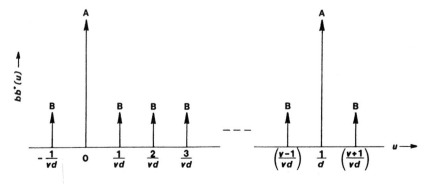

**Figure 3.15** Difference set array power pattern.

Next, let $R$ denote the ratio of sidelobe amplitude to mainlobe amplitude

$$R = \frac{B}{A} = \frac{(k-\lambda)/vd}{[(k-\lambda)/vd] + (\lambda/d)}$$

$$= \frac{k-\lambda}{k+(v-1)\lambda}. \qquad (3.93)$$

For the special class of difference set $\lambda = 1$, we have

$$R = \frac{k-1}{k+v-1}, \qquad (3.94)$$

or

$$R(dB) = 10 \log(k-1) - 10 \log(k+v-1). \qquad (3.95)$$

(3.95) is evaluated in the range of $v$ and $k$ from 10 to 1,000, as shown in Figure 3.16, which indicates the control of sidelobe level in terms of the difference set parameters.

For other ($\lambda \neq 1$) difference sets in general, we have, from Chapter 2, Section 2.11.4 on the properties of planar projective geometry difference sets,

$$\lambda = \frac{k(k-1)}{v-1}.$$

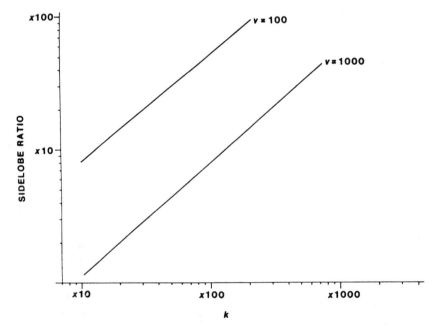

**Figure 3.16** Sidelobe ratio vs. set parameter $k$ for $\lambda = 1$, $v \geq k$.

If we use the $\lambda$ expression in (3.93), we get

$$R = \frac{v - k}{k(v - 1)},\tag{3.96}$$

$$R(dB) = 10 \log (v-k) - 10 \log k - 10 \log (v-1).\tag{3.97}$$

(3.97) is evaluated in Figure 3.17. Once again we see how the sidelobe level can be controlled by the parameters of a difference set. However, it must be pointed out here that difference sets exist for only certain values of parameters.

From (3.78) if we limit the summations to $k$, corresponding to $k$ array elements, then we have

$$b\,b^*(u) = \sum_{m=0}^{k-1} \sum_{\ell=0}^{k-1} \exp[\,j2\pi(x_m - x_\ell)u].\tag{3.98}$$

(3.98) may be considered as the sum of $k^2$ unit phasors, with one phasor for each of the $k^2$ possible differences $(x_m - x_\ell)$. Thus

$$b\,b^*(u) = k^2\tag{3.99}$$

for the aperture points $u=0,\ \pm v,\ \pm 2v,\ \ldots$ . Within the aperture,

$$b\,b^*(u) = k^2\,R\tag{3.100}$$

at aperture points $u = m/vd$ for all $m$.

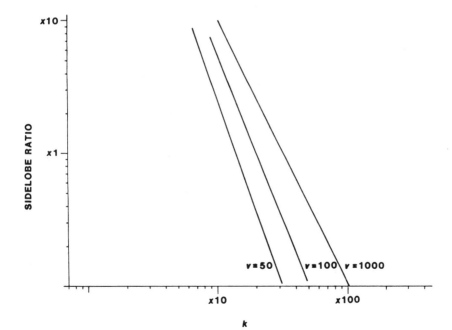

**Figure 3.17** Sidelobe ratio vs. set parameter $k$ for $\lambda \neq 1$, $v \geq k$.

with $R$ as given in (3.93). Now we return to (3.77) and recognize that

$$|b(t/vd)| = \sqrt{b\, b^*(t/vd)} = k \text{ (for } t = 0, \pm v, \pm 2v, \ldots)$$

$$= k \sqrt{R} \; (\text{for } t = \text{noninteger multiples of } v)$$

$$= k \sqrt{\frac{k-\lambda}{k+(v-1)\lambda}}. \tag{3.101}$$

(3.77) may be written as (with $\Theta_\ell$ as the phase of the $\ell$th sample)

$$b(m) = \sum_{\ell=0}^{v-1} \frac{\sin \pi\,(m+0.5-\ell)}{v \sin [\pi\,(m+0.5-\ell)/v]} \cdot b\left(\frac{\ell}{vd}\right)$$

$$= k \left\{ \frac{\sin \pi\,(m+0.5)}{v \sin [\pi\,(m+0.5)/v]} \right.$$

$$\left. + \sum_{\ell=1}^{v-1} \frac{\sin \pi\,(m+0.5-\ell)\,\sqrt{R}\;e^{j\theta\ell}}{v \sin [\pi\,(m+0.5-\ell)/v]} \right\}. \tag{3.102}$$

If we use the relationship

$$\sin [\pi\,(m-\ell+0.5)] = (-1)^{m-\ell}, \tag{3.103}$$

(3.102) becomes

$$|b(m)| = k \left\{ \frac{(-1)^m}{v \sin [\pi\,(m+0.5)/v]} \right.$$

$$\left. + \sum_{\ell=1}^{v-1} \frac{(-1)^{m-\ell}\,\sqrt{R}\;e^{j\theta\ell}}{v \sin [\pi\,(m+0.5-\ell)/v]} \right\}. \tag{3.104}$$

Now let us define the summation term of (3.104) as the sidelobe region and denote it as $Z$. We then have $|b(m)|$ distributed as $k\,\sqrt{R}\,Z$. Let $Y = |Z|^2$ and let $Y_v$ be the largest of $v$ samples of $Y$. We use these relationships in the beam power pattern as

$$b\, b^*(m) \rightarrow k^2\, R|Z|^2 \tag{3.105}$$

$$\overline{b\, b^*(m)} \rightarrow R\,|Z|^2$$

$$\rightarrow R\, Y_v \tag{3.106}$$

where $\rightarrow$ denotes the approximation in distribution, and $\overline{b\, b^*(m)}$ is the normalized peak sidelobe level, which is used as a criterion in the next section for comparison with random arrays.

### 3.7.1   Comparison of Random and Difference Set Arrays

In this section the expected peak sidelobe levels of random and difference set arrays are compared. Examples are given to show the amount of improvement by using the difference set arrays.

Let us define $\overline{B}_v$ as the average amount by which the peak sidelobe level will exceed the theoretical average level of 10 log $1/k$ in dB for a random array. Thus the expected value of the peak sidelobe level for a random array is, in dB,

$$P(R) \simeq 10 \log (1/k) + 10 \log \overline{B}_v. \qquad (3.107)$$

From (3.97) and (3.106), the expected peak sidelobe level is

$$P(D) = 10 \log (v-k) - 10 \log k - 10 \log (v-1) + 10 \log \overline{Y}_v. \qquad (3.108)$$

The values of $\overline{B}_v$ and $\overline{Y}_v$ can be obtained either by computation or by simulation.

*Example 3.7:* Using the values by computation and simulation

$$\overline{B}_v = B + 1 + 2/B$$

$$B = - \ln[1 - (0.5)^{1/v}]$$

and

$$\overline{Y}_v = .848 + 1.28 \log_{10} v$$

in (3.107) and (3.108), Leeper [20] made a comparison of the DSA peak sidelobe level above that of an equivalent random array, as shown in Figure 3.18, where $\beta = k/v$ denotes the thinning factor. We note that the peak sidelobe improvement is a function of the thinning factor. Even with $\beta = 0$, the DSA shows an

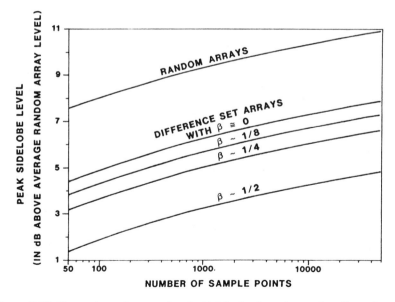

**Figure 3.18** Comparison of expected peak sidelobe level vs. the number of sample points for random and difference set arrays [20].

improvement of 3.0 dB over the random array. For larger values of β corresponding to typical difference set parameters, the amount of improvement is close to 6.0 dB over the comparable random arrays as shown.

*Example 3.8:* In terms of the array power $b\,b^*(u)$ of the beam pattern versus $u$ as (3.78) and with the elements of the $v = 63$, $k = 32$, and $\lambda = 1$ difference set as array element locations, the comparison of magnitude between DSA and random array is shown in Figure 3.19. The elements of the difference set used in the figure are:

|    |    |    |    |    |    |    |    |    |    |
|----|----|----|----|----|----|----|----|----|----|
| 0  | 5  | 5  | 10 | 12 | 15 | 16 | 17 | 18 | 20 |
| 24 | 25 | 25 | 29 | 32 | 34 | 35 | 37 | 38 | 39 |
| 41 | 42 | 45 | 46 | 48 | 50 | 52 | 53 | 54 | 55 |
| 56 | 57 |    |    |    |    |    |    |    |    |

In this example, Leeper placed difference set array elements on a grid with $d = 1/2$ wavelength. The elements for the random arrays were not restricted; they could appear anywhere within the aperture. For this reason they do not show a 0 dB lobe at $u = 2$.

The advantage of DSA is clearly demonstrated in the above two examples. The reason for this advantage as analyzed in Section 3.6 is due to the uniform sidelobe level caused by the DSA structure.

## 3.8   CONTROL OF ALIASING IN SIGNAL PROCESSING

Sub-Nyquist sampling is attractive for baseband signal processing to conserve bandwidth utilization in digital satellite communication. But sub-Nyquist sampling with a uniform rate produces an aliasing effect which degrades the quality of signal transmission. The aliasing effect is a well-known and much-discussed subject in basic communication theory [22, 23]. Aliasing is caused by the spectrum of a sampling function creating a significant amount of signal energy beyond the sampling interval. How to minimize aliasing and maintain a sub-Nyquist sampling rate is a well-known problem in signal processing, filter design, and transmission analysis. In this section we show how the problem may be transformed so that the aliasing effect can be effectively minimized. The method of transformation involves, once again, the application of difference sets.

The sampling process of any time-continuous function is the function itself multiplied by a sequence of samples, i.e., the delta functions $\delta(\cdot)$. In time domain analysis this sequence of delta functions with $k$ sample instants may be represented as

$$s(t) = \sum_{\ell=0}^{k-1} \delta(t - t_\ell). \tag{3.109}$$

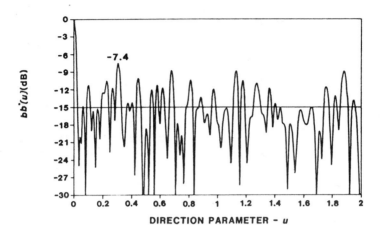

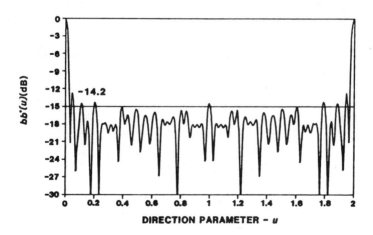

**Figure 3.19** Array power beam pattern comparison [20].

The spectrum of $s(t)$ is

$$S(f) = \sum_{\ell=0}^{k-1} \exp[-j2\pi f t_\ell]. \tag{3.110}$$

Let

$$t_\ell = \frac{(\ell+1)\,d}{\tau} \tag{3.111}$$

where $\tau$ is the smallest interval (or highest frequency content) of the function to be sampled and $d$ is the interval between impulses. Now compare (3.110) and (3.111) with (3.67) and (3.68) and we can establish the following duality: $u \leftrightarrow f$ $S(f) \leftrightarrow b(u)$, $t_\ell \leftrightarrow x_\ell$, $\tau \leftrightarrow \lambda$, and $d, x_\ell$ in DSA are distances measured in meters; $d, t_\ell$ in sampling are times measured in seconds. Once the above duality is recognized, the problem of controlling the array power of the beam pattern $b$ $b^*(u)$ becomes the same as the problem of controlling the aliasing effect $S\,S^*(f)$ in sampling. As a consequence, the solution to one problem becomes the solution to the other. The following result gives another reason for our interest in the analysis and the application of difference sets.

*Example 3.9:* Similar to $b\,b^*(u)$ of (3.98) we have the equivalent $S\,S^*(f)$ from (3.110). Figure 3.20(a) shows the sampling power pattern $S\,S^*(f)$ with uniform deletion of samples. In this power pattern a large single sidelobe impulse is shown at $f = 1$, and the aliasing is generated by the spectra of the signal function. The noise spectrum produced from aliasing is closely related to the spectra of the signal functions.

Figure 3.20(b) shows the power pattern of the same signal function with samples randomly distributed over $\ell = 0$ to $v - 1$; the energy from the impulse of $f = 1$ is separated into $v - 1 = 62$ small impulses and scattered across between $f = 0$ and $f = 2$. When this spectrum is convolved with the signal function, there exist 62 smaller spectra scattered across the same range.

Figure 3.20(c) shows the $S\,S^*(f)$, after the application of difference set sampling with $v = 63$, $k = 32$, and $\lambda = 1$. We can see in this figure that all the noise spectra from aliasing have the same amplitude, but lower magnitude. Leeper has evaluated all three cases under the same conditions.

We must point out that the difference set technique does not eliminate the effect of aliasing, but the technique makes the effect less sensitive by transforming a large sharp spike to a number of smaller spikes. In many cases of satellite signal transmission through nonlinear transponders' constant envelope spectrum is desirable in order to minimize signal degradation.

## 3.9 TRANSPONDER FREQUENCY ASSIGNMENT

Although frequency division transmission schemes are basically analog, the problem of frequency spacing to reduce interference effects is of a discrete nature.

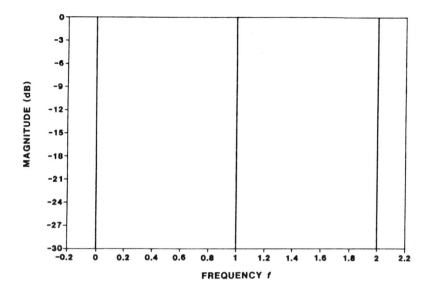

**Figure 3.20** Sampling power pattern—S S*(f) [20].
(a) Uniform sampling.

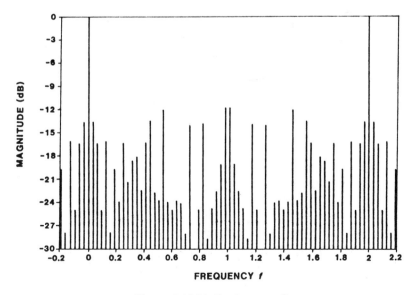

**Figure 3.20**(b) Random sampling.

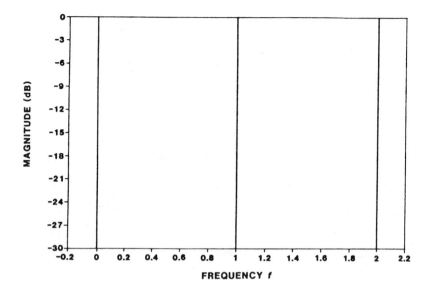

**Figure 3.20** Sampling power pattern—S S*(f) [20].
(a) Uniform sampling.

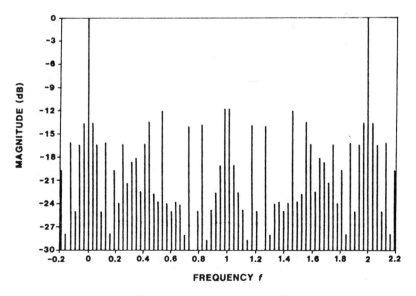

**Figure 3.20**(b) Random sampling.

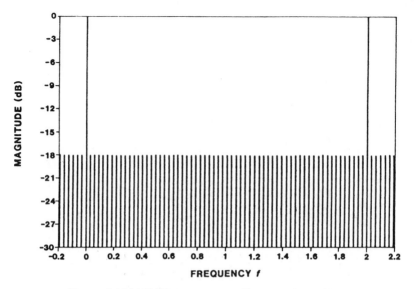

**Figure 3.20**(c)  Difference set sampling, $v = 63$, $k = 32$.

When multiple carrier frequencies pass through any nonlinear device, signal products other than the original frequencies can be produced. These undesirable signals are called intermodulation products and they can cause serious interference in both inband and out-of-band channels. The result is the degradation of the transmitted signal. The effect of this degradation in satellite channels was discussed in Chapter 2, Volume I. In this section we shall present the technique of using difference sets to eliminate or minimize the most common types of intermodulation products when multiple carriers are signaling through a nonlinear transponder.

Long before satellite communication came into being the problem of intermodulation products was well known in terrestrial radio communication. The most common intermodulation occurrence is the third-order intermodulation product in which three carrier frequencies $f_a$, $f_b$, and $f_c$ produce an interference at frequency

$$f_i = f_a + f_b - f_c. \tag{3.112}$$

A third-order intermodulation product can also be produced when the second harmonic of $f_a$ intermodulates with $f_b$ to create

$$f_i' = 2f_a - f_b. \tag{3.113}$$

Babcock observed that by selecting frequencies such that the difference between any pair is unlike that between any other pair, the third-order intermodulation product can be readily reduced [24]. Fang and Sandrin recognize the above requirement for the unlikeness (or distinctiveness) of the differences of all the pairwise frequencies and the definition of a difference set [25]. As described in

Chapter 2, when the set parameter $\lambda = 1$, the corresponding difference set is referred to as a simple or perfect difference set.

The interesting result is that not only do difference sets provide a systematic method of assigning frequency spacings, but all Babcock frequency selections can be derived from difference sets; for larger numbers of frequencies, the difference set approach provides better results than Babcock spacings.

*Example 3.9:* For the Babcock frequency selections of three and four frequencies we have $(f_0, f_1, f_3)$ and $(f_0, f_1, f_4, f_6)$. These correspond to the difference sets $\{D\} = \{0, 1, 3\}$, and $\{D\} = \{0, 1, 4, 6\}$. For five frequencies the Babcock selection is $(f_0, f_1, f_4, f_9, f_{11})$ and the existing difference set is $\{D\} = \{0, 3, 4, 9, 11\}$. For 10 frequencies the Babcock selection is $(0, 1, 7, 11, 26, 39, 47, 56, 59, 61)$, and the difference set method gives $\{D\} = \{0, 1, 6, 10, 23, 26, 34, 41, 53, 55\}$. We note that for the 10-frequency case the difference set solution yields $61 - 55 = 6$ fewer frequency spacings.

*Example 3.10:* Fang and Sandrin made a comparison of the third-order intermodulation spectrum for ten carrier frequencies. As shown in Figure 3.21, the comparison is made of the uniformly spaced frequency, randomly spaced frequency, and spacing according to the difference set arrangement. In terms of the relative power levels in dB, the uniformly spaced case (shown in Figure 3.21(a)) gives the worst result. Some improvement can be made, as seen in

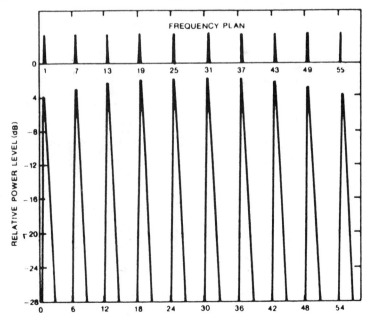

**Figure 3.21(a)** Comparison of the third-order intermodulation spectrum for ten carrier frequencies—Uniform frequency spacing [25].

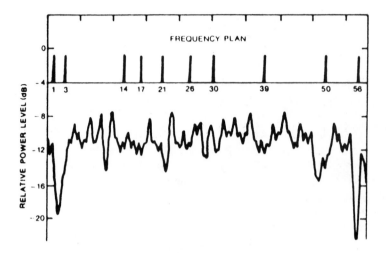

**Figure 3.21(b)** Comparison of third-order intermodulation spectrum for ten carrier frequencies—Randomly spaced frequencies [25].

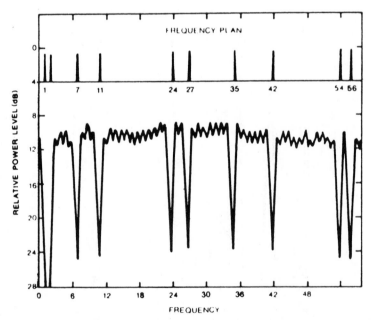

**Figure 3.21(c)** Comparison of third-order intermodulation spectrum for ten carrier frequencies—Difference set arrangement [25].

Figure 3.21(b), in the randomly spaced case. The difference set case, as shown in Figure 3.21(c), clearly exhibits a lower intermodulation level.

It is worthwhile to point out here that it is not necessary to assign the carrier frequency through a difference triangle. The theorem by Fang and Sandrin states: "Assume that there are $L$ consecutive channels numbered 1, 2, 3, ..., $L$, respectively. Let $N$ carriers at the input of a nonlinear repeater be assigned frequencies at the midbands of $N$ channels ($N<L$). Then the necessary and sufficient condition to preclude third-order intermodulation products on the assigned channels is that the difference triangle of these assigned carrier frequencies must have distinct elements." In the corollary to the theorem they further state that the problem of finding a third-order intermodulation-free channel assignment plan for a nonlinear repeater is equivalent to the mathematical problem of searching for a single difference triangle with distinct elements. The fact of the matter is that there is no need to involve the difference triangle. By the property of a difference set (see Chapter 2) the distinctive characteristics of difference triangle elements are implied. As we have seen from Chapter 2, a difference set exists if the elements of the set are known, and the converse is also true. Since a difference triangle is constructed from the set elements, a difference triangle with distinctive elements exists if and only if a simple ($\lambda = 1$) difference set exists. Therefore, the corrected statement without the difference triangle should be:

THEOREM 3.3: Minimization of third-order intermodulation products for optimal frequency spacing is possible if and only if there exists a simple difference set $\{D\} = \{d_0, d_1, ..., d_{k-1}\}$ with parameters $v$, $k$, and $\lambda = 1$. The total frequency spacing needed is $v$, the number of frequencies that can be assigned is $k$, and the frequency assignment is precisely according to the elements of $\{D\}$. That is, $f_0 = d_0, f_1 = d_1, ..., f_{k-1} = d_{k-1}$.

The proof of Theorem 3.3 is obvious once one is thoroughly familiar with the properties of a difference set and the requirement for frequency spacing as described in this section. The validity of Theorem 3.3 can be revealed by reexamination of Example 3.10. The following example demonstrates that difference triangles are not needed in the application of transponder frequency assignment.

*Example 3.12:*   From the difference set $\{D\} = \{1, 2, 5, 10, 16, 23, 33, 35\}$, the corresponding difference triangle is

```
1    3    5    6    7    10--2
4    8    11   13   17--12
9    14   18   23--19
15   21   28--25
22   31--30
32--33
34
```

Subtracting the last diagonal entries and 0 from 35, we have

$$35 - \boxed{34 \quad 33 \quad 30 \quad 25 \quad 19 \quad 12 \quad 2 \quad 0}$$
$$= 1, 2, 5, 10, 16, 23, 33, 35,$$

which are the frequency assignments for $f_1, f_2, \ldots, f_8$, according to Fang and Sandrin. We note that the assignment derived this way is identical to the elements of the original difference set, and we can have the frequency plan immediately written down in general from the elements (or residues) of any difference set without computation and construction of the difference triangles.

*Example 3.13:*   MARISAT has been operational since 1976 to serve mobile maritime users [26]. The system has served the U.S. Navy and connected shore stations with small commercial shipboard terminals which may move continually within the coverage of the satellite. The frequency plan of MARISAT, shown in Figure 3.22, is optimized for minimal third-order intermodulation products. The final frequency plan is derived from the difference set

$$\{D\} = \{6, 11, 21, 24, 30, 48, 59, 80, 92, 100, 114, 131, 150, 154, 157, 180\}.$$

Although Theorem 3.3 supersedes all previous work in this area and puts the problem to rest once and for all, the significant earlier findings by Fang and Sandrin, Edwards, Durkin, and Green, Babcock, Hirata, and Welti are indispensable [24, 25, 27–29]. Without their work the present concise solution could not have been found. However, the problem of using sets with $\lambda \neq 1$ to conserve bandwidth remains unsolved.

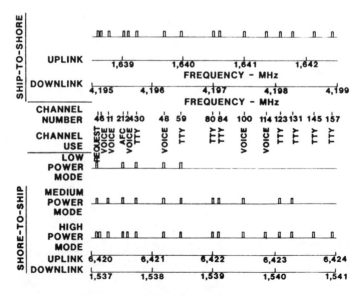

Figure 3.22 MARISAT frequency plan [26].

## 3.10 COLLISION CONTROL IN PACKET TRANSMISSION

In an ALOHA type multiple access transmission system as described in Chapter 3, Volume I, whenever two packets of messages collide in the transmission channel, both must be retransmitted. Very often when the channel is busy, the retransmitted packets collide again and again with other packets. Thus methods to control or to minimize such collisions are important. One of the most effective methods is the application of difference sets suggested by Massey [30]. To understand his scheme, let us begin with a simple example.

*Example 3.14:*   Using the same difference set $\{D\} = \{0, 1, 3\}$ as described in Chapter 2, the corresponding incidence matrix is

$$Q = \begin{bmatrix} 1 & 1 & 0 & 1 & 0 & 0 & 0 \\ 0 & 1 & 1 & 0 & 1 & 0 & 0 \\ 0 & 0 & 1 & 1 & 0 & 1 & 0 \\ 0 & 0 & 0 & 1 & 1 & 0 & 1 \\ 1 & 0 & 0 & 0 & 1 & 1 & 0 \\ 0 & 1 & 0 & 0 & 0 & 1 & 1 \\ 1 & 0 & 1 & 0 & 0 & 0 & 1 \end{bmatrix}.$$

Let each element in this matrix represent a packet, and each row of the matrix be a superpacket. Thus in this example we have seven superpackets and each superpacket contains seven regular packets. The transmission time is based on the incidence matrix, i.e., the $j$th packet of superpacket $t$ starts at time $t + j$, for $t, j = 0, 1, ..., 6$. Transmission is performed on superpackets. Note that if any two superpackets collide in a transmission channel, only a single regular packet can be in collision. This can be seen from the incidence matrix: there is only one coincidence of 1's in any pair of rows in the matrix. By defining an index vector $(d_0, d_1, ..., d_{v-1})$ as a superpacket, Massey gives the following general result.

THEOREM 3.4: Every collision is either a direct hit or an indirect hit on only one packet if and only if the index vector $(d_0, d_1, ..., d_{v-1})$ is a simple difference set (s.d.s.), which is a $\{D\}$ with parameters $v$, $k$, and $\lambda = 1$.

*Proof:* Suppose there exist $p$th and $r$th packets of superpacket $t$, and $q$th and $s$th packets of superpackets $t'$, then we have

$$t + d_p = t' + d_{q'} \tag{3.114}$$

$$t + d_r = t' + d_s. \tag{3.115}$$

Subtracting (3.115) from (3.114), we have

$$d_p - d_r = d_q - d_s. \tag{3.116}$$

Since $d_p$, $d_r$, $d_q$, and $d_s$ are the residues of a $\{D\}$ with $\lambda = 1$, (3.116) can hold if and only if $d_p = d_q$, and $d_r = d_s$. Massey refers to this condition as a direct hit.

Conversely, if the $d$'s are not an s.d.s., then the equality of (3.116) may hold for some $d_p \neq d_q$. That is, $\lambda > 1$ implies the collision of more than one packet. Thus by varying the value of $\lambda$, the amount of packet collision can be controlled.

### 3.11  MINIMIZATION OF BEAM INTERFERENCE

In active phase array multiple beam antennas in which each element is amplified, there exist undesirable beams caused by the intermodulation products of nonlinear amplification. In the ideal case with linear amplification each main beam behaves as if other beams do not exist. In the phase array situation there are two factors that produce beam interference when signaling through a nonlinear device. The first is the intermodulation product of frequencies, as seen in Section 3.9; the second is caused by the phase variations of the array elements. This phase variation directly affects the direction of a beam. For this reason it is important to mention the subject.

For illustration we assume a linear array with only three elements. Each element has three carriers with frequencies $f_a, f_b, f_c$, and phases $\phi_a, \phi_b, \phi_c$. The third-order intermodulation product caused by the frequency at each array element is $f_i = f_a + f_b - f_c$, and the third-order intermodulation product caused by the phases at each element is: $\phi_i = \phi_a + \phi_b - \phi_c$. If the characteristics of the array elements are identical, then each element contributes to the same amount of $\phi_i$. Since the main beam is formed by collecting all the phases from all the elements in the array, there is no doubt that the set of intermodulated phases $\phi_i$ is also collected. The sum of the $\phi_i$'s will form a different beam (the interference beam). The direction of the interference beam depends on the magnitude of the intermodulation phase variations $\phi_i$. In advocation of phase array antennas for satellite communication, Sandrin analyzed, experimented, and observed the phenomena of the intermodulation phase variations [31]. It is natural to use difference sets once again to minimize the intermodulation phase variation, which has exactly the same structure as the intermodulation effect for frequency spacing, as described in Section 3.9.

### 3.12  GENERATION OF ERROR-CORRECTING CODES

Weldon has used projective geometry difference sets to generate one-step, majority logic decodable, random error correction, cyclic block codes [32]. The same sets can be used to obtain multiple-burst error-correcting codes, as illustrated by Mandelbaum [33]. Robinson and Bernstein and Wu used difference sets to generate threshold-decodable convolutional self-orthogonal codes, which are discussed in this section [34, 35].

One-step threshold decodable convolutional codes have the property of self-orthogonality. That is, an error digit is checked by the set of parity check sums and no other digit is checked by more than one such sum. Massey has formulated this criterion for threshold decoding of convolutional codes [36]. Wu described a difference-set coding procedure and calculated a large number of codes with coding rates from 1/2 to 49/50. This unified procedure is described as follows:

Let us assume the existence of a difference set with parameters $v$, $k$, and $\lambda = 1$. The set elements are denoted as $\{d_0, d_1, \ldots, d_{k-1}\}$. We can form the array with the set elements as:

$$
A = \begin{matrix}
d_1 - d_0 & d_2 - d_1 & d_3 - d_2 & \ldots & v + d_0 - d_{k-1} & d_1 - d_0 \\
d_2 - d_0 & d_3 - d_1 & d_4 - d_2 & \ldots & v + d_1 - d_{k-1} & d_2 - d_0 \\
d_3 - d_0 & d_4 - d_1 & d_5 - d_2 & \ldots & v + d_2 - d_{k-1} & d_3 - d_0 \\
\vdots & \vdots & \vdots & & \vdots & \vdots \\
d_{k-2} - d_0 & d_{k-1} - d_1 & d_{k-2} - d_2 & \ldots & v + d_{k-3} - d_{k-1} & d_{k-2} - d_0 \\
d_{k-1} - d_0 & d_{k-2} - d_1 & d_{k-3} - d_2 & \ldots & v + d_{k-2} - d_{k-1} & d_{k-1} - d_0,
\end{matrix}
\tag{3.117}
$$

where the last column is a repeat of the first. The dotted lines indicate how the entries of the array are obtained by successive additions of the entries in the first row. The entries are formed by taking the adjacent differences of the difference set residues.

The total number of columns in $A$ is $k$. To construct a rate $k_0/(k_0 + 1)$ self-orthogonal convolutional code with $d_{min} = J + 1$, partition the columns of $A$ into $k_0$ sections; i.e., $k = J k_0$, where $J$ is the number of columns per section. Next, locate the $J$th row of $A$, and find the set of corresponding entries in each section such that the least of the maximum values is obtained simultaneously in the $J$th row. These numbers are the highest powers of the subgenerator polynomials. That is,

$$
\text{Deg}^{max}\{G^{(1)}(D), \ldots, G^{(k_0)}(D)\} = \min \max\{(\Delta_0, \Delta_1, \ldots \Delta_{J-1}),
$$

$$
(\Delta_J, \Delta_{J+1}, \ldots, \Delta_{2J-1}),
$$

$$
(\Delta_{(k_0-1)J}, \ldots, \Delta_{k_0 J - 1})\}
$$

$$
= \{\Delta_\xi, \Delta_{\xi+J}, \Delta_{\xi+2J}, \ldots, \Delta_{(k_0-1)J+\xi}\}
\tag{3.118}
$$

where $\Delta_\sigma = d_\alpha - d_\beta$ at the $J$th row and $\xi$th column ($\sigma = 0, 1, \ldots J-1$) for $d_\alpha, d_\beta \in \{D\}$. $\text{Deg}^{max}\{x\}$ denotes the maximum degree of $x$. The min max set is always obtained at the end of $J$ comparisons.

Let the column $\Delta$'s corresponding to $\{\Delta_\sigma, \Delta_{\sigma+J}, \ldots, \Delta_{(k_0-1)J+\sigma}\}$ above the $J$th row of $A$ be denoted as $\Delta$ (row, column). Then the complete code generator is obtained:

$$
G(D) = \begin{cases}
G^{(1)}(D) = 1 + D^{\Delta(1,\sigma)} + \ldots + D^{\Delta(J,\sigma)} \\
G^{(2)}(D) = 1 + D^{\Delta(1,\sigma+J)} + \ldots + D^{\Delta(J,\sigma+J)} \\
\vdots \qquad\qquad \vdots \qquad\qquad \vdots \\
G^{(k_0)}(D) = 1 + D^{\Delta(1,\sigma+(k_0-1)J)} + \ldots + D^{(J,\sigma+(k_0-1)J)}.
\end{cases}
\tag{3.119}
$$

Once $G(D)$ is obtained, the implementation is straightforward [35]. When $k$ is sufficiently large, say 8 or 10, then multiple codes can be generated from the same difference set.

*Example 3.15:*   To generate the rate 14/15 ($n_0 = 15$, $k_0 = 14$, $d_{min} = 8$) convolutional code, we choose the difference set $v = 9507$, $k = 98$, and $\lambda = 1$, and generate $A$ from the residues of

$$\{D\} = \quad \{ 1 \quad 13 \quad 68 \quad 97 \quad 137 \quad 360 \quad 568 \quad 611 \quad 657 \quad 670$$

| | | | | | | | | | |
|---|---|---|---|---|---|---|---|---|---|
| 696 | 717 | 833 | 889 | 963 | 1070 | 1071 | 1073 | 1107 | 1122 |
| 1261 | 1378 | 1402 | 1503 | 1984 | 1989 | 2054 | 2163 | 2225 | 2301 |
| 2308 | 2670 | 2748 | 2793 | 2802 | 2825 | 2843 | 2896 | 3000 | 3008 |
| 3169 | 3186 | 3211 | 3527 | 3782 | 3929 | 4128 | 4257 | 4536 | 4594 |
| 4725 | 4745 | 4818 | 5209 | 5215 | 5253 | 5367 | 5371 | 5588 | 5598 |
| 5670 | 5790 | 5847 | 6034 | 6113 | 6124 | 6246 | 6338 | 6399 | 6426 |
| 6566 | 6596 | 6671 | 6687 | 6921 | 7221 | 7243 | 7561 | 7609 | 7829 |
| 7848 | 7862 | 7948 | 8091 | 8233 | 8296 | 8360 | 8560 | 8720 | 8817 |
| 8930 | 9011 | 9126 | 9224 | 9374 | 9409 | 9440}. | | | |

Since $k = 98$, $k_0 = 14$, and $J = 7$, the $\Delta$ entries in the 7th row are:

| | | | | | | | | | | | | | |
|---|---|---|---|---|---|---|---|---|---|---|---|---|---|
| {( 610 | 644 | 602 | 599 | 580 | 473 | 321 | 352 | 413 | 401 | 377 | 390 | 289 | 372), |
| ( 415 | 332 | 432 | 911 | 882 | 932 | 902 | 847 | 899 | 805 | 686 | 759 | 739 | 639), |
| ( 600 | 542 | 588 | 330 | 260 | 376 | 384 | 386 | 684 | 886 | 929 | 1120 | 1088 | 1350), |
| (1383 | 1198 | 963 | 889 | 1081 | 958 | 717 | 773 | 646 | 843 | 780 | 461 | 575 | 594), |
| ( 667 | 742 | 536 | 648 | 668 | 609 | 579 | 532 | 483 | 547 | 441 | 583 | 822 | 817), |
| ( 995 | 1013 | 1158 | 1161 | 941 | 727 | 848 | 672 | 687 | 531 | 712 | 858 | 869 | 839), |
| ( 778 | 802 | 766 | 664 | 654 | 592 | 510 | 497 | 422 | 449 | 380 | 270 | 458 | 635)}. |

After seven comparisons the min max set is {473, 289, 932, 739, 376, 1088, 958, 575, 609, 822, 727, 869, 510, 635}. These are the highest power terms of the set of subgenerators $G^{(j)}(D)$, $j = 1, 2, ..., 14$.

Since $\Delta_{\sigma=5} = \Delta(7, 5) = 473$, then $\Delta(6, 5) = 357$, $\Delta(5, 5) = 336$, $\Delta(4, 5) = 310$, $\Delta(3, 5) = 297$, $\Delta(2, 5) = 251$, and $\Delta(1, 5) = 208$. These are the exponents of the subgenerator $G^{(1)}(D)$. The other subgenerator polynomials are constructed in exactly the same way. As a result, the code generator polynomial is obtained as

$$
\begin{aligned}
& G^{(1)}(D) = 1 + D^{208} + D^{251} + D^{297} + D^{310} + D^{336} + D^{357} + D^{473} \\
& G^{(2)}(D) = 1 + D^{56} + D^{130} + D^{237} + D^{238} + D^{240} + D^{274} + D^{289} \\
& G^{(3)}(D) = 1 + D^{139} + D^{256} + D^{280} + D^{381} + D^{862} + D^{867} + D^{932} \\
& G^{(4)}(D) = 1 + D^{109} + D^{171} + D^{247} + D^{254} + D^{616} + D^{694} + D^{739} \\
& G^{(5)}(D) = 1 + D^9 + D^{32} + D^{50} + D^{103} + D^{207} + D^{215} + D^{376} \\
& G^{(6)}(D) = 1 + D^{17} + D^{42} + D^{358} + D^{613} + D^{760} + D^{959} + D^{1088} \\
G(D) = \quad & G^{(7)}(D) = 1 + D^{279} + D^{337} + D^{468} + D^{488} + D^{561} + D^{952} + D^{958} \qquad (3.120) \\
& G^{(8)}(D) = 1 + D^{38} + D^{152} + D^{156} + D^{373} + D^{383} + D^{455} + D^{575} \\
& G^{(9)}(D) = 1 + D^{57} + D^{244} + D^{323} + D^{334} + D^{456} + D^{548} + D^{609} \\
& G^{(10)}(D) = 1 + D^{27} + D^{167} + D^{197} + D^{272} + D^{288} + D^{522} + D^{822} \\
& G^{(11)}(D) = 1 + D^{22} + D^{340} + D^{388} + D^{608} + D^{627} + D^{641} + D^{727} \\
& G^{(12)}(D) = 1 + D^{143} + D^{258} + D^{348} + D^{412} + D^{612} + D^{772} + D^{869} \\
& G^{(13)}(D) = 1 + D^{81} + D^{168} + D^{196} + D^{294} + D^{444} + D^{479} + D^{510} \\
& G^{(14)}(D) = 1 + D^{68} + D^{80} + D^{135} + D^{164} + D^{204} + D^{427} + D^{635}.
\end{aligned}
$$

The codes so constructed do not propagate errors and are simple to implement, and thus are ideal for high-speed and code-concatenation applications.

## 3.13 CONCLUDING REMARKS

In this chapter we have demonstrated the usefulness of combinatorial sets. As mentioned in the beginning, few simple analytical techniques provide so many possible applications for satellite communication. The beauty of the combinatorial sets is the fact that an almost effortless familiarity with the sets can produce a large number of rewards, and in many cases they turn out to be the optimal solutions. We like to bring out simple techniques that have vast impact in applications, instead of a great deal of analyses with little usefulness. This chapter appears to meet our goal.

## 3.14 REFERENCES

1. Bluestein, L. and Greenspan, R. "Efficient Approximation of Orthogonal Waveforms." Lincoln Lab Report 1964–48, November 1964.
2. Schuleter, G.A. "Effects of Interference for Time-Frequency Coded Communication Systems." Ph.D. Dissertation, Department of Electrical Engineering, Purdue University, August 1972.
3. Solomon, G. "Optimal Frequency Hopping Sequences for Multiple Access. System Engineering Laboratory, TRW Systems Group, Redondo Beach, California, 1972.
4. Kasami, T. and Lin, S. "Multiple Access Codes," *IEEE Transactions on Information Theory* (March 1976).
5. Mersereau, R. and Seay, T. "Multiple Access Frequency Hopping Patterns with Low Ambiguity." *IEEE Transactions on Aerospace and Electronic Systems*, AES-17, 4 (July 1981):571–78.
6. Wu, W. "Random Multiple Access Codes." Ph.D. Dissertation, Department of Electrical Engineering, The Johns Hopkins University, 1976.
7. Ricardi, L. "Communication Satellite Antennas." *Proceedings of the IEEE*, 65, 3 (March 1977):356–69.
8. Kreutel, R.; Difonzo, D.; English, W.; and Grunner, R. "Antenna Technology for Frequency Reuse Satellite Communications." *Proceedings of the IEEE*, 65, 3 (March 1977):370–87.
9. Collin, R. and Zucker, F. *Antenna Theory*. New York: McGraw-Hill, 1969.
10. Reudink, D. and Yeh, Y. "A Scanning Spot-Beam Satellite System." *Bell Systems Technical Journal* (October 1977):1549–60.
11. Stark, L. "Microwave Theory of Phased-Array Antennas—A Review." *Proceedings of the IEEE*, 62, 12 (December 1974):1661–66.
12. Amitay, N.; Galindo, V.; and Wu, C. *Theory and Analysis of Phase Array Antennas*. New York: John Wiley & Sons, 1972.
13. Ma, M. *Theory and Application of Antenna Arrays*. New York: John Wiley & Sons, 1974.
14. Wu, W. and Baker, R. "Aspect Sensing of the Sunblazer." M.I.T. Center For Space Research Report, June 1967.
15. Dolph, C. "A Current Distribution for Broadside Arrays which Optimizes the Relationship between Beam Width and Side-Lobe Level." *Proceedings of the IRE*, 34 (June 1946):335–48.

16. Ishimaru, M. and Chen, Y. S. "Thinning and Broad-banding Antenna Arrays by Unequal Spacings." *IEEE Transactions on Antennas and Propagation*, 13 (January 1965):34–42.

17. Skolnik, M.; Nemhauser, G.; and Sherman, J. "Dynamic Programming Applied to Unequally Spaced Arrays." *IEEE Transactions on Antennas and Propagation*, AP-12 (January 1964):35–43.

18. Lo, Y. "A Mathematical Theory of Antenna Arrays with Randomly Spaced Elements." *IEEE Transactions on Antennas and Propagation*, AP-12 (May 1964):257–68.

19. Steinberg, B. *Principles of Aperture and Array System Design*. New York: John Wiley & Sons, 1976.

20. Leeper, D. *Sidelobe Control in Sub-Nyquist Sampling*. Ph.D. Dissertation, Department of Electrical Engineering and Science, The University of Pennsylvania, 1977.

21. Leeper, D. "Thinned Aperiodic Antenna Arrays with Improved Peak Sidelobe Level Control." U.S. Patent 4071848, January 31, 1978.

22. Peeble, P. *Communication System Principles*. Reading, Mass.: Addison-Wesley, 1976.

23. Oppenheim, A. and Schafer, R. *Digital Signal Processing*. Englewood Cliffs, N.J.: Prentice-Hall, 1975.

24. Babcock, W.C. "Intermodulation Interference in Radio Systems. *Bell Systems Technical Journal* (January 1953):63–73.

25. Fang, R. and Sandrin, W. "Carrier Frequency Assignment for Nonlinear Repeaters." *COMSAT Technical Review*, 7, 1 (Spring 1977):227–44.

26. Lipke, D.W. and Swearingen, D.W. "Communications System Planning for MAR-ISAT." *Conference Record, IEEE International Conference on Communications, ICC 1974*, Session 29, Minneapolis, Minn.: 29B-1—29B-5.

27. Edwards, R.; Durkin, J.; and Green, D. H. "Selection of Intermodulation-Free Frequencies for Multiple-Channel Mobile Radio System." *Proceedings of the IEEE*, 116, 8 (August 1969).

28. Hirata, Y. "A Bound on the Relationship Between Intermodulation Noise and Carrier Frequency Assignment." *COMSAT Technical Review* (Spring 1978):141–54.

29. Welti, G. R. "Butler Matrix Transponder Improvements." U.S. Patent Application, Serial No.1 412399, November 2, 1972.

30. Massey, J. "Superpacket Transmissions." An unscheduled invited lecture presented at the International Telemetering Conference, November 1978.

31. Sandrin, W. A. "Spatial Distribution of Intermodulation Products in Active Phased Array Antennas." *IEEE Transactions on Antennas and Propagation* (November 1973):864–68.

32. Weldon, E., Jr. "Difference-Set Cyclic Codes." *Bell Systems Technical Journal*, 45 (1966).

33. Mandelbaum, D. "Some Classes of Multiple-Burst Error-Correcting Codes Using Threshold Decoding." *IEEE Transactions on Information Theory* (March 1972):285–92.

34. Robinson, J. P. and Bernstein, A. J. "A Class of Binary Recurrent Codes with Limited Error Propagation." *IEEE Transactions on Information Theory*, IT-13, 1 (January 1967):106–113.

35. Wu, W. "New Convolutional Codes—Part II." *IEEE Transactions on Communications* (September 1976).

36. Massey, J. *Threshold Decoding*. Cambridge, Mass.: M.I.T. Press, 1963.

## 3.15 PROBLEMS

1. Design a random multiple access system for $n = 4$ in accordance with the difference set $\{D_E\} = \{1, 3, 4, 12\}$ (Example 3.2), and Appendix 3.

2. Evaluate the performance in terms of error probability for the $n = 4$ random multiple access system of Problem 1.

3. Given a difference set $\{D_E\} = \{d_0, d_1, \ldots, d_{k-1}\}$, draw a logic flow diagram for the random multiple access sequence generation procedure.

4. Formulate the observations of Figure 3.1—the $\alpha$ lines of $A_j$—in analytical expressions.

5. Assuming $P(e_\ell) = 10^{-(\ell+1)}$ for $\ell = 0, 1, 2, 3$ in (3.55), evaluate the miss-detection probability $P_m(\lambda_Q)$ for a random multiple access system with four time divisions and sixteen frequency divisions.

6. Given a $v$-element filled array antenna with the desired sidelobe-to-mainlobe ratio $r$, derive the radiation pattern through a Chebyshev polynomial of degree $v - 1$ and element spacing $d$, $\lambda/2 \leq d < \lambda$. What is the element excitation value if $v = 8$ and $v = 0.1$?

7. Show how to place five array elements in an aperture of eleven such that the sidelobe level is minimal. What is the sidelobe-to-mainlobe ratio?

8. Show how to construct a two-dimensional planar array from the difference set $\{D_P\} = \{0, 1, 3\}$.

9. For a mainlobe magnitude that does not include the magnitude of a sidelobe, reformulate the ratio $R$ of sidelobe to mainlobe in terms of the difference set parameters. How does this new ratio affect the expected peak sidelobe level?

10. To reduce third-order intermodulation products in transponder frequency assignment, use the simple difference set $\{D_P\}$ with parameters of $v = 21$, $k = 5$, and $\lambda = 1$ to construct the corresponding difference triangle, derive the frequency assignment, and verify that the solution can be obtained directly from the elements of the difference set.

11. Using the difference set $\{D_P\} = \{0, 1, 3, 9\}$ with $v = 13$, $k = 4$, and $\lambda = 1$, design a packet satellite transmission system with controlled message collision.

12. Investigate how the direction $(x, y, z)$ of an interference beam depends on the magnitude of the intermodulation phase variations $\phi_i$ and how a difference set with parameters $v$, $k$, $\lambda$ may be used to reduce such an interference beam.

13. From the difference set $\{D_P\} = \{0, 1, 3, 9\}$, generate a self-orthogonal convolutional code of encoding rate 2/3 and a maximum of eight stage shift register code subgenerators.

14. From the same difference set $\{D_P\} = \{0, 1, 3, 9\}$, derive the corresponding cyclic black code. Compare it with the results from Problem 13 in terms of code rate, error-correcting capability, and implementation complexity.

15. From the difference set $\{D_P\} = \{0, 1, 3\}$, derive the rate 1/2 convolutional self-orthogonal code and the rate 3/7 cyclic block code. What are the corresponding code generator polynomials?

# Chapter 4

# CRYPTOLOGY AND MESSAGE SECURITY

## 4.1 MESSAGE SECURITY AND SATELLITE COMMUNICATION

The advantages of satellite communication over terrestrial communication are that messages can be transmitted over a long distance and information can be disseminated over a large area. These advantages can turn into a disadvantage if sensitive information gets into the hands of unintended users.

Information transmitted by satellite is also vulnerable to message tampering and identity falsification. As the nature of the information transmitted through a satellite network becomes increasingly complex, message security is no longer limited to the domain of military, diplomatic, or governmental operations.

Message security is related to message privacy and message authenticity. Message privacy refers to the transmission of information protected against undesirable disclosure and destruction. Message authenticity refers to protection against false identity and unauthorized message alteration. For any practical system design authenticity should be checked before checking message privacy.

Message security can be provided through cryptographic techniques. These techniques will be examined in this chapter along with their relevance to satellite communication.

Even in peacetime, cryptographic satellite comunication systems are justified. For example: a satellite cable television distribution organization wants to discourage unauthorized users from receiving programming reserved for subscribers only; an international banking institution wishes to transfer funds across continents; a briefing for a NATO military exercise needs to be transmitted instantaneously to the headquarters of distant member participants; a giant corporation does not wish its competitors to obtain certain sensitive transmitted data; or a sweet old lady simply does not want anyone else other than the intended party to hear her conversation and she is willing to pay for such a guarantee.

As public, private, and government demands increase, the protection of satellite message transmission against passive eavesdropping or active tampering will become an essential part of future satellite communication system design. System designers will need to be able to derive, choose, and provide the various cryptographic alternatives for a specific application.

There is a dichotomy between satellite communication systems that feature secrecy and those that do not. The designer of a satellite communication system not utilizing secrecy seeks to make the message unmistakably clear at the other end of the satellite link, even when the transmitted signal is corrupted by channel noises, degraded by interference, or impaired by cascade nonlinearities. On the other hand, the designer of a satellite communication system that does utilize secrecy seeks to make the message impossible to recover by an unauthorized earth station receiver, even when the transmitted signal is perfectly received. Thus a different orientation and different methods are needed to accomplish the second objective.

In a series of five papers, Fang, Lu, and Lee of COMSAT Laboratories [1-5] first suggested cryptographic algorithms for satellite communications. Fang [6] and Ingmarsson and Wong of IBM Research [7] have suggested using encryption and authentication with onboard satellite processing capability. Both research teams recommend specific approaches for the encryption and decryption of messages transmitted through a satellite. Their work identifies the need for satellite systems to use ciphers.

In this chapter we will not emphasize the specifics presented in the work cited. The reasons are that, first, presenting a specific algorithm for a particular application may mislead the reader into believing that is the only solution to the problem and, second, some schemes presented have been shown subsequently to be relatively easily broken. Therefore, we shall discuss the principles of cryptography and examine the various fundamental techniques without making any specific recommendation. Once the reader has acquired a clear understanding of the fundamentals, the applicability of a particular scheme should be self-evident.

Based upon the concept of the ''open network'' structure (see Chapter 7), one of the salient features of the recently developed EUTELSAT (European Telecommunication Satellite) single-channel-per-carrier business system is the implementation of the encryption and decryption of messages transmitted through the satellite network. This encryption system has been designed to enable the use of the Data Encryption Standard algorithm (Section 4.6), and provides security for the satellite link by inhibiting unauthorized reception of the satellite communication.

The Centre National d'Etudes des Telecommunication has proposed encryption and decryption at 25.0 Mbps with selective link data through the TDMA structure of the French TELECOM-series domestic satellite systems.

Another satellite system that intends to use encryption is the Packet Switching Satellite System participated in by the U.S., U.K., Germany, Norway, and other countries. These are just three examples of unclassified satellite systems that use or plan to use cryptography during message transmission.

Section 4.2 defines terminologies used in the field of cryptology, describes cryptographic functions, outlines basic cryptographic satellite systems, and points out general problems in cryptographic study.

In Section 4.3 we study the various types of secret systems, and in Section 4.4 we consider cryptographic key management in terms of key generation, distribution, and recognition. In Section 4.5 we discuss cryptographic systems derived from or analyzed by information theory. Section 4.6 gives a detailed description of the Data Encryption Standard.

Section 4.7 describes, analyzes, and generalizes multiple access cryptosystems. Section 4.8 defines and discusses the problem of secrecy measures. Section 4.9 considers the Rivest-Shamir-Adleman algorithm. Section 4.10 discusses McEliece's public key method based on Goppa error-correcting codes. Section 4.11 presents all the techniques for constructing secret sharing systems and Section 4.12 provides the new unified method for secret sharing systems.

## 4.2   CRYPTOLOGY

Cryptology refers to the theory and practice of cryptography and cryptanalysis. Cryptography is the art of writing in or deciphering secret code; in satellite communication, cryptography means encoding and decoding transmitted messages, rendering them inaccessible to unauthorized users of the system. In the terminology of cryptography, the unaltered message is called plaintext, or clear text. The process of transforming a plaintext into an encrypted text or ciphertext is called encryption, and the reverse process is called decryption. The secret text or the ciphertext is also sometimes called a cryptogram. Encryption (or enciphering) and decryption (or deciphering) are systematically accomplished by means of cryptographic keys, which can be either privately or publicly provided for the transmission and reception of messages in secrecy.

### 4.2.1   Cryptographic Functions

General cryptographic functions are shown in Figure 4.1 in which an encryption algorithm $E$ transforms the message $M$ or plaintext to be transmitted into a cryptogram $C$ by the cryptographic key $K_E$. The received message $M'$ is obtained through the corresponding decryption algorithm $D$ with the decryption key $K_D$. These functions can be expressed as:

For encryption:

$$C = E\,(M,\,K_E) \tag{4.1}$$

For decryption:

$$M' = D\,[C,\,K_D]$$

$$= D\,[E\,(M,\,K_E),\,K_D] \tag{4.2}$$

where $C$, $M$, $K_E$, $K_D$, $M'$ are assumed to be in sequence with digital components.

The generation, distribution, and reception of the keys is referred to as key management, which is one of the most important elements of any cryptosystem design.

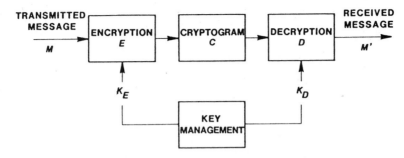

**Figure 4.1** Cryptographic functions.

The converse of designing a cryptosystem to protect secrecy is cryptanalysis, which refers to the theory and practice of uncovering a secret from either ciphertext or encryption keys without the consent or authorization of the sender. The means of doing this is called the method of attack.

It is customary in cryptography to assume (in spite of all the precautions that one might have taken to prevent this) that the cryptanalyst possesses both an encrypter and a decrypter so that any "secret" that resides in the cryptogram is a consequence only of the secret key. It is also the usual practice to use the same key for encrypting several plaintexts. For instance, one might change the key once each week so that all plaintexts encrypted within one week would be encrypted using the same key. It is further customary to classify the kinds of "attacks" that a cryptanalyst might make on a secrecy system as follows:

1. Ciphertext only: The cryptanalyst has available only one or more cryptograms known to have been encrypted with the same key.
2. Known plaintext: The cryptanalyst has available, in addition to the cryptograms to be "solved," one or more plaintexts and their resulting cryptograms known to have been encrypted with the same key.
3. Chosen plaintext: The cryptanalyst can obtain, for any plaintexts that he requests, their resulting cryptograms under the same key that was used for the cryptograms to be "solved."

The designer of a secrecy system would of course desire to make it secure against a chosen-plaintext attack. Failing that, he would wish to make it secure against a known-plaintext attack. But at the very least, he would wish it to be secure against a ciphertext-only attack.

In general, the cryptanalyst will consider that he has "broken" the secrecy system when he can quickly determine the plaintext from the cryptogram no matter what key is used. This usually means that he can quickly find the key itself that was used. He would still regard the system as "broken" if he could quickly determine the plaintext, or large segments of the plaintext, from the cryptogram for a substantial fraction of the possible keys. Or he might have to content himself with a much smaller degree of success, say with quickly determining a small portion of an occasional cryptogram. There is no rigid criterion for saying when a secrecy system has been "broken," but the history of secret communications is replete with cryptanalytic successes that were so complete as to warrant the description "breaking the secret code." Indeed, the record of cryptanalysts on the whole seems much more admirable than that of the designers of secrecy systems.

### 4.2.2 Cryptographic Satellite Communication Systems

The ultimate objective of message security in a satellite network is to transmit a message from user to user. Ciphering equipment can be located either at earth stations or in the satellite. In station-to-station encryption, the satellite appears transparent and the pair of communicating stations shares the key to protect the messages transmitted. Each station may have a number of keys for communicating with different stations. In this example of a TDMA system (see Figure 4.2) the satellite has no role in the cryptographic process.

If we allow the satellite to play an active role in the encryption, decryption processes, many options are available. A possible implementation scheme is shown in Figure 4.3, which indicates that a processing satellite can recognize the keys received from the transmitting uplink stations, and the onboard processor can arrange the generation of different keys in the satellite and distribute the keys in the downlink to the receiving stations. As the system becomes increasingly complex, onboard processing can enhance the level of message security.

Two principal difficulties are encountered in the study of cryptology. The first lies in evaluating a cryptosystem; the second, in study existing systems. Difficult problems that have been studied in depth may not necessarily lead to more secure

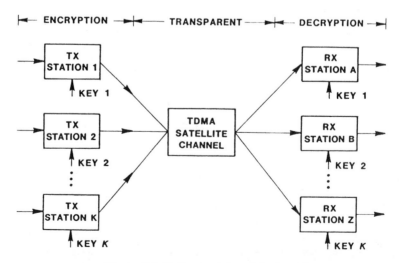

**Figure 4.2** Station-to-station ciphering.

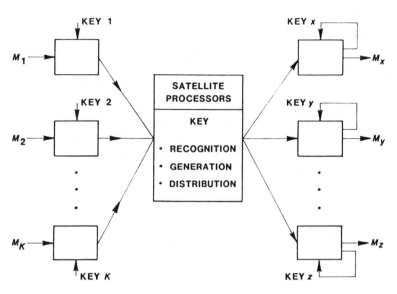

**Figure 4.3** Ciphering with satellite processors.

systems, and longer computation time does not always guarantee a stronger cryptosystem.

Cryptology presents a special difficulty for the student: many good techniques and analyses have been developed but remain the property of organizations whose main business is secrecy. Even algorithms that are supposed to be in the public domain, such as the U.S. Data Encryption Standard, have one essential element—the solution structure—that remains secret.

## 4.3 SECRET SYSTEMS

The secret systems are classified in this section as a special type of encryption system. We shall describe the functions and characteristics of the various secret systems' components—called ciphers—and draw comparisons and discuss trade-offs of some of them. These ciphers are fundamental building blocks to secret system design.

### 4.3.1 Block Ciphering

As the term implies, a block cipher refers to the messages that are encrypted and decrypted in blocks of information digits. Block ciphering has the same basic structure as block coding for error correction (to be described in Chapter 5). Block ciphering consists of block enciphering and block deciphering procedures, just as block coding consists of block encoding and block decoding procedures.

Physically, a coding system consists of an encoder and a decoder, while a ciphering system consists of an encipher and a decipher. Both encoding and enciphering use the various methods of deterministic algebraic transformation.

Both block coding and block ciphering transform a block of message into a block of coded or encrypted message, which depends on the uncoded message or the original block of plaintext. The difference between block coding and block ciphering is that block ciphering is achieved with ciphering keys, while most block coding schemes rely on parity checking. However, some ciphering keyed systems are based on the structure of block codes, as we shall discuss in Section 4.10.

The level of secrecy in block ciphering may be improved by (1) partitioning the plaintext into subblocks, then encrypting and decrypting them separately; (2) repeating the block encryption procedure a number of times; (3) combining (1) and (2). There can be many methods of variation on the above. For half block partition and arbitrary numbers of iteration we provide the following illustration, which happens to be the basic structure of the Data Encryption Standard (to be described in detail in Section 4.6).

The basic function of block ciphering with partition and iteration is shown in Figure 4.4, in which a block of message to be transformed iteratively $i = 0, 1, 2, \ldots, r$ times is divided equally into the left and right halves, denoted as $L(i)$ and $R(i)$. If the block of message is $n$ bits long, then $L(i)$ and $R(i)$ have $n/2$ bits each. Encryption and decryption are accomplished by the set of iteration-dependent keys $K(i+1)$ and a transformation function $f$, which is a function of $R(i)$ and $K(i+1)$ for encryption, and $L(i+1)$ and $K(i+1)$ for decryption. With this arrangement it can be easily verified that the iteration $i + 1$ from the $i$th iteration is:

$$L(i + 1) = R(i)$$

$$R(i + 1) = L(i) \oplus f[k(i + 1), R(i)]$$

where $\oplus$ denotes modulo-2 addition. For decryption, the order of $K(i + 1)$ is reversed, as shown in Figure 4.4.

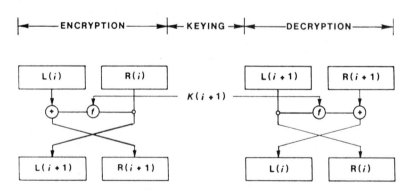

Figure 4.4 Block ciphering with partition and iteration.

### 4.3.2  Blockless Ciphering

In contrast to block ciphering, blockless ciphering has no block structure. The processes of encryption and decryption are performed continuously, on the same principle as convolutional encoding for error correction. However, the method of blockless ciphering is much simpler than encoding and decoding convolutional codes. With the addition of an initiation vector for recognition, the encryption and decryption processes can be accomplished by simple modulo-2 addition, as shown in Figure 4.5. (Such an arrangement should not be new to us because when we discussed the baseband spectrum-spreading multiple access techniques in Chapter 3, Volume I, modulo-2 additions were used in connection with maximum length sequence generators in both the transmitter and receiver. However, we shall not enter into a historic dispute here as to where it was applied first.)

In blockless ciphering, an initializing vector $I$ must be introduced to establish the required synchronization. $I$ can be introduced in more than one way: for example, let the transmitter generate $I$ and then transmit it to the receiver, or let the receiver provide its own $I$ and then inform the transmitter. Each method has its pros and cons from the standpoint of efficiency and cryptographic strength.

In general, initializing vectors are not secret; this degrades the strength of ciphering. There are a number of ways to enhance the situation, such as:

1. varying the length of the initializing vector
2. varying the patterns of the initializing vector either periodically or nonperiodically
3. giving each message its own initializing vector
4. providing feedback information either from the cryptogram or from the message itself.

The next function in blockless ciphering is the key $K$, which is used to encrypt $x$ and to decrypt $y$ through algorithm $G$. When the ciphering sequence $c$ is modulo-2 added to the plaintext $x$, we have the blockless ciphertext $y = x \oplus c$. The reverse process $x = y \oplus c$ is performed at the deciphering end. If $G$ is a linear pseudorandom (maximum length) sequence generator, we have the most

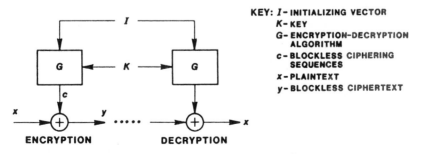

KEY: $I$ – INITIALIZING VECTOR
$K$ – KEY
$G$ – ENCRYPTION-DECRYPTION ALGORITHM
$c$ – BLOCKLESS CIPHERING SEQUENCES
$x$ – PLAINTEXT
$y$ – BLOCKLESS CIPHERTEXT

**Figure 4.5** The function of blockless ciphering.

simple blockless ciphering arrangement, in which case both the *I* and *K* functions may be omitted. If *G* is a nonlinear maximum length sequence generator, then *K* may also be used to extend the period of the generator beyond $2^m - 1$, as described in Chapter 3 of Volume I.

### 4.3.3 Feedback Ciphering

The strength of a cryptogram may be improved by providing a feedback mechanism which may be introduced along the path of an algorithm. Without distinguishing the types of ciphering (block or blockless), a general feedback from the output arrangement is shown in Figure 4.6, in which *G* is the ciphering algorithm, the mapping $G(K_1)$ is used for encryption, and $G(K_2)$ is used for decryption. $K_1$, $K_2$ are the keys. With feedback function *F* the encipher output is:

$$y = [x \pm y F] G(K_1)$$

from which

$$y = \frac{x \, G(K_1)}{1 \mp F \, G(K_1)} \tag{4.3}$$

where *x* is the plaintext or message sequence. On the deciphering end we have

$$[y \pm x F] \cdot G(K_2) = x \tag{4.4}$$

$$y = \frac{x \, [1 \mp F \cdot G(K_2)]}{G(K_2)}. \tag{4.5}$$

For the same sequence *y*, (4.3) must equal (4.5), and we have the conditions

$$G(K_1) = 1 \mp F \, G(K_2)$$

$$G(K_2) = 1 \mp F \, G(K_1)$$

relating the feedback function to the keys. If the feedback function is unity, $G(K_1)$ and $G(K_2)$ are additive inverses of each other. In general *F* and $G(\cdot)$ are linear functions in feedback ciphering.

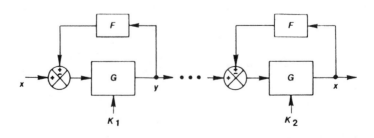

(ENCIPHER)                                  (DECIPHER)

**Figure 4.6** Feedback ciphering.

### 4.3.4  Cascade Ciphering

Each of the ciphering techniques described so far may be connected in cascade, as shown in Figure 4.7, where $E(i)$, $D(i)$ denote the component enciphers and deciphers for $i = 1, 2, \ldots, c$. For each $i$ the $\{E(i), D(i)\}$ pair may be identical or different. Thus, the keys in the set of $K(i)$ can be identical or different. The concatenations can occur anywhere along the ciphering algorithm, but for effectiveness they should be connected from the output of an encipher to the input of the next encipher.

Ciphers are concatenated to increase the cryptographic strength of the overall ciphering system and to reduce possible repetitiveness in the ciphertext, because two identical blocks or short streams of message will result in different ciphertext if cascade ciphering is used. A cascade ciphering system may be implemented in a variety of ways, depending on its application.

### 4.3.5  Transmission Ciphering

Up to this point we assume that all the encryption has been performed. Regardless of which method of ciphering is used, we can further assume that the ciphertext is in the form of binary digits, either blocked or blockless. In this section we suggest one more level of encryption that can be added just before transmitting the ciphertext. The scheme is like another stage of cascade ciphering, except that the level of protection of the transmission ciphering is limited only to casual attackers. Because the implementation of the scheme is different from others, we discuss it separately here.

This method of ciphering, called transmission ciphering, basically transforms a binary digit into a sequence of $m$-ary symbols and each symbol physically represents a segment of time and frequency combination for transmission. In Chapter 3 of Volume I we discussed and illustrated the method for achieving random multiple access, and we showed as an example how each bit of a binary sequence may be transformed into three nonbinary symbols. The symbols come from a matrix whose elements are presented in terms of both time and frequency divisions. In the present situation, for the purpose of ciphering, the binary digits are assumed to be encrypted by any means described so far, and the transmission ciphering key is the coding or transformation technique from the binary digits to the corresponding nonbinary sequences. The receiving station must have the correct key or sequence combination to unlock even a single bit, which is going to be further decrypted.

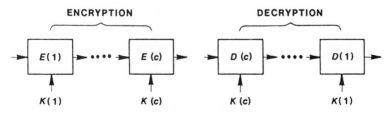

**Figure 4.7** Cascade ciphering.

The algorithm for generation of such nonbinary sequences (or keys in the present application) is described step by step in Section 3.1 of Volume II. Since the number of such sequences is proportional to the 4th power of the sequence length, the longer the sequence length, the better the system will be protected as far as secrecy is concerned. In contrast to the random multiple access application, secrecy is better served with fewer users in the system for a given sequence length.

## 4.4  KEY MANAGEMENT

Just as synchronization establishes meaningful information in a receiver, a strong cipher key provides meaningless information to an unauthorized receiver. A weak key can give away the entire security of the cryptosystem even if the data encryption algorithm is strong (i.e., computationally infeasible to break). If the cryptographic algorithm is assumed to be nonsecret, then the secrecy level of a cryptosystem relies solely upon how the cryptographic keys can be safeguarded. For this reason key management becomes a very important element in the design and operation of a cryptosystem. The subject of key management addresses the issues of the types of keys and considers the principles and techniques of key recognition, generation, and distribution. In addition to the discussion in this section, we shall elaborate on these issues throughout the chapter.

A number of types of keys can contribute to the various levels of cryptographic strength. A master key is a communication key that is permanently stored. A session key (a collection of message keys) is also a communication key but it changes from session to session. A working key is used in conjunction with the cryptographic algorithm. The working key may change from operation to operation. When a key or a set of keys can be derived from other keys, the derived keys are referred to as variants. If a key consists of a number of subkeys, we shall call the subkeys component keys.

### 4.4.1  Key Generation

The purpose of a cipher key is to provide message protection to the ciphertext against clever cryptographic attackers. To a casual attacker what can be more confusing than random numbers that have been used in various applications as key generation required by a cryptographic system? But encryption random number generation sometimes can be predictable. Predictability needs to be avoided by all means by the designers of any cryptosystem. Thus a combination of several random number generators may be used to increase the cryptographic strength of a key. The following example illustrates how a key may be generated.

Let us assume that a cryptosystem is implemented at the communicating earth stations $S(i)$ and $S(j)$ and each station is capable of generating a large set of random numbers. For a given message session the random numbers generated by $S(i)$ and $S(j)$ are denoted as $RN(i)$ and $RN(j)$. Next we proceed in steps as shown in Figure 4.8.

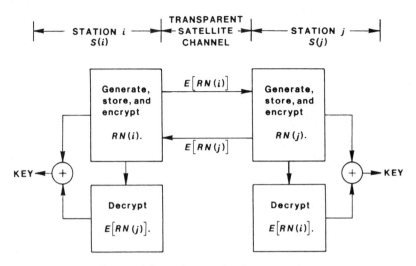

**Figure 4.8** Random number key generation.

Step 1:  Let $S(i)$ generate and store $RN(i)$, and transmit the encrypted $RN(i)$ to $S(j)$.

Step 2:  Let $S(j)$ generate $RN(j)$ and perform modulo-2 addition $RN(i) \oplus RN(j)$ to obtain the session key. Transmit the encrypted $RN(j)$ to $S(i)$.

Step 3:  In $S(i)$ retrieve $RN(i)$ and modulo-2 add with $RN(j)$ to obtain the same session key.

### 4.4.2  Key Distribution

In cryptography there are two types of key distribution: secret key distribution and public key distribution. Secret key distribution is an essential feature of the classical Shannon secrecy system. Secret keys may be distributed through various means such as special transmission channels, registered mail, or trustworthy messengers. As satellite communication does not easily accommodate these secret key distribution methods, we are more interested in public key distribution.

A public key distribution system is the procedure by which any two users who wish to communicate secretly with one another can form a secret key for this purpose without any prior exchange of secret information. These users would then use this key in a "secure" conventional cryptosystem for the actual exchange of their secret messages.

Thus a public key cryptosystem utilizes two related keys, one for encryption and one for decryption. The decryption key must be so difficult to derive from the encryption key that the encryption key can be publicly distributed. Anyone can encrypt, but not everyone can decrypt. This is the whole idea behind the public key cryptosystem.

Diffie and Hellman [8] suggested that practical secrecy could be achieved this way without any exchange of keys as in the classical case. Their results, which will be discussed in more detail in Section 4.7, have stimulated recent interest in modern cryptography.

### 4.4.3  Conditional Key Distribution

Key distribution may be arranged through statistical dependency, which means that distribution of a specific key may depend on a number of successive other keys. This key-dependent relation or conditional key distribution scheme further safeguards the cryptographic key.

Let a cryptographic master key of $r$ component keys be denoted as $(k_1, k_2, \ldots, k_r)$ and let $E_i$, $i = 1, 2, \ldots, r$ denote the event that the component key $k_i$ is transmitted without being attacked or compromised. $\overline{E}_i$ is the complement of $E_i$, that is, $\overline{E}_i$ is the event that key $k_i$ is compromised. The probability of such event is $P(\overline{E}_i)$. The system is no longer considered to be secret only if all the component keys are successfully attacked. The probability of this occurring is:

$$P(A) = P(\overline{E}_1, \overline{E}_2, \ldots, \overline{E}_r) \qquad (4.6)$$

where $A$ denotes the event that the system is nonsecret and $(\overline{E}_1, \overline{E}_2, \ldots, \overline{E}_r)$ denote the joint event of the $r$ component keys being compromised. By probabilistic identity this probability of the system becoming insecure is:

$$P(A) = P(\overline{E}_1) \, P(\overline{E}_2|\overline{E}_1) \, P(\overline{E}_3|\overline{E}_1, \overline{E}_2) \ldots.$$
$$P(\overline{E}_r|\overline{E}_1, \overline{E}_2, \ldots, \overline{E}_{r-1}). \qquad (4.7)$$

The mutual information derived from the joint ensembles is:

$$I(A) = \log \frac{1}{P(A)}$$

$$= \log \frac{1}{P(\overline{E}_1) \, P(\overline{E}_2|\overline{E}_1) \ldots P(\overline{E}_r|\overline{E}_1, \ldots \overline{E}_{r-1})}$$

$$= \log \frac{1}{P(\overline{E}_1)} + \log \frac{1}{P(\overline{E}_2|\overline{E}_1)} + \ldots$$

$$+ \log \frac{1}{P(\overline{E}_n|\overline{E}_1, \overline{E}_2, \ldots, \overline{E}_{r-1})}$$

$$= I(\overline{E}_1) + I(\overline{E}_2|\overline{E}_1) + \ldots$$

$$+ I(\overline{E}_r|\overline{E}_1, \overline{E}_2, \ldots, \overline{E}_{r-1}). \qquad (4.8)$$

The above result indicates that the mutual information based on the $r$-dimensional key ensembles is equal to the sum of the self-information as shown in the last expression of (4.8). All terms of self-information in the expression except the first are conditional on the previous events.

In contrast, if the distribution of the component keys is statistically independent, then the mutual information of the ciphering system becomes

$$I'(A) = \sum_{i=l}^{r} I(\overline{E_i}), \; 1 \leq i \leq r. \tag{4.9}$$

In order to have a more secure system, we wish to have a larger value of mutual information, which means, from (4.8) and (4.9), $I(A) > I'(A)$. This is the justification for distribution of keys by conditional means. The level of implementation of this scheme depends on the relative values of the set of conditional and marginal probabilities under consideration.

The desirability of a larger value of mutual information among the set of component keys is the opposite of the case between plaintext and ciphertext. In the latter situation it is desirable to have the mutual information approach zero in order for a cipher system to be very secure.

### 4.4.4 Key Recognition

Assume that an onboard processor can manage the cryptographic procedures of a satellite network. The network consists of a number of stations which all play passive roles and only respond to the requests of the onboard processor. If the onboard processor provides a session key that is encrypted under the station master key, a practical procedure for recognizing or recovering the session key at each station can be described as follows:

Step 1: When a station receives the ciphertext from the satellite, the station master key is retrieved separately from master key storage. The decryption procedure begins by the use of the working key in order to recognize the session key.

Step 2: When the session key is recognized, it is transferred to working key storage and replaces the master key. The station is then ready to recover plaintext using the recognized session key.

This procedure, which has been suggested primarily for encryption between host computers and terminals [10], is diagramed in Figure 4.9.

### 4.5 INFORMATION-THEORETIC SYSTEMS

The use of secrecy techniques is older than Greek civilization. The secret systems of ancient China were so good that present-day archeologists are still unable to figure them out. Pig Latin and Nimbus talk, codes that children use, are called syllable ciphers. To encode a message in Pig Latin the first letter of a word is switched from the beginning to the end and the syllable AY is added. Thus GO becomes O-GAY. If the word begins with a vowel, AY is added at the end. "Good night Irene" becomes "ood-GAY ight-NAY Irene-AY." To encode a

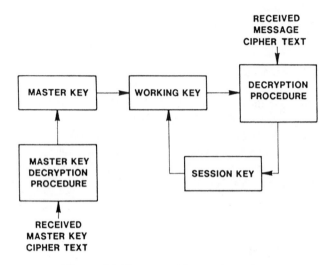

**Figure 4.9** Key recognitions at the stations.

message in Nimbus talk one inserts B after every vowel, then repeats the vowel. Capital Children's Museum becomes ca-BA-pi-BI-ta-BA-l chi-Bl-ldre-BE-n's mu-Bu-se-BE-u-Bu-m.

Modern-day secret system analysis began with Shannon [11], and in this section, which is based on Massey's lecture notes [9], we shall discuss the essence of Shannon's work and its extensions by Hellman and Lu [13, 14].

### 4.5.1  Shannon's Model

The model of a secrecy system proposed by Shannon is shown in Figure 4.10. Here the encryption and decryption keys are the same. The $m$-tuple $X = (x_1, x_2, \ldots, x_m]$ (where we will usually suppose that each $x_i$ is a binary digit) is the

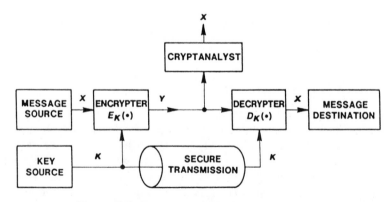

**Figure 4.10** Shannon's model of a secrecy system.

message or plaintext that is to be delivered to the destination. The $s$-tuple $Y = [y_1, y_2, \ldots, y_s]$ is the cryptogram or ciphertext that is transmitted over an insecure channel to the destination. (By "insecure" we mean that the cryptanalyst has access to this cryptogram.) The $t$-tuple $K = [k_1, k_2, \ldots, k_t]$ is the secret key that is sent to the destination over a secure channel. (We suppose that the cryptanalyst has no access to this key.) In general, the transmission of the key $K$, takes place at a much earlier time than that of the cryptogram $Y$.

We can view the key $K$ as determining the particular "scrambling" that the encrypter performs on the plaintext to produce the cryptogram. In other words, the encrypter implements a particular function $E_K(\cdot)$ determined by the value of $K$. We can thus express the operation of the encrypter as

$$Y = E_K(X). \tag{4.10}$$

Of course, $Y$ is actually a function of both $X$ and $K$, but we have chosen to write the right side of (4.10) as $E_K(X)$ rather than $E(X,K)$ to emphasize that we wish to think of the choice of the key as determining the particular functional relation between $X$ and $Y$.

The task of the decrypter is to reproduce the plaintext from the cryptogram and key. We demand that the decrypter implement a function $D_K(\cdot)$ such that

$$X = D_K(Y) \tag{4.11}$$

or, equivalently, such that for every possible key $K$

$$X = D_K [E_K(x)] \tag{4.12}$$

for all possible plaintexts $X$.

### 4.5.2   Ideal Secrecy Systems

From Shannon's model of a secrecy system as given in Figure 4.10 we have already noted that in actual systems the same key is used for many encryptions. To simplify the notation that would otherwise be needed to describe such multiple encipherings in one key, we now suppose that the plaintext $X$ in Figure 4.10 is the concatenation of all the individual plaintexts that are encoded in the same key. Thus, in this section we assume that in Figure 4.10 the key is changed after each encryption.

We begin our analysis by noting that the relation (4.11) merely states that $Y$ and $K$ together uniquely determine $X$. It then follows from fundamental information theory [12] that (4.11) is equivalent to

$$H(X|Y\,K) = 0. \tag{4.13}$$

The relation (4.13) is as obvious as it is general. To obtain a less obvious relation, we first introduce a definition. We will say that a secrecy system is ideal (or ideally secure) if

$$I(X;Y) = 0. \tag{4.14}$$

From the definition of mutual information (4.14) implies that $X$ and $Y$ are statistically independent. When this is the case, the cryptanalyst can estimate $X$ no better from $Y$ than he could from a "cryptogram" that was randomly chosen from the set of possible cryptograms without regard to any plaintext. But it should be emphasized that the security of which we are speaking here is security against a ciphertext-only attack. An ideal secrecy system might not be at all secure from even a known-plaintext attack. Moreover, it is not at all obvious that ideal secrecy systems exist. We defer answering this question until after we have proved the following result.

THEOREM 4.1: In an ideal secrecy system,

$$H(K) \geq H(X). \tag{4.15}$$

*Proof.* We begin by observing that

$$H(X) = H(X|Y) - I(X,Y) = H(X|Y) \tag{4.16}$$

where we have made use of (4.14). Next, we note that

$$H(X|Y) \leq H(X\ K|Y) = H(K|Y) + H(X|Y\ K)$$

$$= H(K|Y) \leq H(K) \tag{4.17}$$

where we have made use of (4.13). (4.16) and (4.17) now imply (4.15).

The import of Theorem 4.1 is that ideal secrecy systems necessarily require a large amount of key. In fact, if the digits in the key $K$ are binary, then

$$H(X) \leq H(K) = H(k_1\ k_2\ \ldots\ k_t) \leq k \text{ bits} \tag{4.18}$$

so that we would need at least as many binary digits in the key as there are bits of information in the plaintext. Our proof of Theorem 4.1 actually implies a more general result from which we could have deduced Theorem 4.1 as a corollary, namely:

THEOREM 4.2: In any secrecy system,

$$I(X;Y) \geq H(X) - H(K). \tag{4.19}$$

*Proof.* By definition,

$$I(X;Y) = H(X) - H(X|Y). \tag{4.20}$$

But inequality (4.17) above was obtained using only (4.13), which applies to any secrecy system, not only an ideal one. Thus, we can use (4.17) in (4.20) to obtain the desired inequality (4.19).

The import of Theorem 4.2 is that in secrecy systems with a small amount of key, the cryptogram necessarily gives much information about the plaintext—whether or not this information is useful to a cryptanalyst is another question!

We are now ready to answer the question of whether or not ideal secrecy systems exist.

THEOREM 4.3: *Ideal secrecy systems exist.*

Our proof of this rather startling (in the light of the disastrous history of secret communications) claim will be by way of construction. We assume with no loss of generality that the cryptogram $X = [x_1, x_2, \ldots, x_m]$ is a binary message, i.e., that each $x_i$ takes values in GF(2). We now choose the key $K = [k_1, k_2, \ldots, k_t]$ and the cryptogram $Y = [y_1, y_2, \ldots, y_s]$ also to have binary components and we also choose

$$t = s = m$$

so that $X$, $Y$ and $K$ all contain the same number of binary components. For our key generator, we choose a binary symmetric source (BSS); the components of $K$ are thus statistically independent and equally likely to be 0 or 1. For our encrypter, we choose simply a modulo-2 adder (Exclusive-Or gate) that we apply component-by-component to $X$ and $K$, i.e,

$$Y = E_K(X) = X + K. \tag{4.21}$$

We now see that we can regard $K$ as the error pattern added by a BSC with crossover probability $\epsilon = \frac{1}{2}$ to the transmitted word $X$ to give the received word $Y$. But

$$I(X;Y) \le sC = 0,$$

since this BSC has capacity $C = 0$. Since always $I(X;Y) \ge 0$, we must conclude that

$$I(X;Y) = 0$$

and, hence, that this simple secrecy system is ideal.

It should be noted, however, that the above ideal secrecy system is completely insecure against a known plaintext attack. For if we know any plaintext-cryptogram pair $(x', y')$ in the same key $k$ as the cryptogram $y$ to be solved, then, because $y' = x' + k$, we can solve for the key as $k = x' + y'$. We can use the now known key $k$ to find the plaintext of the cryptogram $y$ as $x = y + k$.

Ideal secrecy systems of the above type have long been used in diplomatic and espionage circles, although their "perfect" security was first proved by Shannon. The classical name for these systems is the "one-time pad." The name derives from the fact that a secret agent might generate such a key by tossing a "binary coin" $s$ times. He then writes this same sequence on two slips of paper (or "pads"), one of which he hands to a friend and the other he keeps for himself. When later he wishes to send a secret message $X$ to his friend, he encrypts it as described above with his copy of the pad. His friend can then decrypt the cryptogram by using the other copy of the pad, but no one else can do better from the cryptogram than randomly guess as to what $X$ is. The "one-time" portion of the name derives from the early realization by security experts that such a system is disastrously insecure if the same pad is used to encrypt more than one plaintext. The secret agent is always advised to burn his copy of the key immediately after he forms the cryptogram!

What prevents the one-time pad, or any ideal secrecy system for that matter, from being a practical solution for general secret communications is not its vulnerability to a known plaintext attack but rather its requirement for enormous amounts of key when large amounts of plaintext are to be sent. The generation, distribution, and protection of the required amount of secret random key (the so-called "key management problem") is generally not feasible except in situations (such as in diplomacy and espionage) characterized by infrequent short messages whose secrecy is of paramount importance.

Before concluding our study of ideal secrecy systems, we wish to consider the tightness of the bound (4.15) of Theorem 4.1. Recall that in our one-time pad secrecy system, the key and plaintext have the same length, so that (4.15) gives

$$t = s \geqslant H(X). \tag{4.22}$$

Thus, if the message is highly redundant (i.e., if $H(X) \ll s$), the key length will be much greater than implied by the lower bound (4.22). The question then arises whether there exist ideal secrecy systems (with keys having binary components) that are minimal in the sense that

$$t \approx H(X). \tag{4.23}$$

The answer to this question is also yes. To obtain such a minimal ideal system, we need simply (in principle!) augment our one-time pad system with a data compressor and its inverse. We would first compress the actual plaintext $X = [x_1, x_2, \ldots, x_m]$ to obtain a compressed plaintext $X' = [x_1', x_2', \ldots, x_s']$; then we would encrypt according to (4.21), i.e.,

$$Y = X' + K.$$

After decrypting $Y$ to obtain $X'$, we would then use the inverse of the data compressor on $X'$ to obtain the actual plaintext $X$. In principle, we can always build a data compressor so that the length of the compressed message satisfies

$$s \approx H(X) \quad \text{(bits)}$$

and thus, because $t = s$ in our one-time pad system, (4.23) will be satisfied.

We have just shown that the bound of Theorem 4.1 for ideal secrecy systems can be approached as closely as desired. In one respect, this is not very interesting because the bound still implies that impractically long keys must be used in most situations of interest. However, something else of practical interest can also be deduced; namely, that compressing data before encryption enhances the security of the system. This fact, pointed out by Shannon in 1948, is still not widely appreciated. The very redundancy of natural plaintext is the best weapon in the cryptanalyst's arsenal. Denying him the use of this weapon is to put him at a great disadvantage.

We now consider some theoretical aspects of a ciphertext-only cryptanalytic attack on a cryptosystem. In particular, we examine the question of how much ciphertext the cryptanalyst must possess before he can, in principle at least, "break" the cryptosystem. Toward this end, Shannon considered the uncertainty about the key, given the first $n$ digits of the ciphertext, $H(K|y_1 y_2 \ldots y_n)$, which is called the key equivocation. For $n = 0$, the key equivocation is just $H(K)$. The key equivocation can only decrease (or remain constant) as $n$ increases because

$$H(K|y_1 \ldots y_n y_{n+1}) \leqslant H(K|y_1 \ldots y_n).$$

Shannon defined the unicity distance of a cryptosystem and its associated message source as the smallest value of $n$ for which there is only one, or a very small number, of the possible keys consistent with the first $n$ digits of the cryptogram. This is the point where the system can, in principle, be broken (if there are only a small number of keys to consider, one can easily try each in the decrypter to see which gives sensible plaintext when applied to the ciphertext). But this is also the point where the key equivocation first reaches 0, or a small number. In other words, the unicity distance is the smallest $n$ such that

$$H(K|y_1 y_2 \ldots y_n) \approx 0. \tag{4.24}$$

To find the unicity distance, we present Massey's unpublished new derivation and begin by considering how the key equivocation decreases with $n$ in a "typical cryptosystem" [9]. Virtually all cryptosystems and their associated message sources have the property that, for sufficiently small $n$ (how small we leave open for the moment), the first $n$ digits of the cryptogram are virtually "completely random" binary digits. (By completely random, we mean that the digits are statistically independent and equally likely to be 0 or 1, i.e., they are equivalent to the output sequence of a BSS.) For such $n$, we thus have

$$H(y_1 y_2 \ldots y_n) \approx n \text{ bits.} \tag{4.25}$$

We now consider the identities (based on the information measures from Chapter 2)

$$\begin{aligned} H(X Y|K) &= H(X|K) + H(Y|XK) \\ &= H(Y|K) + H(X|YK). \end{aligned} \tag{4.26}$$

In all cryptosystems, the key is chosen independently of the plaintext, i.e.,

$$H(X|K) = H(X), \tag{4.27}$$

and of course the cryptogram is uniquely determined by the plaintext and key, i.e.,

$$H(Y|XK) = 0. \tag{4.28}$$

Using (4.27), (4.28), and (4.13) in (4.26), we conclude that, for all cryptosystems,

$$H(Y|K) = H(X). \tag{4.29}$$

For most cryptosystems and their associated message sources, the uncertainty $H(y_1\, y_2\, \ldots\, y_n | K)$ grows almost linearly with $n$; thus, because this uncertainty by (4.29) equals $H(X)$ when $n \leq s$, we have

$$H(y_1\, y_2\, \ldots\, y_n | K) \approx \frac{n}{s} H(X), \quad 1 \leq n \leq s \qquad (4.30)$$

for most cryptosystems and their associated message sources. Suppose now that (4.30) holds and, also, for sufficiently small $n$, that (4.25) holds. We can use these two approximations in

$$I(K; y_1\, \ldots\, y_n) = H(y_1\, \ldots\, y_n) - H(y_1\, \ldots\, y_n | K)$$

to obtain

$$I(K; y_1\, \ldots\, y_n) \approx n - \frac{n}{s} H(X) = n\rho \qquad (4.31)$$

where we have defined

$$\rho = 1 - \frac{1}{s} H(X). \qquad (4.32)$$

The quantity $\rho$ can be interpreted as the plaintext redundancy per digit of the cryptogram. To see this interpretation, we note that, when $K$ is known, it must be possible to solve for $X$ from $Y$. Hence, the number of binary digits in $Y$ must satisfy

$$s \geq H(X). \qquad (4.33)$$

Moreover, (near) equality will hold in (4.33) if and only if $Y$ is the representation of $X$ obtained by an (almost) perfect data compressor. Inequality (4.33) together with (4.32) implies

$$0 \leq \rho \leq 1 \qquad (4.34)$$

with (near) equality on the left when and only when the system incorporates such (nearly) perfect compression of the plaintext, in addition to supplying secrecy.

We are finally ready to determine how the key equivocation decreases with the length $n$ of plaintext examined. We need only use (4.31) in the relation

$$H(K | y_1\, y_2\, \ldots\, y_n) = H(K) - I(K; y_1\, y_2\, \ldots\, y_n)$$

to conclude that

$$H(K | y_1\, y_2\, \ldots\, y_n) \approx H(K) - \rho n \qquad (4.35)$$

whenever (4.30) holds and $n$ is sufficiently small so that (4.25) also holds. From (4.35) we see that (in most cryptosystems) the key equivocation decreases linearly with the length $n$ of ciphertext examined when $n$ is sufficiently small. The question we now must answer is "How small must $n$ be?" The rather surprising answer is that, in many cases, (4.35) holds until the key equivocation is quite small. When this is so, it follows by setting the left side of (4.35) to 0 that the unicity distance is given by

$$\text{unicity distance} \approx \frac{H(\mathbf{K})}{\rho}. \tag{4.36}$$

In the usual case where the $t$ binary digits in the key are completely random, (4.36) becomes

$$\text{unicity distance} \approx \frac{t}{\rho}. \tag{4.37}$$

By considering an ensemble of appropriately defined cryptosystems and message sources, Shannon showed that (4.35) would indeed remain valid until the key equivocation was quite small so that (4.36) [or (4.37)] is an accurate estimate of the unicity distance. Shannon gave examples which indicate that actual cryptographic schemes and real (as opposed to mathematically defined) message sources conform closely to (4.35) and (4.36).

In Figure 4.11, we sketch the key equivocation as a function of the length $n$ of the cryptotext examined for Shannon's ensemble of cryptosystems and message sources. This can be considered a "canonical" relationship to which most actual systems will conform. There are two points worth emphasizing about this relationship. First, we see that the unicity distance apparently becomes infinite when $\rho = 0$, i.e., when the cryptosystem acts also as a perfect data compressor for its associated message source. Such is indeed the case, but this does not imply perfect secrecy (which, if it did, would contradict Theorem 4.1); rather it implies that, no matter how large $n$ becomes, there will always be about $2^{H(K)}$ possible solutions for the key consistent with the first $n$ digits of the cryptogram and hence also about $2^{H(K)}$ valid plaintext messages consistent with the entire cryptogram. This is the best that one can do with a key whose length is much less than $H(X)$, but it is not "perfect secrecy." For real message sources, zero

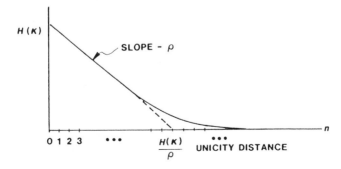

**Figure 4.11** A sketch of the key equivocation given the first $n$ digits of the cryptogram for Shannon's ensemble of randomly chosen cryptosystems and message sources.

redundancy (i.e., $\rho = 0$) can of course never be achieved. In fact, many cryptosystems expand the plaintext ($s > m$) so that $\rho$ is even greater than the per letter redundancy $1 - H(X)/m$ of the plaintext itself. Even when $s = m$ so that $\rho = 1 - H(X)/m$, real message sources (say, of direct binary coded English text or German text) typically give $\rho \approx .8$ so that the unicity distance is only slightly greater than the key length. The second point to remember about Figure 4.11 is that it describes only what is theoretically possible to achieve by cryptanalysis; it says nothing about how much effort will be required to solve the cryptogram. This effort may be so large that, from a practical standpoint, it is impossible to solve the cryptogram even though its length $s$ is far greater than the unicity distance. This failure to take computational effort into account is the fundamental weakness of the information-theoretic approach to cryptography that we have followed to this point. Shannon was quite aware of this weakness and in 1949 gave the outlines of an approach that would remedy this deficiency.

### 4.5.3  Practical Secrecy Systems

Shannon used the terminology "practical secrecy" to describe the kind of secrecy that results when the computational effort to break the cryptographic system is so large as to make successful cryptanalysis impractically difficult. Defining $W_n$ as the average computational work (in some appropriate units, say "person-years" of effort) to solve for (all of) the key(s) consistent with the cryptogram, Shannon suggested that this function would behave as shown in Figure 4.12. (The dotted portion of this curve indicates that the solution for the key is not unique for $n$ in this range.) Naturally, one would choose $s$ (the number of encrypted digits before the key is changed) as some value of $n$ beyond the unicity distance but for which $W_n$ is still unrewardingly large for a cryptographer. But how does one find the function $W_n$? Shannon suggested ways, which we shall now briefly examine.

The first way that one might calculate $W_n$ is to make a list of all the customary cryptanalytic approaches to solving a cryptogram, then to estimate how much work will be required on the average to break the system using this arsenal of

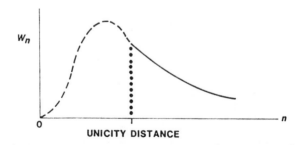

**Figure 4.12**  Shannon's sketch of the average computational work to solve for the set of keys consistent with the first $n$ letters of the cryptogram.

weapons. This approach requires considerable cryptanalytic experience, but it is certainly the method most often used in cryptographic practice.

The second method suggested by Shannon to determine (or bound) $W_n$ was to prove mathematically that breaking the secrecy system must require on the average at least a certain amount of work, regardless of the method used by the cryptanalyst. In fact every sorting algorithm must use an average of at least $\log_2(n!)$ comparisons to order a list of $n$ unequal real numbers in random order. It seems clear that this second approach to practical secrecy would be much superior to the first, if, indeed, one could succeed in proving a lower bound on $W_n$ for some secrecy system.

When Shannon wrote in 1949, such proofs were well-nigh impossible. But the rapid evolution of digital computers in the intervening years has greatly stimulated interest in the general subject of computational complexity. There is now a large body of knowledge in this field that can potentially be applied to assess the practical security of a secrecy system or to suggest new forms of secrecy systems.

### 4.5.4 The Results of Hellman and Lu

Let us consider the output of a message source as a stationary random sequence, each component taking values from a finite alphabet set. If a key is generated from a key source with an alphabet size of $e^{nr}$, then $n$ is the keyword length and $r$ is defined as the key rate.

For uniformly distributed random ciphers, Shannon's result states that a sufficient condition for the existence of a good cipher is that the key rate be larger than the message redundancy. Shannon also showed that it is possible to construct good secrecy systems through source coding techniques. These approaches have been extended by Hellman [13], Lu [14], and others. Hellman considered uniformly distributed ciphers and used the criterion of the correct decryption probability. He showed that this probability can be made small when the key rate is greater than the message redundancy. On the other hand, under the conditions of a binary memoryless source and with additive block ciphering, Lu showed that, in principle, good ciphers exist in which the key rate is less than the message redundancy.

Let $c_k = (c_{k,1}, c_{k,2}, \ldots, c_{k,n})$ be the $k$th keyword and $k$ ranges from 1 to $e^{nr}$. The set of keys is the set of key words $c_1, c_2, \ldots, c_{e^{nr}}$, which defines a cipher with block length $n$ and key rate $r$. For each digit, the corresponding message digit is combined with the same digit of the key word by using the combiner function $f(m_i, c_{k, i})$ for $i = 1, 2, \ldots, n$ where $m_i$ denotes the $i$th digit of the message sequence and $c_{k,i}$ is the $i$th digit of the key word $c_k$. The combiner function is a mapping that produces a cryptogram. When the combiner function is additive-like, Lu refers to it as an additive-like instantaneous block (ALIB) cipher, which can be uniquely encrypted as well as decrypted.

The decipher has two inputs, the key and the cryptogram word. The receiving station generates the key and performs the inverse mapping $f^{-1}(m_i, c_{k, i})$ on each digit in order to recover the message word from the received cryptogram.

The simplest form of Lu's model is the binary additive cipher, which consists of a modulo-2 adder as combiner and a set of $2^{nr}$ key words with the cryptogram formed as $m_i \oplus c_{k,i}$ for $i = 1, 2, \ldots, n$.

## 4.6 THE DATA ENCRYPTION STANDARD

Under Shannon's general concept of "mixing transformations" the Data Encryption Standard (DES) became effective in 1977 [15] for use in unclassified U.S. Government applications. This standard, based on the cryptographic cipher designed by Feistel [16], is a block cipher with both partition and iteration methods as described in Section 4.3.1. The algorithmic standard encrypts a 64-bit block of plaintext into a 64-bit block of ciphertext under the control of a set of 56-bit cryptographic keys, using one key for each iteration. The number of iterations is determined by the combination of the ciphertext, the plaintext, and the corresponding key.

The essential functions of DES are partition, iteration, permutation, shifting, selection, and modulo-2 addition. The essential mechanism of DES, key generation, and cryptographic transformation will be treated separately in detail below.

With the number of iterations to be 16 let $X = x_1, x_2, \ldots, x_{64}$ and $Y = y_1, y_2, \ldots, y_{64}$ denote the input (plaintext) and output (ciphertext) of the algorithm, respectively. With the same notations as in Section 4.3.1, let the left-hand and right-hand partitions be denoted as $L(i)$ and $R(i)$, $i = 0, 1, \ldots, 16$. The components of $L(i)$ and $R(i)$ are denoted as $\ell_1(i), \ell_2(i), \ldots, \ell_{32}(i)$ and $r_1(i), r_2(i), \ldots, r_{32}(i)$. Let the key of the $i$th iteration be denoted as $K(i)$, as shown in Figure 4.13. With a simple initial permutation and inverse initial permutation we have the basic DES encryption algorithm, as shown, where $f$ is the transformation function to be described.

### 4.6.1  Key Generation

For each iteration $i$ in the DES algorithm a different key vector $K(i)$ of 48 bits needs to be generated. The basic key generation function of DES is shown in Figure 4.14, which consists of bit reduction, bit permutation, bit shifting, and bit selection procedures. Each procedure is elaborated in the following sections.

#### 4.6.1.1  Bit Reduction

Bit reduction refers to the removal of eight parity check bits from the 64-bit key input, which is denoted as $K = k_1, k_2, \ldots, k_{64}$. One bit in each 8-bit byte of $K$ is used for error detection through parity checking. These parity check bits are $k_8, k_{16}, k_{24}, k_{32}, k_{40}, k_{48}, k_{56}$, and $k_{64}$. After successful parity checking, the bit reduction procedure removes the parity check digits. What remains after bit reduction is a 56-bit combination. We load this combination into two 28-bit shift registers.

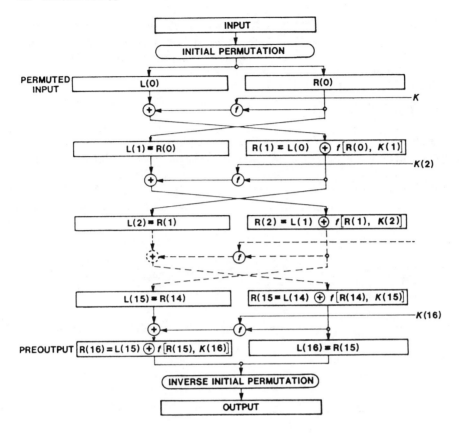

**Figure 4.13** The Data Encryption Standard.

### 4.6.1.2   Bit Permutation

The permutation procedure is specified by DES as

$$C_0 = k_{57}, k_{49}, k_{41}, k_{33}, k_{25}, k_{17}, k_9,$$
$$k_1, k_{58}, k_{50}, k_{42}, k_{34}, k_{26}, k_{18},$$
$$k_{10}, k_2, k_{59}, k_{51}, k_{43}, k_{35}, k_{27},$$
$$k_{19}, k_{11}, k_3, k_{60}, k_{52}, k_{44}, k_{36}, \tag{4.38}$$

$$D_0 = k_{63}, k_{55}, k_{47}, k_{39}, k_{31}, k_{23}, k_{15},$$
$$k_7, k_{62}, k_{54}, k_{46}, k_{38}, k_{30}, k_{22},$$
$$k_{14}, k_6, k_{61}, k_{53}, k_{45}, k_{37}, k_{29},$$
$$k_{21}, k_{13}, k_5, k_{28}, k_{20}, k_{12}, k_4 \tag{4.39}$$

where $C_0$ and $D_0$ represent the $o$th iteration of the shift register contents. Note that the eight parity check elements disappear from $C_0$ and $D_0$. Note also that in DES the combination of bit reduction and bit permutation is called the Permuted Choice 1.

### 4.6.1.3   Bit Shifting

With the register contents in $C_0$ and $D_0$, we shift the registers to the left for encryption and to the right for decryption. The number of shifts for the two registers at each iteration is not the same. That is, the number of shifts for $(C_i, D_i)$ is not necessarily the same for $(C_{i+1}, D_{i+1})$. This shifting number is either 1 or 2 depending on the number of iterations, as specified in Table 4.1.

**Table 4.1**
Number of shifts in key generation.

| Iteration Number | 1, 2, 9, 16 | 3–8, 10–15 |
|---|---|---|
| Number of Shifts | 1 | 2 |

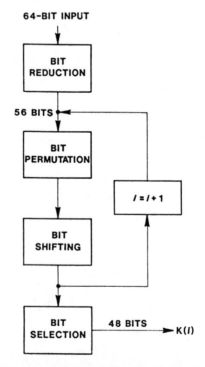

**Figure 4.14** Key generation functions of DES.

Based on the number of shifts for each iteration, as in Table 4.1, we perform the cyclic shifting process beginning with $C_0$ and $D_0$. Since $C_0$ begins with $k_{57}$, $k_{49}$, . . . and ends with $k_{44}$, $k_{36}$, as shown in (4.38), the first shift (a single shift) gives $C_1$ the combination of $k_{49}$, $k_{41}$, . . ., $k_{36}$, $k_{57}$. Note that the first bit of $C_0$ becomes the last bit of $C_1$ and all other bits of $C_0$ have pushed toward the left to form $C_1$ as a consequence of the first shift. The complete shifting processes for both $C_i$ and $D_i$ with $i = 1, 2, . . .$ 16 are also provided by Meyer and Matyas [10], as shown in Tables 4.2 and 4.3. We note that for $i = 3$ to 8 and 10 to 15, double shifts take place.

### 4.6.1.4  Bit Selection

Once the elements of $C_i$ and $D_i$ are obtained from bit shifting, the elements of the key $K(i)$ can be readily derived from the bit selection process, as shown in Figure 4.14. In DES, the selection process of the $i$th iteration for $i \geqslant 2$ is called the Permuted Choice 2 (PC-2). The Permuted Choice 2 defines how the key vector $K(i)$ can be generated from the contents of the shift register pairs $(C_i, D_i)$ for $i = 1, 2, . . .$ 16. Specifically, $K(i)$ is derived from $(C_i, D_i)$ by selecting the key component positions as:

$$14,17,11,24, \; 1,5, \; 3,28,15, \; 6,21,10$$

$$23,19,12, \; 4,26,8,16, \; 7,27,20,13, \; 2 \qquad (4.40)$$

$$41,52,31,37,47,55,30,40,51,45,33,48$$

$$44,49,39,56,34,53,46,42,50,36,29,32. \qquad (4.41)$$

With the key positions as in (4.40) and (4.41) for each iteration, we go to Tables 4.4 and 4.5 to obtain the corresponding position elements for $K(i)$. $K(i)$ is generated simply by combining the selected key components from both $C_i$ and $D_i$. Since each selection contains 24 bits, $K(i)$ has 48 bits.

*Example 4.1:* From (4.40) the positions selected from $C_i$ are 14, 17, 11, . . ., 20, 13, 2, and from (4.41) the positions selected from $D_i$ are 41, 52, 31, . . ., 36, 29, and 32. For $i = 1$, and from Tables 4.4 and 4.5, we identify the corresponding position elements from $C_1$ and $D_1$ as

$$K(C_1) = 10, 51, 34, . . ., 27, 18, 41$$

$$K(D_1) = 22, 28, 39, . . ., 62, 55, 31.$$

By combining the elements of $K(C_1)$ and $K(D_1)$, we have the elements of the key vector

$$K(1) = K(C_1) + K(D_1)$$

$$= 10, 51, 34, . . ., 27, 18, 41,$$

$$22, 28, 39, . . ., 62, 55, 31.$$

**Table 4.2**

Key bits stored in register C for each individual round [10].

| (i) | 1 | 2 | 3 | 4 | 5 | 6 | 7 | 8 | 9 | 10 | 11 | 12 | 13 | 14 | 15 | 16 | 17 | 18 | 19 | 20 | 21 | 22 | 23 | 24 | 25 | 26 | 27 | 28 |
|---|---|---|---|---|---|---|---|---|---|---|---|---|---|---|---|---|---|---|---|---|---|---|---|---|---|---|---|---|
| | | | | | | | | | | | | | Elements in $C_i$ | | | | | | | | | | | | | | | |
| 1 | 49 | 41 | 33 | 25 | 17 | 9 | 1 | 58 | 50 | 42 | 34 | 26 | 18 | 10 | 2 | 59 | 51 | 43 | 35 | 27 | 19 | 11 | 3 | 60 | 52 | 44 | 36 | 57 |
| 2 | 41 | 33 | 25 | 17 | 9 | 1 | 58 | 50 | 42 | 34 | 26 | 18 | 10 | 2 | 59 | 51 | 43 | 35 | 27 | 19 | 11 | 3 | 60 | 52 | 44 | 36 | 57 | 49 |
| 3 | 25 | 17 | 9 | 1 | 58 | 50 | 42 | 34 | 26 | 18 | 10 | 2 | 59 | 51 | 43 | 35 | 27 | 19 | 11 | 3 | 60 | 52 | 44 | 36 | 57 | 49 | 41 | 33 |
| 4 | 9 | 1 | 58 | 50 | 42 | 34 | 26 | 18 | 10 | 2 | 59 | 51 | 43 | 35 | 27 | 19 | 11 | 3 | 60 | 52 | 44 | 36 | 57 | 49 | 41 | 33 | 25 | 17 |
| 5 | 58 | 50 | 42 | 34 | 26 | 18 | 10 | 2 | 59 | 51 | 43 | 35 | 27 | 19 | 11 | 3 | 60 | 52 | 44 | 36 | 57 | 49 | 41 | 33 | 25 | 17 | 9 | 1 |
| 6 | 42 | 34 | 26 | 18 | 10 | 2 | 59 | 51 | 43 | 35 | 27 | 19 | 11 | 3 | 60 | 52 | 44 | 36 | 57 | 49 | 41 | 33 | 25 | 17 | 9 | 1 | 58 | 50 |
| 7 | 26 | 18 | 10 | 2 | 59 | 51 | 43 | 35 | 27 | 19 | 11 | 3 | 60 | 52 | 44 | 36 | 57 | 49 | 41 | 33 | 25 | 17 | 9 | 1 | 58 | 50 | 42 | 34 |
| 8 | 10 | 2 | 59 | 51 | 43 | 35 | 27 | 19 | 11 | 3 | 60 | 52 | 44 | 36 | 57 | 49 | 41 | 33 | 25 | 17 | 9 | 1 | 58 | 50 | 42 | 34 | 26 | 18 |
| 9 | 2 | 59 | 51 | 43 | 35 | 27 | 19 | 11 | 3 | 60 | 52 | 44 | 36 | 57 | 49 | 41 | 33 | 25 | 17 | 9 | 1 | 58 | 50 | 42 | 34 | 26 | 18 | 10 |
| 10 | 51 | 43 | 35 | 27 | 19 | 11 | 3 | 60 | 52 | 44 | 36 | 57 | 49 | 41 | 33 | 25 | 17 | 9 | 1 | 58 | 50 | 42 | 34 | 26 | 18 | 10 | 2 | 59 |
| 11 | 35 | 27 | 19 | 11 | 3 | 60 | 52 | 44 | 36 | 57 | 49 | 41 | 33 | 25 | 17 | 9 | 1 | 58 | 50 | 42 | 34 | 26 | 18 | 10 | 2 | 59 | 51 | 43 |
| 12 | 19 | 11 | 3 | 60 | 52 | 44 | 36 | 57 | 49 | 41 | 33 | 25 | 17 | 9 | 1 | 58 | 50 | 42 | 34 | 26 | 18 | 10 | 2 | 59 | 51 | 43 | 35 | 27 |
| 13 | 3 | 60 | 52 | 44 | 36 | 57 | 49 | 41 | 33 | 25 | 17 | 9 | 1 | 58 | 50 | 42 | 34 | 26 | 18 | 10 | 2 | 59 | 51 | 43 | 35 | 27 | 19 | 11 |
| 14 | 52 | 44 | 36 | 57 | 49 | 41 | 33 | 25 | 17 | 9 | 1 | 58 | 50 | 42 | 34 | 26 | 18 | 10 | 2 | 59 | 51 | 43 | 35 | 27 | 19 | 11 | 3 | 60 |
| 15 | 36 | 57 | 49 | 41 | 33 | 25 | 17 | 9 | 1 | 58 | 50 | 42 | 34 | 26 | 18 | 10 | 2 | 59 | 51 | 43 | 35 | 27 | 19 | 11 | 3 | 60 | 52 | 44 |
| 16 | 57 | 49 | 41 | 33 | 25 | 17 | 9 | 1 | 58 | 50 | 42 | 34 | 26 | 18 | 10 | 2 | 59 | 51 | 43 | 35 | 27 | 19 | 11 | 3 | 60 | 52 | 44 | 36 |

$k_{49}$, $k_{41}$, $k_{33}$, . . ., etc. are the 1st, 2nd, 3rd, . . ., etc. key bits in $C_1$, i.e., key bits in register C during the 1st round.

**Table 4.3**

Key bits stored in register $D$ for each individual round [10].

| $(i)$ | 29 | 30 | 31 | 32 | 33 | 34 | 35 | 36 | 37 | 38 | 39 | 40 | 41 | 42 | 43 | 44 | 45 | 46 | 47 | 48 | 49 | 50 | 51 | 52 | 53 | 54 | 55 | 56 |
|---|---|---|---|---|---|---|---|---|---|---|---|---|---|---|---|---|---|---|---|---|---|---|---|---|---|---|---|---|
| 1 | 55 | 47 | 39 | 31 | 23 | 15 | 7 | 62 | 54 | 46 | 38 | 30 | 22 | 14 | 6 | 61 | 53 | 45 | 37 | 29 | 21 | 13 | 5 | 28 | 20 | 12 | 4 | 63 |
| 2 | 47 | 39 | 31 | 23 | 15 | 7 | 62 | 54 | 46 | 38 | 30 | 22 | 14 | 6 | 61 | 53 | 45 | 37 | 29 | 21 | 13 | 5 | 28 | 20 | 12 | 4 | 63 | 55 |
| 3 | 31 | 23 | 15 | 7 | 62 | 54 | 46 | 38 | 30 | 22 | 14 | 6 | 61 | 53 | 45 | 37 | 29 | 21 | 13 | 5 | 28 | 20 | 12 | 4 | 63 | 55 | 47 | 39 |
| 4 | 15 | 7 | 62 | 54 | 46 | 38 | 30 | 22 | 14 | 6 | 61 | 53 | 45 | 37 | 29 | 21 | 13 | 5 | 28 | 20 | 12 | 4 | 63 | 55 | 47 | 39 | 31 | 23 |
| 5 | 62 | 54 | 46 | 38 | 30 | 22 | 14 | 6 | 61 | 53 | 45 | 37 | 29 | 21 | 13 | 5 | 28 | 20 | 12 | 4 | 63 | 55 | 47 | 39 | 31 | 23 | 15 | 7 |
| 6 | 46 | 38 | 30 | 22 | 14 | 6 | 61 | 53 | 45 | 37 | 29 | 21 | 13 | 5 | 28 | 20 | 12 | 4 | 63 | 55 | 47 | 39 | 31 | 23 | 15 | 7 | 62 | 54 |
| 7 | 30 | 22 | 14 | 6 | 61 | 53 | 45 | 37 | 29 | 21 | 13 | 5 | 28 | 20 | 12 | 4 | 63 | 55 | 47 | 39 | 31 | 23 | 15 | 7 | 62 | 54 | 46 | 38 |
| 8 | 14 | 6 | 61 | 53 | 45 | 37 | 29 | 21 | 13 | 5 | 28 | 20 | 12 | 4 | 63 | 55 | 47 | 39 | 31 | 23 | 15 | 7 | 62 | 54 | 46 | 38 | 30 | 22 |
| 9 | 6 | 61 | 53 | 45 | 37 | 29 | 21 | 13 | 5 | 28 | 20 | 12 | 4 | 63 | 55 | 47 | 39 | 31 | 23 | 15 | 7 | 62 | 54 | 46 | 38 | 30 | 22 | 14 |
| 10 | 53 | 45 | 37 | 29 | 21 | 13 | 5 | 28 | 20 | 12 | 4 | 63 | 55 | 47 | 39 | 31 | 23 | 15 | 7 | 62 | 54 | 46 | 38 | 30 | 22 | 14 | 6 | 61 |
| 11 | 37 | 29 | 21 | 13 | 5 | 28 | 20 | 12 | 4 | 63 | 55 | 47 | 39 | 31 | 23 | 15 | 7 | 62 | 54 | 46 | 38 | 30 | 22 | 14 | 6 | 61 | 53 | 45 |
| 12 | 21 | 13 | 5 | 28 | 20 | 12 | 4 | 63 | 55 | 47 | 39 | 31 | 23 | 15 | 7 | 62 | 54 | 46 | 38 | 30 | 22 | 14 | 6 | 61 | 53 | 45 | 37 | 29 |
| 13 | 5 | 28 | 20 | 12 | 4 | 63 | 55 | 47 | 39 | 31 | 23 | 15 | 7 | 62 | 54 | 46 | 38 | 30 | 22 | 14 | 6 | 61 | 53 | 45 | 37 | 29 | 21 | 13 |
| 14 | 20 | 12 | 4 | 63 | 55 | 47 | 39 | 31 | 23 | 15 | 7 | 62 | 54 | 46 | 38 | 30 | 22 | 14 | 6 | 61 | 53 | 45 | 37 | 29 | 21 | 13 | 5 | 28 |
| 15 | 4 | 63 | 55 | 47 | 39 | 31 | 23 | 15 | 7 | 62 | 54 | 46 | 38 | 30 | 22 | 14 | 6 | 61 | 53 | 45 | 37 | 29 | 21 | 13 | 5 | 28 | 20 | 12 |
| 16 | 63 | 55 | 47 | 39 | 31 | 23 | 15 | 7 | 62 | 54 | 46 | 38 | 30 | 22 | 14 | 6 | 61 | 53 | 45 | 37 | 29 | 21 | 13 | 5 | 28 | 20 | 12 | 4 |

**Table 4.4**

First set of 24 key bits in $K(i)$, the key used at iteration $(i)$ [10].

| $(i)$ | | | | | | | | | | | | Elements in $K(i)$ | | | | | | | | | | | | |
|---|---|---|---|---|---|---|---|---|---|---|---|---|---|---|---|---|---|---|---|---|---|---|---|---|
| | 1 | 2 | 3 | 4 | 5 | 6 | 7 | 8 | 9 | 10 | 11 | 12 | 13 | 14 | 15 | 16 | 17 | 18 | 19 | 20 | 21 | 22 | 23 | 24 |
| Selected Element in Vector $C_i$ | 14 | 17 | 11 | 24 | 1 | 5 | 3 | 28 | 15 | 6 | 21 | 10 | 23 | 19 | 12 | 4 | 26 | 8 | 16 | 7 | 27 | 20 | 13 | 2 |
| 1 | 10 | 51 | 34 | 60 | 49 | 17 | 33 | 57 | 2 | 9 | 19 | 42 | 3 | 35 | 26 | 25 | 44 | 58 | 59 | 1 | 36 | 27 | 18 | 41 |
| 2 | 2 | 43 | 26 | 52 | 41 | 9 | 25 | 49 | 59 | 1 | 11 | 34 | 60 | 27 | 18 | 17 | 36 | 50 | 51 | 58 | 57 | 19 | 10 | 33 |
| 3 | 51 | 27 | 10 | 36 | 25 | 58 | 9 | 33 | 43 | 50 | 60 | 18 | 44 | 11 | 2 | 1 | 49 | 34 | 35 | 42 | 41 | 3 | 59 | 17 |
| 4 | 35 | 11 | 59 | 49 | 9 | 42 | 58 | 17 | 27 | 34 | 44 | 2 | 57 | 60 | 51 | 50 | 33 | 18 | 19 | 26 | 25 | 52 | 43 | 1 |
| 5 | 19 | 60 | 43 | 33 | 58 | 26 | 42 | 1 | 11 | 18 | 57 | 51 | 41 | 44 | 35 | 34 | 17 | 2 | 3 | 10 | 9 | 36 | 27 | 50 |
| 6 | 3 | 44 | 27 | 17 | 42 | 10 | 26 | 50 | 60 | 2 | 41 | 35 | 25 | 57 | 19 | 18 | 1 | 51 | 52 | 59 | 58 | 49 | 11 | 34 |
| 7 | 52 | 57 | 11 | 1 | 26 | 59 | 10 | 34 | 44 | 51 | 25 | 19 | 9 | 41 | 3 | 2 | 50 | 35 | 36 | 43 | 42 | 33 | 60 | 18 |
| 8 | 36 | 41 | 60 | 50 | 10 | 43 | 59 | 18 | 57 | 35 | 9 | 3 | 58 | 25 | 52 | 51 | 34 | 19 | 49 | 27 | 26 | 17 | 44 | 2 |
| 9 | 57 | 33 | 52 | 42 | 2 | 35 | 51 | 10 | 49 | 27 | 1 | 60 | 50 | 17 | 44 | 43 | 26 | 11 | 41 | 19 | 18 | 9 | 36 | 59 |
| 10 | 41 | 17 | 36 | 26 | 51 | 19 | 35 | 59 | 33 | 11 | 50 | 44 | 34 | 1 | 57 | 27 | 10 | 60 | 25 | 3 | 2 | 58 | 49 | 43 |
| 11 | 25 | 1 | 49 | 10 | 35 | 3 | 19 | 43 | 17 | 60 | 34 | 57 | 18 | 50 | 41 | 11 | 59 | 44 | 9 | 52 | 51 | 42 | 33 | 27 |
| 12 | 9 | 50 | 33 | 59 | 19 | 52 | 3 | 27 | 1 | 44 | 18 | 41 | 2 | 34 | 25 | 60 | 43 | 57 | 58 | 36 | 35 | 26 | 17 | 11 |
| 13 | 58 | 34 | 17 | 43 | 3 | 36 | 52 | 11 | 50 | 57 | 2 | 25 | 51 | 18 | 9 | 44 | 27 | 41 | 42 | 49 | 19 | 10 | 1 | 60 |
| 14 | 42 | 18 | 1 | 27 | 52 | 49 | 36 | 60 | 34 | 41 | 51 | 9 | 35 | 2 | 58 | 57 | 11 | 25 | 26 | 33 | 3 | 59 | 50 | 44 |
| 15 | 26 | 2 | 50 | 11 | 36 | 33 | 49 | 44 | 18 | 25 | 35 | 58 | 19 | 51 | 42 | 41 | 60 | 9 | 10 | 17 | 52 | 43 | 34 | 57 |
| 16 | 18 | 59 | 42 | 3 | 57 | 25 | 41 | 36 | 10 | 17 | 27 | 50 | 11 | 43 | 34 | 33 | 52 | 1 | 2 | 9 | 44 | 35 | 26 | 49 |

**Table 4.5**

Second set of 24 key bits in $K(i)$, the key used at iteration ($i$) [10].

| ($i$) | Elements in $K(i)$ | | | | | | | | | | | | | | | | | | | | | | | |
|---|---|---|---|---|---|---|---|---|---|---|---|---|---|---|---|---|---|---|---|---|---|---|---|---|
| | 25 | 26 | 27 | 28 | 29 | 30 | 31 | 32 | 33 | 34 | 35 | 36 | 37 | 38 | 39 | 40 | 41 | 42 | 43 | 44 | 45 | 46 | 47 | 48 |
| Selected Element in Vector $D_i$ | 41 | 52 | 31 | 37 | 47 | 55 | 30 | 40 | 51 | 45 | 33 | 48 | 44 | 49 | 39 | 56 | 34 | 53 | 46 | 42 | 50 | 36 | 29 | 32 |
| 1 | 22 | 28 | 39 | 54 | 37 | 4 | 47 | 30 | 5 | 53 | 23 | 29 | 61 | 21 | 38 | 63 | 15 | 20 | 45 | 14 | 13 | 62 | 55 | 31 |
| 2 | 14 | 20 | 31 | 46 | 29 | 63 | 39 | 22 | 28 | 45 | 15 | 21 | 53 | 13 | 30 | 55 | 7 | 12 | 37 | 6 | 5 | 54 | 47 | 23 |
| 3 | 61 | 4 | 15 | 30 | 13 | 47 | 23 | 6 | 12 | 29 | 62 | 5 | 37 | 28 | 14 | 39 | 54 | 63 | 21 | 53 | 20 | 38 | 31 | 7 |
| 4 | 45 | 55 | 62 | 14 | 28 | 31 | 7 | 53 | 63 | 13 | 46 | 20 | 21 | 12 | 61 | 23 | 38 | 47 | 5 | 37 | 4 | 22 | 15 | 54 |
| 5 | 29 | 39 | 46 | 61 | 12 | 15 | 54 | 37 | 47 | 28 | 30 | 4 | 5 | 63 | 45 | 7 | 22 | 31 | 20 | 21 | 55 | 6 | 62 | 38 |
| 6 | 13 | 23 | 30 | 45 | 63 | 62 | 38 | 21 | 31 | 12 | 14 | 55 | 20 | 47 | 29 | 54 | 6 | 15 | 4 | 5 | 39 | 53 | 46 | 22 |
| 7 | 28 | 7 | 14 | 29 | 47 | 46 | 22 | 5 | 15 | 63 | 61 | 39 | 4 | 31 | 13 | 38 | 53 | 62 | 55 | 20 | 23 | 37 | 30 | 6 |
| 8 | 12 | 54 | 61 | 13 | 31 | 30 | 6 | 20 | 62 | 47 | 45 | 23 | 55 | 15 | 28 | 22 | 37 | 46 | 39 | 4 | 7 | 21 | 14 | 53 |
| 9 | 4 | 46 | 53 | 5 | 23 | 22 | 61 | 12 | 54 | 39 | 37 | 15 | 47 | 7 | 20 | 14 | 29 | 38 | 31 | 63 | 62 | 13 | 6 | 45 |
| 10 | 55 | 30 | 37 | 20 | 7 | 6 | 45 | 63 | 38 | 23 | 21 | 62 | 31 | 54 | 4 | 61 | 13 | 22 | 15 | 47 | 46 | 28 | 53 | 29 |
| 11 | 39 | 14 | 21 | 4 | 54 | 53 | 29 | 47 | 22 | 7 | 5 | 46 | 15 | 38 | 55 | 45 | 28 | 6 | 62 | 31 | 30 | 12 | 37 | 13 |
| 12 | 23 | 61 | 5 | 55 | 38 | 37 | 13 | 31 | 6 | 54 | 20 | 30 | 62 | 22 | 39 | 29 | 12 | 53 | 46 | 15 | 14 | 63 | 21 | 28 |
| 13 | 7 | 45 | 20 | 39 | 22 | 21 | 28 | 15 | 53 | 38 | 4 | 14 | 46 | 6 | 23 | 13 | 63 | 37 | 30 | 62 | 61 | 47 | 5 | 12 |
| 14 | 54 | 29 | 4 | 23 | 6 | 5 | 12 | 62 | 37 | 22 | 55 | 61 | 30 | 53 | 7 | 28 | 47 | 21 | 14 | 46 | 45 | 31 | 20 | 63 |
| 15 | 38 | 13 | 55 | 7 | 53 | 20 | 63 | 46 | 21 | 6 | 39 | 45 | 14 | 37 | 54 | 12 | 31 | 5 | 61 | 30 | 29 | 15 | 4 | 47 |
| 16 | 30 | 5 | 47 | 62 | 45 | 12 | 55 | 38 | 13 | 61 | 31 | 37 | 6 | 29 | 46 | 4 | 23 | 28 | 53 | 22 | 21 | 7 | 63 | 39 |

The complete 16 iterations provide the keys $K(1)$, $K(2)$, . . ., $K(16)$, as shown in Tables 4.4 and 4.5. As a consequence of the above descriptions, the original key generation function for encryption DES as shown in Figure 4.15 should become apparent.

### 4.6.2 The Transformation

In DES, the transformation $f[R(i), K(i+1)]$ consists of bit expansion, key modulo-2 addition, and selection (substitution) and permutation operations, as shown in Figure 4.16. The whole transformation procedure will become apparent after we discuss each operation separately.

#### 4.6.2.1 Bit Expansion

The function of bit expansion in DES is to convert a 32-bit block into a 48-bit block in accordance with the ordering sequence $E$ which assigns and expands the 32-bit block of $R(i)$ into the 48-bit block $U(i)$:

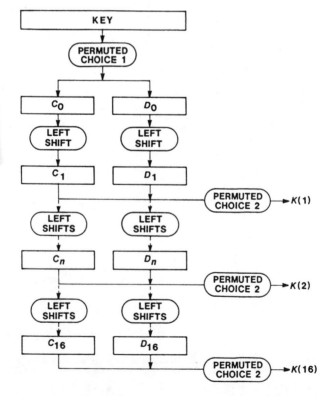

**Figure 4.15** Key generation of DES.

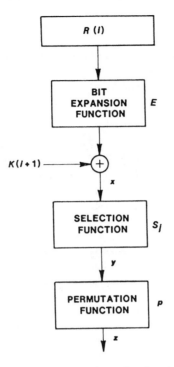

**Figure 4.16** The transformation functions.

$$E = \begin{pmatrix} 32 & 1 & 2 & 3 & 4 & 5 \\ 4 & 5 & 6 & 7 & 8 & 9 \\ 8 & 9 & 10 & 11 & 12 & 13 \\ 12 & 13 & 14 & 15 & 16 & 17 \\ 16 & 17 & 18 & 19 & 20 & 21 \\ 20 & 21 & 22 & 23 & 24 & 25 \\ 24 & 25 & 26 & 27 & 28 & 29 \\ 28 & 29 & 30 & 31 & 32 & 1 \end{pmatrix} \qquad (4.42)$$

Thus the bit expansion function may be expressed as

$$U(i) = E[R(i)]. \qquad (4.43)$$

*Example 4.2:* Let $R(i) = 10101010101010101010111100011011$. Upon bit expansion based on $E$ we have from (4.43)

$$U(i) = \begin{pmatrix} 110101 \\ \vdots \\ 110111 \end{pmatrix}$$

The bit expansion function may be implemented by further partitioning the 32-bit $R(i)$ into eight segments with four bits in each segment. Except for the first and the 32nd bit, each of the end bits of the segment is allocated to two positions, as shown in Figure 4.17. As a consequence, the 32-bit $R(i)$ has expanded to 48 bits, which matches the number of bits in $K(i)$.

### 4.6.2.2 The Selection Function

In DES there exist eight selection functions $S_1, S_2, \ldots, S_8$. Each function $S_j$, $j = 1, 2, \ldots, 8$, has four rows and sixteen columns. The elements of $S_j$ are a specific arrangement of the set of integers from 0 to 15, as shown in Table 4.6.

Each function $S_j$ takes a 6-bit block as its input and provides a 4-bit block output. If we let the input and output of a function $S_j$ be denoted as $x = x_1, x_2, x_3, x_4, x_5, x_6,$ and $y = y_1, y_2, y_3, y_4$, then the purpose of the selection function is to provide the following transformation.

Let $r_j, c_j$ denote a particular row and column of $S_j$; $r_j$ is determined by the pair of first and the sixth digits $(x_1, x_6)$. Since $x_1$ and $x_6$ are binary there can be only four outcomes to indicate the four rows of $S_j$; the other four binary digits $x_2, x_3, x_4, x_5$ provide numbers between 0 and 15, determine $c_j$. The intersection $r_j$ and $c_j$ in $S_j$ gives a specific integer between 0 to 15. When this integer converts to its binary representation, we have $y_1, y_2, y_3, y_4$, as shown in Figure 4.18.

*Example 4.3:* Let $X = (x_1, x_2, x_3, x_4, x_5, x_6) = 101010$; this gives $r_j = (x_1, x_6) = 10$, and $c_j = 0101 = 5$. If we let $j = 7$, the element of $S_7$ in the 3rd row and the 5th column is 12, which has the binary representation $1100 = (y_1, y_2, y_3, y_4) = y$ as the output at $S_7$ due to $X$ as its input.

### 4.6.2.3 The Permutation Function

As described in the last section, each selection function $S_j$ has four bits output. For the total of eight selection functions there is a total of 32 bits. The purpose of the permutation function is to take these 32 bits as input, permute the digits, and produce a 32-bit block output. The permutation is performed according to the permutation function

$$P = \begin{matrix} 16 & 7 & 20 & 21 \\ 29 & 12 & 28 & 17 \\ 1 & 15 & 23 & 26 \\ 5 & 18 & 31 & 10 \\ 2 & 8 & 24 & 14 \\ 32 & 27 & 3 & 9 \\ 19 & 13 & 30 & 6 \\ 22 & 11 & 4 & 25 \end{matrix}$$

which is a condensed way of writing the row vector

$$P = (16, 7, 20, 21, 29, \ldots, 25).$$

Let the 32-bit block input to $P$ be denoted as

$$Y = (y_1, y_2, \ldots, y_{32}).$$

## Table 4.6
The selection functions of DES.

$S_1$

| 14 | 4 | 13 | 1 | 2 | 15 | 11 | 8 | 3 | 10 | 6 | 12 | 5 | 9 | 0 | 7 |
|---|---|---|---|---|---|---|---|---|---|---|---|---|---|---|---|
| 0 | 15 | 7 | 4 | 14 | 2 | 13 | 1 | 10 | 6 | 12 | 11 | 9 | 5 | 3 | 8 |
| 4 | 1 | 14 | 8 | 13 | 6 | 2 | 11 | 15 | 12 | 9 | 7 | 3 | 10 | 5 | 0 |
| 15 | 12 | 8 | 2 | 4 | 9 | 1 | 7 | 5 | 11 | 3 | 14 | 10 | 0 | 6 | 13 |

$S_2$

| 15 | 1 | 8 | 14 | 6 | 11 | 3 | 4 | 9 | 7 | 2 | 13 | 12 | 0 | 5 | 10 |
|---|---|---|---|---|---|---|---|---|---|---|---|---|---|---|---|
| 3 | 13 | 4 | 7 | 15 | 2 | 8 | 14 | 12 | 0 | 1 | 10 | 6 | 9 | 11 | 5 |
| 0 | 14 | 7 | 11 | 10 | 4 | 13 | 1 | 5 | 8 | 12 | 6 | 9 | 3 | 2 | 15 |
| 13 | 8 | 10 | 1 | 3 | 15 | 4 | 2 | 11 | 6 | 7 | 12 | 0 | 5 | 14 | 9 |

$S_3$

| 10 | 0 | 9 | 14 | 6 | 3 | 15 | 5 | 1 | 13 | 12 | 7 | 11 | 4 | 2 | 8 |
|---|---|---|---|---|---|---|---|---|---|---|---|---|---|---|---|
| 13 | 7 | 0 | 9 | 3 | 4 | 6 | 10 | 2 | 8 | 5 | 14 | 12 | 11 | 15 | 1 |
| 13 | 6 | 4 | 9 | 8 | 15 | 3 | 0 | 11 | 1 | 2 | 12 | 5 | 10 | 14 | 7 |
| 1 | 10 | 13 | 0 | 6 | 9 | 8 | 7 | 4 | 15 | 14 | 3 | 11 | 5 | 2 | 12 |

$S_4$

| 7 | 13 | 14 | 3 | 0 | 6 | 9 | 10 | 1 | 2 | 8 | 5 | 11 | 12 | 4 | 15 |
|---|---|---|---|---|---|---|---|---|---|---|---|---|---|---|---|
| 13 | 8 | 11 | 5 | 6 | 15 | 0 | 3 | 4 | 7 | 2 | 12 | 1 | 10 | 14 | 9 |
| 10 | 6 | 9 | 0 | 12 | 11 | 7 | 13 | 15 | 1 | 3 | 14 | 5 | 2 | 8 | 4 |
| 3 | 15 | 0 | 6 | 10 | 1 | 13 | 8 | 9 | 4 | 5 | 11 | 12 | 7 | 2 | 14 |

$S_5$

| 2 | 12 | 4 | 1 | 7 | 10 | 11 | 6 | 8 | 5 | 3 | 15 | 13 | 0 | 14 | 9 |
|---|---|---|---|---|---|---|---|---|---|---|---|---|---|---|---|
| 14 | 11 | 2 | 12 | 4 | 7 | 13 | 1 | 5 | 0 | 15 | 10 | 3 | 9 | 8 | 6 |
| 4 | 2 | 1 | 11 | 10 | 13 | 7 | 8 | 15 | 9 | 12 | 5 | 6 | 3 | 0 | 14 |
| 11 | 8 | 12 | 7 | 1 | 14 | 2 | 13 | 6 | 15 | 0 | 9 | 10 | 4 | 5 | 3 |

$S_6$

| 12 | 1 | 10 | 15 | 9 | 2 | 6 | 8 | 0 | 13 | 3 | 4 | 14 | 7 | 5 | 11 |
|---|---|---|---|---|---|---|---|---|---|---|---|---|---|---|---|
| 10 | 15 | 4 | 2 | 7 | 12 | 9 | 5 | 6 | 1 | 13 | 14 | 0 | 11 | 3 | 8 |
| 9 | 14 | 15 | 5 | 2 | 8 | 12 | 3 | 7 | 0 | 4 | 10 | 1 | 13 | 11 | 6 |
| 4 | 3 | 2 | 12 | 9 | 5 | 15 | 10 | 11 | 14 | 1 | 7 | 6 | 0 | 8 | 13 |

$S_7$

| 4 | 11 | 2 | 14 | 15 | 0 | 8 | 13 | 3 | 12 | 9 | 7 | 5 | 10 | 6 | 1 |
|---|---|---|---|---|---|---|---|---|---|---|---|---|---|---|---|
| 13 | 0 | 11 | 7 | 4 | 9 | 1 | 10 | 14 | 3 | 5 | 12 | 2 | 15 | 8 | 6 |
| 1 | 4 | 11 | 13 | 12 | 3 | 7 | 14 | 10 | 15 | 6 | 8 | 0 | 5 | 9 | 2 |
| 6 | 11 | 13 | 8 | 1 | 4 | 10 | 7 | 9 | 5 | 0 | 15 | 14 | 2 | 3 | 12 |

$S_8$

| 13 | 2 | 8 | 4 | 6 | 15 | 11 | 1 | 10 | 9 | 3 | 14 | 5 | 0 | 12 | 7 |
|---|---|---|---|---|---|---|---|---|---|---|---|---|---|---|---|
| 1 | 15 | 13 | 8 | 10 | 3 | 7 | 4 | 12 | 5 | 6 | 11 | 0 | 14 | 9 | 2 |
| 7 | 11 | 4 | 1 | 9 | 12 | 14 | 2 | 0 | 6 | 10 | 13 | 15 | 3 | 5 | 8 |
| 2 | 1 | 14 | 7 | 4 | 10 | 8 | 13 | 15 | 12 | 9 | 0 | 3 | 5 | 6 | 11 |

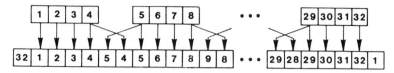

**Figure 4.17** Expansion ($E$) of right half of input to each iteration [10].

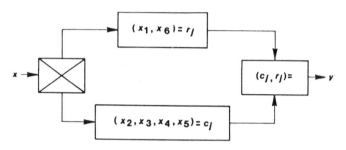

**Figure 4.18** The function of $S_j$.

The permutation function simply performs

$$Z = P(Y) = y_{16}, y_7, y_{20}, \ldots, y_{25}$$

$$= z_1, y_2, y_3, \ldots, y_{32}$$

where $Z$ is the output as a result of $P$. In other words, the order of $Y$ is rearranged (or permuted) according to $P$.

*Example 4.4:* Let $Y = 1100100111101010101111\ 0\ 0110000101$ be the 32-bit input block to $P$; rearranging $Y$ according to $P$, we have

$$Z = P(Y) = 00110001110010011110100111100101.$$

From the above descriptions and examples the functions of transformation $f[R(i), K(i+1)]$ are clear.

## 4.7   MULTIPLE ACCESS CRYPTOSYSTEMS

A multiple access cryptosystem is a cryptographic system in which the keys are distributed publicly. The principle of public key distribution has been mentioned in Section 4.4.2. After defining the one-way functions in Section 4.7.1, we shall be in a position to study the multiple access cryptosystems. In these systems the keys are generated through either the exponentiation or logarithmic process over a finite field. The security for each technique depends crucially on the difficulty of computing the class of functions that has products of exponents.

### 4.7.1 One-Way Functions

A function $f$ that maps the set $A$ into the set $B$, denoted $f:A{\rightarrow}B$, is said to be invertible if $x_1$ and $x_2$ in $A$ with $x_1 \neq x_2$ implies $f(x_1) \neq f(x_2)$. When and only when $f$ is invertible, there exists a function $g$, $g:B{\rightarrow}A$, such that $g[f(x)] = x$ for all $x$ in $A$. Such mappings between two sets form the foundation of modern cryptography. For each key $K$ the encrypting function $E_K$ is invertible where $A$ is the set of possible plaintexts.

The invertible function $f$, $f:A{\rightarrow}B$, is said to be a one-way function if it is "easy" to compute $f(x)$ for all $x$ in $A$, but it is so "very difficult" as to be "practically impossible" when given $f(x)$ to find $x$ for "almost all" $x$ in $A$. In other words, there is a "simple" algorithm (in the sense that only a "small" amount of storage and advance calculation is required by the algorithm) that "quickly" computes (in the sense of using a "small number" of "easily implemented" operations) $f(x)$ when presented with $x$; but every algorithm that computes $x$ from $f(x)$ will either be "very complex" (in the sense of requiring a "very large" amount of storage or a "very large" amount of advance calculation) or will only "very slowly" compute (in the sense of a "very large" number of any "easily" implemented operation) $x$, for "almost all" $x$ in $A$.

In our definition of a one-way function and its explanation, we have placed within quotation marks those words that are not intended to have a precise mathematical meaning but whose intuitive meaning is quite clear. We are deliberately imprecise here for two reasons. The first is that there are many different ways to attach precise meanings to these terms, some of which are much more convenient in certain cases than others so that specification of one way destroys useful freedom. The second and more persuasive reason is that it is not possible to make these terms more precise in the context of apparent difficulty as opposed to essential difficulty. As the reader will soon see, the security of the public key secrecy systems that have been devised up to now depends on the apparent difficulty of some problems, not their essential difficulty.

### 4.7.2 The Diffie-Hellman System

The Diffie-Hellman Scheme relies on discrete exponentiation functions over a finite field as one-way functions [8]. All users of the system agree on some large prime $p$, for which $p-1$ has a large prime factor, and on a primitive element $\alpha$ of GF($p$). Each user then secretly chooses an integer $x$ in the range $0 \leqslant x < p-1$ by a random choice (with the uniform probability distribution). Let $x_i$ denote the secret integer chosen by station $i$. Each station then forms the element $y = \alpha^x$ in GF($p$) and places $y$ in a public directory; thus station $i$ places $y_i = \alpha^{x_i}$ beside his own name in this directory. It is important that only authorized users be allowed to place numbers in this directory, and that the integrity of the directory be maintained in the sense that anyone who later requests information from the public directory receives exactly that information placed there by the authorized stations. There is no need, however, to keep the directory secret.

When station $i$ and station $j$ wish to communicate secretly, they do so by using the key

$$K_{ij} = \alpha^{x_i x_j}, \qquad (4.44)$$

where we are now for simplicity letting the key be an element of $\mathrm{GF}(p)$ but in practice the "key" would be the binary representation of $K_{ij}$. Both station $i$ and station $j$ can easily form this key by exponentiation in $\mathrm{GF}(p)$ because

$$(y_i)^{x_j} = \alpha^{x_i x_j} = (y_j)^{x_i} = K_{ij} \qquad (4.45)$$

so that each of these users requires only the publicly available information from the other. This key generation process for multiple access of pairwise stations $i$ and $j$ is diagrammed in Figure 4.19. However, for the reasons we now state, it seems extremely difficult for anyone else to determine this key.

Exponentiation in such $\mathrm{GF}(p)$ is apparently a one-way function. Thus unless the cryptanalyst knows more than we do about taking logarithms in $\mathrm{GF}(p)$, no unauthorized person can compute $x_i$ from $y_i$ or $x_j$ from $y_j$ as he would need to do in order to form $K_{ij}$ in the manner $(y_j)^{x_i}$ or $(y_i)^{x_j}$. But are there ways to compute $\alpha^{x_i x_j}$ from $y_i$ and $y_j$ without equivalently finding either $x_i$ or $x_j$? By "equivalently finding" we mean here computing some quantity or quantities from which $x_i$ or $x_j$ can easily be found. With present knowledge, it is indeed very difficult to calculate $\alpha^{x_i x_j}$ from $y_i$ and $y_j$ without first obtaining $x_i$ or $x_j$. By contrast, $\alpha^{x_i + x_j}$ is easy to calculate from $y_i$ and $y_j$ without finding $x_i$ or $x_j$ because $y_i y_j = \alpha^{x_i} \alpha^{x_j} = \alpha^{x_i + x_j}$; this suggests that one should not too quickly rule out

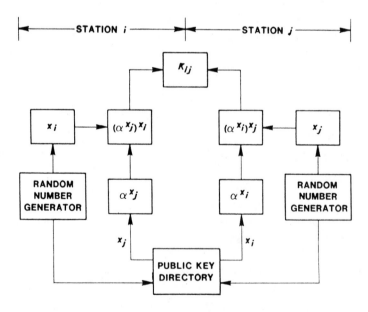

**Figure 4.19** Function of multiple access cryptosystem (all computations modulo-$p$).

the possibility of similarly finding $\alpha^{x_i x_j}$, which determines the strength of the multiple access cryptosystems.

The Diffie-Hellman System is a very neat solution to the key management problem in multiple access cryptosystems. For if there are $S$ stations on the network, the classical solution of distributing keys over secure channels would require

$$\binom{S}{2} = \frac{1}{2} S(S-1)$$

different secret keys so that each station can communicate secretly with each other station. Moreover, each station must store and protect $S-1$ of these secret keys. In the Diffie-Hellman System, however, each station stores and protects only one "secret key" (its own randomly chosen $x_i$). The public directory of $S$ public keys $(y_i, y_2, \ldots, y_S)$ does not even have to be stored by each station but could be accessed on demand, provided there was a mechanism that prevented everyone except the authorized public directory service from sending information purported to be from the public directory. This great simplification of the key management problem in a network when the Diffie-Hellman System is used has generated immense interest in this scheme.

Finally, we wish to emphasize that in the Diffie-Hellman System, the authorized users need never do the difficult operation of "taking logarithms" but only the easy operation of "exponentiation." This is essential since if exponentiation in GF($p$) is truly a one-way function, then no one (authorized or not) could perform the difficult operation.

*Example 4.5:* Let $p = 7$. For a finite field based on the irreducible polynomial $1 + x + x^3$, we have the corresponding GF(7) as:

$$\alpha^0 = 1\ 0\ 0$$

$$\alpha = 0\ 1\ 0$$

$$\alpha^2 = 0\ 0\ 1$$

$$\alpha^3 = 1\ 1\ 0$$

$$\alpha^4 = 0\ 1\ 1$$

$$\alpha^5 = 1\ 1\ 1$$

$$\alpha^6 = 1\ 0\ 1$$

Station $i$ selects the random value $x_i = 4$, and station $j$ chooses $x_j = 5$. Station $i$ places $y_i = \alpha^4 = 011$ and station $j$ places $y_j = \alpha^5 = 111$ in the common key directory. If stations $i$ and $j$ wish to communicate, station $i$ sends to station $j$ the value $y_i = \alpha^4 \bmod(7)$ and station $j$ sends to station $i$ the value $y_j = \alpha^5 \bmod(7)$. Then the key becomes

$$K_{ij} = \alpha^{x_i x_j} = \alpha^{20} = \alpha^6 = 101.$$

### 4.7.3   Trapdoor One-Way Functions

The concept of a "trapdoor one-way function" is another of the innovations in the 1976 paper by Diffie and Hellman. Roughly speaking, we can say that an invertible function is a trapdoor one-way function if this function is easy to compute, but its inverse is usually very difficult to compute unless one knows some additional information (the "trapdoor") that makes the inverse also easy to compute. Just as a combination lock is easy to open only when one knows the combination, so a trapdoor one-way function is easy to invert only if one knows the trapdoor. When we attempt to be more precise, however, we encounter certain subtleties in the concept of a trapdoor one-way function. First, we note that such a function cannot be a one-way function at all because a one-way function has no easy inverse under any conditions. Second, we note that there would have to be many possible trapdoors, since otherwise, by trying each possible trapdoor in turn to see if it works, we could easily invert the function. These two considerations lead us to the following definition:

A trapdoor one-way function is a family of invertible functions $f_k$, $f_k : A_k \rightarrow B_k$, indexed by $k \in K$, such that:

(a) For all $k \in K$, given $k$, it is "easy" to find a pair of algorithms $E_k$ and $D_k$ that, for all $x$ in $A_k$ "easily compute" $f_k$ and its inverse, respectively, such that

$$f_k(x) = E_k(x)$$

$$D_k[f_k(x)] = x.$$

Moreover, given $k$ it is easy to find an algorithm $F_k$ that "easily computes" for all $x$ in some easily identified set $A$ that contains each set $A_k$ whether $x$ is in $A_k$.

(b) For "almost all" $k$ in $K$, given only $E_k$ and $F_k$ and knowledge of the easily identified set $A$, it is so very difficult as to be "practically impossible" to compute $x$ from $y = f_k(x)$ for almost all $x$ in $A_x$.

The first part of this definition should be reasonably clear without further explanation, except for the role of the algorithm $F_k$. The need for this algorithm follows from the fact that the "easily computed" algorithm $E_k$ would not be of much use to anyone unless it was likewise "easy" to identify the numbers $x$ that are allowable arguments for the function that this algorithm computes. We did not encounter this situation in our previous discussion of one-way functions because then there was only a single set $A$ of allowed arguments that we tacitly assumed were easily recognized. Now, when the set of arguments $A_k$ can depend on $k$, we must make sure that we can tell whether a number $x$ known to be in the larger set $A$ (that we again suppose is "easily identified") is also in $A_k$ without having to know $k$ if we are going to expect someone to use the algorithm $E_k$ without knowledge of $k$.

The second part of our definition specifies the sense in which a trapdoor one-way function is "one-way." Notice that it must not be possible (at least for "most" $k$), when given the algorithms $E_k$ and $F_k$, to "compute easily" an algorithm that for "most" $x$ in $A_k$ "easily inverts" the function $f_k$ computed by $E_k$. Of course, given $E_k$ and $F_k$, we can "easily compute" $y_i = f(x_i)$ for any "reasonably small" set of values $\{x_1, x_2, \ldots, x_n\} \in A_k$ that we wish. Thus (b) implies that knowledge of these pairs $(y_i, x_i)$, $i = 1, 2, \ldots, n$ must not be "easily incorporated" into an algorithm that "easily computes" $x$ from $f(x)$ for more than those $x$'s in a "small subset" of $A_k$.

We are about to consider secrecy systems built upon one-way functions in which the set $A_k$ (for the chosen trapdoor $k$) is the set of possible plaintexts. This requires us to put another constraint on trapdoor one-way functions, namely:

(c) The set $A_k$, for every $k$, must be a "convenient" set of numbers for use as the set of plaintexts in a secrecy system. This means, for instance, that there must be an "easily implemented" mapping between the "natural" plaintexts that one might wish to send (e.g., binary sequences of length 1000) and the numbers in the set $A_k$ that will represent those plaintexts. This third condition seems to have been tacitly understood in the literature on public key cryptography, and so we now make it a part of our full definition of a trapdoor one-way function.

We remark that, in their original formulation of trapdoor one-way functions, Diffie and Hellman took $A_k = A$ for all $k$. However, the most interesting of the apparently similar functions that have since been suggested require the dependency of $A_k$ on $k$, so Massey has used the more general definition at the outset.

### 4.7.4   Public Key Cryptosystems

Diffie and Hellman introduced the concept of a trapdoor one-way function to use it as the building block for a public key cryptosystem, a secrecy system that requires no exchange of secret keys and makes no use of any conventional cryptosystem.

In a public key cryptosystem, all the stations agree on a trapdoor one-way function. Each station, say station $i$, then secretly chooses its trapdoor $k(i)$ by a random choice (with the uniform probability distribution) from the set $K$ of possible trapdoors. Station $i$ derives the algorithms $E_{k(i)}$ and $F_{k(i)}$, and then publishes in the public directory. Any station wishing to send information secretly to station $i$ sends the plaintext $x$ as the cryptogram $y = E_{k(i)}(x)$ after first checking with $F_{k(i)}$ to be sure that $x$ is an allowed plaintext. From the definition of a one-way trapdoor function, it follows that it will almost always be a practical impossibility for anyone to find $x$ and $y$ who knows only the information in the public directory, i.e., who knows only the algorithms $E_{k(i)}$ and $F_{k(i)}$, but station $i$ can use its own secret knowledge of $k_i$ to easily derive a simple algorithm $D_{k(i)}$ that can be used to decrypt the cryptogram and obtain $x$.

Diffie and Hellman showed that these trapdoor one-way functions do exist if certain other objects exist whose existence is generally accepted without question by cryptographers. More precisely, they showed that a conventional secrecy system that is secure against a chosen-plaintext attack specifies a one-way trapdoor function when the key is chosen as the trapdoor and the function is implemented by the encrypter. Because the capability to make a chosen-plaintext attack is equivalent to having an encrypter loaded with the secret key so that one can easily form the cryptogram for any desired plaintext, we can clearly see that such a conventional cryptosystem satisfies the three conditions that define a trapdoor one-way function.

### 4.7.5 The Pohlig-Hellman Scheme

For those primes where $p - 1$ has no large prime factor, Pohlig and Hellman [17] have shown a much superior algorithm for finding logarithms in GF($p$). For primes of the pathological form $p = 2^n + 1$, their algorithm (which requires little storage and little advance calculation) computes $x$ from $\alpha^x$ with only (log $p)^2$ multiplications, or in only about 10 milliseconds for $p \approx 2^{100}$ on the same fast computer. The Pohlig-Hellman algorithm is proportionately fast whenever $p - 1$ has no large prime factor; thus, exponentiation in GF($p$) is certainly not a one-way function for such $p$. But when $p - 1$ has a large prime factor, there is no presently known better algorithm than the general one mentioned above. We conclude then that exponentiation in GF($p$) is an apparent one-way function for primes $p$ in which $p - 1$ has a large prime factor. But our confidence that calculating logarithms in GF($p$) is an essentially difficult problem should perhaps be tempered by the fact that not many people have tried to solve this problem in recent years.

In Figure 4.20, we give the best presently known algorithm (from Knuth [18]) for finding logarithms in an arbitrary GF($p$), i.e., for solving for $x$ ($0 \leqslant x < p - 1$) given $y = \alpha^x$ where $\alpha$ is a given primitive element of GF($p$) and $p$ is a given prime. The algorithm exploits the fact that, for any positive integer $d$, there exist unique integers $Q$ and $r$ such that

$$x = Qd + r, \qquad 0 \leqslant r < d. \qquad (4.46)$$

Thus, for any chosen $d$, solving $y = \alpha^x = \alpha^{Qd+r}$ is equivalent to finding integers $r$ and $Q$ such that

$$y\alpha^{-dQ} = \alpha^r \qquad (4.47)$$

where

$$0 \leqslant r < d \qquad (4.48)$$

and

$$0 \leqslant Q \leqslant \left[\frac{p-1}{d}\right]. \qquad (4.49)$$

That $Q$ is restricted to the range given in (4.49) follows from (4.46) and the fact that $0 \leqslant x < p - 1$. The algorithm begins by choosing a "convenient" $d$ (say a power of 2) such that

$$d \approx \sqrt{p}.$$

As we shall shortly see, this choice roughly equalizes the necessary memory in bits and the required number of GF($p$) operations. One can generally trade computation for storage, so the case when both are roughly equal gives perhaps the best measure of the complexity of the algorithm. Alternatively, the product

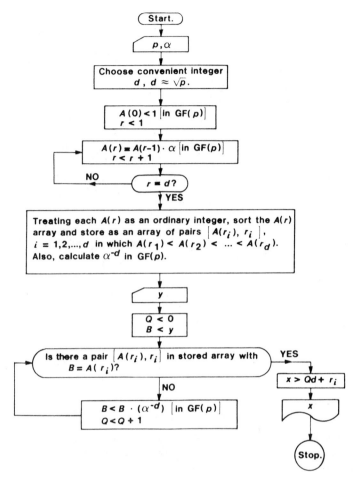

**Figure 4.20** Knuth's algorithm for computing logarithms in GF($p$); $x = \log \alpha y$.

of computational steps and memory generally is a good single parameter measure of complexity.

The algorithm begins by computing and storing $A(r) = \alpha^r$ for $0 \leq r < d$; this requires about $\sqrt{p}$ multiplications in GF($p$) and about $\sqrt{p} \log p$ bits of storage. So that it will be easy later to check whether some number is among these $d$ stored numbers, the list is then treated as a list of real integers and sorted into increasing order $A(r_1) < A(r_2) < \ldots < A(r_d)$, and each entry is marked with the corresponding integer $r_i$. An efficient sort (see Volume I, Chapter 6) requires about $(d \log d) \approx \left(\frac{1}{2}\right) \sqrt{p} \log p$ comparisons, and $d \log d \approx \left(\frac{1}{2}\right) \sqrt{p} \log p$ further bits of storage are needed for the marking integers. In all, we require about $M \approx \left(\dfrac{3}{2}\right) \sqrt{p} \log p$ bits of storage. Because the computation to set up memory is on the same order, we shall not further interest ourselves in this "precomputation."

The element $y$ whose logarithm is to be taken is next entered into the algorithm. The algorithm then computes $B = y\alpha^{-dQ}$ for $Q = 0, 1, 2, \ldots$ until a matching entry $A(\alpha^{r_i})$ is found in storage. This must happen, according to (4.49), in at most $\left\lceil \dfrac{p-1}{d} \right\rceil \approx \sqrt{p}$ steps and, on the average, in about half that many steps.

Thus we need about $\sqrt{p}$ multiplications and, for each multiplication, about $\log d \approx \dfrac{1}{2} \log p$ comparisons. To find whether the number is in the list we first compare $B$ with the number $A(r_i)$ at the middle of the list to find whether the matching number, if any, is in the first half or the second half of the list, etc. Thus, we need about

$$C \approx \sqrt{p} + \frac{1}{2}\sqrt{p} \log p \approx \frac{1}{2}\sqrt{p}\log p$$

operations in $GF(p)$ in order to determine $x$. The computation-memory product is thus seen to be

$$CM \approx \frac{3}{4} p \, (\log p)^2.$$

### 4.7.6  General Multiple Access Cryptosystems

At the heart of the multiple access cryptosystems are the computations of exponentiation and logarithmic functions over a finite field, as we have seen in the previous sections. In this section we shall see what will happen if we break away from these specific functions and generalize to other arbitrary functions.

Let $z$ be a common variable of a multiple access cryptosystem. Let $k_i$, $k_j$ denote the private keys from the two stations $S_i$ and $S_j$, respectively. If $S_i$ desires to communicate with $S_j$ secretly, $S_i$ first needs to extract $f(z,k_j)$ from the public key directory $f(z,\cdot)$. After obtaining $f(z,k_j)$, $S_i$ inserts $f(z,k_j)$ into its own public key variable as $f(f(z\,,\,k_j),\,k_i)$. Similarly $S_j$ obtains $f(f(z,\,k_i),\,k_j)$. For meaningful communication we demand

$$f(f(z,\,k_i),\,k_j) = f(f(z,\,k_j),\,k_i). \qquad (4.50)$$

It turns out that such a general solution exists in order for (4.50) to hold. The solution as suggested by Henze [19] is in the form

$$f(z,\,k) = g^{-1}[g(z) + h(k)] \qquad (4.51)$$

where $k$ is a key in the whole key set, and $g$ is an arbitrary continuous and strictly monotonic function. Here we assume that the discrete encryptic transformations meet such generalization. To show that $f(z,\,k)$ is a solution to (4.50), we may be convinced thrrough the following verification for $g\,g^{-1} = 1$, i.e., $g^{-1}$ is the inverse of $g$, and

$$f(\alpha\,k_i) = g^{-1}[g(\alpha) + h(k_i)] \qquad (4.52)$$

$$\begin{aligned}
f(f(z,\,k_i),\,k_j) &= g^{-1}\{g[f(z,k_i)] + h(k_j)\} \\
&= g^{-1}\{g[g^{-1}(g(\alpha) + h(k_i))] + h(k_j)\} \\
&= g^{-1}\{g(\alpha) + h(k_i) + h(k_j)\} \\
&= g^{-1}\{g(\alpha) + h(k_j) + h(k_i)\}. \qquad (4.53)
\end{aligned}$$

Now we reverse the process from the last expression in (4.53)

$$\begin{aligned}
g^{-1}&\{g(\alpha) + h(k_j) + h(k_i)\} \\
&= g^{-1}\{g[g^{-1}(g(\alpha) + h(k_j))] + h(k_i)\} \\
&= g^{-1}\{g[f(\alpha,\,k_j)] + h(k_i)\} \\
&= f(f(\alpha,\,k_j),\,k_i), \qquad (4.54)
\end{aligned}$$

which is identical to the right-hand side of (4.50) when $\alpha = z$. This general cryptosystem analysis is shown in Figure 4.21. The functions $g(\alpha)$ and $h(k_i)$ can be very general. But there is specific interest for these functions to be logarithmic. When $z = \alpha$,

$$f(\alpha,\,k_i) = \alpha^{k_i},$$
$$f(f(\alpha,\,k_j),\,k_i) = \alpha^{k_j\,k_i} = k_{ij},$$

the case reduces to the Diffie and Hellman system of Section 4.7.2.

## 4.8   A MEASURE OF SECRECY

Before we consider other more sophisticated cryptographic algorithms, we pause in this section to mention the other side of cryptography. It has been said that knowing not only oneself but also the opponent wins every battle. Before the design of any cryptosystem it is paramount to know what means are available at the opponent's disposal. In theory, any cryptographic algorithm using a finite length key can always be broken. The guiding principle of a cipher designer, therefore, would be to make the task of breaking the key so difficult that it seems infeasible. Infeasibility may be measured in terms of the amount of effort required, or the "work factor," as it is called in the cryptographic literature. Thus work factor, a measure of secrecy, is the amount of effort needed to try to break a cryptographic algorithm, and a good cryptographic algorithm design should maximize the work factor. There is no standard expression for the work factor, which may be measured in terms of the number of mathematical computations, cryptanalyst time, the relative degree of difficulty of a problem, or the cost necessary to break the algorithm.

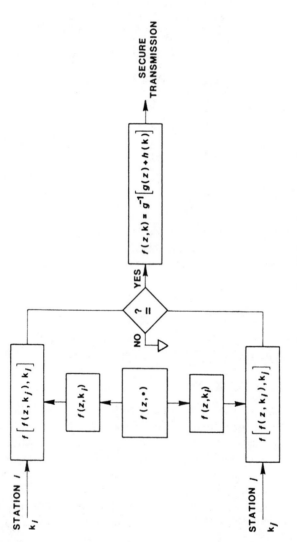

**Figure 4.21** A general multiple access cryptosystem.

In general, there are two types of cryptographic attacks. The exhaustive attack uses the brute force search technique of trial and error. The other type of attack, in contrast, is analytical. The cryptanalyst first has to guess the underlying mathematical formulation used in the cryptographic algorithm; next, the cryptanalyst needs to solve the variable or variables in the formulation, based upon the information available.

The work factor or the success of a cryptographic attack depends on the available information, the type of ciphers, the nature of a plaintext, and the time and storage element tradeoffs. Some specific methods of cryptographic attack and work factors are discussed by Meyer and Matyas [10]. Methods of attack for computer database with regard to inference control are well treated by Denning [20]. We shall not provide more specifics other than the following example.

*Example 4.6:* Diffie and Hellman suggested exponentiation in GF($p$) where $p$ is a large prime such that $p-1$ has a large prime factor. The function in this case is

$$f(x) = \alpha^x$$

where $\alpha$ is a given primitive element of GF($p$), and where $x$ is an integer satisfying $0 \le x < p-1$. Taking the logarithm of the function $f(x)$ we can write

$$\log_\alpha(\alpha^x) = x, \qquad 0 \le x < p-1$$

and call the operation of calculating $x$ from $\alpha^x$ to be taking the logarithm in GF($p$). It is easy to exponentiate in GF($p$) even when $p$ is extremely large (say $p \approx 2^{100}$). An exponentiation algorithm requires only $\log p$ (where all logs here and hereafter are to the base 2 and where we ignore small differences in counting computations or storage) multiplication and only $(\log p)^2$ binary digits of storage and calculates $\alpha^x$ from $x$ with only $\log p$ GF($p$) multiplications. For $p \approx 2^{100}$, only $10^2$ multiplications are required. Thus, assuming a fast computer that could do such a multiplication in 1 microsecond, we could compute $\alpha^x$ from $x$ in $\dfrac{1}{10}$ millisecond. On the other hand, the best general algorithm for computing logarithms in GF($p$), which we described above, requires about $\dfrac{3}{2} \sqrt{p} \times \log p$ bits of storage and calculates $x$ from $\alpha^x$ in about $\dfrac{1}{2} \approx \sqrt{p} \log p$ GF($p$) operations. For the same $p \approx 2^{100}$, assuming that the vast necessary memory was available, calculating $x$ from $\alpha^x$ would require $\dfrac{1}{2} \times 2^{50} \times 100 \approx 10^{16.7}$ operations, or about $10^{10.7}$ seconds on the same fast computer. (One year $= 10^{7.5}$ seconds, so that $10^{10.7}$ seconds $\approx 1600$ years. This might give a better picture of the enormity of the required computation described by Massey [9].)

## 4.9   THE RIVEST-SHAMIR-ADLEMAN ALGORITHM

Shortly after the appearance of the Diffie-Hellman paper, Rivest, Shamir, and Adleman (RSA) announced their discovery of an apparent trapdoor one-way function for use in a public key cryptosystem [21].

In the RSA algorithm, each station independently and randomly chooses two large primes, say $p_1$ and $p_2$. We shall not append a subscript to remind us that these are the primes chosen, say, by station $i$—as the notation becomes too cumbersome—but the reader should keep in mind that we are indeed considering the actions of a single station.

First the station forms

$$m = p_1 p_2 \tag{4.55}$$

and the Euler function (Chapter 2, Section 2.4)

$$\phi(m) = (p_1 - 1)(p_2 - 1). \tag{4.56}$$

Next, the station randomly selects an integer $e$, $1 \leq e \leq \phi(m)$, such that the greatest common divisor is

$$\gcd[\phi(m), e] = 1, \tag{4.57}$$

after which the station computes the unique inverse of $e$,

$$d = e^{-1} \text{ [inverse modulo } \phi(m)] \tag{4.58}$$

whose existence is guaranteed. This station then publishes $(m, e)$ in the public directory. The set of allowed plaintext is just the set of positive integers smaller than, and relatively prime to, $m$, i.e.,

$$A_k = \{x : 1 \leq x < m \text{ and } \gcd(x, m) = 1\}. \tag{4.59}$$

The encrypting function, by which other stations send secret messages to this station, is just

$$y = R_m(x^e) = x^{e(\text{mod } m)} \tag{4.60}$$

where $y$ is the cryptogram for the plaintext $x$. This station decrypts such cryptograms in the manner

$$x = R_m(y^d) = y^{d(\text{mod } m)}. \tag{4.61}$$

The RSA algorithm, as shown in Figure 4.22, is summarized as follows.

Step 1: Select two prime numbers $p_1$ and $p_2$. Calculate $m = p_1 p_2$.
Step 2: Calculate $\phi(m) = (p_1 - 1)(p_2 - 1)$.
Step 3: Select an integer $e$ such that $\gcd[\phi(m), e] = 1$.
Step 4: Using the Euclidean algorithm, calculate $d$, the inverse of $e$, modulo $\phi(m)$.
Step 5: For encryption raise the message to the power of $e$ modulo-$m$; for decryption raise the cryptogram to the power of $d$ modulo-$m$.

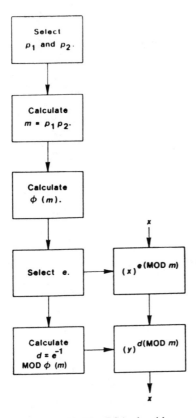

**Figure 4.22** The RSA algorithm.

*Example 4.7*: Proceeding with the RSA algorithm in steps, we have:

Step 1: Let $p_1 = 5$ and $p_2 = 7$, $m = 5 \times 7 = 35$.
Step 2: $\phi(m) = \phi(p_1) \phi(p_2) = (p_1 - 1)(p_2 - 1) = 24$.
Step 3: Select $e = 5$, because gcd $(24, 5) = 1$.
Step 4: $d = e^{-1} = 5$, because $e \times d = 5 \times 5 = 25 = 1$ mod 24.
Step 5: Raise the message $x$ to the power 5 for both encryption and decryption. If $x = 2$, $2^5$ mod 35 $= 32$ is then given for the corresponding encrypted text. For decryption we perform

$$32^d \bmod 35 = 32^1 \times 32^2 \times 32^2$$

$$= 32 \times 1024 \times 1024$$

$$= 33{,}554{,}432 \bmod 35$$

$$= 2$$

as the original message.

### 4.9.1    Analyses of the RSA Algorithm

First, we verify that the plaintext can indeed be decrypted according to (4.61).
From (4.60), it follows that

$$R_m(y^d) = R_m[(x^e)^d]$$

$$= R_m(x^{de}). \tag{4.62}$$

But (4.58) implies

$$de = Q \, \phi(m) + 1 \tag{4.63}$$

for some integer $Q$, and hence

$$R_m(x^{de}) = R_m(x^{Q\phi(m)}x)$$

$$= R_m[\{R_m(x^{\phi(m)})\}^Q x] \tag{4.64}$$

But the fact that $\gcd(x,m) = 1$, as specified by (4.59) ensures that we can
apply Euler's theorem [actually, because of (4.55) and (4.56), we need only
apply the special case of Euler's theorem] to obtain

$$R_m[x^{\phi(m)}] = 1 \tag{4.65}$$

so that (4.64) then gives

$$R_m(x^{de}) = R_m(x) = x \tag{4.66}$$

where the last equality follows from the fact that $1 \leq x < m$. Thus, we have
shown the validity of the decrypting operation (4.61).

We now consider more explicitly the trapdoor one-way function on which the
RSA algorithm rests. The trapdoor may be taken to be the three-tuple

$$k = (p_1, p_2, e). \tag{4.67}$$

Note that the third component of the trapdoor is publicly revealed, but we include
it in the trapdoor nonetheless, for otherwise the trapdoor would not suffice to
specify the encrypting and decrypting functions as our definition demands. The
first two components of the trapdoor are, however, kept secret, as is the number
$d$ determined from them.

As the fast and simple algorithm for encryption, $E_k$, we can choose the easy
exponentiation (by repeated squaring) algorithm that will compute $R_m(x^e)$ in
about 2 log $m$ multiplications modulo-$m$. RSA recommend choosing $p_1$ and $p_2$
as primes on the size of $10^{100} \approx 2^{332}$ so that $m \approx 2^{664}$, and hence encryption
takes about 664 multiplications of 330 binary digit numbers. Note that the
plaintext $x$ is also represented by about 664 binary digits so that we need one
such multiplication for each bit of the plaintext. As the fast and simple algorithm,
$F_k$, for determining whether an integer $x$, $0 < x < m$, is valid plaintext, i.e.,
whether $\gcd(x, m) = 1$, we can use Euclid's greatest common divisor algorithm
that requires about 2 log $m$ operations of taking remainders. However, the
probability that an $x$ randomly chosen (with the uniform probability distribution)

in the range $1 \leq x < m$ satisfies $\gcd(x, m) = 1$ is just

$$\frac{\phi(m)}{m} = \frac{(p_1 - 1)(p)_2 - 1)}{p_1 p_2} = 1 - \frac{1}{p_1} - \frac{1}{p_2} + \frac{1}{p_1 p_2}, \quad (4.68)$$

which is so close to 1 that one need not bother checking whether $x$ is a valid plaintext. In fact, choosing an $x$ such that $\gcd(m,x) \neq 1$ will allow the sender to break the system since $m = p_1 p_2$ then implies either $\gcd(m,x) = p_1$ or $\gcd(m,x) = p_2$. Both $p_1$ and $p_2$ could now be found by dividing the publicly known $m$ by $\gcd(m,x)$. The transmitting station then knows the entire trapdoor! Thus, if the RSA algorithm is secure so that the chances of breaking it by a random guess are virtually nil, it will be a waste of time to check whether $\gcd(x,m) = 1$.

For the simple decrypting algorithm, $D_k$, we can also use the easy exponentiation algorithm. Thus, decrypting will require about $2 \log m$ multiplications modulo-$m$.

We now must show that the user can easily enough choose his trapdoor $k = (p_1, p_2, e)$ and determine $d$. The determination of $d$ and $e$ is easily done by Euclid's greatest common divisor algorithm, but how does one randomly choose large primes easily? Let $\pi(n)$ denote the number of primes less than $n$. It is well known in number theory (Tchebycheff's Theorem) that

$$\pi(n) \approx \frac{n}{\ell n \ n}, \ n \text{ large}. \quad (4.69)$$

Thus, the probability of a randomly chosen $m$, $\frac{1}{2} 2^{332} \leq m < 2^{332}$, being a prime is about

$$\frac{\pi(2^{332}) - \pi(2^{331})}{2^{332} - 2^{331}} \approx \frac{1}{2 \ \ell n(2^{332})} \approx \frac{1}{115}$$

so that on the average we would need to choose about 115 such numbers before we find a prime. Thus, choosing a prime $p$, $p \approx 2^{332}$, is easy provided that it is easy to check whether a number of this size is a prime. For this purpose, RSA suggest using a probabilistic test proposed by R. Solovay and V. Strassen that works as follows. Let $m$ be the integer to be tested for primality. Randomly choose an integer $n$ in the interval $1 \leq n < m$ (with the uniform probability distribution) and then check to see whether

$$\gcd(m,n) = 1 \text{ and } J(n,m) \equiv n^{(m-1)/2} \ (\text{mod } m). \quad (4.70)$$

Here $J(n,m)$ is the "Jacobi symbol" defined recursively by

$$J(n,m) = \begin{cases} 1, \text{ if } n = 1 \\ J(\frac{n}{2}, m)(-1)^{(m^2-1)/8}, \text{ if } n \text{ is even} \\ J(R_n(m), n)(-1)^{(m-1)\ (n-1)/4}, \text{ otherwise}. \end{cases} \quad (4.71)$$

The condition (4.70) must hold if $m$ is a prime, but has probability at most $\frac{1}{2}$ of holding if $m$ is not prime. Thus, if (4.70) holds for, say 100 independently randomly chosen values $n$, the probability that $m$ is not a prime is less than $2^{-100} \approx 10^{-30}$ and therefore we could confidently decide that $m$ is a prime.

The security of the RSA algorithm rests on the fact that factoring $m = p_1 p_2$, where $p_1$ and $p_2$ are large primes such that $p_1 - 1$ and $p_2 - 1$ each have large prime factors and $\gcd(p_1 - 1\ p_2 - 1)$ is small, is an apparently difficult problem. (Note that the above fast test for primeness did not yield a factorization of $m$ when $m$ was nonprime.) Because $m = p_1 p_2$ and $e$ are publicly known in the RSA system, we could break the system (i.e., find $k = p_1, p_2, e$) if we could factor $m$. Rivest, Shamir, and Adleman cite an unpublished algorithm of Schroeppel that factors a large positive number $n$ in about

$$e^{\sqrt{\ell n(n)\ \ell n\ \ell n\ (n)}}$$

operations as the best general factoring algorithm known. For $m \approx 2^{664}$, this algorithm requires about $3.8 \times 10^9$ years to factor $m$, assuming a very fast computer that can perform one operation per microsecond. Thus, it is unfeasible by the best-known methods to break the RSA algorithm by attempting to factor $m$.

It would be desirable to show that any way of breaking the RSA algorithm is equivalent to factoring $m$. The system could be broken if the opponent could determine $\phi(m) = (p_1 - 1)(p_2 - 1)$ from $m = p_1 p_2$ and $e$, since the unauthorized station could then find $d$ just as did the originator of the system. But finding $\phi(m)$ is equivalent to factoring $m$ in this case because

$$m - \phi(m) + 1 = p_1 + p_2$$

and

$$\sqrt{(p_1 + p_2)^2 - 4m} = p_1 - p_2.$$

Since the quantities on the left of these two equations can be found in order without difficulty, we can easily get those on the right and hence easily get $p_1$ and $p_2$. The unauthorized user could also break the system by finding $d$, since he could then do decryption in the same way as the authorized user. Rivest, Shamir, and Adleman stated that finding $d$ is also equivalent to factoring $m$. A modification of the RSA algorithm has been suggested by Williams [22]. For our purpose, however, we shall limit our discussions to the original algorithm.

## 4.10   McELIECE'S PUBLIC KEY SYSTEM

The rich structures of error-correction codes have been explored for cryptosystem designs. McEliece [23] first suggested error-correcting codes for public key distributions; Lu, Lee, and Fang [15] proposed sequential decoding of convolutional codes for satellite cryptosystem design; and Karnin, Greene, and Hellman [24] connected maximum distance codes to secret sharing systems. In this section we discuss only the merit of the approach suggested by McEliece because of its potential usefulness and far-range implications. The maximum distance codes applied to secret sharing systems will be treated separately in Section 4.11.

To generate a public key in McEliece's scheme, a station first selects a Goppa code chosen at random from a large set of codes. The code generator matrix is the encryption key. The codeword of the chosen code is then permuted and scrambled. Errors are then introduced on purpose. McEliece suggests the use of fast algorithm for decoding and decryption. The key becomes public after the permutation and selection of the Goppa code generator matrix. Any transmitting station can permute and select a Goppa code, but only the communicating receiving station knows the inverse permutation, which is used for deciphering the ciphertext in addition to the error-correcting algorithm. The reason for choosing Goppa codes is that there is a large number of such codes with good code minimum distance. The existence of this large number of Goppa codes and the large number of codewords from each code add an extra security measure in cryptosystem design.

## 4.10.1   The Goppa Codes

In 1970 and 1971 V.D. Goppa, the information theorist from the U.S.S.R., published two papers in *Problemy Peredachi Informatsi* [25,26]. Berlekamp [27] first evaluated and examined the properties of this class of generally noncyclic, nonbinary block codes, and called them Goppa codes. These codes happen to be closely related to the Reed-Solomon (RS), Bose-Chaudhuri-Hocquenghem (BCH), and even cyclic codes [28]. For example, Retter showed that Goppa codes can be decoded by BCH decoders. Delsarte demonstrated that Goppa codes are subfield subcodes of modified RS codes and Tzeng showed that Goppa codes become cyclic when their overall parity check matrix is extended [29,30,31]. Upon investigation of the applicability of Goppa codes to satellite communication, I concluded that they are not desirable from the standpoints of efficiency, error-correcting capability, and ease of implementation. For this reason, Goppa codes will not be discussed in Chapter 5 but we shall discuss the encryption properties of Goppa codes in this section.

The best way to describe Goppa codes is to start with some well-known codes such as the BCH codes. A binary BCH code of $n = 2^m - 1$ nonzero elements can be defined over the finite field GF($2^m$). A binary BCH code of block length $n$ that can correct up to $t$ random errors must contain $\alpha, \alpha^2, \ldots, \alpha^{2t}$ consecutive roots of a codeword polynomial. Decoding BCH code is based on the syndrome polynomial $S(x)$, which is the result of the multiplication of a received codeword $R = \{R_i, i = 0, 1, 2, \ldots, n-1\}$ to the code parity check matrix whose elements are the roots $\alpha^j, j = 1, 2, \ldots, 2t$. Let

$$S(x) = S_1 + S_2 x + \ldots + S_{2t} x^{2t-1}$$

$$= \sum_{j=1}^{2t} S_j x^{j-1}$$

$$= \sum_{j=1}^{2t} x^{j-1} \sum_{i=0}^{n-1} R_i \alpha_i^{-j} \qquad (4.72)$$

$$= \sum_{i=0}^{n-1} R_i \sum_{j=1}^{2t} x^{j-1} \alpha_i^{-j}$$

The summation under $j = 1$ to $2t$ is

$$\sum_{j=1}^{2t} \frac{x^{j-1}}{\alpha_i^j} = \frac{x^0}{\alpha_i} + \frac{x}{\alpha_i^2} + \frac{x^2}{\alpha_i^3}$$

$$+ \ldots \cdot \frac{x^{2t-1}}{\alpha_i^{2t}}$$

$$= \frac{1}{\alpha} \left( 1 + \frac{x}{\alpha_i} + \frac{x^2}{\alpha_i^2} + \ldots + \frac{x^{2t-1}}{\alpha_i^{2t-1}} \right)$$

$$= \frac{1}{\alpha} \left[ 1 + (Y) + (Y)^2 + \ldots + (Y)^{2t-1} \right]$$

$$= \frac{1}{\alpha} \left[ \frac{Y^{2t}-1}{Y-1} \right]$$

$$= \frac{1}{\alpha} \left[ \frac{\left(\dfrac{x}{\alpha_i}\right)^{2t} - 1}{\left(\dfrac{x}{\alpha_i}\right) - 1} \right] = \frac{\left(\dfrac{x}{\alpha_i}\right)^{2t} - 1}{x - \alpha_i}$$

$$= \frac{1}{\alpha_i^{2t}} (x^{2t-1} + \alpha_i x^{2t-2} + \alpha_i^2 x^{2t-3} + \ldots$$

$$+ \alpha_i^{2t-1}) \bmod x^{2t} \qquad (4.73)$$

Substituting (4.73) back into (4.72) we have

$$S(x) = \sum_{i=0}^{n-1} R_i \left[ \frac{\left(\dfrac{x}{\alpha_i}\right)^{2t} - 1}{x - \alpha_i} \right]. \qquad (4.74)$$

For $x \neq \alpha_i$, $(x/\alpha_i)^{2t} \neq 1$, the numerator cannot be zero. Thus, for the syndrome $S(x)$ to be zero,

$$\sum_{i=0}^{n-1} \frac{R_i}{x - \alpha_i} \equiv 0 \bmod x^{2t}. \qquad (4.75)$$

If $x^{2t}$ in (4.75) is replaced by a Goppa polynomial $g(x)$, then the Goppa code is defined as

$$\sum_{i=0}^{n-1} \frac{G_i}{x - \alpha_i} \equiv 0 \bmod g(x) \tag{4.76}$$

where $G_i$, $i = 0, 1, \ldots, n-1$ denotes a received Goppa codeword and $g(x)$ is the code polynomial of degree $s$ with coefficients from $GF(q^m)$. With the above definition, Goppa codes are shown [28] to have code minimum distance $d_{\min} \geq s + 1$, with the number of information digits $k \geq n - ms$ where $s$ is the number of distinct roots of the code generator polynomial [see (4.78)]. Note that from the congruence of (4.76) it can be observed that a Goppa code is determined by $g(x)$ and the set of $\alpha_i$'s.

As any other parity check codes, Goppa codes can also be expressed in terms of the corresponding parity check matrix $H$, which contains the set of roots of the Goppa polynomial as well as the set of $\alpha_i$'s. Let $g(x)$ in general be expressed as

$$g(x) = (x - \beta_1)^{r_1} (x - \beta_2)^{r_2} \ldots . (x - \beta_s)^{r_s} \tag{4.77}$$

where $\beta_1, \beta_2, \ldots, \beta_s$ are the roots of $g(x)$ and $r_1, r_2, \ldots, r_s$ denote the number of particular repeated roots. Then the parity check matrix of a Goppa code is

$$H = \begin{bmatrix} \dfrac{1}{\beta_1 - \alpha_1} & \dfrac{1}{\beta_1 - \alpha_2} & \cdots & \dfrac{1}{\beta_1 - \alpha_n} \\[2mm] \dfrac{1}{(\beta_1 - \alpha_1)^2} & \dfrac{1}{(\beta_1 - \alpha_2)^2} & \dfrac{1}{(\beta_1 - \alpha_n)^2} \\[1mm] \vdots & \vdots & \vdots \\[1mm] \dfrac{1}{(\beta_1 - \alpha_1)^{r_1}} & \dfrac{1}{(\beta_1 - \alpha_2)^{r_1}} & \dfrac{1}{(\beta_1 - \alpha_n)^{r_1}} \\[2mm] \vdots & \vdots & \vdots \\[1mm] \dfrac{1}{\beta_s - \alpha_1} & \dfrac{1}{\beta_s - \alpha_2} & \dfrac{1}{\beta_s - \alpha_n} \\[2mm] \dfrac{1}{(\beta_s - \alpha_1)^2} & \dfrac{1}{(\beta_s - \alpha_2)^2} & \dfrac{1}{(\beta_s - \alpha_n)^2} \\[1mm] \vdots & \vdots & \vdots \\[1mm] \dfrac{1}{(\beta_s - \alpha_1)^{r_s}} & \dfrac{1}{(\beta_s - \alpha_2)^{r_s}} & \dfrac{1}{(\beta_s - \alpha_n)^{r_s}} \end{bmatrix} \tag{4.78}$$

For $R$ to be a valid received Goppa coded word, we must have

$$R \times H^T = 0. \tag{4.79}$$

*Example 4.8*: Given a Goppa polynomial $g(x) = 2 + x + x^2 = (x - \alpha)(x - \alpha^3)$, where $\alpha$ is primitive in $GF(3^2)$ and $2 + \alpha + \alpha^2 = 0$, then $\alpha^2 = 2\alpha$

$+ 1, \alpha^3 = 2\alpha + 2, \alpha^4 = 2, \alpha^5 = 2\alpha, \alpha^6 = \alpha + 2, \alpha^7 = \alpha + 1$, and $\alpha^8$
$= 1$. Let $\alpha_1 = 0, \alpha_2 = 1$, and $\alpha_3 = 2$, then

$$
H = \begin{bmatrix} \dfrac{1}{(\alpha - 0)} & \dfrac{1}{\alpha - 1} & \dfrac{1}{\alpha - 2} \\ \dfrac{1}{\alpha^3 - 0} & \dfrac{1}{\alpha^3 - 1} & \dfrac{1}{\alpha^3 - \alpha} \end{bmatrix}
$$

$$
= \begin{bmatrix} \alpha^7 & \alpha^2 & \alpha \\ \alpha^5 & \alpha^6 & \alpha^3 \end{bmatrix}
$$

$$
= \begin{bmatrix} \alpha^7 & \alpha^2 & \alpha \\ \alpha & \alpha^2 & \alpha^7 \end{bmatrix}.
$$

This Goppa code, which comes from [31], is the $n = 3, k = 1$ linear code over
GF(3) that has three codewords (0, 0, 0), (1, 2, 1), and (2, 1, 2).

## 4.11  SECRET SHARING SYSTEMS

To prevent the loss or disappearance of the common cryptographic keys due to
station failure in a satellite system, a secret $S$ can be divided into $n$ segments
$s_1, s_2, \ldots, s_n$. Each segment may represent a station chosen from a set $\{s\}$ and
$s_i < S$ for all $i = 1, 2, \ldots, n$. A secret sharing system has the properties that
the secret $S$ is recoverable from any $k$ segments, while $k \leq n$ and knowledge of
$k - 1$ or fewer segments provides no clue, or is irrecoverable on $S$. That is to
say that secret $S$ of $n$ stations can only be reconstructed from any $k$ number of
stations. Such a secret sharing system is also called a threshold scheme, with $k$
being the threshold. Shamir [32] first realized the need for such a system and
provided a systematic procedure for its construction. Almost immediately after,
Blakely [33] extended Shamir's work from modulo a prime number to modulo
a primitive polynomial with the finite field operation from GF($p$) to specific
GF($2^n$). Recently, Karnin, Greene, and Hellman [24] applied the properties of
maximum distance codes to the secret sharing system. A modular approach to
key safeguarding in such secret sharing systems is suggested by Asmuth and
Bloom [34]. McEliece and Sarwate [35] observed that Shamir's scheme is closely
related to the Reed-Solomon coding scheme for an error-and-erasure decoding
algorithm.

In this section, we discuss these various methods for constructing secret sharing
systems. We should note here that a common property of such systems is that
the degree of secrecy depends on the amount of information and, specifically,
on the number of predetermined components. This is different from the cryptosystems described in previous sections, in which the secrecy depends on the
degree of computational difficulty. In a secret sharing security system, a segment
with a small number of bits can provide the same security as a segment with a
large number of bits, because the information on secret sharing is the same in
both cases.

### 4.11.1 Shamir's Polynomial Method

Using numerical analysis with prime numbers as a point of departure, Shamir originated the idea of enumerating two polynomials to construct a secret system. For a $k$ out of $n$ secret system, his scheme first chooses a prime number $p$ which is larger than both $S$ and $n$, then determines the interpolating polynomial $g(x)$ of degree $k-1$ with the property $q(x_i) = y_i$ for $i = 1, 2, \ldots, k$ and all $y_i$'s distinct. If we let $q(x)$ be expressed as

$$q(x) = a_0 + a_1 x + \ldots + a_{k-1} x^{k-1} \tag{4.80}$$

then the secret $S = a_0$. The $n$ components of the secret, i.e., $s_1, s_2, \ldots, s_n$, are obtained by solving $q(x)$ on the set of $x_1, x_2, \ldots, x_n$ as

$$s_i = q(x_i) \bmod p \tag{4.81}$$
$$i = 1, 2, \ldots, n.$$

If we consider $q(x)$ as a hypergeometrical line, then the pairs $(x_i, s_i)$ can be considered as points on the line. Since $k$ points determine $q(x)$, the secret can be determined and recovered from $k$ components. However, $q(x)$ cannot be determined from fewer than $k$ components.

To reconstruct $q(x)$ from the set of $k$ components, Shamir suggests the use of the Lagrange polynomial

$$q'(x) = \sum_{c=1}^{k} \frac{s_c(x - x_1)(x - x_2) \ldots (x - x_k)}{(x_c - x_1)(x_c - x_2) \ldots (x_c - x_k)} \tag{4.82}$$

while the Lagrange polynomial enumeration is also carried out in modulo $p$. The following example is from Denning [20].

*Example 4.9*: For $s = 13, p = 17$,

$$q(x) = (2x^2 + 10 x + 13) \bmod 17$$

can be used for $k = 3$ (because $q(x)$ has the degree of $k-1$). For $n = 5$, we need to evaluate $q(x)$ for $x = 1, 2, \ldots, 5$ to obtain the set of components as

$$s_1 = q(1) = 25 \bmod 17 = 8$$

$$s_2 = q(2) = 41 \bmod 17 = 7$$

$$s_3 = q(3) = 61 \bmod 17 = 10$$

$$s_4 = q(4) = 85 \bmod 17 = 0$$

$$s_5 = q(5) = 113 \bmod 17 = 11.$$

Thus far, the representation of the secret 13 by the set of five components 8, 7, 10, 0, and 11 is complete. To show that any $k = 3$ out of the five components can recover $q(x)$, we use (4.82). Assuming $s_1, s_3, and s_5$ are selected, we have

$$q'(x) = \left\{ 8 \frac{(x - 3)(x - 5)}{(1 - 3)(1 - 5)} + 10 \frac{(x - 1)(x - 5)}{(3 - 1)(3 - 5)} \right.$$

$$\left. + 11 \frac{(x - 1)(x - 3)}{(5 - 1)(5 - 3)} \right\} \bmod 17$$

$$= (19 x^2 - 92 x + 81) \bmod 17$$

$$= 2 x^2 + 10 x + 13 = q(x).$$

The result is obtained by using both additive and multiplicative inverses with the prime 17.

The problems with Shamir's method are the generation of $q(x)$ and the enumeration of $q'(x)$ with inverses. But the method provides the significant link between the algebraic structure and secret sharing systems, and allows other developments to follow.

### 4.11.2   The Method of Matrices by Karnin, Greene, and Hellman

Karnin, Greene, and Hellman have proved a series of theorems for the same type of secret sharing system, which establish the following interesting results.

1. The problem is equivalent to finding a set of $n + 1$ matrices, $A_i$, $i = 1, 2,$ . . ., $n$ GF($p$), each matrix has dimension km × m, where $m$ is the size of the secret and $k$ is the number of the key segments required to construct the secret.
2. The problem of finding a $k$ out of $n$ secret sharing system is the same as finding a $k \times (n + 1)$ size matrix $G$, in which any $k$ columns of $G$ are linearly independent. Furthermore, $G$ can be assumed to contain a $k \times k$ identity matrix in the first $k$ columns of $G$. For a reconstructable $k$ out of $n$ secret sharing system, $G$ must be the generator matrix of an error-correcting code that can correct any $(n + 1) - k$ erasures and cannot correct $(n + 1) - (k + 1)$ or more number of erasures.
3. The maximum distance code with code minimum distance $n - k + 1$ is recommended.
4. The required generator matrix $G$ with matrix elements from a primitive element of GF($p^m = n$) has the form

$$G = \begin{bmatrix} 1 & 0 & 1 & 1 & . & . & . & 1 \\ 0 & 0 & \alpha & \alpha^2 & . & . & . & \alpha^{n-1} \\ 0 & 0 & \alpha^2 & \alpha^4 & . & . & . & \alpha^{(n-1)2} \\ \vdots & \vdots & \vdots & \vdots & & & & \vdots \\ 0 & 1 & \alpha^{k-1} & \alpha^{2(k-1)} & & & & \alpha^{(n-1)(k-1)} \end{bmatrix}$$

$$(4.83)$$

with a size of $k \times (n + 1)$. $G$ has the property that any of its $k$ columns are linearly independent.

5. If we delete the second column (dotted) in $G$, the resulting matrix is equivalent to Shamir's polynomial interpolation scheme.

To appreciate the inner working of this scheme, we need to know what a maximum distance code is. A maximum distance code (also called maximum-distance separable code, or maximum code) is a class of error-correcting block codes. Let $n$, $k$, and $d$ denote the codeword length, the number of information digits, and the minimum distance of a maximum distance code. The most significant characteristic of the codes is that the code distance is $d = n - k + 1$, which is the maximum attainable minimum distance value of any code.

Maximum distance codes were first studied by Singleton [36]. Additional analyses, such as the weight distribution of the code, can be found in [37, 38, 39].

Maximum distance codes can be constructed from finite projective geometries or orthogonal arrays, as well as from Reed-Solomon codes. First of all we recognize immediately that both the Reed-Solomon code and the maximum distance code have the same code distance property, i.e., $d = n - k + 1$. Although Karnin, Greene, and Hellman's result is based on the generator matrix $k$ of size $\times$ $(n + 1)$, a normal Reed-Solomon code generator matrix has a size of $k \times n$, but an extended Reed-Solomon code generator matrix has a size of $k \times (n + 1)$.

### 4.11.3  The Modular Method of Asmuth and Bloom

Based on the first-order congruence of the Chinese Remainder Theorem (see Chapter 2) Asmuth and Bloom use the congruence classes of a number associated with the secret $S$ as the set components. To construct a $k$ out of $n$ secret sharing system their method needs a prime $p$ and $n$ integers $m_1$, $m_2$, . . ., $m_n$ with the following constraints:

(a) $m_1 < m_{i+1}$ for all $i$,

(b) $(m_i, m_j) = 1$ for $i \neq j$,

(c) $(p, m_i) = 1$ for all $i$,

(d) $\displaystyle\prod_{i=1}^{k} m_i > p \prod_{i=1}^{K-1} m_{n-i+1}$

Any $k$ number of the $n$ components will be sufficient to recover the secret $S$.

To derive the set of components, let $y = S + Ap$ where $0 \leq S < p$, and $A$ is an arbitrary integer subject to the condition $0 \leq y < \displaystyle\prod_{i=1}^{k} m_i$. Then the components can be obtained by calculating

$$s_i \equiv y \pmod{m_i}. \tag{4.84}$$

The steps for generating the set of $s_i$'s are shown in Figure 4.23.

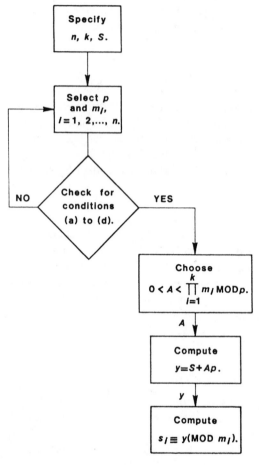

**Figure 4.23** Generation of secret components by congruence classes.

To reconstruct $S$ from the $k$ components, knowing $y$ is sufficient and $y$ can be determined by the Chinese Remainder Theorem, let

$$W_j = \prod_{k \leq j} m_k \qquad (4.85)$$

and

$$z_j = s_j \qquad (4.86)$$

$$z_{j+1} = z_j + a_j W_j$$

$$= S_{j+1}(\bmod\ m_{j+1}) \qquad (4.87)$$

where

$$a_j \equiv (s_{j+1} - z_j)\ W_j^{-1}\ (\bmod\ m_{j+1}). \qquad (4.88)$$

Euclidean algorithm can be used to calculate $W_j^{-1}$ from $W_j$. Once $a_j$ is found, we can calculate

$$z_k \equiv y(\mathrm{mod}\ W_k) \qquad (4.89)$$

and $y$ is obtained as the remainder by dividing $z_k$ by $W_j$.

The modular approach to secret sharing systems is mathematically elegant. The secret recovery process is shown to be $0(k)$ number of computations compared with $0(k \log^2 k)$ more computations by the method described in Section 4.11.1. But because of the restrictive conditions and the required inverse computations, the practicality of the scheme remains to be further investigated.

## 4.12   THE UNIFIED METHOD FOR CONSTRUCTION OF SECRET SHARING SYSTEMS

In this section, we demonstrate that the results of the last section can be unified and extended into a simple procedure for the construction of secret sharing systems. The tool used for this unification is orthogonal Latin squares, which are defined and discussed in Chapter 2. Here we relate the set of orthogonal Latin squares as a common source of origin to all the schemes previously described. Unification is obtained by demonstrating that orthogonal Latin squares are the roots of the polynomial approach by Shamir, the maximum distance code approach by Karnin, Greene, and Hellman, and the Reed-Solomon error-and-erasure coding observation by McEliece and Sarwate.

### 4.12.1   Maximum Distance Codes and Reed-Solomon Codes

It is well known in coding theory [38, 39] that all Reed-Solomon codes having parameters $[n = q^m, k, d = n - k + 1]$ with symbols over $GF(q^m)$ are maximum distance codes. The square brackets around the code parameters $n$, $k$, $d$ indicate that they denote number of characters in contrast to $(n, k, d)$, in which binary digits are denoted. Both maximum distance and Reed-Solomon code generator matrices have the dimension of $k \times n$. Reed-Solomon codes with maximum distance will be further discussed in Chapters 5 and 6.

### 4.12.2   Latin Squares and Reed-Solomon Codes

In this section we shall demonstrate that Reed-Solomon codes can be generated from orthogonal Latin squares. As we have seen from the last section, maximum distance codes thus can also be generated this way.

In general, from the set of elements $0$, $\alpha^0 = 1$, $\alpha$, $\alpha^2$, . . ., $\alpha^{n-1}$ the general matrix of the corresponding $[n, k, d]$ Reed-Solomon codes over $GF(n = q^m)$ looks like:

$$G_n = \begin{bmatrix} 1 & 1 & 1 & 1 \\ 0 & \alpha & \alpha^2 & \alpha^{n-1} \\ 0 & \alpha^2 & \alpha^4 & \alpha^{2(n-1)} \\ \vdots & \vdots & \vdots & \vdots \\ 0 & \alpha^{k-1} & \alpha^{2(k-1)} & \alpha^{(k-1)(n-1)} \end{bmatrix} \qquad (4.90)$$

Through a number of examples, we shall see how Latin squares generate such Reed-Solomon codes and their extensions.

*Example 4.10*: A Reed-Solomon code over GF(4) has the element 0, $\alpha^0$, $\alpha$, and $\alpha^2$. The generator polynomial of the code is

$$g(x) = x(x-1)(x-\alpha)(x-\alpha^2).$$

With $\alpha^2 = \beta$, the multiplication and addition operations over the finite field are shown in Table 4.7.

**Table 4.7**
Operation in GF(4).

| $X$ | 0 | 1 | $\alpha$ | $\beta$ | | $+$ | 0 | 1 | $\alpha$ | $\beta$ |
|---|---|---|---|---|---|---|---|---|---|---|
| 0 | 0 | 0 | 0 | 0 | | 0 | 0 | 1 | $\alpha$ | $\beta$ |
| 1 | 0 | 1 | $\alpha$ | $\beta$ | | 1 | 1 | 0 | $\beta$ | $\alpha$ |
| $\alpha$ | 0 | $\alpha$ | $\beta$ | 1 | | $\alpha$ | $\alpha$ | $\beta$ | 0 | 1 |
| $\beta$ | 0 | $\beta$ | 1 | $\alpha$ | | $\beta$ | $\beta$ | $\alpha$ | 1 | 0 |
| (a) Multiplication | | | | | | (b) Addition | | | | |

The generator matrix of the [3, 2, 2] Reed-Solomon code over GF(4) is

$$G = \begin{bmatrix} 1 & 1 & 1 \\ 0 & \alpha & \beta \end{bmatrix}.$$

The codeword length has three characters of which two are information characters, and the code has a code minimum distance of 2. In accordance with the operation in GF(4), the set of codewords are the linear combinations of the code generator matrix, and they are listed in Table 4.8 [39].

**Table 4.8**
[3, 2, 2] Reed-Solomon codewords.

| | | |
|---|---|---|
| 0 | 0 | 0 |
| 0 | 1 | $\alpha$ |
| 0 | $\alpha$ | $\beta$ |
| 0 | $\beta$ | 1 |
| 1 | $\alpha$ | 0 |
| $\alpha$ | $\beta$ | 0 |
| $\beta$ | 1 | 0 |
| $\alpha$ | 0 | 1 |
| $\beta$ | 0 | $\alpha$ |
| 1 | 0 | $\beta$ |
| 1 | $\beta$ | $\alpha$ |
| $\alpha$ | 1 | $\beta$ |
| $\beta$ | $\alpha$ | 1 |
| 1 | 1 | 1 |
| $\alpha$ | $\alpha$ | $\alpha$ |
| $\beta$ | $\beta$ | $\beta$ |

*Example 4.11*: Let a Latin square of order 4 corresponding to the four elements of GF(4) be as shown in Table 4.9. Note that, in this table, each entry of the square is designated by the row and column from 0, 1, $\alpha$ to $\beta$. For example, we have the Latin square entry 0 corresponding to row 0 and column 0. For the $\alpha$ row and column $\beta$ we also have 0 as the entry. Now referring back to the set of codewords of Table 4.8, let the first and second character of the codeword be the row and column designation of Table 4.9, and the third character be the corresponding entry in the Latin square. From the above two examples we have the corresponding two words 000 and $\alpha\beta0$.

*Example 4.12*: Expressed in terms of the elements in GF(2), the codeword symbols of the [3, 2, 2] Reed-Solomon code are listed in Table 4.10, which is obtained by the following symbol transformation from GF(4) to GF(2):

| GF(4) | 0 | 1 | $\alpha$ | $\alpha^2 = \beta$ |
|---|---|---|---|---|
| GF(2) | 00 | 10 | 01 | 11 |

The binary codewords can also be directly generated from

$$G = \begin{bmatrix} 1 & 0 & 1 & 0 & 1 & 0 \\ 0 & 0 & 0 & 1 & 1 & 1 \end{bmatrix}$$

over GF(2). Thus the Latin square is equivalent to the generator matrix of a Reed-Solomon code, which provides the structure for Shamir's polynomial interpolation scheme.

### 4.12.3   Construction of Extended Reed-Solomon Codes from Mutually Orthogonal Latin Squares

We have shown in Section 4.11.2 that a set of general required matrices for a secret sharing system is not exactly the same as the code generator matrix of a Reed-Solomon code.

Karnin, Greene, and Hellman's scheme requires a set of $n + 1$ matrices over $GF(2^m)$, each matrix having the size of $k\,m \times m$ over GF(2). A Reed-Solomon code generator matrix of size $k \times (n + 1)$ is needed to preserve the code distance property, as it turns out that the required system matrices can be derived from an extended Reed-Solomon code generator matrix. In this section we show how mutually orthogonal Latin squares can generate extended Reed-Solomon codes.

**Table 4.9**

A Latin square $L_i$ with entries from GF(4).

|  | 0 | 1 | $\alpha$ | $\beta$ |
|---|---|---|---|---|
| 0 | 0 | $\alpha$ | $\beta$ | 1 |
| 1 | $\beta$ | 1 | 0 | $\alpha$ |
| $\alpha$ | 1 | $\beta$ | $\alpha$ | 0 |
| $\beta$ | $\alpha$ | 0 | 1 | $\beta$ |

**Table 4.10**
Binary codewords of the [3, 2, 2] Reed-Solomon code.

| | | | | | |
|---|---|---|---|---|---|
| 0 | 0 | 0 | 0 | 0 | 0 |
| 0 | 0 | 1 | 0 | 0 | 1 |
| 0 | 0 | 0 | 1 | 1 | 1 |
| 0 | 0 | 1 | 1 | 1 | 0 |
| 1 | 0 | 0 | 1 | 0 | 0 |
| 0 | 1 | 1 | 1 | 0 | 0 |
| 1 | 1 | 1 | 0 | 0 | 0 |
| 0 | 1 | 0 | 0 | 1 | 0 |
| 1 | 1 | 0 | 0 | 0 | 1 |
| 1 | 0 | 0 | 0 | 1 | 1 |
| 1 | 0 | 1 | 1 | 0 | 1 |
| 0 | 1 | 1 | 0 | 1 | 1 |
| 1 | 1 | 0 | 1 | 1 | 0 |
| 1 | 0 | 1 | 0 | 1 | 0 |
| 0 | 1 | 0 | 1 | 0 | 1 |
| 1 | 1 | 1 | 1 | 1 | 1 |

*Example 4.13*: Referring to Table 4.9, which shows the Latin square $L_1$ with entries from GF(4), we now need another Latin square of order 4 and mutually orthogonal to $L_1$. There are a number of ways to generate mutually orthogonal Latin squares. In this example, the other orthogonal square can be obtained by simply taking the transpose of $L_1$, i.e., converting row to column or column to row in $L_1$ we get $L_2$, as shown in Table 4.11.

The orthogonality of $L_1$ and $L_2$ can be easily verified, as in Table 4.12. Using the same technique as before, we can add another character to each codeword in accordance with the entries of $L_2$. For example, the entry corresponding to row 0 and column 1 is $\beta$, which is added to the previous codeword (0 1 $\alpha$) to become (0 1 $\alpha$ $\beta$). The complete set of new codewords based on the entries of $L_2$ is shown in Table 4.13. This new set of codewords can be generated from the generator matrix

$$G' = \begin{bmatrix} 1 & 0 & 1 & 1 \\ 0 & 1 & \alpha & \beta \end{bmatrix}$$

$$= \begin{bmatrix} 1 & 0 & 1 & 1 \\ 0 & 1 & \alpha & \alpha^2 \end{bmatrix}$$

We start from $G$ of $n = 3$ columns and obtain $n + 1 = 4$ columns in $G'$ which contains the same field elements as in $G$. In fact, $G'$ is known as an extended Reed-Solomon code generator matrix of $G$ [38, 39]. Each column (as shown before) can be extended $m = 2$ columns of binary symbols.

The total eight columns in $G'$ can be partitioned into four matrices as required by the method of Karnin, Greene, and Hellman.

Since $G'$ exists for $n = 3$, without loss of any generality from the set of elements 0, $\alpha^0$, $\alpha^1$, $\alpha^2$, . . ., $\alpha^{n-1}$ of GF($q^m$), we immediately have

$$
G_{n+1} = \begin{bmatrix}
1 & 0 & 1 & 1 & & . & . & . & & 1 \\
0 & 0 & \alpha & \alpha^2 & & . & . & . & & \alpha^{n-1} \\
0 & 0 & \alpha^2 & \alpha^4 & & . & . & . & & \alpha^{2(n-1)} \\
\vdots & \vdots & \vdots & \vdots & & & & & & \vdots \\
0 & 1 & \alpha^{k-1} & \alpha^{2(k-1)} & & . & . & . & & \alpha^{(k-1)(n-1)}
\end{bmatrix} \tag{4.91}
$$

which is the foundation of Karnin, Greene, and Hellman's secret sharing system. The existence of $G_{n+1}$ from $G_n$ in general will be proved in Chapter 5.

In this section, we have linked all secret sharing systems known today to mutually orthogonal Latin squares. In the next section, we show how a set of such squares can be systematically constructed.

**Table 4.11**

The orthogonal Latin square $L_2$.

|   | 0 | 1 | $\alpha$ | $\beta$ |
|---|---|---|----------|---------|
| 0 | 0 | $\beta$ | 1 | $\alpha$ |
| 1 | $\alpha$ | 1 | $\beta$ | 0 |
| $\alpha$ | $\beta$ | 0 | $\alpha$ | 1 |
| $\beta$ | 1 | $\alpha$ | 0 | $\beta$ |

**Table 4.12**

The orthogonality of $L_1$ and $L_2$.

| (0, 0) | ($\alpha$, $\beta$) | ($\beta$, 1) | (1, $\alpha$) |
|--------|------|------|------|
| ($\beta$, $\alpha$) | (1, 1) | (0, $\beta$) | ($\alpha$, 0) |
| (1, $\beta$) | ($\beta$, 0) | ($\alpha$, $\alpha$) | (0, 1) |
| ($\alpha$, 1) | (0, $\alpha$) | (1, 0) | ($\beta$, $\beta$) |

**Table 4.13**

The extended [4, 2, 3] Reed-Solomon code from the [3, 2, 2] Reed-Solomon code.

| | | | |
|---|---|---|---|
| 0 | 0 | 0 | 0 |
| 0 | 1 | $\alpha$ | $\beta$ |
| 0 | $\alpha$ | $\beta$ | 1 |
| 0 | $\beta$ | 1 | $\alpha$ |
| 1 | $\alpha$ | 0 | $\beta$ |
| $\alpha$ | $\beta$ | 0 | 1 |
| $\beta$ | 1 | 0 | $\alpha$ |
| $\alpha$ | 0 | 1 | $\beta$ |
| $\beta$ | 0 | $\alpha$ | 1 |
| 1 | 0 | $\beta$ | $\alpha$ |
| 1 | $\beta$ | $\alpha$ | 0 |
| $\alpha$ | 1 | $\beta$ | 0 |
| $\beta$ | $\alpha$ | 1 | 0 |
| 1 | 1 | 1 | 1 |
| $\alpha$ | $\alpha$ | $\alpha$ | $\alpha$ |
| $\beta$ | $\beta$ | $\beta$ | $\beta$ |

### 4.12.4    The Construction of Mutually Orthogonal Latin Squares

From Sections 4.12.1 to 4.12.3 we have demonstrated that all the secret sharing systems described in Sections 4.11.1 to 4.11.2 can be traced to orthogonal Latin squares. Thus, the problem of constructing a secret sharing system reduces to the problem of constructing orthogonal Latin squares. The subject of the construction of orthogonal Latin squares has strong ties to the mathematics of combinatorial theory [40, 41, 42]. A set of mutually orthogonal Latin squares can be constructed using finite fields, resolvable incomplete block designs, magic squares, and planar projective geometrices. Except for planar geometry, discussed in Chapter 2 of Volume II, all these topics are defined and presented in Volume I. In this section, we only present a theorem and its proof to explain how the squares can be derived from projective geometry planes, and describe an algorithm for constructing the squares when the order of the square is a power of a prime.

THEOREM 4.4: Every finite projective geometry plane of order $n$ defines a set
   of $n-1$ mutually orthogonal Latin squares.
*Proof.* As defined in Chapter 2, a finite planar projective geometry PG($t$, $n$) has $t$ dimension over GF($n$). The number of points of a line in PG($t$, $n$) is shown to be $n + 1$. Choose any line in PG($t$, $n$) and denote the $n + 1$ points on this line as $A$, $E$, $B_1$, $B_2$, . . . , $B_{n-1}$. We also know from Chapter 2 that there are $n$ lines (other than the one chosen) passing through each of the above $n + 1$ points. Let the lines passing through point $A$ be denoted as $a_1$, $a_2$, . . . , $a_n$, the lines passing through point $E$ be denoted as $e_1$, $e_2$, . . . . , $e_n$, and the lines passing through point $B_j$, $j = 1, 2, . . . , n-1$, be denoted as $b_{j1}$, $b_{j2}$, . . . , $b_{jn}$. Then every geometric point $P(i, j)$ can be expressed in terms of the $n + 1$ lines as $e_i$, $a_j$, $b_{2k_2}$, . . . , $b_{(n-1)k_{n-1}}$, where $i, j = 1, 2, . . . , n$ are the row and column designations of a square. The set of orthogonal Latin squares can be constructed by placing the value of $k_\ell$ in the entry position $(i, j)$ of the Latin square $L_\ell$. By numeration of both $i$ and $j$ we fill the $n - 1$ squares of size $n \times n$. The theorem is proved by recognizing the fact that this construction procedure prevents $k_\ell$ in position $(i, j)$ of $L_\ell$ from being in the same position in another Latin square $L_m \neq L_\ell$. Hence the theorem is proved.

Since projective planar geometry of order $n$ exists for every integer $n$ in which $n$ is a power of a prime number, for every such integer $n$ there exists a set of $n - 1$ mutually orthogonal Latin squares of order $n$. This proof is a modified version from Theorem 4.3.

### 4.12.5    Prime Powered Orthogonal Latin Squares

When the order of a set of orthogonal Latin squares is a power of a prime number, the squares can be constructed through the elements of a finite field. The result is summarized in the following algorithm.

Step 1: From a GF($P^m = n$) where $m$ is any positive integer and $p$ is a prime number, we determine the field elements as $\alpha_0 = 0$, $\alpha_1 = 1$, $\alpha_2 = \alpha$, $\alpha_3 = \alpha^2, \ldots, \alpha_{n-1} = \alpha^{n-2}$. GF($n$) can be generated by an irreducible polynomial of degree $m$ with the primitive element $\alpha$.

Step 2: Denote the entry in the $i$th row and $j$th column of the first Latin square as $L_1(i, j)$, where $i, j = 0, 1, \ldots n-1$. The entries of the 0th row and 0th column of $L_1$ are in natural order, i.e., $\alpha_0, \alpha_1, \alpha_2, \ldots, \alpha_{n-1}$, which is $L_1(i, 0) = \alpha_i$ and $L_1(0, j) = \alpha_j$. The other entries of $L_1$ are separately determined as follows.

Step 3: For $i \neq j$,

$$L_1(i, j) = \alpha_i + \alpha_j \tag{4.92}$$

For $i = j$

$$L_1(i+1, j+1) = \begin{cases} 0, & \text{if } (i, j)_1 = 0 \\ \alpha_{s+1}, & \text{if } (i, j)_1 = \alpha_s \\ \alpha_1, & \text{if } (i, j)_1 = \alpha_{n-1} \end{cases} \tag{4.93}$$

where $0 < s < n - 1$, and the construction of $L_1$ is completed.

Step 4: The other squares are obtained by cyclic permutation of the rows of $L_1$ except the 0th row.

*Example 4.14*:

Step 1: Let $m = p = 2$, $n = p^m = 4$. Let the elements of GF(4) be denoted as $\alpha_0 = 0$, $\alpha_1 = 1$, $\alpha_2 = \alpha$, and $\alpha_3 = \alpha^2$. GF(4) can be generated by the irreducible polynomial $1 + x + x^2$.

Step 2:

$$L_1 = \begin{array}{|cccc|} \hline \alpha_0 & \alpha_1 & \alpha_2 & \alpha_3 \\ \alpha_1 & & & \\ \alpha_2 & & & \\ \alpha_3 & & & \\ \hline \end{array}$$

Step 3: $L_1(1, 1) = \alpha_1 + \alpha_1 = 0$

$L_1(1, 2) = \alpha_1 + \alpha_2 = 1 + \alpha = \alpha^2$

$L_1(1, 3) = 1 + \alpha^2 = \alpha$

$L_1(2, 1) = \alpha_2 + \alpha_1 = \alpha + 1 = \alpha^2$

$L_1(2, 2) = 0$

$L_1(2, 3) = \alpha_2 + \alpha_3 = \alpha + \alpha^2 = 1$

$L_1(3, 1) = \alpha_3 + \alpha_1 = \alpha^2 + 1 = \alpha$

$L_1(3, 2) = \alpha_3 + \alpha_2 = \alpha^2 + \alpha = 1$

$L_1(3, 3) = 0$.

Combining the above entries and Step 2 we have

$$L_1 = \begin{vmatrix} \alpha_0 & \alpha_1 & \alpha_2 & \alpha_3 \\ \alpha_1 & \alpha_0 & \alpha_3 & \alpha_2 \\ \alpha_2 & \alpha_3 & \alpha_0 & \alpha_1 \\ \alpha_3 & \alpha_2 & \alpha_1 & \alpha_0 \end{vmatrix}$$

$$= \begin{vmatrix} 0 & 1 & \alpha & \alpha^2 \\ 1 & 0 & \alpha^2 & \alpha \\ \alpha & \alpha^2 & 0 & 1 \\ \alpha^2 & \alpha & 1 & 0 \end{vmatrix}$$

$$= \begin{vmatrix} 0 & 1 & 2 & 3 \\ 1 & 0 & 3 & 2 \\ 2 & 3 & 0 & 1 \\ 3 & 2 & 1 & 0 \end{vmatrix}$$

The entries of the last square are just the subscripts of the first square of $L_1$.

Step 4: Letting the 0th row of $L_1$ stay the same we do cyclic permutation of the other three rows and obtain

$$L_2 = \begin{vmatrix} 0 & 1 & 2 & 3 \\ 3 & 2 & 1 & 0 \\ 1 & 0 & 3 & 2 \\ 2 & 3 & 0 & 1 \end{vmatrix}$$

and

$$L_3 = \begin{vmatrix} 0 & 1 & 2 & 3 \\ 2 & 3 & 0 & 1 \\ 3 & 2 & 1 & 0 \\ 1 & 0 & 3 & 2 \end{vmatrix}$$

As we state in Chapter 2, interchanging columns in a Latin square yields a Latin square. In this example, if we interchange the columns of $L_2$ to

$$L_2' = \begin{vmatrix} 0 & 2 & 3 & 1 \\ 3 & 1 & 0 & 2 \\ 1 & 3 & 2 & 0 \\ 2 & 0 & 1 & 3 \end{vmatrix}$$

and interchange the columns of $L_3$ to

$$L_3' = \begin{vmatrix} 0 & 3 & 1 & 2 \\ 2 & 1 & 3 & 0 \\ 3 & 0 & 2 & 1 \\ 1 & 2 & 0 & 3 \end{vmatrix}$$

with $\alpha = 2$ and $\alpha^2 = \beta = 3$, $L_2'$ and $L_3'$ are the two mutually orthogonal Latin squares of order 4 used for demonstration earlier.

The construction algorithm presented in this subsection provides the foundation of the following theorem, which is stated here without the proof. The proof can be seen, for example, in [40, 41].

THEOREM 4.5: For $n \geq 3$ and $n$ a prime or a power of a prime, there exists a set of $(n - 1)$ orthogonal Latin squares of order $n$.

When $n$ is not a prime or power of a prime, then $n$ may be decomposed into prime powers as described in Chapter 2. The number of orthogonal Latin squares of nonprime order is no more than the minimum of the factored prime power minus one.

## 4.13   CONCLUDING REMARKS

In this chapter, we have related cryptology to digital satellite communication. We have elucidated the needs, justifications, and examples of cryptosystems that use satellites and the characteristic distinction between systems that use encryption/decryption and those that do not.

Cryptosystems render transmitted messages unintelligible to all but the intended receivers. Cryptology refers to the theory and practice of cryptography and cryptanalysis and is concerned with the degree of unrecognizability of a message. In this chapter, we have discussed most significant cryptographic algorithms and cryptosystems, but we have omitted techniques that are known to be potentially easy to break. These include algorithms based on first-order congruence classes and other algorithms. The first-order congruence class applied to secret sharing systems produces different results, however, and thus has been included.

Multiple access cryptosystems were discussed in Section 4.6. Such systems were based on the discrete exponential and logarithmic one-way function over a finite field $GF(p)$, where $p$ is a large prime. If we use $GF(2^m = p)$ for any positive integer $m$, the advantage of binary field operation for implementation of encryption and decryption units becomes a disadvantage in terms of the strength of the cryptosystem. For example, if $m = 127$, the data base required to break the multiple access cryptosystem can take two weeks for the Adleman's algorithm [43]. A modified algorithm of Blake, Fuji-Hara, Mullin, and Vanstone takes about nine hours [44], and further reduction to eleven minutes has proved to be possible by Coppersmith [45]. However, for general $p \neq 2^m$ the system remains secure.

We have also mentioned some problems involved in studying cryptology: the difficulty of evaluating the strength of a cryptosystem and the unavailability of some of the secrets in a cryptographic algorithm.

We have devoted the last two sections to secret sharing systems because of their importance to satellite communication. Section 4.11 surveys, for the first time, all the work done in this area; Section 4.12 presents a unified construction method with proofs and verifications, also for the first time.

## 4.14   REFERENCES

1. Fang, R. and Lu, S. "Authentication Over Low-Data-Rate Channels with Feedback." *COMSAT Technical Review* 9, 1 (Spring 1979): 1–13.
2. Lu, S. and Lee, L. "A Simple and Effective Public-Key Cryptosystem." *COMSAT Technical Review* 9, 1 (Spring 1979): 15–24.
3. Lee, L. and Lu, S. "A Multiple-Destination Cryptosystem for Broadcast Networks." *COMSAT Technical Review* 9, 1 (Spring 1979): 25–35.
4. Lu, S. and Lee, L. "Message Redundancy Reduction by Multiple Substitution: A Preprocessing Scheme for Secure Communication." *COMSAT Technical Review* 9, 1 (Spring 1979): 37–47.
5. Lu, S.; Lee, L.; and Fang, R. "An Integrated System for Secure and Reliable Communications Over Noisy Channels." *COMSAT Technical Review* 9, 1 (Spring 1979): 49–60.
6. Fang, R. "Application of Cryptographic Techniques to Commercial Digital Satellite Communications." In *New Concepts in Multi-User Communication.* Proceedings of the NATO Advanced Study Institute, Edited by J. Skwirzynski, 677–96. Alphen aan den Rijn, The Netherlands: Sijthoff & Noordhoff, 1981.
7. Ingemarsson, I. and Wong, C. "Encryption and Authentication in On-board Processing Satellite Communication Systems." *IEEE Transactions on Communications Technology* COM-29, 11 (November 1981): 1684–87.
8. Diffie, W. and Hellman, M. "New Directions in Cryptography." *IEEE Transactions on Information Theory* 22 (November 1976): 644–54.
9. Massey, J. Lecture Notes on Digital Communication, September 1981.
10. Meyer, C. and Matyas, S. *Cryptography: A New Dimension in Computer Data Security.* New York: John Wiley & Sons, 1982.
11. Shannon, C. "Communication Theory of Secrecy Systems." *Bell Systems Technical Journal* 28 (1949): 656–715.
12. Gallager, R. *Information Theory and Reliable Communication.* New York: John Wiley & Sons, 1968.
13. Hellman, M. "An Extension of the Shannon Theory Approach to Cryptography." *IEEE Transactions on Information Theory,* IT-23, 3 (May 1977): 289–94.
14. Lu, S. "Random Ciphering Bounds on a Class of Secrecy Systems and Discrete Message Sources." *IEEE Transactions on Information Theory,* IT-25, 4 (July 1979): 405–414.
15. *Data Encryption Standard,* Federal Information Processing Standard (FIPS) Publication 46, National Bureau of Standards, U.S. Department of Commerce, Washington, D.C., January 1977.
16. Feistel, H. "Cryptography and Computer Privacy." *Scientific American* 228 (May 1973): 15–23.
17. Pohlig, S. and Hellman, M. "An Improved Algorithm for Computing Logarithms over $GF(p)$ and Its Cryptographic Significance." *IEEE Transactions on Information Theory* IT-24 (January 1978): 106–111.
18. Knuth, D. *The Art of Computer Programming, Vol. 2: Seminumerical Algorithms.* Reading, Mass.: Addison-Wesley, 1969.
19. Henze, E. "The Solution of the General Equation for Public Key Distribution Systems." *IEEE Transactions on Information Theory* IT-28, 6 (November 1982).
20. Denning, D. *Cryptography and Data Security.* Reading, Mass.: Addison-Wesley, 1982.

**21.** Rivest, R., Shamir, A., and Adleman, L. "A Method for Obtaining Digital Signatures and Public-Key Cryptosystems." *Communications of the ACM* 21, 2 (1978): 120–126.

**22.** Williams, H. "A Modification of the RSA Public Key Encryption Algorithm." *IEEE Transactions on Information Theory* IT-26, 6 (November 1980): 726–29.

**23.** McEliece, R. "A Public-Key Cryptosystem Based on Algebraic Coding Theory." *JPL-DSN Progress Report 42-44*, California Institute of Technology (Jan.–Feb. 1978): 114–16.

**24.** Karnin, E.; Greene, J.; and Hellman, M. "On Secret Sharing Systems." *IEEE Transactions on Information Theory*, IT-29, 1 (January 1983): 35–41.

**25.** Goppa, V. "A New Class of Linear Error-Correcting Codes." *Problemy Peredachi Informatsiy* (September 1970).

**26.** Goppa, V. "Rational Representation of Codes and (L, g)-Codes." *Problemy Peredachi Informatsiy* (September 1971).

**27.** Berlekamp, E. "Goppa Codes." *IEEE Transactions on Information Theory* 19 (1973): 590–92.

**28.** McEliece, R. "The Theory of Information and Coding." In *Encyclopedia of Mathematics and Its Applications*, Vol. 3. Reading, Mass.: Addison-Wesley, 1977.

**29.** Retter, C. "Goppa Decoding with BCH Decoder." *IEEE Transactions on Information Theory* (January 1975).

**30.** Delsarte, P. "Generalized BCH and Goppa Codes as Subfield Subcodes of Modified Reed-Solomon Codes." *IEEE Transactions on Information Theory* (September 1975).

**31.** Tzeng, K. "Cyclic Codes Obtained When Goppa Codes Are Extended by Overall Parity Check." *IEEE Transactions on Information Theory* IT-21, 6 (November 1975): 712–16.

**32.** Shamir, A. "How To Share a Secret." *Communications of the ACM* 22 (November 1979): 612–13.

**33.** Blakely, G. "Safeguarding Cryptographic Keys." *Proceedings of the National Computer Conference* (AFIPS) 48 (1979): 313–17.

**34.** Asmuth, C. and Bloom, J. "A Modular Approach to Key Safeguarding." *IEEE Transactions on Information Theory* IT-30, 2 (March 1983): 208–210.

**35.** McEliece, R. and Sarwate, D., "On Sharing Secrets and Reed-Solomon Codes." *Communications of the ACM* 24, 9 (September 1981): 583–84.

**36.** Singleton, R. "Maximum Distance q-nary Codes." *IEEE Transactions on Information Theory* IT-10 (April 1964): 116–18.

**37.** Peterson, W. and Weldon, E. *Error-Correcting Codes*. Cambridge, Mass.: M.I.T. Press, 1972.

**38.** Berlekamp, E. *Algebraic Coding Theory*. New York: McGraw-Hill, 1968.

**39.** MacWilliams, F. and Sloane, N. *The Theory of Error-Correcting Codes*. Amsterdam: Elsevier North-Holland, 1978.

**40.** Mann, H. *Analysis and Design of Experiments*. New York: Dover, 1949.

**41.** Liu, C. *Introduction to Combinatorial Mathematics*. New York: McGraw-Hill, 1968.

**42.** Denes, J. and Keedwell, A. *Latin Squares and Their Applications*. English Universities Press, 1974 (Distributed by Academic Press, New York).

**43.** Adleman, L., "A Sub-exponential Algorithm for the Discrete Logarithm Problem with Applications to Cryptography." *Proceedings of the IEEE 20th Annual Symposium on Foundations of Computer Science*, 55–60 (1979).

**44.** Blake, I.; Fuji-Hara, R.; Mullin, R.; and Vanstone, S. "Computing Logarithms in Finite Field of Characteristic Two." *SIAM Journal on Algebraic and Discrete Methods*, in press.

**45.** Coppersmith, D. "Fast Evaluation of Logarithms in Fields of Characteristic Two." *IEEE Transactions on Information Theory*, in press.

### 4.15  PROBLEMS

1. Single channel per carrier (SCPC) is the simplest form of international commercial digital satellite communication system. Show in a block diagram where and how message security can be provided by using the existing Data Encryption Standard.

2. In a nonlinear substitution (or selection) transformation a block of $n$ bits input is first represented as one of $2^n$ different characters. A selection is then made of one of the other characters. The selected character is then made to convert back to an $n$-bit block output. How many such selections are possible?

3. In designing a satellite switched TDMA system with onboard processing capability, show the most compatible ciphering algorithms for controlling information and the message itself.

4. How can the random number key generation procedure described in Section 4.4.1 for station-to-station transmission be applied to multiple stations?

5. In a conditional cryptographic key distribution system, how is the level of secrecy affected by the number of component keys? How can one achieve the maximum value of the mutual information $I(A)$ which is derived from the joint ensemble $P(A)$ as in Section 4.4.3?

6. Let $p = 4$; calculate $\log_\alpha (\alpha^x)$ by using Knuth's algorithm (Section 4.7).

7. Let $p = 15$; generate a pair of keys for the multiple access cryptosystem described in Section 4.8.

8. Let $p_1 = 3$ and $p_2 = 11$; apply the Rivest-Shamir-Adleman algorithm to encrypt and to decrypt message 2.

9. Use the Goppa code with the generator polynomial $g(x) = 2 + x + x^2$ over GF($3^2$) to design a McEliece public key system.

10. For the construction of a secret sharing system among $n$ stations show that a Latin square of order $n$ can always be constructed for any positive integer $n$.

11. Show how to construct a 3 out of 5 secret sharing system using the unified method of Section 4.11.

12. From the irreducible polynomial $1 + x + x^3$ over GF(2), construct the corresponding practical secret system by first generating the seven mutually orthogonal $8 \times 8$ Latin squares.

# Chapter 5

# CHANNEL CODING: PERFORMANCE AND VARIATIONS

## 5.1 INTRODUCTION

One of the important advantages of digital communication over analog transmission is the ease with which error control can be implemented. A significant improvement in the quality of a transmission channel is feasible if codes are properly chosen and decoding techniques are efficiently applied. Except for limited variations in modulation index, such equivalent means for channel improvement are not available in analog transmission.

Although error controls in digital satellite communication share common problems in general with all digital communication, with or without satellites, satellite systems present additional constraints on error coding such as path delay, interference, nonlinearity, and multidestinational characteristics. Depending on the specific application, selection of a code or a decoding algorithm can be a complicated task, because the choice is affected by overall system design parameters. To determine whether an error control scheme is justified for a particular application, the following factors need to be considered:

- Transmission channel characteristics
- Decoding performance
- Nature of message format
- Speed of the codec operation
- Coding efficiency
- Availability of a feedback channel
- Delay throughput
- The location of the codec in the transmission link and the code transparent capability
- Multidestination decoding capability
- Complexity and cost.

These factors are not necessarily independent. From the standpoint of system design, the most important factor is decoding performance. After all, this is the main purpose for using error control for digital systems in the first place. Therefore, this chapter will emphasize the relative performance of different error coding schemes. Other factors are briefly described in this section only.

Before selecting any encoding scheme or a specific code, it is highly desirable to know the characteristics of the transmission channel. A code that is closely matched with the transmission channel can provide the most coding gain and transmission efficiency with the least codec complexity. The channel characteristics for digital satellite communication are investigated in Volume I, Chapter 2. For coding purposes the channel of interest includes modem, TWTA, and link filters.

The format of the message can affect the choice of the type of codec (encoder-decoder pair). If the messages have block structures, then block codes may be preferred. Otherwise tree codes can be considered. In many practical situations messages do not have fixed block sizes, nor are they indefinitely long in comparison to the code constraint length. In such cases the choice of a code, or a set of codes, becomes less obvious.

The message speed at the decoder is an important factor. Very high speed data (over 100 Mbps for the present state of the art) will rule out many decoding schemes which otherwise would be attractive in terms of performance.

Encoding rates and processing delays affect system efficiency. One must often choose between performance and system efficiency, because in general low rate codes and longer decoding delays give better performances than codes of the same complexity. Without a feedback channel, error detection and retransmission will be impossible.

Multidestination decoding capability refers to the ability of a single decoder to decode messages from encoders corresponding to multiple earth stations. Different multiple access transmission systems impose different problems for encoding and decoding. For example, in a burst mode TDMA system, neither block codes nor tree codes are ideal because all bursts are transmitted with variable length to accommodate traffic variation, and because all bursts begin and end abruptly without any tolerance for sequence memory. With performance, speed, efficiency, complexity, location, nature of channel errors, and other factors all taken into consideration, no single code, or class of codes, is optimal when it comes to application. This includes satellite communication system designs. Therefore, after the discussion of the justification, limitation, and performance interpretation of coding in Section 5.2, a classification of useful codes and error coding techniques is addressed in Section 5.3.

Section 5.4 concerns error detection and retransmission problems. Besides the discussion on the types and tradeoffs of the ARQ schemes, analyses and derivations of throughput efficiencies and performance measures are presented. Examples of actual ARQ codes used in the cited satellite systems are described. Section 5.5 contains the error rate calculation, performance evaluation, and a brief description of the most useful classes of block codes. Section 5.6 discusses convolutional codes with threshold, Viterbi, and sequential decoding algorithms. Error rate expressions and representative code performances are included for each of the decoding schemes. The multiple stack algorithm for sequential decoding appears to be most promising for future high-speed digital satellite communication. In Section 5.7, the limit of soft decision is derived. Performance improvements are shown for channel measurement, maximum radius, Viterbi, weighted erasure, and *a posteriori* probability (APP) decoding techniques. Section 5.8 gives the reason for code concatenation in multiple message transmission. Again, performance improvements of sample code concatenation are indicated. Section 5.9 concludes and summarizes the purpose of Chapter 5.

## 5.2   JUSTIFICATION OF ERROR CONTROLS IN SATELLITE COMMUNICATION LINKS

In Chapter 2 of Volume I, the elements of available signal-to-noise ratio ($(C/N)_{\text{AVA}}$ and required signal-to-noise ratio $(C/N)_{\text{REQ}}$) were discussed. For economic design $(C/N)_{\text{AVA}}$ is fixed or optimized to a specific value for a set of given satellite system parameters. But $(C/N)_{\text{REQ}}$ must be less than $(C/N)_{\text{AVA}}$. If the signaling rate, system bandwidth, and the combined losses cannot be reduced further, the one way to reduce $(C/N)_{\text{REQ}}$ is to reduce the information bit-energy-to-noise-density ratio ($E_b/N_0$) through error correction or detection schemes or encoding schemes and maintaining a constant error rate. This is the justification of error coding from the viewpoint of satellite system design. After a discussion of coding gains and losses with respect to error rates, the question of the limiting value of coding gain will be discussed.

### 5.2.1   Coding Gains and Losses

An error-correcting code with parity check digits may be used in order to improve the link quality; i.e., by decreasing the link error rate for the same required $E_b/N_0$, or by maintaining the same error rate with reduced $E_b/N_0$ in the power-limited situation. Without referring to a specific code, a general illustration of coding gain with respect to $E_b/N_0$ is presented in Figure 5.1. Where

$$snr(o) = \left(\frac{E_b}{N_0}\right)_2$$

$$snr(c) = \left(\frac{E_b}{N_0}\right)_c$$

$$snr(i) = \left(\frac{E_b}{N_0}\right)_i$$

$$snr(L) = \left(\frac{E_b}{N_0}\right)_L.$$

For a given transmission channel bit error rate (BER) $P_i$ and a decoded bit error rate $P_o$ the required signal-to-noise ratios (SNR) are $snr(i)$ for the uncoded case and $snr(c)$ for the coded case. In terms of transmission bit energy $E_b$, the

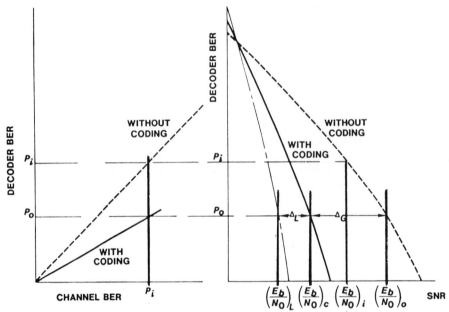

**Figure 5.1** Performance of parity check codes.

required SNR reduction or coding gain from the uncoded condition is $\Delta_G = snr(o) - snr(c)$, where $snr(o)$ is the required $E_b/N_0$ to achieve $P_o$ if coding is not used. In terms of carrier frequency energy $C$ and noise $N = N_0 W_b$, we have

$$\left(\frac{C}{N}\right) \text{ dB} = \left(\frac{E_b}{N_0}\right) \text{ dB} + 10 \log\left(\frac{R_b}{W_b}\right) \tag{5.1}$$

where $R_b$ and $W_b$ are the signaling bit rate and the transmission bandwidth. To simplify our discussion here let $R_b = W_b$; we have from (5.1), $(C/N) = (E_b/N_0)$. Although the coding gain $\Delta_G$ in $E_b/N_0$ is the difference between $snr(o)$ and $snr(c)$, if we let $snr(o) = (E_b/N_0)_o$ and $snr(c) = (E_b/N_0)_c$, we have $(C/N)_o = (E_b/N_0)_o$ but we do not have $(C/N)_c = (E_b/N_0)_c$. The fact is $(C/N)$ for the coded case changed due to $N$ changed because the transmission rate changed, since the transmission rate depends on the encoding rate $R$. Thus, even if the value of $C$ remains the same, the carrier-to-noise ratio after decoding is no longer the same. For constant duration encoding, that is, encoding accomplished by speeding, the transmission speed is increased by $1/R$. Because an increase in speed can be obtained only through bandwidth expansion, as a consequence the noise $N$ is also increased proportionally. Expressing the SNR loss due to encoding rate $R$ as $\Delta_L = 10 \log_{10}(1/R)$ in dB we have the carrier-to-noise ratio as

$$\left(\frac{C}{N}\right)_c = snr(c) - \Delta_L = snr(N) \tag{5.2}$$

expressed in dB. To minimize the required $(C/N)_c$, $\Delta_L$ needs to be large, which implies the low code rate $R$. Hence the objectives of low signaling power and high efficiency transmission in satellite communication are working against each other. This is an expected result as information theory indicates.

## 5.2.2  Performance Bounds

If the space segment of the communication satellite link can be modeled as an additive white Gaussian noise (AWGN) channel, then the capacity of this channel is given by Shannon's well-known formula:

$$C_T = W_B \log_2 \left( 1 + \frac{C}{N} \right), \tag{5.3}$$

where $C_T$ is in bits per second and $W_B$ is the system bandwidth. Since

$$\log_x A = \frac{\log_y A}{\log_y X}$$

$$C_T = W_B (\log_2 e) \, \ell n \left( 1 + \frac{C}{N_0 W_B} \right) \tag{5.4}$$

for $N = N_0 W_B$, and $N_0$ is the noise density. But

$$\ell n \, Z \begin{cases} = Z - 1 & \text{for } Z = 1 \\ < Z - 1 & \text{for } Z > 0, Z \neq 1. \end{cases} \tag{5.5}$$

Substituting the above inequality (5.5) into (5.4), we have:

$$C_T \leq \frac{C}{N_0} (\log_2 e) = \frac{E_b}{N_0} R_b (\log_2 e) \tag{5.6}$$

where $R_b$ is the bit rate of transmission. As this rate approaches its limit, $R_b = C_T$ and (5.6) becomes

$$\frac{E_b}{N_0} > \frac{1}{\log_2 e} = -1.6 \text{ dB.} \tag{5.7}$$

That is, in terms of signal-to-noise ratio $E_b/N_0$ for binary signaling in an additive Gaussian channel, the ultimate limit of the required signal-to-noise ratio is $-1.6$ dB for any coding scheme, of any encoding rate, being transmitted at any speed.

In terms of information bit energy $E_b$ per noise density $N_0$, (5.3) can be written as

$$C_T = W_B \log_2 \left[ 1 + \left( \frac{E_b}{N_0} \right) \left( \frac{R_b}{W_B} \right) \right]. \tag{5.8}$$

As we have said, the system performance bound is obtained when $C_T$ approaches $R_b$ as a limit. Then (5.8) becomes

$$\frac{R_b}{W_B} = \log_2 \left[ 1 + \left( \frac{E_b}{N_0} \right) \left( \frac{R_b}{W_B} \right) \right] \tag{5.9}$$

or

$$2 \exp\left(\frac{R_b}{W_B}\right) = 1 + \left(\frac{E_b}{N_0}\right)\left(\frac{R_b}{W_B}\right) \tag{5.10}$$

gives

$$\frac{E_b}{N_0} = \frac{2 \exp\left(\dfrac{R_b}{W_B}\right) - 1}{\dfrac{R_b}{W_B}}. \tag{5.11}$$

Since $R_b = R \cdot R_e$, we have in terms of $R_e$, the bit rate after coding,

$$\frac{E_b}{N_0} = \frac{2^{R\,R_e/W_B} - 1}{\dfrac{R\,R_e}{W_B}}. \tag{5.12}$$

(5.12) is evaluated in terms of $R_e/W_B$ versus $E_b/N_0$ with coding rates as parameters. With both coordinates expressed in dB, the curves in Figure 5.2 clearly show the limit of $-1.6$ dB for $E_b/N_0$.

As shown in Figure 5.2, the region below $E_b/N_0 = 3.0$ dB may be considered as a power-limited region, while the region $E_b/N_0 > 6.0$ dB may be considered

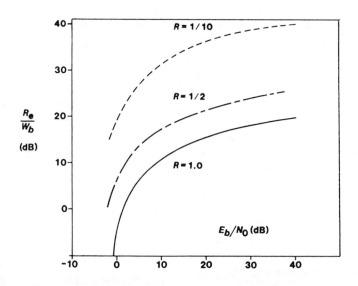

**Figure 5.2** Capacity bound with error coding rates.

as a bandwidth-limited region. The bandwidth and power variation are clearly affected by the coding rate, as shown in the figure. $R_b/W_B$ is normally expressed in bits per cycle.

Most early communication satellite systems are power-limited. As improvements are made on launch vehicles and satellites become larger, the tendency has shifted from power-limited to bandwidth limited. Meanwhile, it is commonly recognized that error coding is useful only in a power-limited transmission environment. One of the intentions of this discussion is to point out that error coding can also be useful in bandwidth-limited digital satellite communication systems by employing high-rate low-redundancy codes.

## 5.3   A CLASSIFICATION OF ERROR CONTROL TECHNIQUES

One of the primary reasons that digital communication is preferred over analog transmission is the feasibility of channel coding, which is also referred to as error control. Error control can be classified into two basic categories: (1) error detection and retransmission, which presupposes the availability of an automatic repeat request (ARQ) scheme, and (2) forward error correction (FEC).

A classification of error control techniques is shown in Figure 5.3. Except for the cases of repeated word transmission or simple even-odd nonzero symbol checking, almost all error control techniques use error-correcting codes.

A general discussion is provided in this section based on the classification presented in Figure 5.3. Detailed discussions will be provided in Sections 5.4 through 5.8 of this chapter. Throughout this chapter, all tree codes are convolutional codes. For hard decisions, all block codes can be decoded algebraically. Tree codes have less mathematical structure than block codes, and tree codes can be decoded either algebraically or probabilistically. Both block and tree codes can be designed to handle burst errors and to employ soft decisions.

The choice of a type of code for a particular application depends on system parameters such as coding gain, code efficiency, nature of the message formation, transmission speed, implementation complexity, and cost. Both types of codes are powerful and useful for digital satellite communications. The most useful decoding techniques for block codes are Bose Chaudhuri-Hocquenghem (BCH), Reed-Solomon (RS), majority logic, Fire codes or interleaving for burst-error correction, and compound channel codes for both random-error and burst-error correction. The principal decoding techniques for tree codes are threshold, sequential, and maximum likelihood (Viterbi). The latter two are probabilistic. Burst or burst/random errors for tree codes can be handled either by burst trapping or interleaving.

Whether the error control is ARQ or FEC, block or tree code encoding is required. Encoding for block codes is a division process, while encoding for tree codes is a multiplication process. The block encoders can be implemented with feedback shift registers, while tree code encoders can be implemented with feedforward shift registers. For an $(n, k, d_m)$ block code the encoding process is a mapping from $k$ information digits to an $n$-digit codeword with the detection or correction capability of the code measured by its minimum distance $d_m$.

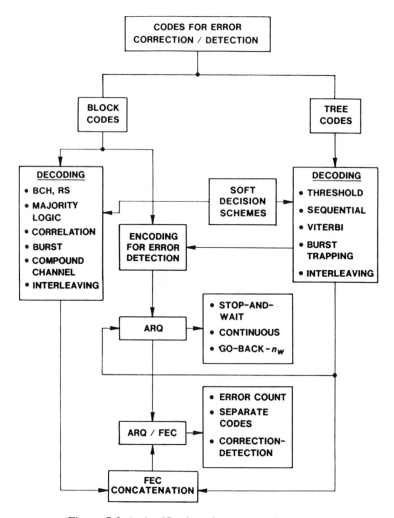

**Figure 5.3** A classification of error control techniques.

The encoding of a tree code also contains an inherent block structure. For an $(n_o, k_o, d)$ tree code within each bit time a block of $k_o$ digits is encoded and $n_o$ digits (differing by a constraint length delay) are available at the encoder output. In general, $n_o < n$ and $k_o < k$. For tree codes, $d$ is either the minimum feedback distance or the free distance. For a tree code with an $m_s$-stage shift register, the $n_o$ encoded symbols of the present block depend not only on the $k_0$ digits of the present input block, but also on the past $m_s$ blocks, or $m_s k_o$ digits.

Unfortunately, because of differences in decoding schemes, there are two definitions of constraint length for tree codes. For Massey's syndrome decoding scheme, the constraint length of a rate $k_o/n_o$ tree code is defined as

$$n_A = (m_s + 1)k_0n_0. \tag{5.13}$$

For Viterbi's decoding scheme, the constraint length of a tree code of the same rate is defined at the encoder input as

$$K = (m_s + 1)k_o. \tag{5.14}$$

As both definitions are used in the literature, care must be exercised to follow the correct one.

   With syndrome decoding, ARQ can be obtained from either block codes or tree codes for simple detection of channel errors. ARQ does not require any block decoding procedure as for error correction, but it does require a tree code decoding procedure for probabilistic decoding. The types of ARQ, their tradeoffs, hybrid ARQ-FEC, and their efficiency improvements are investigated in detail in Section 5.4

   Forward error correction may be used to reduce the power requirement at the expense of increased signal bandwidth. Hence, when adequate bandwidth allocation is available to accommodate the redundancy introduced by the code, the use of FEC coding can trade off the capacity of a power-limited digital communication system. As shown in Section 5.2.1, coding performance is given in terms of average probability $P_o$ of bit error versus $E_b/N_0$ for an ideal additive white Gaussian noise channel. It is important, however, to evaluate the coding performance over a more realistic channel that includes other effects, such as nonlinearity, interference, and imperfect synchronization.

   As shown in Figure 5.3, both block and tree codes can be employed in FEC. The performance of these types of code will be discussed in Section 5.5 and 5.6, respectively. For hard decisions there are basically two types of FEC decodings, algebraic and probabilistic. Useful algebraic decoding schemes basically consist of majority-logic and Reed-Solomon decoding algorithms of which the BCH decoding algorithm is a subclass. Majority-logic decoding schemes rely on the existence of a set of independent estimates. Majority-logic decoding can be either a one-step or multiple-step operation. For tree codes, only one-step majority-logic (or threshold) decoding is known at present. For algebraic decoding of the Reed-Solomon or nonbinary BCH codes, error magnitudes must be determined and error locations must be identified. The decoding algorithm essentially solves a set of simultaneous equations based on the syndrome of parity checks from a received codeword.

   In probabilistic decoding, a metric is computed for each contending bit sequence, and the sequence with the winning metric determines the bit decisions. The metric for any bit sequence is related to the conditional probability of that sequence after the received code sequence has been observed.

   In sequential decoding, the idea is to trace the correct path of the code tree without going down too many wrong branches and having to back up. Since it attempts to trace only the one correct sequence of many possible code sequences, the decoder complexity of sequential decoding does not increase rapidly with constraint length. Hence, moderately long constraint lengths may be employed for convolutional codes. The result is a very large coding gain at the expense of speed. Sequential decoding is thus most useful for applications that do not require real-time decoding, such as deep-space telemetry. Because of its trial-and-error method, sequential decoding requires considerable buffering and storage.

Viterbi decoding uses parallel processing to retain all possible code sequences that could be the correct path. The required processing grows exponentially with constraint length. Viterbi decoding is most suitable for soft detection of the code symbols; hence, large coding gains can be achieved even with small constraint lengths. Sequential decoding and Viterbi decoding are both maximum-likelihood methods of decoding that employ cumulative likelihood metrics. From a practical standpoint, these are essentially recursive algorithms for implementing correlation decoding.

The analog values of the received voltages for the code symbols contain channel reliability information that is lost when hard decision is performed. Consequently, some of the potential coding gain cannot be achieved with decoding techniques that are restricted to hard-decision inputs. Soft decision allows most of the channel information of the analog voltages to be retained. Therefore, FEC decoding techniques that can utilize the channel information provided by soft decision will have increased coding gains when soft decision is applied. Soft decision techniques can be applied to both block and tree decoders, either algebraic or probabilistic. The improvements of coding gain from soft decision techniques will be illustrated in Section 5.7.

For occasional burst errors in a satellite channel, either a burst-error FEC or an interleaving random-error FEC can be used. Both block and tree codes are available for burst-error FEC. For example, Fire codes and Reed-Solomon codes are burst-error-correcting block codes; Iwadare-Massey codes and Gallager's diffused codes are burst-error-correcting tree codes. Interleaving techniques may be used to disperse bursty channel errors.

A real channel usually exhibits both random and burst errors. In such a case, a limited number of block compound channel codes and burst-trapping tree codes are available.

When two or more FEC codes are connected in cascade, the combined encoding scheme is referred to as concatenation, which produces powerful results. In most cases, concatenations are limited to two codes, called inner and outer codes. The details of code concatenation are included in Section 5.8.

When an unspecified transmission channel produces random errors, error control performance is often expressed in terms of the decoder output error rate as a function of the channel error rate. When the satellite links can be modeled as an AWGN channel, the appropriate performance measure is the $E_b/N_0$ ratio required to maintain the decoder output bit error rate. Thus almost all code performances included in this chapter are so expressed.

Appendix 4 lists 1,100 useful codes in 20 categories and shows how they can be stored in a computer data bank for selection.

## 5.4   ERROR DETECTION AND RETRANSMISSION WITH CORRECTION

One of the most primitive, but powerful, error-coding techniques for improving message reliability in transmission is error detection and retransmission, which has been referred to as automatic repeat request (ARQ). Compared with coding through repetition or odd-even parity checking, ARQ schemes are very simple to implement. The general block diagram of a simple ARQ arrangement, shown in Figure 5.4, consists essentially of an encoder, a storage buffer, and an error detector.

The encoding processes for both block and convolutional codes are well known and well documented [1–5]. In general the encoder consists of a multistage shift register with modulo adders. In addition to being a replica of the encoder, the error detector can be as simple as a modulo-2 adder for a code syndrome formation. Acknowledgments are provided at the output of the detector. Positive acknowledgment (ACK) means no error was detected. Negative acknowledgment (NCK) refers to the contrary. A buffer is required at the transmitter to store every transmitted word until an ACK or NCK signal is received from the feedback channel.

All ARQ employing parity check codes can utilize the single modulo-2 adder detection scheme shown in Figure 5.5, where the encoder or the replica of the encoder can be implemented in a $k$ or an $n-k$ configuration [1]. Coding for ARQ depends on the length and rate of the code and specific information storage problem. As shown in Figure 5.5, the detector is simply a modulo-2 adder, which adds modulo 2, the received parity check sequence to the locally generated parity check sequence.

Thus for error detection only the encoding process is utilized. Encoding-decoding processes will be presented in Chapter 6. For the present purpose it is sufficient to state that encoding is determined by the corresponding code generator polynomial $g(x)$.

Section 5.4.1 discusses the commonly available ARQ schemes and codes. Section 5.4.2 outlines the advantages and disadvantages of ARQ techniques. Section 5.4.3 describes the three types of hybrid ARQ-FEC combinations. Sec-

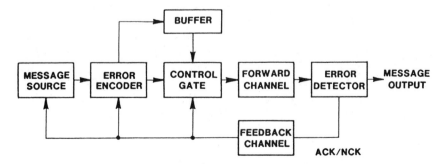

**Figure 5.4** Block diagram of simple error-detection arrangement.

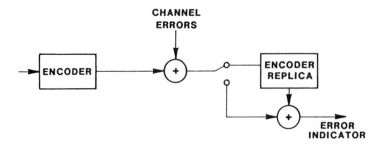

**Figure 5.5** Error detections in ARQ.

tion 5.4.4 establishes simple but general efficiency measures for all the ARQ and hybrid ARQ-FEC techniques discussed. The efficiency improvement of the hybrid scheme over the ARQ scheme alone can be readily seen through the expression derived. As expected, the efficiency improvement depends on the power of FEC employed. Asymptotic verification is also discussed. Section 5.4.5 provides samples of practical efficiency evaluations. Section 5.4.6 illustrates examples of ARQs in either operational or experimental satellite systems.

### 5.4.1 Types of ARQ

From an operational point of view, ARQ has been classically classified as stop-and-wait, continuous, go-back-$n_w$, and selective repeat schemes [6, 7]. Stop-and-wait ARQ indicates that each transmitted word requires an acknowledgment before the next word can be sent. The stop-and-wait scheme makes sure that each word is reliably received. The advantages of such a scheme are that the same forward message channel may be used for acknowledgment during feedback and a one-word buffer is sufficient. The disadvantage is that it is inefficient. The continuous ARQ provides better efficiency than the stop-and-wait scheme; but a feedback channel becomes essential. As the name implies, the transmitter of a continuous ARQ system continuously transmits coded words. When a particular word is acknowledged negatively, that word and all subsequently transmitted code words are retransmitted. In such an arrangement, sufficient buffer must be provided to store all the transmitted messages for which acknowledgments have not been received.

Go-back-$n_w$ ARQ systems transmit codewords without waiting for a positive acknowledgment. When the transmitter receives a negative acknowledgment, it retransmits all the words including the one that contained errors. The selective repeat ARQ is just a simplified version of the go-back-$n_w$ scheme, in which retransmissions are limited only to those words that contained errors; $n_w$ can be determined by the channel and detecting error rates, signaling speed, length of the word, and the code rate.

The most common ARQ systems, referred to by the practitioners as cyclic redundancy checks (CRC), simply use cyclic block codes with parity checking for error detection. The commonly used ones are CRC-12, CRC-16, and CRC-

CCITT. The CRC-12 uses 6-bit characters and two characters per block. The code generator polynomial is

$$g(\text{CRC-12}) = 1 + x + x^2 + x^3 + x^{11} + x^{12}$$
$$= (1+x)(1+x^2+x^{11}).$$

The CRC-16 uses 8-bit characters, also with two characters per codeword.

$$g(\text{CRC-16}) = 1 + x^2 + x^{15} + x^{16}$$
$$= (1+x)(1+x+x^{15}),$$

which can detect a burst of errors up to 16 bits long. CRC-CCITT is just another code of the same code length as CRC-16 with code polynomial

$$g(\text{CRC-CCITT}) = 1 + x^5 + x^{12} + x^{16}.$$

Although most error-detection systems employ block codes, the use of ARQ is not restricted to block codes. In the early 1960s the Lincoln Laboratory at the Massachusetts Institute of Technology implemented several hardware sequential decoders with convolutional codes for ARQ. The decoder was called SECO and actually was used in a hybrid error control system. The sequential decoder was designed to detect hard-to-correct error patterns. When error correction was in doubt, the sequential decoder was switched off and it requested retransmission at a lower code rate. The code rate varied from 1/6, 1/5, . . ., 1/2, 3/5, to 3/4. The decoder operated at a maximum signaling rate of 50 kbps. The performance of ARQ by sequential decoding and Viterbi decoding methods are compared by Drukarev and Costello [8]. Thus any error-correction code can be used for error detection.

## 5.4.2    ARQ Tradeoffs

This section provides a brief guideline to the consideration of ARQ schemes in digital satellite communication systems. The decision to incorporate ARQ, and, if so, which type and which code, is based on the particular system configuration, the nature of the messages, and other system constraints. Almost all the factors need to be satisfactory in order to suggest the use of ARQ. However, a single factor may revoke its use. Detailed tradeoffs have to be considered separately.

### 5.4.2.1    Some Advantages of Using ARQ

- ARQ techniques are useful when the characteristics of all or part of the transmission channel are unpredictable and when the transmission requires great accuracy. ARQ is less sensitive to channel variations.
- In the case of block codes, the same code can provide better error-detecting capability than error-correcting capability.
- ARQ is extremely simple to implement.
- Since error detection is calculated from $n-k$ parity check digits, every $(n, k)$ cyclic code detects any burst errors of length $n-k$ or less.

- Although most ARQ schemes assume noiseless feedback channels, acknowledgments can also be coded to guard against errors occurring in the feedback channel.

### 5.4.2.2  Some Disadvantages of Using ARQ

- Uninterruptible messages such as real-time digital television or telephone transmissions cannot use ARQ.
- Transmissions that require storage, which can be a source of error, cannot use ARQ.
- ARQ cannot be used where a feedback link is neither available nor feasible.
- ARQ leads to inefficient use of the channel if messages are often repeated. Throughput is reduced.
- Transmission delay such as in communication satellite channels makes certain ARQ systems ineffective.

For synchronous orbit satellite communication, the round-trip propagation delay is approximately 600 ms, which is the time required for a codeword to be transmitted and the ACK/NCK signal to be received at the transmitter. During this duration no other codeword may be transmitted.

- ARQ cannot be used when each transmission requires certain identifications and protocols of message blocks and when message blocks must be counted continuously.

Selective repeat ARQ is useful for circumventing the unavoidable time delay of the satellite link. This delay makes it almost impossible for other ARQ schemes to be used effectively for satellite communication.

Selective ARQ uses a shorter code than other ARQ schemes. For the same code rate, the code with shorter block length is less powerful and less detectable than longer codes. From the standpoint of implementation, selective ARQ requires additional memory and selective signaling logic.

### 5.4.3  Hybrid ARQ-FEC

By introducing forward error correction (FEC) in combination with ARQ, the throughput of the system can be improved. FEC corrects some of the errors, thus reducing the number of retransmissions.

In most real channels both random and burst errors may occur. Thus compound channel FEC codes are most suitable. The uncorrectable errors after FEC decoding are then treated with ARQ. In order to achieve high throughput in a hybrid system, we should design the system such that FEC operates most of the time.

Basically there are three types of hybrid FEC-ARQ schemes. The first one is actually an FEC with the total number of occurrences of errors being counted. When the number of total errors is close to or exceeds the capability of the decoder, a request for retransmission will be initiated. It is safe to set the error threshold number for retransmission within the minimum or free distance of the code, because the behavior of most decoders is not predictable once the number of channel errors exceeds the distance criteria of the code. For a longer code it is not feasible to check the decoder behavior for all possible large channel error patterns.

The second type of ARQ-FEC scheme uses two separate codes. In the normal operation, only an FEC codec is used. When the channel becomes very noisy, the FEC codec is turned off and an ARQ code is turned on. The process is reversed when the channel is back to normal.

The third type of hybrid scheme uses a single error-correcting block code. It is well known from coding theory that a minimum distance $d_m$ cyclic code can detect up to $d_t = d_m - 1$ errors within a code block. However, the code can also be used to correct $t$ errors and to detect $d$ errors simultaneously. For a $(n, k, d_m)$ block code, the relation of $t$ and $d_t$ is related to the error-correcting capability of the code $E$ by (5.15)

$$d_m > 2E + 1 = E + E + 1 = (E - \alpha) + (E + \alpha) + 1 = t + d_t + 1 \quad (5.15)$$

with $t = E - \alpha$, $d_t = E + \alpha$, and $0 \leq \alpha \leq E$. The only restriction in (5.15) is that $d \geq t$. Thus by using codes with large distance, both error correction and error detection can be provided by the same code. However, when $\alpha > 0$ the one-to-one exchange between error correction and error detection makes the practice futile. The advantage of this approach is that both error correction and error detection can be performed simultaneously.

When timing iteration is tolerable, another ARQ-FEC technique from the same code may be applied. The idea of this type of hybrid scheme, which also utilizes the minimum distance of a single code, is to correct first up to $E$ errors in a received word. After the errors are corrected, the received word is then processed for error detection with a detection capability $d_m - 1$. This approach essentially utilizes the code capability twice, first for error detection and then for error correction. Both detection capability $(d_m - 1)$ and correction capability $[(d_m - 1)/2]$ are used separately up to the full strength of the code.

### 5.4.4 Efficiency Measures

The net information throughput of the channel is a useful performance measure for evaluating the effectiveness of an ARQ scheme or a hybrid ARQ-FEC scheme, because there may be substantial differences between the transmission rate of the channel and the overhead factors to be discussed.

#### 5.4.4.1 Throughput Efficiency (ARQ)

The performance of one of the systems employing ARQ is evaluated in terms of throughput efficiency $\eta_A$, which is defined as the ratio of the number of errorless information digits to the total number of digits transmitted, including retransmission. In general $\eta_A$ is the product of three factors.

$$\eta_A = \left(\begin{array}{c}\text{repetition}\\\text{efficiency}\end{array}\right)\left(\begin{array}{c}\text{delay}\\\text{factor}\end{array}\right)\left(\begin{array}{c}\text{coding}\\\text{efficiency}\end{array}\right) = \eta_r \cdot \eta_w \cdot \eta_c \quad (5.16)$$

where

$$\eta_r = \frac{\text{errorless words received}}{\text{total words transmitted}}$$

$$= \frac{\text{total number of words transmitted } - \text{ number of words repeated}}{\text{total number of words transmitted}}$$

$$= \frac{\dfrac{\text{total number of words}}{\text{average number of errors}} - \dfrac{\text{number of repeated words}}{\text{average number of errors}}}{\dfrac{\text{total number of words}}{\text{average number of errors}}}$$

$$= \frac{\dfrac{1}{P_w'} - R_p}{\dfrac{1}{P_w'}} = 1 - P_w'R_p, \quad (5.17)$$

with $P_w'$ being the word error rate, and $R_p$ being the number of codewords that will be repeated for each error detected.

If $\tau_w$ is the round-trip time delay through satellites, or a satellite, then the delay factor with $\tau_w$ expressed in number of codewords is

$$\eta_w = \frac{1}{\tau_w + 1}. \quad (5.18)$$

For either an $(n, k)$ block code or an $(n_0, k_0)$ convolutional code, the code rate is

$$R = \frac{k}{n} \text{ or } \frac{k_0}{n_0} = \eta_c. \quad (5.19)$$

Substituting (5.17), (5.18), and (5.19) into (5.16), we obtain in general the throughput efficiency of an ARQ system

$$\eta_A = (1 - P_w'R_p)\left(\frac{1}{\tau_w + 1}\right) R. \quad (5.20)$$

For a stop-and-wait ARQ scheme in a satellite channel $R_p = \tau_w = 1$, (5.20) reduces to

$$\eta_A(\text{S.W.}) = \left(\frac{1 - P_w'}{2}\right) R. \quad (5.21)$$

For selective repeat ARQ $\tau_w = 0$ and $R_p = 1$, (5.20) becomes

$$\eta_A(\text{S.R.}) = (1 - P_w')R. \quad (5.22)$$

Thus, with the same code and error rate, selective repeat ARQ is twice as efficient as stop-and-wait ARQ in satellite channels, as seen from (5.21) and (5.22).

### 5.4.4.2 Throughput Efficiency (ARQ-FEC)

From (5.20) we shall examine how the throughput efficiency $\eta_A$ may be increased by employing FEC.

Assume, for simplicity, that a random error satellite channel can be modeled as binary symmetric with a channel bit error rate $p$. An FEC ($n$, $k$, $d_m \geq 2E+1$) code can decrease $P_w'$ (word error rate without coding) to

$$P_w = \sum_{i=E+1}^{n} \binom{n}{i} p^i (1-p)^{n-i}. \tag{5.23}$$

For small $p \ll 1$, the first term in (5.23) is most significant and $(1-p) \sim 1$, then

$$P_w \sim \binom{n}{E+1} p^{E+1}$$

$$= \frac{n!}{[n-(E+1)]!\,(E+1)!} p^{E+1}$$

$$\sim \frac{n^{E+1} p^{E+1}}{(E+1)!}$$

$$= \frac{(n\,p)^{E+1}}{(E+1)!}. \tag{5.24}$$

For a given $p$, FEC reduces the effective error rate from the bit error rate $p$ to a word error rate $P_w < p$. Note that $P_w'$ relates to $p$ through word partitions. Since $P_w'$ is the average word error rate without coding, $P_w' = np$ for a word of length $n$. Substituting $np$ for $P_w'$ in (5.24), we have

$$P_w \sim \frac{(P_w')^{E+1}}{(E+1)!}. \tag{5.25}$$

It is easy to observe that asymptotically for any $P_w' < 1$

$$\lim P_w \to 0 \tag{5.26}$$

as the FEC correction capability $E$ gets large.

Next we examine how FEC affects $R_p$. Let

$$R_p' = \begin{bmatrix} \text{number of repeated words} \\ \text{without using FEC} \end{bmatrix}$$

$$- \begin{bmatrix} \text{number of codewords need not} \\ \text{repeat by using FEC} \end{bmatrix}. \tag{5.27}$$

From coding theory, an ($n$, $k$, $d_m$) $q$-ary linear block code has $q^k$ codewords and $q^{n-k}$ cosets. If a code is to correct $E$ or fewer errors, all $n$-tuples of weight $E$

or less must be coset leaders. Thus the number of cosets must be greater than or equal to the number of sequences of weight $E$ or less. This establishes the well-known sphere-packing bound [1].

$$q^{n-k} \geq 1 + \binom{n}{1}(q-1) + \binom{n}{2}(q-1)^2 + \ldots \binom{n}{E}(q-1)^E. \quad (5.28)$$

For binary block code $q = 2$, (5.28) becomes

$$2^{n-k} \geq \sum_{j=0}^{E} \binom{n}{j}, \quad (5.29)$$

or for at least single-error words

$$2^{n-k} - 1 \geq \sum_{j=1}^{E} \binom{n}{j}. \quad (5.30)$$

Equality holds in (5.30) for perfect codes.

Since the number of repeated codewords either with or without FEC in (5.27) based on word errors has been detected, the total number of error-detectable words in a $(n, k, d_m)$ code is

$$\binom{n}{1} + \binom{n}{2} + \ldots + \binom{n}{2E} = \sum_{i=1}^{2E} \binom{n}{i}. \quad (5.31)$$

Thus, in (5.27)

$$\begin{bmatrix} \text{the number of codewords} \\ \text{not necessary to repeat by} \\ \text{using FEC} \end{bmatrix} = \frac{\begin{bmatrix} \text{the number of words} \\ \text{correctable by FEC} \end{bmatrix}}{\begin{bmatrix} \text{the number of error words} \\ \text{detectable by the code} \end{bmatrix}}$$

$$= \frac{2^{n-k} - 1}{\sum\limits_{i=1}^{2E} \binom{n}{i}} \geq \frac{\sum\limits_{j=1}^{E} \binom{n}{j}}{\sum\limits_{i=1}^{2E} \binom{n}{i}}. \quad (5.32)$$

Substituting (5.30) and (5.32) into (5.27), we have

$$R_p' = R_p - \left[ \frac{2^{n-k} - 1}{\sum\limits_{i=1}^{2E} \binom{n}{i}} \right] \quad (5.33)$$

or

$$R_p' \leq R_p - \left[ \frac{\sum\limits_{j=1}^{E} \binom{n}{j}}{\sum\limits_{i=1}^{2E} \binom{n}{i}} \right]. \quad (5.34)$$

When FEC is introduced in ARQ, $\tau'_w \leq \tau_w$. When different codes are used in the hybrid system, $R' \neq R$. Combining $P_w$ and $R'_p$, the hybrid throughput efficiency becomes

$$\eta_{AF} = 1 - P_w R'_p \left( \frac{1}{\tau'_w + 1} \right) R', \tag{5.35}$$

where $P_w$ and $R'_p$ are the dominant factors for the improvement of efficiencies.

### 5.4.4.3   An Asymptotic Verification of ARQ-FEC Efficiency

Without FEC, the throughput efficiency is determined by $P'_w$, $R_p$, $\tau_w$ and $R$, as in (5.20). With FEC, the efficiency is determined by $P_w$, $R'_p$, $\tau'_w$ and $R'$ as in (5.35). In this subsection the asymptotic throughput efficiency improvements are investigated in terms of the ratio of $\eta_{AF}$ to $\eta_A$.

Assume for the case of a hybrid ARQ-FEC scheme that a code is used twice both for detection and correction; then $R' = R$, $\tau'_w = \tau_w$. In this case, let $\eta_w \eta_c = \eta$, then

$$\eta_A = (1 - P'_w R_p)\, \eta \tag{5.36}$$

$$\eta_{AF} = (1 - P_w R'_p)\, \eta. \tag{5.37}$$

We define the ratio as

$$r = \frac{1 - \dfrac{\eta_{AF}}{\eta}}{1 - \dfrac{\eta_A}{\eta}}. \tag{5.38}$$

Substituting (5.36) and (5.37) into (5.38) we have, with the help of (5.25), (5.27), and (5.32)

$$r = \frac{P_w R'_p}{P'_w R_p} = \frac{1}{P'_w R_p} \left\{ \frac{(P'_w)^{E+1}}{(E+1)!} \left[ R_p - \frac{2^{n-k} - 1}{\displaystyle\sum_{i=1}^{2e} \binom{n}{i}} \right] \right\}. \tag{5.39}$$

For either stop-and-wait or selective repeat ARQ schemes $R_p = 1$, (5.39) simplifies to

$$r = \frac{(P'_w)^E}{(E+1)!} \left[ 1 - \frac{2^{n-k} - 1}{\displaystyle\sum_{i=1}^{2E} \binom{n}{i}} \right]. \tag{5.40}$$

Since for perfect codes we know from (5.30) that

$$2^{n-k} - 1 = \sum_{j=1}^{E} \binom{n}{j} \ll \sum_{j=1}^{2E} \binom{n}{j}, \tag{5.41}$$

(5.40) can be further simplified as

$$r \sim \frac{(P_w')^E}{(E + 1)!} = \frac{(P_w')^{E+1}}{(E + 1)!P_w'} = \frac{P_w}{P_w'}. \tag{5.42}$$

However, from (5.38) we can obtain $\eta_{AF}$ in terms of $r$, $\eta_A$, and $\eta$ as

$$\eta_{AF} = (1 - r)\eta + r\eta_A. \tag{5.43}$$

Substituting (5.42) into (5.43), we have

$$\eta_{AF} = \left(1 - \frac{P_w}{P_w'}\right)\eta + \frac{P_w}{P_w'}\eta_A. \tag{5.44}$$

But for $R_p = 1$ ARQ

$$\eta_A = \eta_r\eta_w\eta_c = (1 - P_w'R_p)\eta = (1 - P_w')\eta. \tag{5.45}$$

Substituting (5.45) into (5.44), we obtain

$$\eta_{AF} = \left(1 - \frac{P_w}{P_w'}\right)\eta + \frac{P_w}{P_w'}(1 - P_w')\eta = (1 - P_w)\eta. \tag{5.46}$$

Asymptotically, (5.46) becomes

$$\lim_{P_w \to 0} \eta_{AF} \to \eta \tag{5.47}$$

which checks with the original expression for $\eta_{AF}$. That is, for $P_w \to 0$ the repetition part of throughput efficiency $\eta_{AF}$ approaches unity. The overall throughput efficiency approaches the product of $\eta_w$ and $\eta_c$.

*Example 5.1:* Consider the simple (7, 3, 3) Hamming code in a binary symmetric channel with its bit error rate $p = 0.1$. Without FEC, the word error rate is $P_w' = np = 7 \times 0.1 = 0.7$. Using ARQ only, $\eta_A = (1 - 0.7)(3/7) = 0.128$. With FEC $P_w = .21$. $\eta_{AF} = [1 - (0.21 \times 0.464)](0.428) = 0.387$ and it increases to more than double the original throughput efficiency with the simple application of FEC.

### 5.4.5 Performance Measures

As early as 1961, Peterson [1] addressed the issue of the probability of an undetected error $P_{ud}$ for block codes used in error detection. He defined $P_{ud}$ as the probability that the received word differs from the transmitted word. By the weights of codewords in Hamming codes he established the undetected error probability in a binary symmetric channel with transitional probability $p$ as

$$P_{ud} = \sum_{j=1}^{n} W(j)p^j (1 - p)^{n-j}$$

$$= (1 - p)^n \sum_{j=1}^{n} W(j) \left(\frac{p}{1 - p}\right)^j$$

$$= (1 - p)^n \left[f(\frac{p}{1 - p}) - 1\right], \tag{5.48}$$

where

$$f(x) = \frac{1}{n + 1} \left[ (1 + x)^n + n(1 + x)^{\frac{n-2}{2}} (1 - x)^{\frac{n+1}{2}} \right] \qquad (5.49)$$

is the weight-enumerating polynomial and $W(j)$ is the number of codewords of weight $j$. The collection of $W(j)$, i.e.,

$$W(z) = \sum_{j=0}^{n} W(j)z^j, \qquad (5.50)$$

is referred to as the code weight enumerator, which has been extensively investigated in coding theory. However, because of the practical difficulty in obtaining the weight enumerators even with the help of MacWilliams' and Pless's power-moment identities [2, 3], the following approximate approach is presented instead.

### 5.4.5.1 Undetected Error Rate

The performance of an ARQ system may be analyzed by means of classical detection theory with error coding included in the detection process. For a linear $(n, k, d_m)$ code with a parity check matrix $H$, the key to successful error detection is through the formation and recognition of the syndrome sequence

$$\overline{S} = (\overline{I} + \overline{E})H^T, \qquad (5.51)$$

where $\overline{I}$ is the encoded information sequence, $\overline{E}$ is the error sequence, and $H^T$ is the transposition of $H$. For error detection, the detector detects whether $\overline{S} = 0$ or $\overline{S} \neq 0$ to determine the absence or presence of channel errors. As shown in Figure 5.6, the detection error rate consists of false and miss (or undetected) error events. The false detected error $P_{fd}$ occurs when the error sequence $\overline{E}$ is zero but a nonzero syndrome sequence has been indicated.

$$P_{fd} = P_r(\overline{S} \neq 0, |\overline{E} = 0)$$

$$= P_r(\overline{S} \neq 0|\overline{E} = 0) \cdot P_r(\overline{E} = 0). \qquad (5.52)$$

| $\overline{S}$ \ $\overline{E}$ | $= 0$ | $> 0$ | |
|---|---|---|---|
| | | $\leq d_m - 1$ | $\geq d_m$ |
| $= 0$ | normal operation (no error) | abnormal (decoder) failure | $P_{ud}$ |
| $\neq 0$ | $P_{fd}$ | normal operation | normal operation |
| | | (error detected) | |

**Figure 5.6** Detection error events in ARQ systems.

$P_r(\overline{E} = 0)$ is inversely proportional to the noise level of a transmission channel, which may exhibit low error rate in satellite channels. Nevertheless $P_r(\overline{E} = 0)$ is small but nonzero for most practical channels. It is not possible to have a nonzero syndrome ($\overline{S} \neq 0$) in (5.52) because from (5.51) when $\overline{E} = 0$, $\overline{E}H^T$ must be zero for a nonsingular $H$. Since $\overline{E} = 0$, $\overline{I}H^T$ must be zero from the structure of the code. Thus $\overline{S}$ has to be zero for $\overline{E} = 0$. As a consequence, $P_r(\overline{S} \neq 0/\overline{E} = 0) = 0$, and the false detection error rate in an ARQ system using syndrome detection is zero. Therefore we need not consider $P_{fd}$ further.

The undetected error event occurs as no detection of errors is made ($\overline{S} = 0$) when one or more errors occurred; i.e., $\overline{E} > 0$. The joint event of $\{\overline{S} = 0, \overline{E} > 0\}$ thus makes up the probability of undetected error $P_r (\overline{S} = 0, \overline{E} > 0)$. The other two combinations of $\overline{S}$ and $\overline{E}$ are normal operations. Since $\overline{E} > 0$ consists of two parts, $0 \leq \overline{E} \leq d_m - 1$ and $\overline{E} > d_m$, we have

$$
\begin{aligned}
P_{ud} &= P_r(\overline{S} = 0, \overline{E} > 0) \\
&= P_r(\overline{S} = 0, 0 < \overline{E} \leq d_m - 1) + P_r(\overline{S} = 0, |\overline{E} \geq d_m) \\
&= P_r(\overline{I}H^T = 0, 0 < \overline{E} \leq d_m - 1) \cdot P_r(\overline{E}H^T = 0, 0 < \overline{E} \leq d_m - 1) \\
&\quad + P_r(\overline{S} = 0, \overline{E} \geq d_m) \\
&= P_r(\overline{S} = 0, \overline{E} \geq d_m) \\
&= P_r(\overline{S} = 0|\overline{E} \geq d_m) \cdot P_r(\overline{E} \geq d_m).
\end{aligned}
\tag{5.53}
$$

In arriving at (5.53), the fact that

$$
P_r(\overline{I}H^T = 0, 0 < \overline{E} \leq d_m - 1) = 0 \tag{5.54}
$$

has been utilized because this fact is guaranteed by the property of the code. The condition $0 \leq \overline{E} \leq d_m - 1$ implies $\overline{E}H^T \neq 0$, which cannot produce a syndrome $\overline{S} = 0$ for an undetected error. An examination of the last expression in (5.53) follows.

In the evaluation of $P_r(\overline{S} = 0/\overline{E} \geq d_m)$, we ask how many syndrome sequence patterns will give the all-zero sequence, conditional on the number of channel errors per code block equal to or exceeding the minimum distance of the code. As it turns out, if an error sequence is equal to any linear combination of the rows of the parity check matrix $H$, then the zero syndrome will result. The number of rows in $H$ is $n - k$. The number of linear combinations of the $n - k$ rows is

$$
\binom{n-k}{2} + \binom{n-k}{3} + \cdots + \binom{n-k}{n-k} = \sum_{i=2}^{n-k} \binom{n-k}{i}. \tag{5.55}
$$

The total number of linear combinations and the original $n - k$ rows of $H$ therefore make $i$ from 1 to $n - k$ under the summation in (5.55). By the identity of binomial coefficient, $L_c = 2^{n-k} - 1$. From the sphere-packing bound for a linear binary $(n, k, d_m)$ code,

$$2^{n-k} \geq \sum_{i=0}^{E} \binom{n}{i}. \tag{5.56}$$

For $2^n$ possible sequences in the sample space

$$P_r(\overline{S} = 0 | \overline{E} \geq d_m) \geq \frac{\left[\sum_{i=0}^{E} \binom{n}{i} - 1\right]}{2^n}. \tag{5.57}$$

Next, for given $d_m$, the probability of the event that the number of errors in $\overline{E}$ is equal to or larger than $d_m$ is obviously channel dependent. Assume again that the channel bit error rate is $p$; then

$$P_r(\overline{E} \geq d_m) = \sum_{j=d_m}^{n} \binom{n}{j} p^j(1 - p)^{n-j}. \tag{5.58}$$

Substituting (5.57) and (5.58) into (5.53), we have

$$P_{ud} \geq 2^{-n} \left[\sum_{i=0}^{E} \binom{n}{i} - 1\right] \left[\sum_{j=d_m}^{n} \binom{n}{j} p^j \left(1 - p\right)^{n-j}\right]. \tag{5.59}$$

### 5.4.6   ARQ in Digital Communication Satellite Systems

ARQ schemes have been widely used in terrestrial networks for a long time. In this section, five ARQ schemes in digital satellite communication systems are presented. Two ARQ schemes are in experimental systems and three were implemented in global operational systems.

### 5.4.6.1   *ARQ in MARISAT*

MARISAT has been operational for global maritime application since 1976. In the data transmission channel of MARISAT there exists a block coding for ARQ [9].

The calculated code performance of the ($n = 63$, $k = 36$, $d_m = 11$) BCH code for the MARISAT System is shown in Figure 5.7. The bit-error rates for the code used for both error correction and error detection are indicated in terms of signal-to-noise ratio ($E_b/N_0$ in dB) in an additive white Gaussian channel.

Laboratory measurements have been performed for several runs. Each run consists of $864 \times 10^6$ bits. At channel error rates below $10^{-2}$, no undetectable errors have been observed. At higher channel error rates, miss detection (or false detection) synchronization (unique word) errors are dominant. The design error rate of the codec is well within the synchronization error rate of $10^{-5}$.

The baseband signaling format of MARISAT shown in Fig. 5.8 consists of a unique word (UW) and its inverse ($\overline{UW}$), ship identification (SID), message type and message, TTY for network interfacing, and channel testing for bit-error evaluation (BEE). The message, message type, and SID are coded by the ($n = 63$, $k = 36$, $d_m = 11$) BCH code.

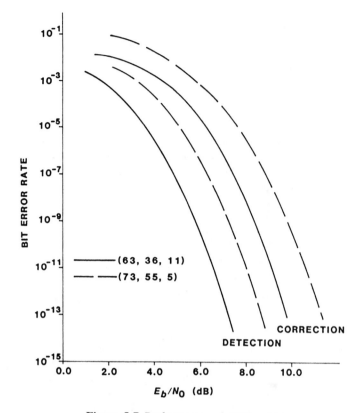

**Figure 5.7** Performances of ARQ codes.

The roots of the (63, 36, 11) BCH codes are $\alpha^1$, $\alpha^3$, $\alpha^5$, $\alpha^7$, and $\alpha^9$. The minimal polynomials are chosen from degree 6:

$$f_1(x) = 1 + x + x^6$$

$$f_2(x) = 1 + x + x^2 + x^5 + x^6$$

$$f_3(x) = 1 + x^3 + x^6$$

$$f_4(x) = 1 + x^2 + x^3, \text{ and}$$

$$f_5(x) = 1 + x + x^2 + x^4 + x^6.$$

The generator polynomial of the code is

$$g(\text{MARISAT}) = 1 + x + x^4 + x^8 + x^{15} + x^{17}$$
$$+ x^{18} + x^{19} + x^{21} + x^{22} + x^{27}.$$

Figure 5.9 shows the corresponding encoder circuit.

| UW, $\overline{\text{UW}}$ | SID TYPE MESSAGE | (63,36,11) CODE | TTY BEE |
|---|---|---|---|
| | INFORMATION BITS | PARITY CHECK BITS | |

**Figure 5.8** MARISAT coded signal format.

### 5.4.6.2 ARQ in INTELSAT SCPC

INTELSAT's single channel per carrier (SCPC) provision is an outgrowth of COMSAT's SPADE system, in which two block-coded ARQ schemes were employed. For the common signaling channel a (73, 55, 5) nonprimitive BCH code was selected primarily to accommodate the number of message digits per block needed to be coded. The ARQ encoder uses an 18-stage shift register as shown in Figure 5.10.

The generator polynomial of the SCPC ARQ code is

$$g(\text{SCPC}) = x^{18} + x^{16} + x^{12} + x^{10} + x^9 + x^6$$
$$+ x^4 + x^3 + x^2 + x + 1.$$

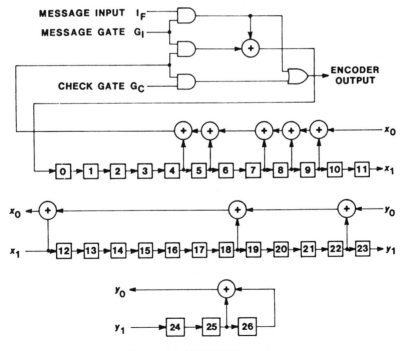

**Figure 5.9** MARISAT encoder.

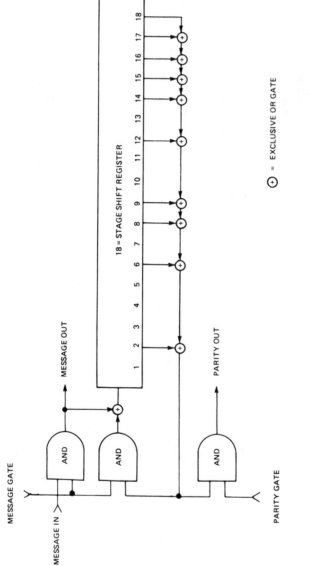

**Figure 5.10** Parity generator of $(73,55)$ BCH code having a generator polynomial $g(x)$ $= X^{18} + X^{16} + X^{12} + X^{10} + X^9 + X^6 + X^4 + X^3 + X^2 + X + 1$.

This system has been implemented by Digital Communications Corporation, Fujitsu, and Nippon Electric Corporations for INTELSAT. A 55-bit . . . 1010 pattern will give a parity check sequence of 100101100010011111.

In the data recorder portion of the same SCPC system there is a powerful (255, 163, 25) BCH code designed for ARQ for the purpose of high degree error protection such as billing. The primitive polynomial for the code was chosen as

$$m_1(x) = x^8 + x^4 + x^3 + x^2 + 1.$$

The generator polynomial was then determined as

$$g(x) = [M_1(x) M_3(x) \ldots\ldots M_{23}(x)]$$

$$= 1 + x^7 + x^8 + x^{11} + x^{12} + x^{15} + x^{17}$$

$$+ x^{20} + x^{22} + x^{24} + x^{25} + x^{26} + x^{29}$$

$$+ x^{30} + x^{31} + x^{32} + x^{33} + x^{35} + x^{36}$$

$$+ x^{39} + x^{41} + x^{42} + x^{44} + x^{46} + x^{52}$$

$$+ x^{53} + x^{54} + x^{56} + x^{57} + x^{58} + x^{59}$$

$$+ x^{66} + x^{69} + x^{71} + x^{72} + x^{74} + x^{75}$$

$$+ x^{80} + x^{87} + x^{89} + x^{90} + x^{91} + x^{92}.$$

The generator polynomial $g(x)$ is obtained as the least common multiple of the minimum polynomials $m_i(x)$, $i = 1, 3, 5, 7, \ldots 23$.

$$g(x) = \text{LCM} [M_1(x), M_2(x), \ldots M_{23}(x)],$$

$$\text{with } M_1(x) = M_2(x) = M_4(x) = M_8(x) = M_{16}(x)$$

$$M_3(x) = M_6(x) = M_{12}(x)$$

$$M_5(x) = M_{10}(x) = M_{20}(x)$$

$$M_7(x) = M_{14}(x)$$

$$M_9(x) = M_{18}(x)$$

$$M_{11}(x) = M_{22}(x)$$

$$M_{13}(x)$$

$$M_{15}(x)$$

$$M_{17}(x)$$

$$M_{19}(x)$$

$$M_{21}(x)$$

$$M_{23}(x)$$

and

$$M_3(x) = 1 + x + x^2 + x^4 + x^5 + x^6 + x^8$$

$$M_5(x) = 1 + x + x^4 + x^5 + x^6 + x^7 + x^8$$

$$M_7(x) = 1 + x^3 + x^5 + x^6 + x^8$$

$$M_9(x) = 1 + x^2 + x^3 + x^4 + x^5 + x^7 + x^8$$

$$M_{11}(x) = 1 + x + x^2 + x^5 + x^6 + x^7 + x^8$$

$$M_{13}(x) = 1 + x + x^3 + x^5 + x^8$$

$$M_{15}(x) = 1 + x + x^2 + x^4 + x^6 + x^7 + x^8$$

$$M_{17}(x) = 1 + x + x^4$$

$$M_{19}(x) = 1 + x^2 + x^5 + x^6 + x^8$$

$$M_{21}(x) = 1 + x + x^3 + x^7 + x^8$$

$$M_{23}(x) = 1 + x + x^5 + x^6 + x^8.$$

### 5.4.6.3 ARQ in Experimental TDMA Systems

COMSAT's 50 Mbps TDMA-1 system and the 120 Mbps TDMA-2 system each contains the (31, 21, 5) block code for ARQ. The block diagram for the encoder, implemented with $k$-stage configuration, is shown in Figure 5.11. A 21-bit 1010 . . . pattern will produce a 10-bit 1110001101 parity check pattern. The reason for $k$-stage encoding is to provide storage during TDMA burst transmission, because messages are transmitted intermittently from one TDMA burst to another.

Ten international manufacturing companies supply the test equipment for

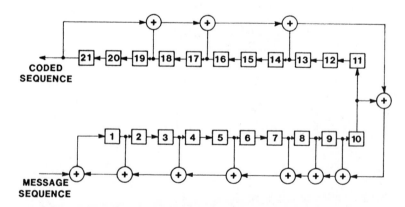

**Figure 5.11** BCH encoder for TDMA ARQ.

INTELSAT's TDMA field trials, which have been conducted since 1978 with the participation of France, the Federal Republic of Germany, Italy, the United States, and the United Kingdom. In the control signaling channel of the testing system, which is designed as the preamble of the future operational unit, there is a (39, 26, 6) cyclic code for ARQ. This code, which is not a BCH code, has a code generator polynomial $g(x) = 1 + x^4 + x^7 + x^{11} + x^{12} + x^{13}$. The calculated code performance is shown in Figure 5.12.

### 5.4.6.4    ARQ in Satellite Packet Switching Experiment

In the most recent Atlantic Packet Satellite Experiment sponsored by ARPA through INTELSAT IV-A with earth stations in Etam, West Virginia; Goonhilly Downs, England; and Tanum, Sweden, there is a 24-bit parity check code. The encoder connection is shown in Figure 5.13, where an $n - k$ configuration is used. Each packet is formatted as shown below for transmission [10].

| DLE | STX | Packet | DLE | ETX |
|-----|-----|--------|-----|-----|

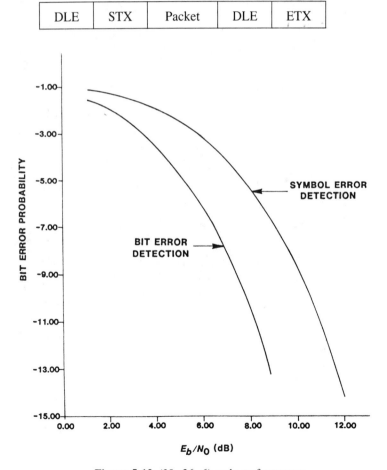

**Figure 5.12** (39, 26, 6) code performances.

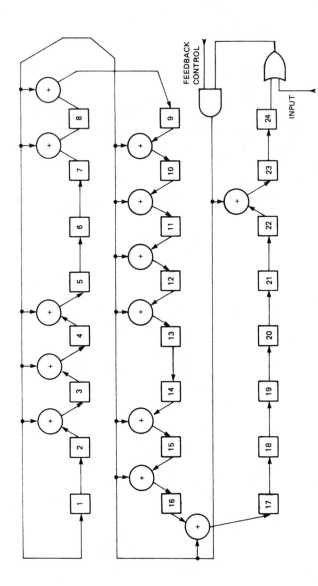

**Figure 5.13** ARQ encoder in satellite packet switching experiment.

where

$$\boxed{\text{DLE}} = \text{data link escape} = 00010000 = 0^3 \, 10^4$$

$$\boxed{\text{STX}} = \text{start of text} = 00000010 = 0^6 \, 10$$

$$\boxed{\text{ETX}} = \text{end} = 00000011 = 0^6 \, 1^2$$

For an impulse packet $0^{21} \, 10^{11}$ of 32 bits long the corresponding parity check sequence is $0^2 \, 10^3 \, 1^2 \, 01^2 \, 0^7 \, 1^2 \, 010^2$.

## 5.5 BLOCK CODES

The usefulness, justification, and performance limitations of error coding are discussed in Section 5.2. As mentioned in Section 5.3, block codes are useful both in ARQ and in FEC for error controls. The theory and implementation of block codes can be found in Chapter 6. For digital satellite communication, block coding can be very useful for control messages or information messages that have block structure such as burst transmission in TDMA systems. Since some block codes are powerful and can also be easily implemented in future satellite applications, in this section we shall discuss three aspects of block coding. From a system standpoint, we shall show in Section 5.5.1 how block coding performance can be calculated. Section 5.5.2 discusses the classes of most useful block codes. In Section 5.5.3, a sample of block coding performances is presented.

### 5.5.1 The Bit Error Rate Calculation

This section gives bit error rate expressions that are used to obtain a set of performance measures for the codes presented in Section 5.5.3. We begin by assuming a memoryless linear channel and two-level signaling in additive Gaussian noise without cochannel or intersymbol interferences. To keep the discussion simple, the effect of delay in the channel is not included. As discussed in Volume I, some of the assumptions made here are valid from a practical viewpoint.

When an $(n, k, d_m \geq 2E + 1)$ block code is used in a discrete memoryless binary symmetric channel (BSC) with channel transitional error rate $p$, the word error rate at the output of the decoder is

$$P_W = \sum_{i=E+1}^{n} \binom{n}{i} P^i (1 - p)^{n-i} = \sum_{i=E+1}^{n} p_w(i), \qquad (5.60)$$

where $p_w(i)$ represents the word error rate of exactly $i$ errors in a codeword that cannot be corrected. The probability of $i \geq E + 1$ errors occurring conditional on a word error is

$$p_i(b/w) = i/n. \tag{5.61}$$

The corresponding bit error rate when the block is used is

$$P_B = P_{(B,\ W)} = P_{(B/W)} \cdot P_W$$

$$= \sum_{i=E+1}^{n} p_i(b/w)\, p_w(i) = \sum_{i=E+1}^{n} \frac{i}{n}\, p_w(i), \tag{5.62}$$

with

$$p_w(i) = \binom{n}{i} p^i\, (1 - p)^{n-i}. \tag{5.63}$$

If the symbol signaling energy is $E_s$ with a symbol signaling rate $R_s$, for symmetric hard decision detection the BSC transition error rate in terms of the complementary error function is:

$$p = \frac{1}{2}\mathrm{erfc}\left(\sqrt{\frac{2\,E_s}{N_0}}\right) = \frac{1}{2}\mathrm{erfc}\left(\sqrt{\frac{2mRE_b}{N_0}}\right) = \frac{1}{2}\mathrm{erfc}(\alpha). \tag{5.64}$$

The transmitting signal amplitudes are assumed to be $\pm \sqrt{E_s}$ and $m$ is the number of coded digits per symbol, $R$ is the coding rate, and $E_b$ is the uncoded information bit energy. For large value of $\alpha$ the function $\mathrm{erfc}(\alpha)$ is known to be bounded [11].

$$\left(1 - \frac{1}{2\alpha^2}\right) E(\alpha) < \mathrm{erfc}(\alpha) < E(\alpha) \tag{5.65}$$

with

$$E(\alpha) = \frac{1}{\sqrt{\pi}\,\alpha}\, e^{-\alpha^2} \tag{5.66}$$

$$\alpha = \left(\frac{2mRE_b}{N_0}\right)^{\!1/2} > 0. \tag{5.67}$$

For $\alpha > 1$ and $\mathrm{erfc}(\alpha) < 0.01$, the bound of (5.65) converges to $E(\alpha)$. Otherwise

$$\mathrm{erfc}(\alpha) \simeq \int_{\alpha}^{\infty} \frac{2}{\sqrt{\pi}}\, e^{-\beta^2/2}\, d\beta. \tag{5.68}$$

Substituting (5.65) or (5.68) into (5.64) and then into (5.62), we have

$$P_B = \frac{1}{n} \sum_{i=E+1}^{n} \binom{n}{i} (i) \left(\frac{1}{2}\right)^{\!n} \mathrm{erfc}^i(\alpha)\, [2 - \mathrm{erfc}(\alpha)]^{n-i}. \tag{5.69}$$

In terms of the corresponding error function $\mathrm{erf}(\alpha)$,

$$\mathrm{erfc}(\alpha) = 1 - \mathrm{erf}(\alpha). \tag{5.70}$$

(5.67) and (5.69) can be used to calculate the performance of any $(n,\ k,\ d_m)$ block code by the parameters $n$, $m$, $R$, $E$, $E_b/N_0$. From (5.66), $E(\alpha)$ may be approximated as

$$\text{erfc}(\alpha) < e^{-\alpha^2/2}. \tag{5.71}$$

Substituting (5.71) into (5.69),

$$P_B < \frac{1}{n} \sum_{i=E+1}^{n} \binom{n}{i} (i) \left(\frac{1}{2}\right)^n e^{-\alpha^2 i/2} \cdot \left[2 - e^{-\alpha^2/2}\right]^{n-i}. \tag{5.72}$$

For small signal-to-noise ratio, $E_b/N_0 \to 0$. This implies $\alpha \to 0$. (5.72) then becomes

$$\lim_{\alpha \to 0} P_B \to \frac{1}{n \, 2^n} \sum_{i=E+1}^{n} \binom{n}{i} (i). \tag{5.73}$$

Recall that

$$(x + 1)^n = 1 + nx + \frac{n(n-1)}{2!} x^2 + \cdots + x^n$$

$$= \binom{n}{0} + \binom{n}{1} x + \binom{n}{2} x^2 + \cdots + \binom{n}{n} x^n, \tag{5.74}$$

and, differentiating (5.74) with respect to $x$, we get

$$n(x + 1)^{n-1} = \binom{n}{1} + 2 \binom{n}{2} x + \cdots + n \binom{n}{n} x^{n-1}. \tag{5.75}$$

If we let $x = 1$ in (5.75),

$$n2^{n-1} = \binom{n}{1} + \binom{n}{2} 2 + \cdots + \binom{n}{n} n$$

$$= \sum_{i=1}^{n} \binom{n}{i} (i)$$

$$= \sum_{i=0}^{n} \binom{n}{i} (i), \tag{5.76}$$

since

$$\sum_{i=E+1}^{n} \binom{n}{i} (i) < \sum_{i=0}^{n} \binom{n}{i} (i). \tag{5.77}$$

Substituting (5.76) and (5.77) into (5.73),

$$\lim_{\alpha \to 0} P_B < \frac{1}{n2^n} n2^{n-1} = \frac{1}{2}. \tag{5.78}$$

(5.78) indicates that when $E_b/N_0$ approaches zero with small values of $E$, the bit error rate approaches 1/2, which may be worse than the error rate without coding.

For large signal-to-noise ratio $E_b/N_0 \to \infty$, and as $\alpha \to \infty$, it is easy to see that

$$\lim_{\alpha \to \infty} P_B \to 0. \tag{5.79}$$

For erfc($\alpha$) $\ll$ 1 in (5.69), and using the first term approximation,

$$P_B \simeq \frac{n^{E+1}}{(E + 1)! \, n} (E + 1) \, [\text{erfc}(\alpha)]^{E+1}. \tag{5.80}$$

Substituting exp($-\alpha^2$), the bounded value of erfc($\alpha$), (5.80) becomes

$$P_B \leqslant \frac{n^E}{E!} [\exp(-\alpha^2)]^{E+1} \left[ \frac{1}{2^{E+1}} \right]$$

$$= \frac{n^E}{E} \exp[-2m(E + 1) \, snr(c)] \tag{5.81}$$

where we denote the required $E_b/N_0$ after coding as snr($c$) in consistency with Section 5.2.1. For the uncoded case, we have

$$P_B^{(1)} < \exp[-snr(o) \, 2m]. \tag{5.82}$$

At the same decoder output bit error rate $p_0 = P_B = P_B^1$. Equating (5.81) to (5.82) we have

$$\frac{n^E}{E} \exp[-2mR(E + 1) \, snr(c)] \leq \exp[-snr(o) \, 2m]. \tag{5.83}$$

Take $\ell n$ on both sides of (5.83)

$$\ell n \left( \frac{n^E}{E!} \right) - 2mR(E + 1) \, snr(c) \leq - snr(o) \, 2m. \tag{5.84}$$

For large $n$, $n^E \gg E!$ because $E \ll n$ always.

$$\ell n \left( \frac{E!}{n^E} \right) \rightarrow 0. \tag{5.85}$$

In terms of coding gain $\Delta_G$ defined in Section 5.2.1, (5.84) becomes

$$\Delta_G = \frac{snr(o)}{snr(c)} \rightarrow 2R \, (E + 1). \tag{5.86}$$

The bound of (5.86) says that coding gain is a function of the product of coding rate $R$ and the error-correcting capability $E$. It is not just $E$.

To see how the bit error rates differ in an AWGN channel with and without coding, let the coded bit error rate be denoted $P_B^{(2)}$ in (5.81). After taking $\ell n$ of both sides, (5.81) becomes

$$snr(c) < - \frac{\ell n \left( \dfrac{E!}{n^E} \right) + \ell n \, P_B^{(2)}}{2 \, mR \, (E + 1)}. \tag{5.87}$$

For the uncoded case, (5.82) gives

$$snr(o) < -\frac{\ell n\, P_B^{(1)}}{2m}. \tag{5.88}$$

To evaluate the bit error rates at the same signal-to-noise ratio, that is, $snr(c)$ $= snr(o)$, we obtain, with the aid of (5.85), from (5.87) and (5.88),

$$R(E + 1)\, \ell n\, [P_B^{(1)}] = \ell n\left(\frac{E!}{n^E}\right) + \ell n\, P_B^{(2)}, \tag{5.89}$$

or

$$\ell n\, P_B^{(2)} = \ell n\, [P_B^{(1)}]^{R(E+1)} - \ell n\left(\frac{E!}{n^E}\right)$$

$$= \ell n\, [P_B^{(1)}]^{R(E+1)}\left(n^E/E!\right),$$

or

$$P_B^{(2)} = [P_B^{(1)}]^{R(E+1)}\, (n^E/E!). \tag{5.90}$$

(5.90) was obtained from previous assumptions including large code length $n$ (thus large $E$). Since $P_B^{(1)} < 1$, $P_B^{(2)}$ decreases exponentially with $E$ and $R$.

From (5.81), the approximated bit error rate after decoding is seen to be a function of snr($c$), or required coded ($E_b/N_0$), code length $n$, number of bits per symbol $m$, encoding rate $R$, and the number of error-correcting capabilities $E$ of the code. A necessary condition for the existence of a block code, linear or nonlinear, with the code parameters $n$, $E$, $R$ is the Gilbert bound in [1]

$$1 - R \geq H\left(\frac{E}{n}\right), \tag{5.91}$$

where

$$H\left(\frac{E}{n}\right) = -\left(\frac{E}{n}\right)\log\left(\frac{E}{n}\right) - \left(1 - \frac{E}{n}\right)\log\left(1 - \frac{E}{n}\right) \tag{5.92}$$

in the entropy of the ratio $E/n$. In other words, the values of $n$, $E$, and $R$ in (5.81) cannot be selected arbitrarily. These values are related precisely from the existence of a particular code.

The above error rate analysis of the channel under consideration applies to most block codes and it is independent of the choice of decoding algorithm. As a consequence, the result is useful for any block code. This is not true for convolutional codes. As we shall see in Section 5.6, a different decoding scheme of a convolutional code determines the corresponding decoding error rate evaluation. Thus they have to be treated separately.

### 5.5.2 Most Useful Block Codes

Although the error rate calculation derived in Section 5.5.1 is common to all block codes, the structure of the codes and the decoding techniques can be quite different. For practical reasons, the following brief discussion will be restricted to the three most useful classes of block codes, so that a sample of code performance can be presented in Section 5.5.3.

### 5.5.2.1 Nonbinary Character Error-Correcting Codes

The nonbinary character error-correcting (NBCEC) codes consist of two subclasses of error-correcting codes, namely the block NBCEC codes, which are represented by Reed-Solomon (RS) codes, and the convolutional NBCEC codes, which are represented by Tong-Ebert (TE) codes. Both are nonbinary character error-correcting codes with different characteristics. In the following the principles of the RS codes are described. The TE codes will be mentioned under convolutional coding in Section 5.6.4.

In Section 4.11.3 of Chapter 4 we saw the usefulness of extended RS codes for the construction of secret sharing systems; in this section we study the basics of RS codes strictly for error correction. An example of RS codec implementation will be described in Chapter 6.

For multiple burst error corrections or code concatenation the RS code is one of the most powerful channel coding techniques. The symbols of the RS code come from the set of integers $\{0, 1, 2, \ldots, (q - 1)\}$, where $q$ is a power of a prime $q = p^m$ for some prime $p$ and positive integer $m$. The length of an RS codeword is equal to $n = q - 1 = p^m - 1$. The minimum distance of an RS code is $d = n - k + 1 > 2t + 1$, which is capable of correcting $t -$ symbol errors in $n$ symbols. Every code symbol from the set of the integers $\{0, 1, 2, \ldots, q - 1\}$ can be expressed as an $m$-tuple with $0, 1, \ldots, (p - 1)$ elements in each tuple. For the most practical case of digital communication $p = 2$, i.e., every symbol of an RS code is represented by an $m$-tuple of binary symbols. Thus the block length of an RS code consists of $mn$ bits. If each binary channel error burst is less than or equal to the length $m$ of a code symbol, then RS codes can be, and have been, used effectively for multiple bursts error correction.

For any primitive block codes there exists a primitive element $\alpha$ in $GF(q^m)$ such that the powers of $\alpha$ exhaust the extension field. The size of the field equals the length of the code.

For a $\nu$th order primitive RS code the parity check polynomial $h(x)$ has as roots $\alpha^0, \alpha^1, \alpha^2, \ldots, \alpha^\nu$ of consecutive order of the element $\alpha$. That is

$$h(x) = (x - \alpha^0)(x - \alpha^1) \ldots (x - \alpha^\nu).$$

The degree of $h(x)$ is $k = \nu + 1$. Each root of $h(x)$ determines a row of the generator matrix $G$.

$$G = \begin{bmatrix} \alpha^0 & \alpha^0 & \cdot & \cdot & \cdot & \cdot & \cdot & \alpha^0 \\ \alpha^0 & \alpha^1 & \alpha^2 & \cdot & \cdot & \cdot & \cdot & \alpha^{n-1} \\ \alpha^0 & \alpha^2 & \alpha^4 & \alpha^6 & \cdot & \cdot & \cdot & \alpha^{2(n-1)} \\ \cdot & & & & & & & \cdot \\ \cdot & & & & & & & \cdot \\ \cdot & & & & & & & \cdot \\ \alpha^0 & \alpha^\nu & \alpha^{2\nu} & \alpha^{3\nu} & \cdot & \cdot & \cdot & \alpha^{\nu(n-1)} \end{bmatrix}$$

We know that for any cyclic code the generator polynomial $g(x) = (x^n - 1)/h(x)$. If $\alpha^0$, $\alpha^1$, $\alpha^2$, . . ., $\alpha^{d-1}$ are the consecutive roots of $g(x)$. Then these roots form the rows of the parity check matrix $H$.

$$H = \begin{bmatrix} \alpha^0 & \alpha^1 & \alpha^2 & \cdot & \cdot & \cdot & \alpha^{n-1} \\ \alpha^0 & & & & & & \alpha^{2(n-1)} \\ \alpha^0 & & & & & & \cdot \\ \cdot & & & & & & \cdot \\ \cdot & & & & & & \cdot \\ \cdot & & & & & & \cdot \\ \alpha^0 & \alpha^{d-1} & \alpha^{2(d-1)} & \cdot & \cdot & \cdot & \alpha^{(d-1)(n-1)} \end{bmatrix}$$

Since $d - 1 = n - k = n - (v + 1)$, which is the last column. With $G$ or $H$ obtained from the respective roots from $h(x)$ or $g(x)$ a primitive Reed-Solomon code is determined.

The properties of RS codes are summarized as follows:

- The symbols of an RS code come from the set of integers $\{0, 1, 2, \ldots (q - 1)\}$, with $q = p^m$, a power of a prime $p$ and positive integer $m$.
- Every code character can be expressed as an $m$-tuple with $0, 1, 2, \ldots,$ $(p - 1)$ elements in each tuple.
- In terms of number of characters, the length of an RS code is $n = q - 1 = p^m - 1$.
- The code is capable of correcting $E$ character errors within $n$.
- The minimum distance of the code is

$$d_m \geq 2E + 1$$

$$= n - k - 1,$$

where $k$ is the number of information characters.
- The number of parity check characters is $n - 1$.
- Let $\alpha$ be an element of $GF(q)$, and let $n$ be the order of $\alpha$. The minimum function of $\alpha^j$ is simply $x - \alpha^j$.
- The generator polynomial of the RS code is in the form $g(x) = (x - \alpha)(x - \alpha^2) \ldots (x - \alpha^{d-1})$.
- For binary transmission, $p = 2$, $q = 2^m$, and each $q$-ary character can be expressed as an $m$-tuple over $GF(2)$. That is, each character can be expressed as a collection of $m$ binary digits. As a consequence, $([2^m]^m - 1, [2^m]^m - 1 - 2t, d)$ RS code over $GF(2^m)$ can be treated as an $(m\{[2^m]^m - 1\}, m\{[2^m]^m - 1 - 2t\}, md)$ code over $GF(2)$.
- The dual of an RS code is also an RS code.
- Every RS code over $GF(2^m)$ is optimum with respect to the Reiger bound.
- The implementation of an RS codec is much simplified if the fast Fourier transform technique is used.
- A simple RS codec can be built and operated up to 120.0 Mbps.

- A large number of RS codes with varied rates and error-correcting capabilities exists.
- RS is the only class of codes known for effective multiple burst-error correction.

### 5.5.2.2  Bose-Chaudhuri-Hocquenghem (BCH) Codes

One of the most powerful classes of useful error control codes for random channels is the class of Bose-Chaudhuri-Hocquenghem (BCH) codes, which can be either algebraically or majority logically decoded. The theory of BCH codes is well known and is available in texts and in the literature. Because of the significance of BCH codes in digital satellite communication, the next chapter discusses separately binary BCH encoding-decoding techniques as well as decoder complexity estimations. In this section we point out that BCH codes are subfield, subcodes of the generalized RS codes. Specifically, we shall show that primitive BCH codes can be derived from primitive RS codes. The corresponding nonprimitive codes can be obtained from the primitive ones.

Let a $\nu$th order information polynomial of an RS code be defined as [12]

$$C(x) = a_0 + a_1 X^j + a_2 X^{2j} + \cdots a_\nu X^{\nu j}. \qquad (5.93)$$

For $j = 0, 1, 2, \ldots$ (5.93) expands to, with $a_i \in GF(q)$,

$$C(1) = a_0 + a_1 + a_2 + \cdots + a_{\nu-1}$$

$$C(\alpha) = a_0 + a_1\alpha + a_2\alpha^2 + \cdots + a_{\nu-1}\alpha^{\nu-1}$$

$$C(\alpha^2) = a_0 + a_1\alpha^2 + a_2\alpha^4 + \cdots + a_{\nu-1}\alpha^{2(\nu-1)}. \qquad (5.94)$$

Take a vector $C$ of $C(x)$ and encode it into a codeword $c = C \cdot G$, where $G$ is the generator matrix of the code.

$$G(x) = c_0 + c_1 X + c_2 X^2 + \cdots c_{n-1} X^{n-1}. \qquad (5.95)$$

With the evaluated $c(\alpha^j)$ as coefficients, the code vector is:

$$[C(1), C(\alpha), \ldots, C(\alpha^{q-2})] \qquad (5.96)$$

$$c(x) = C(1) + C(\alpha) X + C(\alpha^2) X^2$$
$$+ \cdots C(\alpha^{n-1}) X^{n-1}$$
$$= a_0 + a_1 + \cdots + a_{\nu-1}$$
$$+ a_0 X + a_1 \alpha X + a_2 \alpha^2 X$$
$$+ \cdots a_{\nu-1} \alpha^{\nu-1} X + \cdots$$
$$= a_0(1 + X + X^2 + \cdots + X^{n-1})$$
$$+ a_1 [1 + \alpha X + \alpha^2 X^2 + \cdots (\alpha X)^{n-1}] + \cdots$$
$$+ a_{\nu-1} [1 + \alpha^{\nu-1} X + (\alpha^{\nu-1} X)^2$$
$$+ \cdots + (\alpha^{\nu-1} X)^{n-1}]. \qquad (5.97)$$

For arbitrary $i$ $(0 < i < v - 1)$, we have

$$C_i(x) = a_i (1 + \alpha^{j_i} X + \alpha^{2j_i} X^2 + \cdots \alpha^{(n-1)j_i}X^{n-1}). \quad (5.98)$$

Let $X = \alpha^{-j}$. (5.98) becomes

$$C_i(\alpha^j) = a_i (1 + \alpha^{j_i} \alpha^{-j} + \alpha^{2j_i} \alpha^{-2j}$$

$$+ \cdots + \alpha^{(n-1)j_i} \alpha^{-(n-1)j})$$

$$= a_i(\alpha^{\Delta \cdot 0} + \alpha^{\Delta \cdot 1} + \cdots \alpha^{\Delta \cdot k} + \cdots \alpha^{\Delta \cdot (n-1)}), \quad (5.99)$$

where

$$\Delta = j_i - j \quad (5.100)$$

$$c_i(\alpha^{-j_i}) = - a_i. \quad (5.101)$$

Enumerating (5.101) for $j_i$ to $v - 1$, we have

$$c(\alpha) = - a_0$$

$$c(\alpha^{-1}) = - a_1$$

$$\vdots \qquad \vdots$$

$$c(\alpha^{-v}) = - a_v. \quad (5.102)$$

Since $\alpha^q = \alpha$ due to primitivity, $\alpha^{q-1} = 1$. Multiplying both sides by $\alpha^{-j}$, we get $\alpha^{q-1-j} = \alpha^{-j}$. When $n = q - 1$, these codes are BCH codes in GF($q$). Thus BCH codes are subcodes of RS codes as just demonstrated, although they were developed independently. As a consequence some of the properties of BCH codes are similar to those of RS codes. In particular:

- For any positive integer $s$ there exists a $q$-ary BCH code of length $n = q^s - 1$, where $q = p^m$ as before. (For $s = 1$, the code becomes RS.)
- A BCH code is specified by its generator polynomial $g(x)$, which in turn is specified by its roots.
- For binary transmission, let $\alpha$ be a primitive element of GF($2^m$) and let $m_i(x)$ be the minimum polynomial of the roots, $\alpha^i$ for $i = 1, 2, 3, \ldots, 2E$. Then $g(x)$ is the least common multiple of the minimum polynomials $g(x) = $ L.C.M. $\{m_i(x)\}$.
- To obtain $g(x)$, one first chooses a primitive polynomial of order $m$ such as $m_1(x)$. Since $m_1(x)$ contains the 1st, 2nd, 4th, 8th, $\ldots$ powers of $\alpha$ as its roots and $m_2(x) = m_4(x) = m_8(x) \ldots$, one can only choose $m_3(x)$, $m_5(x)$, $m_7(x) \ldots$ as minimum polynomials. An example is presented in the section of ARQ for INTELSAT (Section 5.4.6.2).
- BCH codes are effective for random error correction and are relatively easier to decode than RS codes. However, comparison is difficult because of the difference in error-correcting capabilities.

Both RS and BCH decoders have been known to produce burst errors when the capability of the hard-decision decoder is exceeded. That is, if the channel error rate is $p_b > E$, and $d_m > 2E + 1$ is the minimum distance of the code, then the decoder produces additional bursty errors. The error-producing phenomenon depends on the channel error distribution. Whether the same class of decoders exhibits the same error-producing phenomena in general with the identical channel error statistics is not known at present.

### 5.5.2.3  Quadratic Residue Codes

With regard to the solutions of a second-order congruence relation

$$x^2 \equiv q_r \bmod P \tag{5.103}$$

quadratic residue codes may be derived, where $q_r$ is a quadratic residue of a prime $p$. Let $\alpha$ be a primitive element of $GF(p)$; then

$$q_r^{(p-1)/2} \equiv x^{p-1} \equiv 1 \bmod p. \tag{5.104}$$

If $p = 8m \pm 1$, then it has been shown that 2 is a quadratic residue of $p$. $\alpha^i$ is a quadratic residue if and only if $2\alpha^i$, $4\alpha^i$, $8\alpha^i$, . . . are also, and the set of quadratic residues consists of at least one cycle set, which implies that $x^p - 1$ factors into $(x - 1)\,\alpha_r(x)\,\alpha_n(x)$, where the roots of $\alpha_r(x)$ are quadratic residues of $p$ and the roots of $\alpha_n(x)$ are nonresidues. The codes can be generated by $\alpha_r(x)$, $(x - 1)\,\alpha_r(x)$, $\alpha_n(x)$, $(x - 1)\,\alpha_n(x)$.

The minimum distances of such codes are

$$d^2 > n, \text{ for } p = 8m + 1 \tag{5.105}$$

and

$$d(d - 1) \geq n - 1, \text{ for } p = 8m - 1. \tag{5.106}$$

Most quadratic residue codes are of code rate about 1/2. Most binary quadratic residue codes are tabulated by Berlekamp [2].

A practical example of the quadratic residue code is the QR (17, 9, 5) code, which has been shortened to QR(16, 8, 5) with generator polynomial

$$G(x) = 1 + x^3 + x^4 + x^5 + x^8.$$

This code has been implemented in the all-digital TDMA Satellite Business System (SBS) for U.S. domestic operation. Most QR codes can be decoded by means of the permutation method. However, simple codes may be decoded by looking them up in a table.

### 5.5.2.4  Threshold Decodable Codes

Among known decoding algorithms, majority or threshold decoding techniques are the simplest to implement. Both block and tree codes are threshold decodable, as demonstrated by Massey [40]. The difference between majority logic and threshold decoding for block codes was observed by Gore [12]. Codes which are threshold decodable or which can be decoded by a threshold decoding algorithm which are codes constructed by trial and error or from finite geometries, and Reed-Muller codes, maximum length codes, quasi-cyclic codes, low density parity check codes, and some BCH codes. Depending on the code structure, threshold decoding can be applied in a single step or multiple steps. However, the codes share the following common properties:

- From the parity check matrix $H$ of a threshold-decodable code a syndrome sequence $\bar{S}$ is formed from an error sequence $\bar{E}$ as $\bar{S} = \bar{E} H^T$. The set of syndrome sequences $\bar{S}$ forms a check sum, $\Sigma\bar{S}$, which contains the error digits. A collection of $\Sigma\bar{S}$ is said to be orthogonal on a particular error digit if it appears in each check sum in the collection and no other error digit appears in more than one check sum (for single step decoding).
- Under the above principle of orthogonality, if the number of $\Sigma\bar{S}$ collection is $J$, then the $(n, k, d_m)$ code with parity check matrix $H$ has a minimum distance $d_m = J + 1$, and the code is guaranteed to correct up to $J/2$ number of errors in a block with $n$ bits.

### 5.5.3   Block Code Performance

Figure 5.14 gives the performance in terms of decoder input-output error rates for a sample of block codes. The code length varies from 15 to 1057. These block codes do not belong to the same class. The (23, 12, 7) code is the Golay perfect code, and the (255, 215, 11) and (15, 5, 7) codes are BCH codes. The

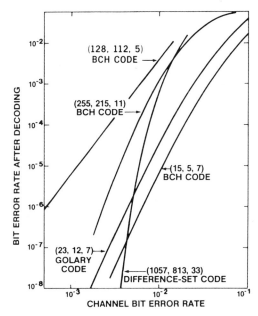

**Figure 5.14** Block code performances in binary symmetric channel with high bit error rate.

(1057, 813, 33) code is a planar projective geometry difference set code. These codes are compared at relatively high channel error rates, i.e., above $10^{-3}$. However, they can perform significantly differently at lower channel error rates. If a channel error rate is $3 \times 10^{-3}$, then, as can be seen from Figure 5.14, the (1057, 813, 33) code outperforms the (15, 5, 7) code with decoding error rates near $10^{-8}$. If the channel error rate is $10^{-2}$, then the (15, 5, 7) code outperforms the (1057, 813, 33) code. If the satellite channel can be modeled as additive Gaussian noise, then in terms of signal-to-noise ratio the performances of the same sample set of block codes are shown in Figure 5.15. This figure shows the required signal-to-noise ratios in dB at a specified decoding bit error rate $P_B$. For $P_B$ less than $10^{-5}$, the (1057, 813, 33) code clearly shows superior performance.

But code error rate performance alone cannot determine the selection of codes. Next, the efficiency, or coding rate, needs to be examined. One way to compare all this information is to look at the code performances of the decoding error rate with respect to the coding rate. For example, with a BCH block code of length 1023, the required $E_b/N_0$ versus code rate $R > 0.5$ is shown in Figure 5.16 for three decoding bit error rates signaling in the AWGN channel. Depending on the error rates, the minimal signal-to-noise ratios occur at 0.575, 0.65, and 0.625.

A more extensive evaluation of other code lengths is shown in Figure 5.17, which gives the signal-to-noise ratios required to achieve a decoding word error rate of $10^{-8}$ for binary BCH codes of lengths from 31 to 1023. The code performance curves indicate that minimum signal-to-noise ratios occur inside the code rate window of 0.5 to 0.6. These calculations were provided to the author by C. E. Holborow. If the quality, in terms of signal-to-noise ratio, is more demanding than bandwidth efficiency, then codes like (31, 16, 7), (63, 36, 11), . . ., (511, 358, 37), and (1023, 573, 101) may be applicable.

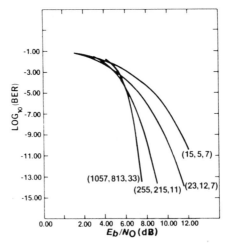

**Figure 5.15** Block code performances in an additive Gaussian channel.

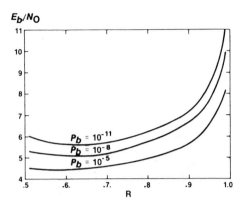

**Figure 5.16** BCH codes $n = 1023$ $E_b/N_0$ (in dB) vs. code rate $R$.

For nonbinary Reed-Solomon codes, similar evaluations are shown in Figure 5.18. The minimal signal-to-noise ratio window shifts toward the higher end of the coding rate scale between 0.6 and 0.7 at the same decoding word error rate of $10^{-8}$. It appears, from the standpoint of performance improvement, that Reed-Solomon codes of the same length and rate outperform the corresponding BCH codes even in random error channels. But this need not always be true. In the introduction of FFT (fast Fourier transform) decoding of RS codes, Gore evaluated a number of BCH and RS codes in the binary channels, as shown in Figure 5.19 [12]. As indicated, RS codes are better at low channel error rates. As noted in the figure, BCH codes can outperform the corresponding RS codes at high channel error rates.

Normally, within the same class of codes, the longer the code length, the higher the code rate; and this may result in a larger number of correctable errors,

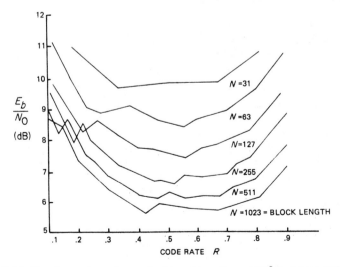

**Figure 5.17** Signal-to-noise ratio required to achieve $P_w = 10^{-8}$ for binary BCH codes.

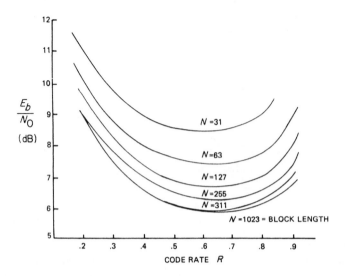

$\dfrac{E_b}{N_0}$ (dB)

CODE RATE $R$

$N = 31$

$N = 63$

$N = 127$

$N = 255$

$N = 311$

$N = 1023 = $ BLOCK LENGTH

**Figure 5.18** Signal-to-noise ratio required to achieve $P_w = 10^{-8}$ for RS codes.

but the codec complexity increases. As shown in Figure 5.14, for channel bit error rate less than $4 \times 10^{-3}$, the (1057, 813, 33) code outperforms most comparable random error-correcting codes of the same complexity. That is, the (1057, 813, 33) code can be majority logically decoded and can be operated at very high speed. Unfortunately, few projective geometry block codes, such as the 1057 block length code, exist.

## 5.6 CONVOLUTIONAL CODES

Convolutional (or tree) codes can be decoded either probabilistically or algebraically. Probabilistic decoding of convolutional codes consists of sequential decoding and maximum-likelihood (Viterbi) decoding. Algebraic decoding of convolutional codes mainly refers to majority or threshold decoding. Since the performance of convolutional codes depends on the decoding techniques, each type of decoding performance needs to be considered separately.

### 5.6.1 Threshold Decoding

Threshold decoding algorithms for convolutional codes in general were formulated by Massey [13]. Hagelberger discussed majority decoding procedures for self-orthogonal convolutional codes [14]. Self-orthogonality in threshold decoding refers to the automatic formation of $J = d_m - 1$ parity check equations to check on an error digit. The error digit is checked by each of the parity check equations; no other error digit is checked by more than one equation.

Wu has demonstrated the usefulness of threshold decoding in digital satellite communication systems, and generated a large number of one-step threshold

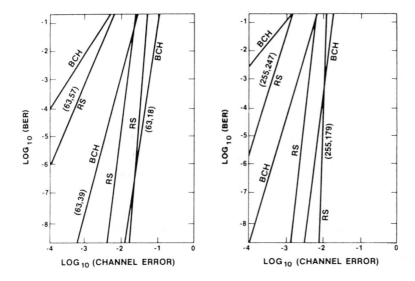

**Figure 5.19** Comparisons of BCH and RS code performances in BSC [12].

decodable self-orthogonal convolutional codes, which are known to have the following advantages [15]:

- They are simple to implement.
- They do not propagate errors.
- They guarantee correction capability beyond code minimum distance.
- They can operate at very high speeds.
- When the channel error condition exceeds the decoding capability, the decoder does not produce bursty errors at its output.
- A large number of high efficiency codes are available.

The disadvantage of this class of decoding algorithms is that the coding gain is not as large as those obtainable with probabilistic decoding techniques.

### 5.6.1.1   Decoding Error Rate Calculation

Through a direct argument, we suggest a simple expression for the bit error rate of this decoding [15]. It has been experimentally verified that the results are remarkably close to this expression. The reason for the simplicity is that error propagation can be ignored and the expression can be applied to all the codes in this class.

Assume that a communication satellite channel can be modeled as an ideal AWGN channel. The bit error rate from an ideal receiver with matched filter and coherent detector is

$$p = \frac{1}{2} \left[ 1 - \text{erf} \sqrt{\frac{E_s(1 - \rho)}{2N_0}} \right]$$

where                                                                                              (5.107)

$$\text{erf}(x) = \frac{2}{\pi} \int_0^x e^{-t^2} \, dt,$$

$E_s$ = average signal energy per symbol,
$N_0$ = noise power density, and
$\rho$ = cross-correlation coefficient of the waveforms
$(-1 \leq \rho \leq +1)$.

For a coded system of encoding rate $R$, $E_s/N_0 = m(E_b/N_0)R$, where $E_b$ is the average signal energy per uncoded bit. If the modulation system is phase-shift keying (PSK), then $\rho = -1$, which minimizes $p$ in (5.107). Then (5.107) becomes

$$p = \frac{1}{2} \left[ 1 - \text{erf} (\sqrt{mR} \sqrt{E_b/N_0}) \right].$$                        (5.108)

For an $(n_0, k_0, d_m)$ self-orthogonal tree code, the bit error probability $P_B$, after threshold decoding in an additive Gaussian channel, can be written as

$$P_B = \min(P_b, P_b')$$                                                                          (5.109)

where

$$P_b = \sum_{i=(d_m+1)/2}^{n_e} \binom{n_e}{i} [2^{n_e}]^{-1} \left[ \frac{1 - X}{1 + X} \right]^i [1 + X]^{n_e}$$   (5.110)

$$P_b' = \frac{1}{n_A 2^{n_A}} \sum_{j=(d_m+1)/2}^{n_A} \binom{n_A}{j} \left[ \frac{1 - X}{1 + X} \right]^j [1 + X]^{n_A}$$   (5.111)

$$X = \text{erf} \sqrt{Rm} \sqrt{E_b/N_0}$$                                                         (5.112)

where $n_A$ = constraint length = $(1 + m_s)n_0$ and $n_e$ = effective constraint length of the same code. Note that $n_e$ is defined for the total number of distinct noise digits checked by the set of $J$ orthogonal parity checks. Actually $n_e$ is the sum of the "size" $(n_i)$ of such a set; i.e.,

$$n_e = 1 + \sum_{i=1}^{J} n_i,$$                                                                  (5.113)

where the "size" $n_i$ is the number of noise digits, excluding the 0th noise digit, which are checked by the set.

The effective constraint length $n_e$ of a threshold decodable convolutional code is related to the code rate $R$, the number of check equations $J$, and the number of inputs per block $k_0$

$$n_e \leq \frac{1}{2} \frac{R}{1 - R} J^2 + \frac{k_0}{2} J + 1 + \frac{1}{2} r \left[ k_0 + r \frac{R}{1 - R} \right], \quad (5.114)$$

where $r = J \mod(n_0 - k_0)$. For high-rate codes under consideration, $r = 0$ for $n_0 - k_0 = 1$. Thus,

$$n_e \leq \frac{1}{2} \left[ \frac{RJ^2}{1 - R} + k_0 J + 2 \right]. \quad (5.115)$$

(5.109) will be used to evaluate a number of codes in Section 5.6.1.2.

### 5.6.1.2 Threshold Decoding Performances in Random Error Satellite Channels

Based on (5.109), the bit error rate performance of the rate 7/8, $n_A = 376$, $n_e = 71$, and $d_m = 5$ code with generator polynomial expressed in terms of its nonzero terms

$$G(7/8) = \begin{cases} 0, & 3, & 19, & 42 \\ 0, & 21, & 34, & 43 \\ 0, & 29, & 33, & 47 \\ 0, & 25, & 36, & 37 \\ 0, & 15, & 20, & 46 \\ 0, & 2, & 8, & 32 \\ 0, & 7, & 17, & 45 \end{cases} \quad (5.116)$$

is shown in Figure 5.20. This codec was built by Digital Communication Corporation, Nippon Electric Corporation, and Fujitsu for INTELSAT's SCPC digital system. The measured data is also shown.

The second performance example is shown in Figure 5.21. The code is a rate 3/4, $n_A = 80$, $n_e = 31$, $d_m = 5$, with generator

$$G(3/4) = \begin{cases} 0, & 3, & 15, & 19 \\ 0, & 8, & 17, & 18 \\ 0, & 6, & 11, & 13. \end{cases} \quad (5.117)$$

The measured results both in Gaussian and binary symmetric channels are indicated. Note that for the Gaussian channel, the scale is in terms of measured carrier-to-noise ratio $(C/N_0)$, which was defined and related to $E_b/N_0$ as described in Section 5.2.

The performance of a larger constraint length $n_A = 1176$, $n_e = 148$, $d_m = 7$ code performance is shown in Figure 5.22. For this codec, the measured difference between with and without feedback syndrome resetting is clearly shown.

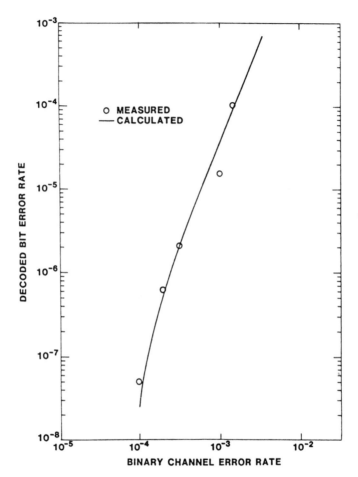

**Figure 5.20** Performance of code $G(7/8)$.

The above three codes were evaluated under assumed random error satellite channel conditions. Performances of two of the three codes in nonlinear satellite channels and channels with interference will be treated in Sections 5.6.1.3 and 5.6.1.4.

### 5.6.1.3 Threshold Decoding of Self-Orthogonal Codes in Nonlinear Satellite Channels

In terms of the decoder input (or channel output) BER versus decoder output BER, the performances of rate 3/4 codec and rate 7/8 codec, described in Section 5.6.1.1, in channels with nonlinearities, are shown in Figure 5.23 and Figure 5.24.

As the carrier-to-interference ratio ($C/I$) changes from infinity to 10 dB, little difference can be observed for both codecs. These results agree with the earlier

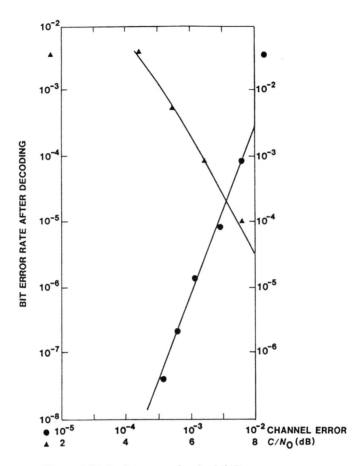

**Figure 5.21** Performance of code $G(3/4)$.

findings that these types of decoders are insensitive to the increase of channel errors [15]. That is, the decoders do not produce additional errors at their outputs when the codec capabilities are exceeded. As long as the interference-producing mechanisms produce random errors, the decoders do not distinguish between the errors from the transmission medium and errors from interference. As for bursty channel errors, the result will also be different for burst interference errors.

These results indicate the fact that, within the range of practical interests, a threshold decoder with convolutional self-orthogonal codes does not degrade codec performance even in a nonlinear satellite channel [16].

### 5.6.1.4 Performance of Threshold Decoding in the Presence of Interference

In terms of carrier-to-noise ratio ($C/N$) the decoding BER in a nonlinear channel of the same two codecs was evaluated (Figures 5.25 to 5.27) for different carrier-to-interference ratios. As interference increases, or as $C/I$ decreases, as shown

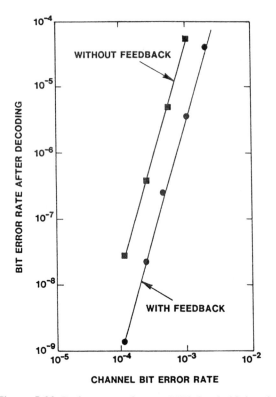

**Figure 5.22** Performance of $n_A = 1176$ threshold decoder.

in the figures, the channel (without coding) degrades significantly. Although interference causes the overall system performance to degrade, the actual coding gain increases as the channel degrades. For example, at $C/I = \infty$, the coding gain is about 4.5 dB in $C/N$ at $10^{-4}$ for the rate 3/4 codec. At $C/I = 20$ dB the coding gain is about 6.0 dB.

At $C/I = 10$ dB, the channel degrades drastically. However, the coded performances do not differ significantly from the channel characteristics, at least in the $C/N$ range measured. The adaptive characteristic of the decoder described in [15] would lead us to expect in general that a codec pushed beyond its performance capability will follow the pathological channel characteristic closely as long as the channel errors are not bursty.

From these measured data, the following conclusions can be drawn:

1. Nonlinearity degrades a transmission channel significantly. For example, at $C/N = 15$ dB, both linear and 6 dB backoff nonlinear channels at $C/I = \infty$ yield a BER about $10^{-5}$. With the nonlinearity of 0 dB backoff the channel degrades to above $2 \times 10^{-4}$.

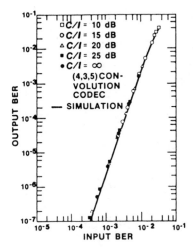

**Figure 5.23**  FEC decoder input BER vs. output BER for (3/4) code (nonlinear channel; 0 dB backoff).

2. As $C/I$ decreases to 10 dB, the uncoded channel behaves so badly that at $C/N = 15$ dB the nonlinearity at 0 dB backoff contributes only to about twice the decoding BER over the linear case. Thus the decoders perform as faithfully in the nonlinear channel as in the linear channel.
3. Again the actual coding gain increases as $C/I$ decreases even in nonlinear channels.

### 5.6.2   Viterbi Decoding

Viterbi originated a decoding algorithm for tree codes, and Forney established the trellis structure for an in-depth analysis of the Viterbi algorithm (VA) [17, 18]. Forney also linked the algorithm to the shortest-path problem in operations research.

Viterbi decodable tree codes are basically nonsystematic codes, which are unlike group block codes that can be transformed into systematic codes. The powerful gains obtainable from such decoding actually result from the nonsystematic encoding arrangement.

In error performance calculation, Viterbi decoding differs significantly from either block codes or threshold decodable tree codes. Calculation of the decoding error rate depends not on the transmission channel alone, but actually on a particular encoder configuration. As a result no systematic procedures exist to generate Viterbi-decodable codes. Codes (through encoder connections) can be determined by trial-and-error of all possible encoder shift register tap connections that yield the minimum or minimal error rates. The criterion for the search is maximization of the minimum free distance for a given code constraint length.

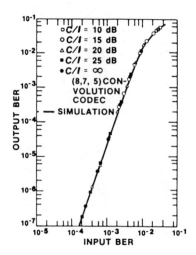

**Figure 5.24** FEC decoder input BER vs. output BER for (8, 7) code (nonlinear channel; 0 dB backoff).

There are two possible measures of decoding performance. The first-event error probability $P_E$, or the decoder initial error rate, refers to the probability that the correct path is excluded at the beginning of the process in an arbitrary decoding step. The other measure is the bit error rate $P_B$, which is the expected ratio of bit errors to the total number of bits transmitted.

The code performance calculations are outlined as follows:

1. Construct a state diagram representation of the encoder operation from encoder shift register tap connections by adding the corresponding trellis to branch distance profiles.

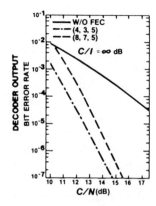

**Figure 5.25** BER vs. $C/N$ for $C/I = \infty$ (nonlinear channel; 0 dB backoff).

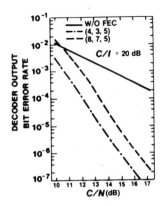

**Figure 5.26** BER vs. $C/N$ for $C/I = 20$ dB (nonlinear channel; 0 dB backoff).

2. Apply linear signal flow graph techniques such as node splitting, loop elimination, or branch combination to the encoder state diagram. This process will lead to an end-to-end transfer function $T(D,L,N)$, where the powers of $D$ represent the weight or distance measure in a branch of the state diagram. The branches of the state diagram are actually the possible encoder output combinations that result from each set of inputs, $L$ is used to determine the length of a given path. $N$ is the number of segments in a branch. The power of $L$ will increase by one every time a branch is passed through. If a branch transition was caused by a channel error, the $N$ segments are then included. The power of $N$ determines the number of bit errors for the path(s) corresponding to the segments in the polynomial form of $T(D,L,N)$.

3. The first-event probability is obtained by letting $L = N = 1$ in $T(D,L,N)$; i.e.,

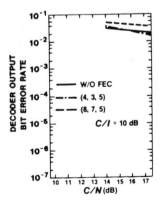

**Figure 5.27** BER vs. $C/N$ for $C/I = 10$ dB (nonlinear channel; 0 dB backoff).

$$P_E = T(D,L,N)\bigg|_{L=N=1} = T(D) = \sum_{k=d} a_k D^k$$

with $D^k = P_k$.

4. To obtain the bit error rate, each term in $T(D,L,N)$ must be weighted by the power of $N$. The weights can be obtained by differentiating $T(D,L,N)$ at $L = N = 1$, i.e.,

$$\frac{d}{dN} T(D,L,N)\bigg|_{L=N=1} = \sum_{k=d} c_k D^k$$

Let $D^k = P_k$; again

$$P_B < \sum_{k=d} c_k P_k.$$

Both $P_E$ and $P_B$ depend on $P_k$, the $k$th pairwise decision error rate. $P_k$ turns out to be a function of the characteristics of the transmission channel.

### 5.6.2.1 Viterbi's Error Rate Derivation

Viterbi's elegant derivation of the decoding error probability can be found in [19]. Each code, in terms of the connection of the modulo-2 adder in the encoder, yields a specific decoding error rate. Since the emphasis here is on classes of codes and different decoding performances for error control rather than a specific code, we shall omit the detailed error rate derivation for which the result applies to a specific code. Instead, the background that leads to the derivation of $P_B$ in general is presented here following Viterbi's original derivation [20].

For any received data sequences, $y$, the minimum error probability decoder chooses the path that maximizes the log-likelihood function (metric)

$$\ell n P(y \mid x^{(m)})$$

over all possible paths $x^{(m)}$. If each symbol is transmitted (or modulates the transmitter) independent of all preceding and succeeding symbols, and the interference corrupts each symbol independent of all the others, then the channel, which includes the modem, is said to be memoryless and the log-likelihood function

$$\ell n P(y \mid x^{(m)}) = \sum_{i=1}^{b} \sum_{j=1}^{n} \ell n P(y_{ij} \mid x_{ij}^{(m)}), \tag{5.118}$$

where $x_{ij}^{(m)}$ is a code symbol of the $m$th path, $y_{ij}$ is the corresponding received (demodulated) symbol, $j$ runs over the $n$ symbols of each branch, and $i$ runs over the $b$ branches in the given path.

Decisions are made after each set of new branch metrics has been added to the previously stored metrics. To analyze performance, we must evaluate $P_k$, the pairwise error probability for an incorrect path that differs in $k$ symbols from the correct path. Letting $x_{ij}$ and $x'_{ij}$ denote symbols of the correct and incorrect paths, respectively, we obtain

$$P_k(x,x') = \Pr\left[ \sum_{i=1}^{b} \sum_{j=1}^{n} \ell n P(y_{ij} \mid x'_{ij}) > \sum_{i=1}^{b} \sum_{j=1}^{n} \ell n P(y_{ij} \mid x_{ij}) \right]$$

$$= \Pr\{ \sum_{r=1}^{k} \ell n P(y_r \mid x'_r)/P(y_r \mid x_r) > 0 \}$$

$$= \Pr\{ \prod_{r=1}^{k} P(y_r \mid x'_r)/P(y_r \mid x_r) > 1 \}, \tag{5.119}$$

where $r$ runs over the $k$ code symbols in which the paths differ. This probability can be rewritten as

$$P_k(x,x') = \sum_{y \in Y_k} \prod_{r=1}^{k} P(y_r \mid x_r), \tag{5.120}$$

where $Y_k$ is the set of all vectors $y = (y_1, y_2 \ldots y_r, \ldots y_k)$ for which

$$\prod_{r=1}^{k} P(y_r \mid x'_r)/P(y_r \mid x_r) > 1. \tag{5.121}$$

But if this is the case, then

$$P_k(x,x') < \sum_{y \in Y_k} \prod_{r=1}^{k} P(y_r \mid x_r) \left[ \frac{P(y_r \mid x'_r)}{P(y_r \mid x_r)} \right]^{1/2}$$

$$< \sum_{\text{all } y \in Y} \prod_{r=1}^{k} P(y_r \mid x_r)^{1/2} P(y_r \mid x'_r)^{1/2}, \tag{5.122}$$

where $Y$ is the entire space of received vectors. The first inequality is valid because we are multiplying the summand by a quantity greater than unity, and because we are merely extending the sum of positive terms over a larger set. Finally, we may break up the $k$-dimensional sum over $y$ into $k$ one-dimensional summations over $y_1, y_2 \ldots y_k$ respectively, and this yields

$$P_k(x,x') < \sum_{y_1} \sum_{y_2} \cdots \sum_{y_k} \prod_{r=1}^{k} P(y_r \mid x_r)^{1/2} P(y_r \mid x'_r)^{1/2}$$

$$= \prod_{r=1}^{k} \sum_{y_r} P(y_r \mid x_r)^{1/2} P(y_r \mid x'_r)^{1/2}. \tag{5.123}$$

To illustrate the use of this bound, let us consider two specific channels. For the BSC, $y_r$ is either equal to $x_r$, the transmitted symbol, or to $\bar{x}_r$, its complement. Now $y_r$ depends on $x_r$ through the channel statistics. Thus

$$P(y_r = x_r) = 1 - p$$

$$P(y_r = \bar{x}_r) = p. \tag{5.124}$$

For each symbol in the set $r = 1, 2, \ldots k$ by definition $x_r \neq x'_r$. Hence for each term in the sum, if $x_r = 0$, then $x'_r = 1$ or vice versa.

Hence, whatever $x_r$ and $x'_r$ may be,

$$\sum_{y_r=0}^{1} P(y_r \mid x_r)^{1/2} P(y_r \mid x'_r)^{1/2} = 2\sqrt{p(1 - p)} = P_1 \tag{5.125}$$

and the product (5.123) of $k$ identical factors is

$$P_k = 2^k p^{k/2} (1 - p)^{k/2} \tag{5.126}$$

for all pairs of correct and incorrect paths. Substituting (5.126) for $p_k$, we have

$$P_E < \sum_{k=d}^{\infty} a_k P_k$$

or

$$P_E < T(D) \Big|_{D^1} = 2\sqrt{p(1 - p)} = P_1. \tag{5.127}$$

In an additive white Gaussian channel, the distance measure is the metric

$$\sum_{i=1}^{b} \sum_{j=1}^{n} x_{ij} Y_{ij},$$

where $x_{ij} = \pm 1$ are the transmitted code symbols, $y_{ij}$ the corresponding received (demodulated) symbols, and $j$ covers the $n$ symbols of each branch while $i$ covers all the branches in a particular path.

Assume, without loss of generality, that the correct (transmitted) path $\mathbf{x}$ has $x_{ij} = +1$ for all $k$ and $j$ (corresponding to the path of all zeros if the input symbols were 0 and 1). Let us consider an incorrect path $\mathbf{x}'$ merging with the correct path at a particular step, which has $k$ negative symbols ($x'_{ij} = -1$), and the remainder, positive symbols. Such a path may be incorrectly chosen only if it has a higher metric than the correct path; i.e.,

$$\sum_{i=1}^{b} \sum_{j=1}^{n} x'_{ij} y_{ij} \geq \sum_{i=1}^{b} \sum_{j=1}^{n} x_{ij} y_{ij} \tag{5.128}$$

or

$$\sum_{i=1}^{b} \sum_{j=1}^{n} (x'_{ij} - x_{ij}) y_{ij} \geq 0,$$

where $i$ runs over all branches in the two paths. But since, as we have assumed, the paths $x$ and $x'$ differ in exactly $k$ symbols, wherein $x_{ij} = 1$ and $x'_{ij} = -1$, the pairwise error probability is just

$$
P_k = \Pr\left[\sum_{i=1}^{b} \sum_{j=1}^{n} (x'_{ij} - x_{ij})y_{ij} \geq 0\right]
$$

$$
= \Pr\left[\sum_{r=1}^{k} (x'_r - x_r)y_r \geq 0\right]
$$

$$
= \Pr\left[-2\sum_{r=1}^{k} y_r \geq 0\right] = \Pr\left[\sum_{r=1}^{k} y_r \leq 0\right], \tag{5.129}
$$

where $r$ runs over the $k$ differing symbols of the two paths.

It can be shown that the $y_{ij}$ are independent Gaussian random variables of variance $N_0/2$ and mean $\sqrt{E_s}x_{ij}$, where $x_{ij}$ is the code symbol actually transmitted. Since we are assuming that the (correct) transmitted path has $x_{ij} = +1$ for all $i$ and $j$, it follows that $y_{ij}$ or $y_r$ has mean $\sqrt{E_s}$ and variance $N_0/2$. Therefore, since the $k$ variables $y_r$ are independent and Gaussian, the sum $\sum_{r=1}^{k} y_r$ is also Gaussian with mean $k\sqrt{E_s}$ and variance $kN_0/2$.

Consequently,

$$
P_k = \Pr(z < 0) = \int_{-\infty}^{0} \exp\left[-\frac{(z - k\sqrt{E_s})^2}{kN_0}\right] \Big/
$$

$$
\sqrt{\pi kN_0}\, dz
$$

$$
= \int_{\sqrt{2kE_s/N_0}}^{\infty} [\exp(-x^2/2)/\sqrt{2\pi}]\, dx
$$

$$
\triangleq \text{erfc}\,\sqrt{2kE_s/N_0} \tag{5.130}
$$

where $E_s$ is the symbol energy. The bound on $P_E$ is

$$
P_E < \sum_{k=d}^{\infty} a_k P_k, \tag{5.131}
$$

where $a_k$ are the coefficients of

$$
T(D) = \sum_{k=d}^{\infty} a_k D^k \tag{5.132}
$$

and $d$ is the minimum distance between any two paths in the code. We may simplify this procedure considerably while loosening the bound only slightly for this channel by observing that for $x \geq 0$, $y \geq 0$,

$$
\text{erfc}\,\sqrt{x + y} \leq \exp(-y/2)\,\text{erfc}\,\sqrt{x}. \tag{5.133}
$$

Consequently, for $k \geq d$, letting $\ell = k - d$, we obtain from (5.130)

$$P_k = \text{erfc}\sqrt{2kE_s/N_0} = \text{erfc}\sqrt{2(d + \ell)E_s/N_0}$$

$$\leq \exp(-\ell E_s/N_0)\,\text{erfc}\sqrt{2d\,E_s/N_0}. \tag{5.134}$$

Hence the bound of (5.131), using (5.132), becomes

$$P_E < \sum_{k=d}^{\infty} a_k P_k \leq \text{erfc}\sqrt{2dE_s/N_0}\sum_{k=d}^{\infty} a_k \exp[-(k-d)E_s/N_0]$$

or

$$P_E < \text{erfc}\sqrt{2dE_s/N_0}\,\exp(dE_s/N_0)\,T(D)\Big|_{D=\exp(-E_s/N_0)}. \tag{5.135}$$

The bit error probability can be obtained in exactly the same way. Just as in the case of BSC, we have

$$P_B < \sum_{k=d}^{\infty} c_k P_k, \tag{5.136}$$

where $c_k$ are the coefficients of

$$\frac{dT(D,N)}{dN}\Big|_{N=1} = \sum_{k=d}^{\infty} c_k D^k. \tag{5.137}$$

Thus, following the same arguments which led from (5.131) to (5.135), we have for a binary-tree code,

$$P_B < \text{erfc}\sqrt{2dE_s/N_0}\,\exp(dE_s/N_0)\,dT(D,N)/dN\Big|_{N=1,D=\exp(-E_s/E_0).} \tag{5.138}$$

Since the symbol energy relates to the uncoded bit energy by

$$E_s = R\,E_b m, \tag{5.139}$$

(5.138) becomes

$$P_B < \text{erfc}\sqrt{\frac{2dRmE_b}{N_0}}\,\exp\left[dRm\left(\frac{E_b}{N_0}\right)\right]\frac{dT(N,D)}{dN}\Big|_{N=1,D=\exp\left[-Rm\left(\frac{E_b}{N_0}\right)\right].} \tag{5.140}$$

Based on either (5.138) or (5.140), the bit error rate performances of Viterbi decoders can be calculated. For rate 1/2 codes of constraint lengths from 3 to 6 the performances are shown in Figure 5.28.

A fixed amount of path history storage is needed in performance determination of Viterbi decoding. The amount of path storage required is equal to the number of states multiplied by the length of the information bit path history per state (or path length for short). It has been demonstrated theoretically and through simulation that a value of five times the code constraint length is sufficient.

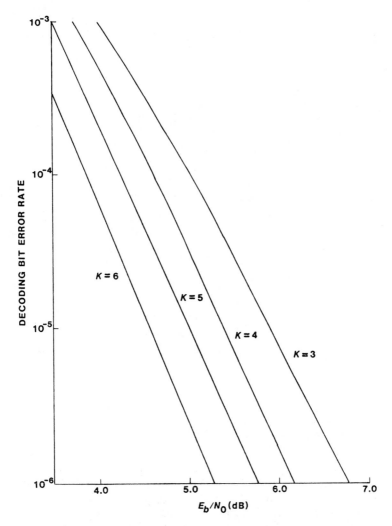

**Figure 5.28** Rate 1/2 bit error rate calculation from equation (5.140). (*Used by permission of Dr. Andrew Viterbi.*)

### 5.6.2.2  Performance Evaluation

Heller and Jacobs and others have simulated and evaluated Viterbi decoder performance [21]. Figure 5.29 shows the decoding bit error rate versus $E_b/N_0$ for rate 1/2 decoders with hard decision and 32-bit decoder path length. It can be observed that each increment in code constraint length in the figure is an improvement in coding gain of about 0.5 dB at a bit error rate of $10^{-5}$. Also, the higher the $E_b/N_0$ value, the larger the bit error rate improvements for larger constraint length codes.

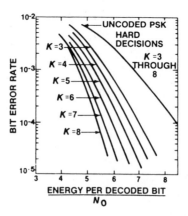

**Figure 5.29** Rate 1/2 Viterbi decoder performance with 32-bit path length [21].

With $K = 6$ and rate 3/4, the performance of the hard decision Viterbi decoder is shown in Figure 5.30. At $E_b/N_0 = 6.0$ dB, the bit error rate is about $3 \times 10^{-4}$. At $E_b/N_0 = 7.0$ dB, the error rate is shown to be $2 \times 10^{-5}$.

All soft-decision performance improvements on the Viterbi decoder will be discussed, along with soft-decision decoding schemes for other codes, in Section 5.7.

### 5.6.3 Sequential Decoding

Sequential decoding, the second probabilistic decoding procedure for convolutional codes, searches sequentially for the most likely message path for a received code sequence. It successively explores the encoded tree in an attempt to find the path whose metric compares with a predetermined but variable running threshold. The concept behind sequential decoding is to find the decoding metric for each message and extend the tree from node to node with the largest value in decoding metric. A decoding metric is a probabilistic quantity that depends on the transmission channel transitional probability, the codeword length, and the probability of the coded message being sent, and that can be maximized for minimal error decoding.

Sequential decoding was first discovered by Wozencraft and later refined by Fano, Zigangirov, and Jelinek [22–25]. The algorithm in the last two papers is referred to as a stack singular algorithm. The term ''stack'' is used to indicate that the decoder stores the previously explored nodes in a stack with decreasing metric down into the stack and, at each step, extends the node at the top of the stack. The top node of the stack has the greatest likelihood of detecting the transmitted codeword. The stack is initially loaded with the original node whose metric is set to zero. The following decoding rules are then executed:

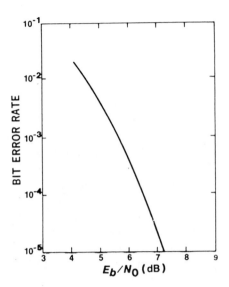

**Figure 5.30** Performance curve for the $K = 6$ Viterbi decoder with hard receiver quantization, code rate $= 3/4$ [21].

1. Compute the metrics of the two successors of the top node and add them to the stack in places determined by their metrics.
2. Delete the node whose successors were just added.
3. If the top node is in the last level of the tree, stop. Otherwise go back to 1.

Sequential decoding is restricted to code rates less than the channel computational cutoff rate $E(\rho,Q)$, which was defined in Section 2.6.2.1, Volume I. $E(\rho,Q)$ is also referred to as a channel quality factor. For an AWGN channel of large bandwidth, the channel cutoff rate is one-half of the channel capacity at low signal-to-noise ratio.

Massey and Costello have evaluated, compared, and suggested rate 1/2 convolutional codes with Fano sequential decoding for the deep space channel [26]. Earlier, Massey virtually compared all the rate 1/2 binary tree codes with equivalent constraint length of 24 and recommended the following codes for the International Ultraviolet Explorer (IUE) spacecraft sponsored by NASA:

$$G(0) = \begin{cases} 74042402 \\ 07121635 \\ 55346125 \end{cases}$$

$$G(1) = \begin{cases} 54042402 \\ 07121635. \\ 75744143 \end{cases}$$

These codes were decoded from 50,000 frames of length 256 on the simulated deep-space channel with $E_b/N_0$ of 3.0 dB.

### 5.6.3.1  Error Probability Bound

There are two types of error possibilities in sequential decoding. The first one is the undetected error, which occurs when the decoder accepts a number of wrong hypotheses but moves forward and ends up on the correct path. The second type of decoding error results from buffer overflow caused by excessive computation needed in the search. More often the limitation of sequential decoding is bounded by the number of steps in the computation $C$, a random variable with Pareto distribution. The probability that the number of computations $C$ exceeds some limit $L$ can be approximated by $\Pr(C > L) \simeq C_f L^{-\alpha(E)}$, where $\alpha(E)$ is the Pareto exponent, which is inversely proportional to the data rate, and $C_f$ is a constant factor. $\alpha(E) = 1$ is considered the capability of sequential decoders. The decoding limit $L$ is determined by decoding speed and buffer size.

For a rate $R = k_o/n_o$ convolutional code with constraint length $n_A = (1 + m_s)n_o$ in sequential decoding, the probability that an arbitrary subblock of $n_o$ bits is decoded in error is derived by Gallager [27].

$$P_e(n_o) \leq A \exp[-n_A E(1, Q)] \tag{5.141}$$

where

$$\Delta A = \left[\frac{m_s + 1}{(1 - z)^2} + \frac{1 + z}{(1 - z)^3}\right] \exp\left[n_o R + \frac{\Delta}{2}\right] \tag{5.142}$$

and

$$z = \exp\{n_o[E(1, Q) - R]\}. \tag{5.143}$$

The bound of (5.141) is valid for

$$R < E(1, Q) = -\ell n \sum_{j=0}^{J-1}\left[\sum_{k=0}^{K-1} Q(k) \sqrt{P(j/k)}\right]^2, \tag{5.144}$$

with $\Delta$ being the running threshold spacing. $Q(k)$ and $P(j/k)$ are the input and transitional probabilities of a discrete memoryless channel with $K$ inputs and $J$ outputs. For binary symmetric channel $Q(0) = Q(1) = 1/2$ and bit error rate $p$, (5.144) simplifies to

$$E(1, 1/2) = \ell n\, 2 - 2\, \ell n\left[\sqrt{P} + \sqrt{(1 - p)}\right]. \tag{5.145}$$

If $p$ can be estimated in a BSC, $E(1, 1/2)$ is known. For a given tree code $n_o$, $R$, $m_s$, and $n_A$ are determined. $z$ can be first obtained from (5.143), then, given a running threshold spacing $\Delta$, $A$ in (5.142) can also be calculated, and in turn the bound of $P_e(n_o)$ can be obtained. A simple observation of (5.141) reveals that

1. $P_e(n_0)$ decreases exponentially with increasing either code constraint length $n_A$ or channel quality factor $E(1, Q)$, or both.
2. $P_e(n_0)$ decreases exponentially with decreasing running threshold spacing $\Delta$.

Since $E(1, Q)$ is a function of $E_s/N_0$ through the channel transition probabilities, the value of $E_b/N_0$ at which $R = E(1, Q)$ may be computed as a function of $m$, the number of bits per symbol, and the number of quantization levels $Q$. For $m \to 0$ and $Q \to \infty$, $E_b/N_0 \to 1.4$ dB, which is considered the performance limit of sequential decoding in BPSK channels [11].

### 5.6.3.2   Measured Decoding Performance

Forney and Bower foresaw the potential gains of sequential decoding [28]. Forney first simulated and later hardware-implemented the rate 1/2 systematic convolutional code of constraint length 47. The generator of this code is 7154737013174652. For phase ambiguity resolution in conjunction with a PSK modem, the code was modified as 7154737013174642 in order to be transparent. The measured performances of the decoder are shown in Figures 5.31 and 5.32, which give the code performance for data rates of 9,600 bps and 5.0 Mbps. The results indicate that

1. The decoding error rate increases with increasing data rate for any given $E_b/N_0$.
2. In the decoder, the decoding error rate decreases with increasing memory capacity.

The back-search limit $(BL)$ was 240 and the metric was $1/-11$, with memory capacity varying from $2.^{10}$ to $2^{16}$ branches.

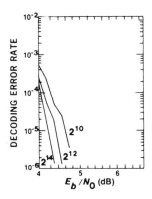

**Figure 5.31** Sequential decoder performance, digitally generated errors. $R_b = 9600$ bit/s, metric ratio—$1/-11$, back-search limit—240. Parameter is number of branches stored in bulk buffer memory [28].

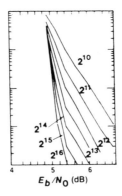

**Figure 5.32** Performance as in Figure 5.31, except with $R_b = 5$ Mbit/s [28].

A state-of-the-art sequential decoder performance from M/A-COM-Linkabit Corporation is shown in Figure 5.33, in which we reconfirm that both the coding rate and signaling speed are factors of sequential decoder performance. The decoder capability at $BL > 160$ and 5.0 Mbps data rate and a 5.0 dB $E_b/N_0$ coding gain at $10^{-5}$ is significantly demonstrated by Forney and Bower.

Cahn, Huth, and Moore simulated sequential decoding performance in conjunction with phase lock demodulation [29]. With a rate 1/2 convolutional code of constraint length 64 and decoding buffer size $1.3 \times 10^5$ bits, their results are shown in Figure 5.34. At $10^{-5}$, more than 4 dB gain appears possible for hard decision ($Q = 1$) over uncoded CPSK channel. An additional 1.0 dB is seen for $Q = 2$ (4-level quantization), and 0.4 dB more improvement for $Q =$

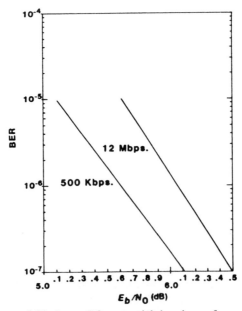

**Figure 5.33** A rate 7/8 sequential decoder performance.

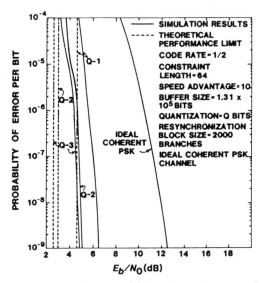

**Figure 5.34** Comparison of sequential decoder performance for 1, 2, and 3 bits of quantization [29].

3. The dotted curves are the corresponding theoretical limitations of sequential decoding with the three $Q$ values considered.

### 5.6.3.3   Performance Comparison of Fano and Stack Algorithms

Performance comparisons of Fano and Stack algorithms were provided by Geist through computer simulations of a nonsystematic convolutional code of memory [30]. This code was constructed by Costello with the code generator polynomials in octal numbers as

$$G(0) = 533533676737$$

$$G(1) = 733533676737.$$

The time to decode a frame was chosen as a measure of decoding effort, because the time required is roughly proportional to the number of computations. If a finite amount of time is allowed for decoding each frame, and any frame cannot be completely decoded in the maximum allocated time $t$, it must be considered as an erasure. Then $P_r\{T > t\}$ is the erasure probability, which is plotted against $t$ for the comparison of the Jelinek and the Fano algorithms. For binary symmetric channels with channel transitional probabilities 0.033, 0.045, and 0.057, the simulation results are shown in Figures 5.35 to 5.37. Figure 5.37 shows the distribution of computation time in Gaussian channel.

It can be observed from the curves that the Jelinek algorithm performs faster than the Fano algorithm for higher channel transition probabilities. At a lower channel transition probability, the Fano algorithm can perform faster at higher values of erasure probabilities. The crossover point depends on the channel conditions as shown.

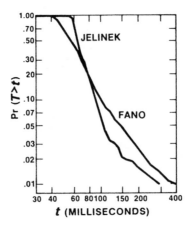

**Figure 5.35** Distribution of computing time, binary symmetric channel, $p = .033$ [30].

In all of the above simulations, Geist defined a computation step for the Fano algorithm as an entry of either the function of the "look forward on best branch" or the function of "look forward on next best branch" [30]. A computation step in the Jelinek algorithm is defined as an execution of step 1 of the algorithm described in Section 5.6.3.

A time-consuming search is required during decoding step 1 for the insertion of two new nodes into the stack. The algorithm is then modified by letting the nodes enter equally spaced bins. Each bin provides nodes of a specified metric value. Hence the placement of new nodes depends only on its metric value and not on the metric values of other nodes in the stack.

### 5.6.3.4  *Performance of the Multiple Stack Algorithm*

For erasure-free sequential decoding, the multiple stack algorithm (MSA) was introduced by Chevillat and Costello [31]. The algorithm is shown to be capable

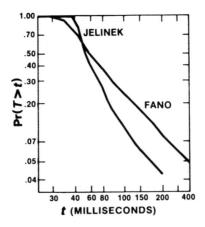

**Figure 5.36** Distribution of computing time, binary symmetric channel, $p = .045$ [30].

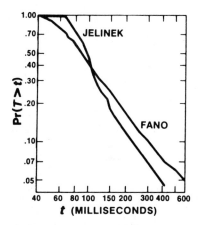

**Figure 5.37** Distribution of computing time, Gaussian channel [30].

of providing relatively low error rates at substantially higher decoding speeds than the Viterbi decoding algorithm. Although this advantage in computational speed is achieved at the expense of an increase in storage complexity, it is done with modest decoding effort. In some high-speed digital satellite communication systems, however, such a tradeoff may be justifiable. An example of a high-speed satellite system is the Tracking Data Relay Satellite System (TDRSS)/ Advanced Westar domestic satellite communication system, which consists of four $K$-band (14/12 GHz) transmission channels, each with a nominal bandwidth of 225 MHz. Each transmission channel supports a digital transmission rate of 250 Mbps using QPSK. Another example is the INTELSAT TDMA, which is operated at 120.0 Mbps. No single Viterbi decoder can operate at these speeds at present.

The merit of MSA lies in the exploitation of the fact that the decoding effort is relatively independent of the code constraint length. The basic idea of MSA is a process of coarse tuning and then fine tuning. The general philosophy of MSA emerges from the use of a number of memory stacks. The first stack is made large enough so that only decoding words with a large number of errors will force the use of other stacks. In contrast to conventional stack algorithms, the MSA advances quickly through the tree and finds an estimate. Only then does the MSA explore in detail all the possible paths in search of the most likely one. As it turns out, the modified scheme provides some very interesting results.

The MSA operates initially exactly like the conventional or single stack algorithm (SSA) for sequential decoding. Starting with the origin node, the path of the original node in the stack is extended. If maximum path metric is not obtained the others are eliminated, the successors are inserted, and the stack is ordered according to the node likelihood values or metrics. Each stack entry consists of a node identifier and its metric. Decoding proceeds by extending the node at the top of the stack again.

If a terminal node in the tree is reached before the first stack fills up, decoding is completed and the path leading from the origin to this terminal node becomes the decoded word. Therefore, if decoding can be completed in the first stack, the MSA executes exactly the same decoding steps as the SSA and thus outputs exactly the same decoding decision; i.e., decoding is asymptotically maximum likelihood for these received words.

However, if the received sequence is one of those few which require extended searches, i.e., a potential erasure, the first stack fills up. In this case, the top nodes of the first stack, i.e., those with the best metrics, are transferred to a second stack. Decoding then proceeds in the second stack using only these transferred nodes.

If the top node in the second stack reaches the end of the tree before the stack fills up, the terminal node is stored as a tentative decision in a special register. The decoder then deletes the remaining nodes in the second stack and returns to the first stack where decoding continues. If the decoder reaches a terminal node before the first stack fills up again, the metric of the new terminal node is compared with that of the tentative decision. The node with the larger metric is retained and becomes the final decoding decision. It can be shown that this decision must be at least as good as that made by the SSA with an arbitrarily large stack.

However, if the first stack fills up again before the end of the tree is reached, a different stack is formed by transferring the top nodes of the first stack. If the second stack fills up also, a third stack is formed by transferring the top nodes from the second stack. Additional stacks are formed in a similar manner until a tentative decision is made. The decoder always compares a new terminal node's metric with the contents of the tentative decision register and retains the node with the best metric. The rest of the nodes in that stack are then deleted and decoding proceeds in the previous stack.

The algorithm terminates if it reaches the end of the tree in the first stack. The only other way decoding can be completed is by exceeding the computational limit $C_{lim}$. In this case the best tentative decision obtained becomes the final decoding decision.

As shown in Figure 5.38, the function of the multiple stack algorithm can be summarized in steps as follows:

Step 1: *Initialization.* Set all memory stacks to zero. Place the origin (or root) node of the estimated code tree into the first stack.

Step 2: *Path Extension.* The path of the decoding tree is extended by identification of the largest path metric computation. As a consequence a new node is merged and the original node is eliminated from this stack.

Step 3: *Stack Entry Reordering.* From the new node found in Step 2, compute the new branch metrics, which place in the stack in order according to the values of the metrics.

Step 4: *Stack Producing.* If the memory of the first stack is full, then transfer its content to other stacks until a temporary decision is reached for a complete path of a block of information digits.

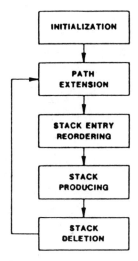

**Figure 5.38** Function of the multiple stack algorithm.

Step 5: *Stack Deletion*. Improve the temporary decision reached in Step 4 by processing stacks in reverse order. The stacks are deleted if they do not provide the largest path metric value.

The MSA terminates when either Step 5 reaches the decoding computational limit or a decision is reached at the first stack.

A performance comparison of the Viterbi algorithm (VA) and MSA is shown in Figure 5.39. Decoding speed and complexity were compared at equal or comparable error rates. Curves 1 and 2 show the error rate performances of constraint lengths 4 and 5 for the Viterbi decoders, and the other curves show those for the MSA decoder with a first stack size of 85. The VA $K = 4$ code performs $C_{VA} = 2^3 = 8$ node extensions per information bit and achieves an error rate $1.5 \times 10^{-3}$ at 5.5 dB $E_b/N_0$. The MSA achieves the same error rate with only $C_{avg} = 3.35$ computations per information bit (2.4 times less). At 6.5 dB, the computational reduction is about six times. Curves 2 and 3 achieve about the same error rate $5 \times 10^{-4}$ at 5.5 dB. The computation of VA is $C_{VA} = 16$ node extension, while the MSA needs $C_{avg} = 1.37$, or about twelve times less. However, VA needs only 32 buffer units while 530 are required for MSA. If the MSA's computational upper limit is lowered from $C_{lim} = 4096$ (curve 3) to $C_{lim} = 2048$ (curve 4), $C_{avg}$ decreases to 1.3 with a slight increase in error rate.

Figure 5.40 shows another decoding error performance for MSA and VA under the same channel noise condition. All three codes are with rate 1/2. The Viterbi decoder has a convolutional code of constraint length 8. The code is represented by $(2, 1)8$. The selected parameters for the $(2, 1)8$ and $(2, 1)15$ MSA are shown in Table 5.1. The reason for such parameter selection is interestingly discussed by Ma [32].

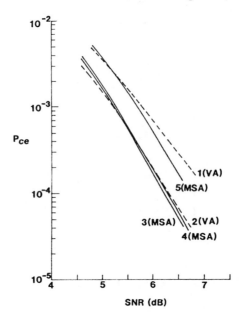

**Figure 5.39** Error probability as a function of SNR, VA, and MSA, $k = 60$ [31].

It can be seen from Figure 5.40 that the (2, 1)8 VA performs better than the (2, 1)8 MSA but performs worse than the (2, 1)15 MSA. The increase of constraint length from 8 to 15 implies a modest increase in complexity for the MSA, as was expected. But the main tradeoff is the speed of the decoder operation, as we stated at the beginning of this section.

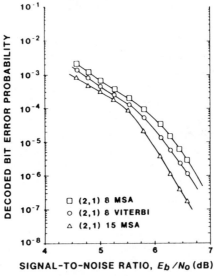

**Figure 5.40** Error performance comparisons of the (2, 1)8 MSA, (2, 1)8 Viterbi algorithm, and (2, 1)15 MSA [32].

**Table 5.1**

MSA parameter selection.

| Parameters / Codes | (2, 1)8 | (2, 1)15 |
|---|---|---|
| The size of the first stack | 1024 | 2900 |
| The size of the $i$th stack | 11 | 11 |
| Decoding constraint length | 64 | 64 |
| Computational limit | 1700 | 3600 |
| Number of transferred nodes due to stack overflow | 3 | 3 |

As the speed of digital satellite communication systems pushes higher and the cost of memory devices becomes lower, MSA will prove to be one of the promising error-control techniques.

### 5.6.4 Nonbinary Character Error-Correcting Convolutional Codes

In order to avoid the sub-burst message length and fixed block code length problem in the nonbinary block codes, convolutional character correcting codes may be used. These codes, developed by Ebert and Tong, have greater error-correcting capability than self-orthogonal threshold decodable codes [40]. The codes are capable of correcting $E$ character errors within a constraint length of $(2E^2 - E + 1)/(1 - R)$ channel characters. While $R$ is the encoding rate, the channel characters must be of a finite field of at least $[R(2E - 1)(E - 1) + 1]/(1 - R)$ elements.

In terms of the minimum distance $d$ of the code the required nonzero field elements are at least $(1/1 - R)[(d - 2)(d - 3) + 2]/2$. For example, for the correction of 2 character errors $(d = 5)$ with a rate of at least 7/8 the code constraint length is $n/(1 - R) = \{[(d - 2)(d - 1)/2] + 1\}/(1 - R) = 56$. Using 4 bits per character the code then has an overall constraint length of 56 $\times$ 4 = 224 bits. This class of codes can be encoded by a standard convolutional encoder or by a number of accumulators that can also perform multiplication at high speed. Decoding can be accomplished by a modified BCH decoder.

## 5.7 SOFT-DECISION DECODING

With the exception of sequential decoding performances of Figure 5.35, almost all coding gain performances presented thus far have been computed on the basis of hard-decision detections in the receiver. That is, the output of the demodulator is sampled and a two-level decision, either 1 or 0, is made on each bit. However, the reliability of the information contained between the two extreme levels is lost when such a hard decision is forced to be made. When the received bit signal level is quantized into $2^Q$ subdivisions and then a threshold level for decision is set, the detection process is called soft decision, which can improve all hard-decision coding performance.

Assume that a demodulator output assembles a block of $n$ numbers from demodulating a received codeword. Each number (or level) has a mean $\pm \sqrt{E_b R m}$ and a variance $N_0/2$. For a code of length $n$ and minimum correcting capability $2d$ in soft decision, the word error rate in such a Gaussian process is

$$P_W = \int_{-\infty}^{0} \frac{1}{\sqrt{2\pi}\,\sigma} \exp\left( -\frac{(x - 2d\sqrt{RE_b m})^2}{2\,\sigma^2} \right) dx \qquad (5.146)$$

with

$$\sigma^2 = \frac{2d\,N_0}{2} = d\,N_0. \qquad (5.147)$$

Let

$$t = \frac{x - 2d\sqrt{RE_b m}}{\sqrt{2d\,N_0}}; \qquad (5.148)$$

then

$$x = -\infty \rightarrow t = -\infty$$

$$x = 0 \rightarrow t = \sqrt{\frac{2d\,RE_b m}{N_0}} \qquad (5.149)$$

and

$$\frac{dx}{dt} = \sqrt{2d\,N_0}. \qquad (5.150)$$

Thus (5.146) becomes

$$P_W = \frac{1}{\sqrt{\pi}} \int_{-\infty}^{-\sqrt{\frac{2d\,RE_b m}{N_0}}} \exp(-t^2\,dt)$$

$$= \frac{1}{2} - \frac{1}{2} \operatorname{erf}\left( \sqrt{\frac{2d\,RE_b m}{N_0}} \right) < \frac{1}{2} \exp(-2dRE_b m/N_0). \qquad (5.151)$$

Then, similar to the argument used for hard-decision decoding, we have

$$P_B' < \frac{E + 1}{n} P_W = \frac{E + 1}{2n} \exp(-2d\,mR(E_b/N_0)_s) \qquad (5.152)$$

with the uncoded bit error rate

$$P_B^{(1)} < \exp(-(E_b/N_0)_2\, 2m),$$

where $(E_b/N_0)_2$, $(E_b/N_0)_s$ denote the required signal-to-noise ratios, both uncoded and coded with soft decision. To obtain the coding gain at the same bit error rate, equate (5.152) to (5.82) and take $\log_e$ of both sides. We have

$$snr(o) = \left(\frac{E_b}{N_0}\right)_2 \le dR\left(\frac{E_b}{N_0}\right)_s - \frac{1}{2m}\,\ell n\left(\frac{E+1}{2n}\right). \tag{5.153}$$

Let $(E_b/N_0)_s$ in (5.153) be denoted $S_D$. For the hard-decision decoding, (5.84) gives

$$\left(\frac{E_b}{N_0}\right)_2 = R(E+1)\left(\frac{E_b}{N_0}\right)_c - \frac{1}{2m}\,\ell n\left(\frac{n^E}{E!}\right). \tag{5.154}$$

Let $(E_b/N_0)_c$ in (5.154) be denoted as $H_D$. Equate (5.153) to (5.154) for the same uncoded $(E_b/N_0)_2$ in order to obtain $S_D$.

$$S_D = \frac{E+1}{d}H_D + \frac{1}{2d\,Rm}\,\ell n\left[\frac{E+1}{2n}\Big/\left(\frac{n^E}{E!}\right)^{1/2m}\right]$$

$$< \frac{E+1}{2E+1}H_D - \frac{1}{2d\,Rm} \tag{5.155}$$

with

$$n \gg \left[E!\left(\frac{E+1}{2n}\right)^m\right]^{1/E}. \tag{5.156}$$

To obtain the soft-decision coding gain over hard-decision coding gain we divide $S_D$ in (5.156) by $H_D$:

$$\Delta'_G = \frac{S_D}{H_D} \simeq \frac{E+1}{2E+1} - \frac{1}{2H_D\,dRm}. \tag{5.157}$$

For large $E$,

$$\Delta'_G \simeq \frac{1}{2} - \frac{1}{2H_D\,dRm}. \tag{5.158}$$

For useful codes, $d \gg 1$; for efficient codes, $R \to 1$. For most digital modulation schemes, $m > 1$. $H_D$ is theoretically bounded by (5.7) of Section 5.2.2 for any code. Most practical values of $H_D$ can range from $|2|$ to 10. When these values are considered in (5.158), it is easy to conclude that $|H_D\,dRm| \gg 1$ and (5.158) reduces to

$$\Delta'_G \simeq \frac{1}{2}, \tag{5.159}$$

which shows that the difference between soft-decision and hard-decision coding gains can be approximately 3.0 dB. Because of implementation losses and finite quantization, (5.159) actually establishes the soft-decision improvement bound in an additive Gaussian channel for either block or convolutional code decoding schemes.

The idea of soft-decision decoding began with Elias's erasure channel [42]. Decoding block code for erasure channels has been studied by Epstein and Forney [43]. An *a posteriori* probability (APP) threshold decoding scheme has been formulated by Massey using channel statistics. Recent development of soft-decision decoding algorithms has been advanced by Chase, Dorsch, Weldon, and Hartmann and Rudolph [35–38]. In the following subsections, we shall discuss the soft-decision algorithms that can provide additional coding performance improvement, such as Chase's algorithm and its performances, and the coding performance improvements of Rudolph. In addition to the soft-decision performance for Viterbi decoding and a modified soft-decision algorithm for threshold decoding, we shall include computer simulation results.

### 5.7.1 Decoding with Channel Measurement Information for Block Codes

Chase constructed a soft-decision decoding algorithm for block codes and analyzed the expected gain [35]. The soft-decision decoding algorithm is optimum in the sense that with increasing signal-to-noise ratio the asymptotic behavior of the probability of error cannot be improved. This result can be shown to be true for both the white Gaussian noise and for fading channels.

The concept of decoding with channel measurement information can be illustrated with the aid of the block diagram in Figure 5.41. The designer of the encoder and decoder typically models a communication system, as shown by the block diagram, with the major modification that no channel measurement

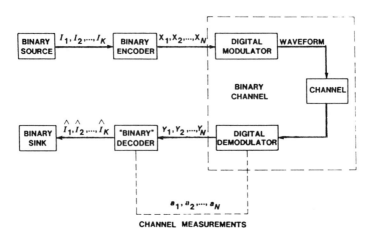

**Figure 5.41** Decoding with Chase's channel measurement information [35].

information is fed into the decoder. Conventionally, the data modulator and demodulator are combined with the actual communication link so that a binary communication channel results, with a true binary decoder following the data demodulator. The use of channel measurement information by the decoder, in conjunction with the algebraic properties of the binary code, represents the nonconventional part of this communication system model.

The type of channel measurement information used is a set of $N$ confidence values for the set of $N$ digits in the received binary word. Each digit is assigned a confidence value which, in some applications, can be chosen so that it is monotonically related to the probability that this binary digit is received correctly. As an example, if the analog decision statistic for a given transmitted digit is $v_i$, the optimum decision rule for a given received digit is

$$Y_i = \begin{cases} 1, & \text{if } \Pr[X_i = 1 \mid v_i] \geq \Pr[X_i = 0 \mid v_i] \\ 0, & \text{otherwise.} \end{cases} \tag{5.160}$$

That is, the most probable value of $Y_i$ is chosen. The confidence value of each $Y_i$ can now be chosen as the value, or monotonically related to the value, of the a posteriori probability given by $\Pr(Y_i \mid v_i)$. For many communication channels this confidence value is equivalent to the magnitudes of the decision statistics of each digit in the codeword.

To describe this decoding algorithm, we start by describing conventional algebraic decoding in the following manner: The received sequence $\vec{Y} = Y_1, Y_2, \ldots, Y_N$ is used to compute a syndrome which, by algebraic decoding, is used to obtain an estimate of the error sequence given by $\vec{Z} = Z_1, Z_2, \ldots, Z_N$. An estimate of the transmitted code is thus given by $\vec{X} = \vec{Y} \oplus \vec{Z}$, where the notation $\oplus$ represents term-by-term modulo-2 addition. Conventionally, for a given syndrome, $\vec{Z}$ is chosen as the error sequence of minimum Hamming weight, where the Hamming weight of $\vec{Z}$ is

$$w(\vec{Z}) = \sum_{i=1}^{N} Z_i. \tag{5.161}$$

The number of possible error sequences that give rise to the same syndrome is equal to the number of codewords. In terms of $\vec{Z}$, these are given by

$$\vec{Z}_m = \vec{Z} \oplus \vec{X}_m \qquad \text{for } m = 1, 2, \ldots, 2^k. \tag{5.162}$$

This is by definition a coset. The conventional decoder picks the coset leader which is that element with minimum Hamming weight.

The soft-decision decoder makes use of the set of channel measurements $\{\alpha_i\}$ related to the received sequence $\vec{Y}$. $\{\alpha_i\}$ can be considered as a measure of the reliability of the received digits $\vec{Y}$. For the Gaussian channel, the $\{\alpha_i\}$ are taken as the analog output of a matched filter.

When we are decoding with channel measurement information, given by the set of $\vec{\alpha} = \alpha_1, \alpha_2, \ldots, \alpha_N$, the $\vec{Z}$ of minimum binary weight is no longer the best choice. We will now choose the error pattern of minimum analog weight from some given set of possible error patterns. The analog weight of a given error pattern $\vec{Z}$ is defined as

$$W_\alpha(\vec{Z}) = \sum_{i=1}^{N} \alpha_i Z_i. \tag{5.163}$$

Thus the choice of an error pattern is based on the weighted sum of potential errors rather than just the number of errors (number of 1s) in a given $\vec{Z}$.

The algorithm then operates as follows: The received vector $\overline{Y}$ is first hard quantized into the estimated binary vector $\overline{X}' = (X_1', X_2' \ldots X_n')$. Next the $c$ least reliable components of $\overline{X}'$ are located. Then $2^c$ binary vectors

$$X'(\boldsymbol{\epsilon}) = X'(\epsilon_1, \epsilon_2, \ldots, \epsilon_c)$$

$$= (\epsilon_1, \epsilon_2, \ldots, \epsilon_c, X_{c+1}', X_{c+2}', \ldots, X_n') \tag{5.164}$$

are generated. Finally, each $X'(\boldsymbol{\epsilon})$ is decoded by the binary decoding algorithm. Since there are $2^c$ $X'(\boldsymbol{\epsilon})$ vectors, $2^c$ applications of the binary decoding algorithm are needed to produce $\{\overline{X}^{(1)}, \overline{X}^{(2)}, \ldots \overline{X}^{(c)}$ number of at most $c$ distinct codewords. The soft-decision decoder selects the codeword for which the inner product $\overline{X}^{(i)} - \overline{Y}$ is largest.

For $c = 0$ the algorithm reduces to hard decision. For $c = n$ the algorithm becomes a maximum-likelihood decoding algorithm. Thus, basically, Chase's algorithm is a quasi-maximum-likelihood decoding algorithm, but we shall show that only small values of $c$ can contribute to significant coding improvement.

Figure 5.42 shows the performances of the (23, 12, 7) Golay code with hard decision ($c = 0$) and soft decisions ($c = 2, c = 4$). The improvement over hard-decision decoding is more than 1.5 dB even at very high decoding error rates. Comparisons are also made in the figure with no coding and Viterbi decoding of rate 1/2, $K = 7$ tree code. Figure 5.43 shows the performance improvements resulting from soft-decision of the rate 1/2 (32, 16) second-order

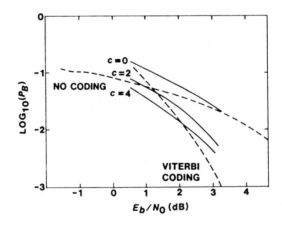

**Figure 5.42** (23, 12) Golay $E = 3$, $Q = 3.65$ [39].

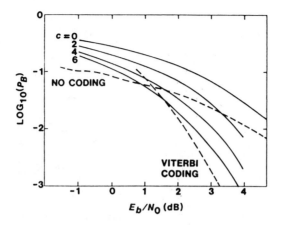

**Figure 5.43** $(32, 16)$ $R - M E = 3, Q = 4$ [39].

Reed-Muller code with $c = 2, 4, 6$. A 2 dB improvement is possible even at decoding error rate of $10^{-2}$. Figures 5.44 and 5.45 give the performance improvements of $(63, 36, 11)$ BCH code and the $(95, 39, 19)$ shortened BCH code. These results were obtained by Baumert and McEliece [39].

### 5.7.2  Optimal Bit-by-Bit and Maximum Radius Decoding

By defining a soft-decision function

$$\rho(x) = \frac{1 - \phi(r_m)}{1 + \phi(r_m)} \tag{5.165}$$

with $\phi(r_m)$ being the likelihood ratio

$$\phi(r_m) = \frac{P_r(r_m/1)}{P_r(r_m/0)}, \tag{5.166}$$

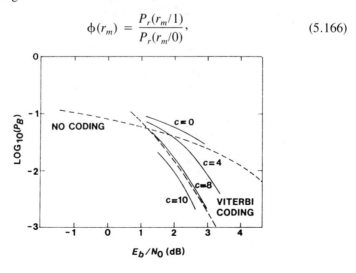

**Figure 5.44** $(63, 36)$ BCH $E = 5, Q = 6.29$ [39].

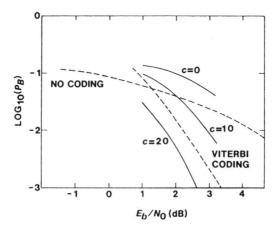

**Figure 5.45** (95, 39) Short BCH $E = 9$, $Q = 7.8$ [39].

where $r_m$ is the $m$th symbol of the received code word, Hartmann and Rudolph constructed the symbol-by-symbol decoding rule for linear codes [38]. For an $(n, k, d_m)$ binary block code the decoding rule is: Set $\hat{c}_m = 0$ if

$$\sum_{j=1}^{2^{n-k}} \prod_{\ell=0}^{n-k} \left(\frac{1 - \phi_\ell}{1 + \phi_\ell}\right)^{c_{j\ell} \oplus \delta_{me}} > 0 \qquad (5.167)$$

and $\hat{c}_m = 1$ otherwise. Here $c_{j\ell}$ denotes the code symbol in the $\ell$th position of the $j$th codeword of the dual $(n, n - k)$ code, and

$$\delta_{m\ell} = \begin{cases} 1, & \text{if } m = \ell \\ 0 & \text{otherwise.} \end{cases} \qquad (5.168)$$

$\oplus$ denotes modulo-2 addition.

By iterative extension of the algebraic analog decoding, a "maximum-radius" decoding (MRD) scheme has been shown to be asymptotically optimum for cyclic codes in AWGN channel. To reduce the complexity, the decoder corrects only all analog error patterns that fall within a Euclidean sphere whose radius is equal to half the minimum Euclidean distance of the code. The performance evaluations employing MRD for the (21, 11) and (73, 45) projective geometry (PG) codes in an AWGN channel are shown in Figures 5.46 and 5.47, respectively. The performances of the same code by means of optimum bit-by-bit decoding are also indicated.

### 5.7.3  Soft-Decision Performance in Viterbi Decoding

The hard-decision code performance of Viterbi decoding is shown in Figures 5.29 and 5.30 of Section 5.6.2. The improvement in performance in Viterbi decoding of a tree code with soft decision is shown in Figure 5.48 with the levels of quantization as parameters. A 2-dB improvement in $E_b/N_0$ can be seen

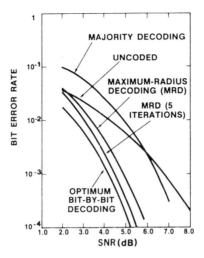

**Figure 5.46** Bit error rate of the (21, 11) PG code over AWGN channel (antipodal signaling) [38].

from hard-decision (2-level) to 8-level quantization almost independently of the decoding error rates.

Figure 5.49 shows the performance improvement due to 3-bit (8-level) quantization of the rate 1/2, $K = 7$ Viterbi decoder. The hard-decision decoder performance was shown in Figure 5.29. Again, we see the 2-dB difference between the soft- and hard-decision decodings. For phase ambiguity resolution, differential encoding may be used. The coding performances with and without differential encoding are also indicated. Figure 5.49 shows the performance of

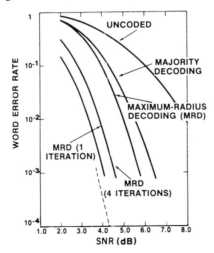

**Figure 5.47** Word error rate of the (73, 45) PG code over AWGN channel (antipodal signaling) [38].

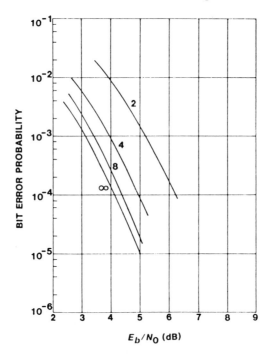

**Figure 5.48** Performance comparison of Viterbi decoding using a rate 1/2, $K = 5$ code with 2, 4, and 8 level quantization, path length = 32 bits [52].

the same rate 3/4 code as in Figure 5.30 but with 8-level soft decision. The codecs in Figure 5.49 were manufactured by the Linkabit Corporation.

The length of path memory is known to affect Viterbi decoder performance as well as quantization level. At a quantization level of 8, Figure 5.50 shows

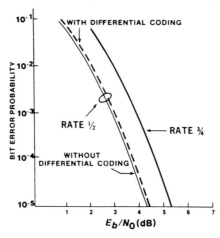

**Figure 5.49** Performance curve for the $K = 7$ code with 3-bit receiver quantization [19, 21].

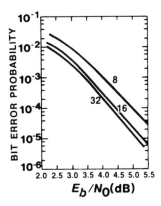

**Figure 5.50** Performance comparison of Viterbi decoder of $K = 5$, rate 1/2 code with bit path lengths of 8, 16, and 32, quantization level = 8 [19, 21].

the decoder output bit error rate performance versus $E_b/N_0$ for path lengths of 8, 16, and 32 for a rate 1/2 code of constraint length $K = 5$. The performance curve for path length of 32 is almost identical to that of an infinite path decoder.

For conventional rate $(n_0 - 1)/n_0$ Viterbi decoding, the state decision, i.e., addition, comparison, and selection function, requires $2^{n_0 - 1}$ number of computations at each state. Thus Viterbi decoding of high rate convolutional codes is a formidable task. But with bit deletion from a low rate code, high rate convolutional codes can be derived with the state decision computations reduced to binary decisions. As a consequence high rate Viterbi decoding becomes practically realizable. An example of such modified code performance with soft decision is shown in Figure 5.51. Both the rate 7/8 and the rate 15/16 convolutional codes are derived from the rate 1/2, $K = 7$ code. The codes are evaluated by the researchers at the KDD Laboratory. More will be discussed about such codec complexity in Chapter 6.

### 5.7.4  Generalized Minimum Distance Decoding

Using likelihood information in algebraic decoding algorithms, Forney proposed generalized minimum distance (GMD) decoding [34]. For hard-decision (HD) decoding in an AWGN channel with antipodal signaling, the probability of decoding error is

$$p_r(\text{HD}) \le \exp\left(-0.5\, dR\, \frac{E_b}{N_0}\right), \tag{5.169}$$

where $d$ and $R$ are the Hamming distance and code rate, respectively. At high $E_b/N_0$ the maximum-likelihood (ML) decoding and GMD decoding have the same approximated expression, that is

$$p_r(\text{GMD}) = p_r(\text{ML}) \simeq \exp\left(-dR\, \frac{E_b}{N_0}\right). \tag{5.170}$$

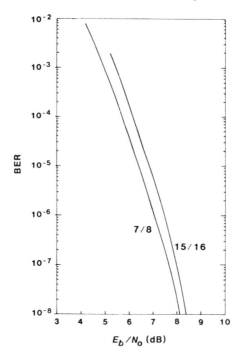

**Figure 5.51** Error rate performance of high rate Viterbi decoders with soft decision.

When decoding with erasure and errors, the probability of error is

$$P_r(\text{E\&E}) \leq \exp\left(- \ 0.686 \ dR \ \frac{E_b}{N_0}\right). \tag{5.171}$$

For small values of $E_b/N_0$, GMD decoding is no better than HD decoding, while for high $E_b/N_0$ channels, GMD decoding approaches the performance of ML decoding. Unfortunately, error controls are needed most where $E_b/N_0$ is small. Thus GMD decoding is not attractive for applications where the channels are noisy.

### 5.7.5  Weighted Erasure Decoding

Weighted erasure decoding (WED) is similar to GMD and uses part of the APP decoding procedure and Reddy's reliability indicator. WED was formulated by Weldon and relies on the assignment of a set of ordered weights

$$0 = W_0 \leq W_1 \leq W_2 \leq \cdots \leq W_{Q-1} = 1 \tag{5.172}$$

to the corresponding $Q$ quantized levels. The condition on the set of weights is

$$W_i + W_{Q-1-i} = 1 \tag{5.173}$$

for $i = 0, 1, \ldots, Q - 1$. The distance between any two quantized levels $L_i$ and $L_j$ is $|W_j - W_i|$. The decoding distance is then defined over the collection of such weights. The decoding rule is to decode the received word as the codeword with the smallest distance. WED can be realized by providing $r$ bits for every decoding symbol where $r$ is related to the quantization level by

$$2^{r-1} < Q \le 2^r. \tag{5.174}$$

The decoder input word is diagrammed first into an $r \times r$ matrix with binary valued elements. Based on the set of $r$ relative weights $V_i$ from each row of the word matrix, the weight of the $i$th digit is

$$W_i = C_{ri} V_r + C_{(r-1)i} V_{r-1} + \cdots + C_{1i} V_1 \tag{5.175}$$

with $C_{ri}$ being 0 or 1, and

$$\sum_{\sigma=1}^{r} V_\sigma = W_{Q-1} = 1. \tag{5.176}$$

Let $S_1$ and $S_0$ denote the index set of the rows corresponding to 1 and 0, respectively, in the first column of a tentatively decoded matrix. The decoding rule is then to choose the first information digit to be 0 if

$$\sum_{S_0} R_\sigma V_\sigma > \sum_{S_1} R_\sigma V_\sigma. \tag{5.177}$$

Otherwise, choose this digit to be a 1. In (5.177), Reddy's reliability indicator is

$$R_\sigma = \max(0, d - 2F_\sigma), \tag{5.178}$$

where the number of changes in row $\sigma$ is

$$F_\sigma = \begin{cases} E_\sigma, E_\sigma < d/2 \\ E_\sigma, \text{(correct decoding)} \ E_\sigma \ge \dfrac{d}{2} \end{cases} \tag{5.179}$$

$$F_\sigma \ge d - E_\sigma, \text{(incorrect decoding)} \ E_\sigma \ge d/2. \tag{5.180}$$

$E_\sigma$ is the number of bit errors in the $\sigma$th row.

For majority logic decoding of either block or tree code, WED can be simplified to provide weights to each row estimates of the word matrix. The sum of the weighted estimates of all rows $S_m$ is then compared with a threshold

$$T = \frac{1}{2} \sum_{i=1}^{r} V_r d, \tag{5.181}$$

for each row containing $d$ estimates. The final decision is obtained from

$$S_m \begin{cases} \ge T \to 1 \\ < T \to 0. \end{cases} \tag{5.182}$$

Computer simulations were carried out to examine the performances of WED for threshold decoding of convolutional codes which have been discussed in

Section 5.6.1 [16]. Performances of three 2-error correcting codecs (all $J = 4$) with rate 1/2, 3/4, and 7/8 are shown in Figures 5.52, 5.53, and 5.54, respectively. As shown in Figure 5.52, in the case of the rate 1/2 code, the gain of the 4-level WED over hard-decision performance is about 0.8 dB. As shown in Figure 5.53, the rate 3/4 code provides a difference of about 0.6 dB. For the rate 7/8 code the difference in gain decreases to 0.4 dB. At least for threshold-decodable self-orthogonal convolutional codes, this seems to indicate that the higher the code rate the less the improvement will be due to WED.

### 5.7.6 *A Posteriori* Probability Decoding

Massey introduced *a posteriori* probability (APP) decoding, which can be used to improve the performance of threshold decoders. The computation of the set of weights necessary in such decoding appears complicated in practice [40]. In the next section some simplifications of the input probability and weight calculations will be analyzed. Simulation results are presented in this section.

Suppose that $V$ is the value of the initial noise bit $e_0$ and $\{A_i\}$ is the parity check set. The algorithm is obtained by maximizing the conditional probability

$$\Pr[(e_0 = V) \mid \{A_i\}]. \tag{5.183}$$

According to Bayes' rule,

$$\Pr[(e_0 = V)\mid\{A_i\}] = \frac{\Pr[\{A_i\} \mid (e_0 = V)]\,\Pr(e_0 = V)}{\Pr[\{A_i\}]}. \tag{5.184}$$

Because of the orthogonality of set $\{A_i\}$ on $e_0$ and the bit independence of the noise sequence,

$$\Pr[\{A_i\}] \mid (e_0 = V)] = \prod_{i=1}^{J} \Pr[A_i \mid (e_0 = V)]. \tag{5.185}$$

Substituting (5.185) into (5.184) yields

$$\Pr[(e_0 = V) \mid \{A_i\}] = \frac{\displaystyle\prod_{i=1}^{J} \Pr[A_i \mid (e_0 = V)]\,\Pr(e_0 = V)}{\Pr[\{A_i\}]}. \tag{5.186}$$

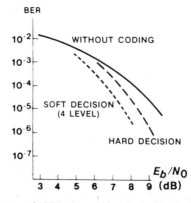

**Figure 5.52** Performance of WED for self-orthogonal code (coding rate: 1/2, $J$:4).

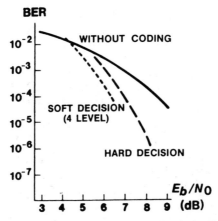

**Figure 5.53** Performance of WED for self-orthogonal code (coding rate: 3/4, $J$:4).

For a fixed set of $\{A_i\}$, the quantity that maximizes $\Pr[(e_0 = V) \mid \{A_i\}]$ is

$$\prod_{i=1}^{J} \Pr[A_i \mid (e_0 = V)] \Pr(e_0 = V). \qquad (5.187)$$

For optimum threshold decoding to determine whether the information bit $i_0$ is in error by maximizing (5.183) from the set of $J$ parity checks, $e_0$ should be chosen as 1 or 0 in (5.183). If $e_0 = 1$,

$$\Pr[(e_0 = 1) \mid \{A_i\}] > \Pr\{(e_0 = 0) \mid \{A_i\}]. \qquad (5.188)$$

Substituting (5.184) into (5.188) yields

$$\Pr[\{A_i\} \mid (e_0 = 1)] \Pr(e_0 = 1) > \Pr[\{A_i\} \mid (e_0 = 0)] \Pr(e_0 = 0). \qquad (5.189)$$

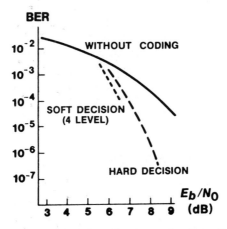

**Figure 5.54** Performance of WED for self-orthogonal code (coding rate: 7/8, $J$:4).

Substituting (5.185) into (5.189) results in

$$\Pr(e_0 = 1) \prod_{i=1}^{J} \Pr[A_i \mid (e_0 = 1)] > \Pr[e_0 = 0) \prod_{i=1}^{J} \Pr(A_i \mid e_0 = 0). \quad (5.190)$$

(5.190) can be rewritten as

$$\prod_{i=1}^{J} \frac{\Pr[A_i \mid (e_0 = 1)]}{\Pr[A_i \mid (e_0 = 0)]} > \frac{\Pr(e_0 = 0)}{\Pr(e_0 = 1)}. \quad (5.191)$$

Let

$$P_i = \Pr[(A_i = 0) \mid (e_0 = 1)] = \Pr[(A_i = 1) \mid (e_0 = 0)]$$

$$q_i = \Pr[(A_i = 1) \mid (e_0 = 1)] = \Pr[(A_i = 0) \mid (e_0 = 0)]; \quad (5.192)$$

and since $A_i$ is either 1 or 0,

$$A_i = 1 \Rightarrow \frac{\Pr[(A_i = 1) \mid (e_0 = 1)]}{\Pr[(A_i = 1) \mid (e_0 = 0)]} = \frac{q_i}{p_i} = \left(\frac{q_i}{p_i}\right)^{2A_i - 1} \quad (5.193)$$

$$A_i = 0 \Rightarrow \frac{\Pr[(A_i = 0) \mid (e_0 = 1)]}{\Pr[(A_i = 0) \mid (e_0 = 0)]} = \left(\frac{q_i}{p_i}\right)^{2A_i - 1}. \quad (5.194)$$

Substituting (5.193) and (5.194) into (5.191) yields

$$\prod_{i=1}^{J} \left(\frac{q_i}{p_i}\right)^{2A_i - 1} > \frac{q_0}{p_0} \quad (5.195)$$

for $A_i = 1$ or 0. Taking the logarithm of both sides of (5.195) results in

$$\sum_{i=1}^{J} (2A_i - 1) \log\left(\frac{q_i}{p_i}\right) > \log\left(\frac{q_0}{p_0}\right)$$

or

$$\sum_{i=1}^{J} A_i \left[2 \log\left(\frac{q_i}{p_i}\right)\right] > \sum_{i=0}^{J} \log\left(\frac{q_i}{p_i}\right). \quad (5.196)$$

(5.196) indicates that, for a binary memoryless channel with additive noise, the decoding rule is as follows: If the sum of the members of the parity check set $\{A_i\}$ weighted by the factor $\omega_i = \log(q_i/p_i)$ exceeds the threshold value $T = \frac{1}{2}\sum_{i=0}^{J} \log(q_i/p_i)$, $e_0 = 1$ should be chosen. This algorithm is summarized in Figure 5.55.

### 5.7.6.1  Performance of APP Decoding

The APP decoding algorithm can be applied to both block and convolutional codes with threshold decoding. Neuman and Lumb were the first to evaluate the performance gains of APP decoding [41]. Figure 5.56 shows the differences between hard-decision (majority decision) and APP decoding with hard-decision feedback (hard-decision APP) performances of the (44, 22, 9) convolutional

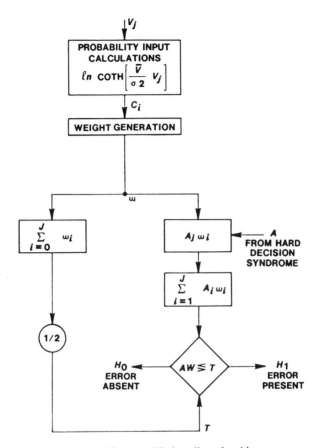

**Figure 5.55** The APP decoding algorithm.

code in BSC and AWGN channels. The improvement resulting from APP in an AWGN channel is shown to be between 1.0 and 1.5 dB, depending on the level of decoding bit error rate. Figure 5.57 indicates the computer simulation performances of (73, 45, 5) block code. Full APP decoding means that the bit error probability after decoding is fed back as its corresponding $c_j^B$, the set of probability inputs.

For the rate 7/8 constraint length 376 tree code described in Section 5.6.1.2, Rhodes and Lebowitz of COMSAT simulated a modified APP addition with 8-level quantization. The modification placed the weighting factors on the received information symbols only. For 8-level quantization, the information symbol is classified into four reliability levels, 1, 3, 5, and 7. The performance of such a soft-decision codec is shown in Figure 5.58.

## 5.8 CODE CONCATENATION IN MULTIPLE MESSAGE TRANSMISSION

Code concatenation means connecting two or more error-correcting codecs in cascade. Code concatenation can be a practical solution to provide unequal error

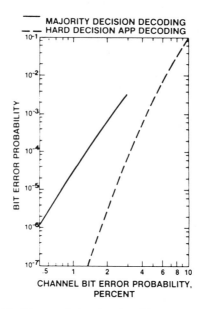

**Figure 5.56** (44, 22) convolutional code—bit error rate performance.

protections for multiple mixed-message transmissions through satellites. For example, the acceptable error rates for telephone and television transmissions are different. The transmissions of computer data, bank statements, and medical records are obviously more important than, say, reporting the number of fish in the Potomac River. For all information to be transmitted, however, the least

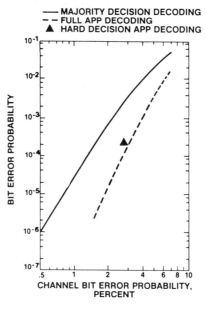

**Figure 5.57** (73, 45) block code—bit error rates [41].

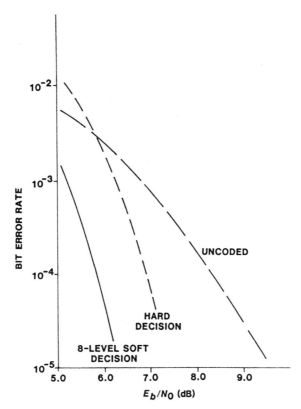

**Figure 5.58** Simulation performance of $(8, 7, 5)$ $n_A = 376$ feedback threshold decoder with modified APP decoding.

demanding user (or type of message) will desire a minimum acceptable quality of the transmission channel. For illustration, a three-stage concatenation is shown in Figure 5.59 for the transmission side only. Such an arrangement provides multilevel error rates in a satellite link. In this case we assume that the transmission speed of the different services is the same. In most practical applications it is not only the speed of the services that varies, but also transmission speed

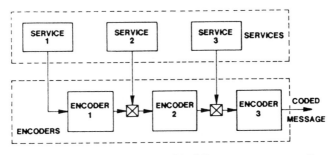

**Figure 5.59** Uniform speed service with different error rate encoding.

changes due to mixed size earth station operation in a satellite network. In this situation the error codecs need to be selected for speed compatibility as well as for error performance desirability. A functional arrangement for such combination is shown in Figure 5.60.

The idea of error-correcting code concatenation began with Elias's iterative procedure [42]. Forney studied block code concatenations extensively [43]. Gore demonstrated a new class of iterative codes [44], calling them the product-generator codes. Pinsker and Stiglitz considered the concatenation of block codes with convolutional codes with sequential decoding [45, 46]. A block code was used as an inner code to increase the computational cutoff rate of the sequential decoder. Odenwalder investigated Reed-Solomon codes in concatenation with Viterbi decoding and produced some impressive expected performances [47] and Ramsey provided both analytical and simulation results on cascaded convolutional codes [48]. In particular, Ramsey provided valuable information about the burst-error statistics of a large number of Viterbi decoders. In terms of actual applications, Baumert and McEliece evaluated a low rate Golay-Viterbi concatenated coding scheme for the 1977 Mariner Jupiter/Saturn mission [49]. For fading channels, Cohn and Levesque evaluated an adaptive decoding technique for concatenated block codes [50]. And Wu has recently shown the advantages of concatenation in identical multistage self-orthogonal tree codes [51].

One of the undesirable properties of concatenated codes is that the output of the inner decoder produces unpredictable or additional errors, usually bursty, when the number of channel errors exceeds the designed capability of the decoder.

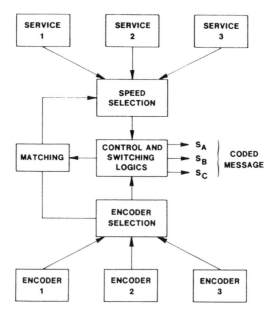

**Figure 5.60** Code concatenations for multiple, mixed-message transmission.

When burst errors are produced by an inner decoder, the problem is traditionally met by a multiple-burst error-correcting outer decoder such as a Reed-Solomon or an interleaved convolutional code. However, such a remedy is inefficient and costly. This could be one of the reasons why concatenated codes of attractive properties are not widely used. Thus a desirable property of an inner decoder of concatenated codes is that it does not produce more errors than it receives from the channel and it does not generate burst errors when the channel errors are nonbursty. As it turns out, the simple threshold decoders for convolutional self-orthogonal codes exhibit such properties. The raw data will be presented in Section 5.8.1. In Section 5.8.2 the powerful performance results of Viterbi/ Reed-Solomon code concatenation will be discussed. Finally, a multistage concatenation scheme will be investigated in Section 5.8.3.

### 5.8.1   Measured Results of Self-Orthogonal Codes for Concatenation

In this section, computer simulation results of two convolutional self-orthogonal codes (CSOC) are presented for the rate 0.75 code of constraint length 80 and the rate 0.875 code of constraint length 376. Both codes were described in Section 5.6.1.

Table 5.2 gives the measured performance of the rate 0.75 code with generator polynomial $G(3/4)$ of (5.117) at high channel error rates. $P$ is the measured channel bit error rate from the IBM standard Gaussian random number generation subroutine. $P_e$ is the measured error rate after decoding. Since the constraint length of the code is relatively short, 1000 bits of data were considered sufficient for each computer run.

Table 5.3 gives the measured performance of the rate 0.875 code with generator polynominal $G(7/8)$ of (5.116). For this longer constraint length code, 10,000 bits are used for each run. The computation time for the noisiest channel condition was 5.88 minutes, of which 2.73 minutes was CPU time on an IBM 360/65 machine.

We may observe from both tables that the values of $P_e$ are always less than $P$. This is even true for channel error rates exceeding 0.5 (which does not

**Table 5.2**

Simulated performance of the rate 0.75 codec.

| Specified channel error rate | Actual channel error rate $P$ | Decoder output error rate $P_e$ |
|---|---|---|
| 0.007 | 0.009 | 0.002 |
| 0.010 | 0.022 | 0.015 |
| 0.030 | 0.036 | 0.027 |
| 0.158 | 0.200 | 0.178 |
| 0.291 | 0.367 | 0.273 |
| 0.383 | 0.473 | 0.355 |
| 0.500 | 0.637 | 0.459 |

**Table 5.3**
Simulated performance of the rate 0.875 codec.

| Specified channel error rate | Actual channel error rate $P$ | Decoder output error rate $P_e$ |
|---|---|---|
| 0.0012 | 0.0006 | 0.0000 |
| 0.0046 | 0.0042 | 0.0022 |
| 0.010 | 0.0096 | 0.0075 |
| 0.158 | 0.178 | 0.162 |
| 0.383 | 0.433 | 0.366 |
| 0.500 | 0.566 | 0.474 |

correspond to any real situation in practice, of course). At no time during the course of simulation were undecodable output errors in burst patterns observed when the channel errors were random.

To investigate the behavior of the decoders, the averaged values of $P_e$ are insufficient because they do not show the error patterns. Run length statistics are presented in Figures 5.61 to 5.65. Figures 5.61 and 5.62 are for the rate 0.75 decoder, while Figures 5.63 and 5.64 are for the rate 0.875 decoder. The run length statistics used herein indicate the frequency of the number of error-free digits between two error digits. Since the point of interest is for high channel error rates, the figures reflect the results of simulation with high channel error rates.

For the shorter constraint length rate 0.75 codec, the results show that the decoder may produce clustered errors if the channel errors are highly clustered. This result confirms the fact that the decoders are useful for correction of random errors. For the rate 0.875 codec, the measurements of the run length statistics uniformly indicate the desirable behavior of CSOCs in code concatenation. The

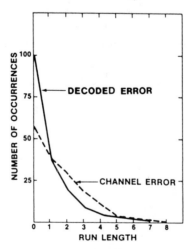

**Figure 5.61** (4, 3, 5) decoded error statistics at channel error rate $P = 0.20$.

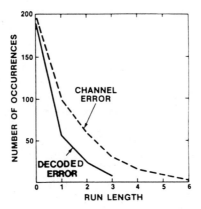

**Figure 5.62** (4, 3, 5) decoded error statistics at channel error rate $P = 0.367$.

performance resulting from the concatenation of two identical rate 0.875 codecs is shown in Figure 5.65.

### 5.8.2   Viterbi/Reed-Solomon Code Concatenation

In their study of the hybrid coding sytem, Odenwalder et al. analyzed and evaluated the concatenation of convolutional coding and Viterbi decoding with Reed-Solomon codes [52]. The concatenated coding system is with the Viterbi code as inner codec and the Reed-Solomon code as outer codec. The results are shown in Figure 5.66 for the inner code with 8 levels of receiver quantization and a path length memory of 32 bits in concatenation with various $2^m$ alphabet

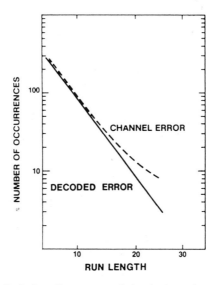

**Figure 5.63** (8, 7, 5) decoding error statistics at channel error rate $P = 0.18$.

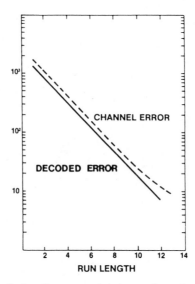

**Figure 5.64** (8, 7, 5) decoding error statistics at channel error rate $P = 0.433$.

sizes for the outer code, where $E$ is the number of Reed-Solomon symbol errors; i.e., an RS block error occurs when more than $E$ symbol errors occur in the block.

The performances of code concatenation are excellent from the standpoint of coding gain. However, making such concatenation applicable to high-efficiency and high-speed satellite communication will be an engineering challenge.

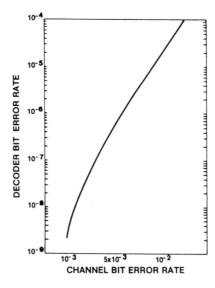

**Figure 5.65** Performance of concatenation of two identical rate 0.875 convolutional self-orthogonal codes.

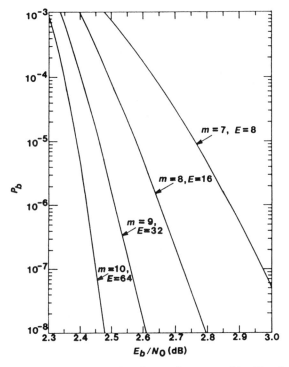

**Figure 5.66** Summary of concatenated coding performance with a $K = 8$, $R = 1/2$ inner code and $2^m$-symbol $E$-error-correcting outer code [52].

### 5.8.3   Multistage Concatenation

From Tables 5.2 and 5.3 and the single decoder output error run length statistics shown in Figures 5.61 to 5.65, it is clear that the feedback decoders in connection with CSOCs do not produce dependent errors at the output of the decoders. It is well known that this is not true for BCH, Reed-Solomon, sequential, and Viterbi decoders. When the number of random channel errors exceeds the capability of such decoders, they produce bursty, dependent errors at the decoder outputs. For feedback decoders, the uncorrectable output errors not only are independent, but appear in dispersion, a desirable property in code concatenation. As an example, let us consider the three-stage concatenation scheme shown in Figure 5.67. The codecs in concatenation are identical: the (8, 7, 5) code used in the INTELSAT SCPC system as previously described. The $E$s denote the encoders and the $D$s, the decoders. The rate decreases from 0.875, 0.76, to 0.67 as a result of the concatenation. For a channel bit error rate of $10^{-2}$, the expected error rate improvements are $2 \times 10^{-3}$, $10^{-4}$, and $2.5 \times 10^{-8}$, respectively. For applications where a moderate decoding delay is tolerable, significant coding gain can be obtained with very simple implementation.

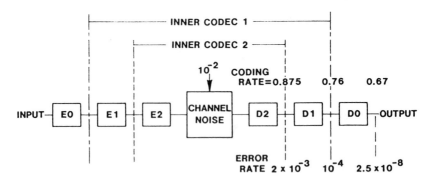

**Figure 5.67** A three-stage concatenation.

Obviously, CSOCs used for concatenation need not be identical. Various codes with different error-correcting capabilities and rates may be concatenated. However, identical codecs are easier to produce and are thus more economical.

The high-rate self-orthogonal tree codes derived in [15] can be applied to code concatenation. Codes of various rates with different degrees of coding gain may be combined. The attractiveness of using CSOCs in concatenated configurations is not only their availability but also the error-suppression characteristics of the decoders even at high channel error rates. Through computer simulation, the results of two representative codecs were presented. The high channel error rates used are for the purpose of testing the codecs only. The robust error characteristics of the decoders suggest their concatenations in practical applications. The large number of this class of codes generated gives the system designer a choice of error controls.

In simulations, the decoders usually correct random channel errors. When the number of channel errors exceeds the designed capabilities of the codes, the decoders do not produce bursty errors. This is in contrast with the behavior of some of the most powerful decoders, such as the Reed-Solomon, BCH, and Viterbi, which exhibit additional uncorrectable errors of burst nature at the outputs of the decoder. Threshold decoders of CSOCs exhibit such interesting properties because of the following characteristics: As the number of channel errors increases, the average weight of the syndrome sequence increases, causing more error indications. However, the nonzero feedback digits always reduce the weight of the syndrome, thus minimizing the effect of false error indication. This phenomenon appears to be responsible for the robust nature of errors at the output of the decoder. A comparison of known high-rate concatenated codes was recently reported by Voukalis [54].

### 5.8.4   The CCSDS Recommendation

NASA's Goddard Space Flight Center (GSFC) Data Systems Requirement Committee was given the authority to establish a coding standard that is primarily intended to ensure compatibility between spacecraft and ground encoding and decoding systems while allowing sufficient flexibility to meet most mission requirements. The standard provides information and guidance for users, designers [53], and potential users of data processing systems supported at GSFC. It provides selectable levels of performance to permit reasonable flexibility in trading off between redundance, deletion rate, and error control while maintaining spacecraft and ground system standardization. It is intended that the compatibility between the encoding and decoding processes will be achieved by agreement to conform to a particular code of this standard without the necessity for additional specifications.

This recommended standard outlines the type of support available from the GSFC Spaceflight Tracking and Data Network (STDN) and data handling and mission operation facilities. The codes described will be supported by these facilities. The codes in this standard are applicable to the deep space missions. The standard recommends that the convolutional codes should be used in low signal-to-noise applications and the block codes in high signal-to-noise applications. In particular, only the modified Hamming codes (63, 56) or a truncated form will be used for the command uplink channel.

The coding standard applies to all spacecraft using GSFC support facilities for forward- or return-link data transmission. In transmitting data over commercial or other telecommunications channels, or in recording and replaying data by magnetic tape systems, additional encoding/decoding processes not covered in this standard may take place without significant effect on the user.

The coding standard recommends the approach of code concatenation. The inner and outer codes suggested for the standard are shown in Tables 5.4 and 5.5, respectively. The implementation design and performance of the codes have been evaluated with the cooperation of the Jet Propulsion Laboratory and the

**Table 5.4**

Inner coding standards.

| Code rate $R$ | Viterbi decoding $G(D)$ | Sequential decoding | |
|---|---|---|---|
| | | Systematic | Nonsystematic |
| ½ | $G = 1111001$ | $G = 1000 \ldots 0$ | $G = 111, 001, 011, 101, 011, 011,$ $110, 111, 110, 111, 011, 111,$ $011, 101, 101, 011$ |
| | $G = 1011011$ | $G = 111, 001, 101, 100,$ $111, 011, 111, 000,$ $001, 011, 001, 111,$ $100, 110, 101, 001$ | $G = 101, 011, 011, 101, 011, 011,$ $110, 111, 110, 111, 011, 111,$ $011, 101, 101, 011$ |
| ½ | $G = 1111001$ $G = 1011011$ $G = 1110101$ | Not applicable | Not applicable |
| Constraint length $K$ | 7 | 21, 30, 40 and 48 | 24, 32, 40 and 48 |

**Table 5.5**
Outer coding standards.

| M | $f(x)$ | $x$ | $R$ | $K$ | $d_m$ | $E$ | |
|---|---|---|---|---|---|---|---|
| 5 | $x^5 + x^3 + 1$ | 31 symbols | 0.484 | 15 | 17 | 8 | symbol |
| | | 155 bits | | 75 | | | bits |
| 8 | $x^8 + x^4 + x^3 + x^2 + 1$ | 255 symbols | 0.937 | 223 | 17 | 8 | symbol |
| | | 2040 bits | | 1912 | | | bits |

following organizations: Centre National D'Etudes Spatiales (CNES)/France, Deutsche Forschungs-und Versuchsanstalt für Luft und Raumfahrt e.V (DFVLR)/ West Germany, European Space Agency (ESA)/Europe, Instituto de Pesquisas Espaciais (INPE)/Brazil, National Aeronautics and Space Administration (NASA)/ USA, and National Space Development Agency of Japan (NASDA)/Japan. The NASA coding standard work has evolved and improved to be part of the Consultative Committee for Space Data Systems (CCSDS) recommendation [55].

With the rates 1/2, 1/3, constraint length 7 convolutional code encoding and Viterbi decoding as inner codec and the (255, 223, 33) symbol, $m = 8$, (2040, 1784, 264) bit, Reed-Solomon codec as outer codec, the concatenated code performance in the recommended standard is shown in Figure 5.68. The rate

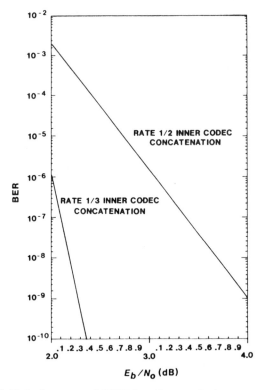

**Figure 5.68** Performance of CCSDS coding standards recommendation.

1/2 codec has been implemented and evaluated. The hardware aspect of this codec will be discussed in the next chapter.

## 5.9 CONCLUDING REMARKS

In this chapter we have provided the justification, variation, classification, performance measure, and limitation of basic error coding applications for satellite communication channels. Justifications have been discussed through power, bandwidth, and efficiency tradeoffs. Variations were illustrated by the diversity of applicable codes and decoding techniques.

This diversity could be unified into a simple classification. For practical purposes critical distinctions and applicabilities among the codes have been discussed. Potentially useful codes and coding techniques have been suggested and their coding gain expressions were provided. Code performance measures were compared on a sample of applicable codes for channels under consideration.

Code and decoding selection criteria were put forward for satellite transmission environment. Several implementable codes are listed in Appendix 4.

Coding limitations were sized up through performance bounds. New formulations were derived for forward error correction, error detection, and their combination. We also included both analytical and experimental results of soft-decision and code concatenations.

The main purpose of this chapter is to provide a unified approach of error control techniques for use in satellite communication system designs, rather than to recommend any particular type of code or any specific decoding algorithm. To use or not to use error coding, or which coding technique should be used in a satellite system, depend on two most important criteria, performance and complexity. In this chapter we provided some answers to the criterion of performance. The issue of codec complexity will be addressed in the next chapter.

## 5.10 REFERENCES

1. Peterson, W. W. and Weldon, E. *Error Correcting Codes,* 2nd ed. Cambridge, Mass.: M.I.T. Press, 1972.
2. Berlekamp, E. R. *Algebraic Coding Theory.* New York: McGraw-Hill, 1968.
3. MacWilliams, F. and Sloane, N. *The Theory of Error-Correcting Codes.* Amsterdam: North-Holland Publishing Co., 1977.
4. McEliece, R. "The Theory of Information and Coding." *Encyclopedia of Mathematics and Its Applications,* Volume 3. Reading, Mass.: Addison-Wesley, 1977.
5. Lin, S. and Costello, D. *Error Control Coding: Fundamentals and Applications.* Englewood Cliffs, N.J.: Prentice-Hall, 1983.
6. Bernice, R. J. and Frey, A. H., Jr. "An Analysis of Retransmission Systems." *IEEE Transactions on Communications Technology* COM-12 (December 1964): 135–45.
7. Gatfield, A. "Error Control on Satellite Channels Using ARQ Techniques." *COMSAT Technical Review* 6,1 (Spring 1976).

8. Drukarev, A. and Costello, D. "Hybrid ARQ Error Control Using Sequential Decoding." *IEEE Transactions on Information Theory* IT-29,4 (July 1983): 521–535.
9. Wu, W. W. "Coding for Multiple-Access Communications Satellites." *Proceedings of 1975 Allerton Conference on Circuit and System Theory.*
10. Kahn, R. "Special Issue on Packet Communications Networks." *Proceedings of the IEEE* (November 1978).
11. Wozencraft, J. and Jacobs, I. *Principles of Communication Engineering.* New York: John Wiley & Sons, 1965.
12. Gore, W. "Transmitting Binary Symbols with Reed-Solomon Codes." *The Johns Hopkins University Department of Electrical Engineering Report,* No. 73-5, April 1973.
13. Massey, J.L. *Threshold Decoding.* Cambridge, Mass.: M.I.T. Press, 1963.
14. Hagelberger, D. "Recurrent Codes for the Binary Symmetric Channel." Lecture Notes on Theory of Codes, University of Michigan Summer Conference, 1962.
15. Wu, W. W. "New Convolutional Codes: Part I." *IEEE Transactions on Communications Technology* 23,9 (September 1975). Part II (January 1976).
16. "High-Speed Forward Error Correcting Codec." *Final Study Report* IS-838 by KDD for INTELSAT, January 1976.
17. Viterbi, A. "Error Bounds for Convolutional Codes and an Asymptotically Optimum Decoding Algorithm." *IEEE Transactions on Information Theory* (April 1967): 260–69.
18. Forney, D., Jr. "The Viterbi Algorithm." *Proceedings of the IEEE* 61 (March 1973).
19. Viterbi, A. "Convolutional Codes and Their Performance in Communication Systems." *IEEE Transactions on Communications Technology* COM-19,5 (October 1971): 751–71.
20. Viterbi, A. and Omura, J. *Principles of Digital Communication and Coding.* New York: McGraw-Hill, 1979.
21. Heller, J. A. and Jacobs, I. M. "Viterbi Decoding for Satellite and Space Communication." *IEEE Transactions on Communications Technology* COM-19,5 (October 1971): 751–71.
22. Wozencraft, J. "Sequential Decoding for Reliable Communications." *M.I.T. Research Laboratory of Electronics Technical Report,* No. 325, August 1957.
23. Fano, R. "A Heuristic Discussion of Probabilistic Decoding." *IEEE Transactions on Information Theory* (April 1963).
24. Zigangirov, K. "Some Sequential Decoding Procedures." *Problemy Peredachi Informatsii* 2,4 (1966).
25. Jelinek, F. "A Fast Sequential Decoding Algorithm Utilizing a Stack." *IBM Journal of Research* 13 (November 1969).
26. Massey, J. and Costello, D., Jr. "Nonsystemmatic Convolutional Codes for Sequential Decoding in Space Applications." *IEEE Transactions on Communications Technology* COM-19,5 (October 1971): 806–812.
27. Gallager, R. *Information Theory and Reliable Communications.* New York: John Wiley & Sons, 1968.
28. Forney, G., Jr. and Bower, E. "A High-Speed Sequential Decoder: Prototype Design and Test." *IEEE Transactions on Communications Technology* (COM-19,5 (October 1971).
29. Cahn, C.; Huth, G.; and Moore, C. "Simulation of Sequential Decoding with Phase Locked Demodulation." *IEEE Transactions on Communications Technology* (February 1973).

**30.** Geist, J. "A Comparison of Fano and Jelinek Sequential Decoding Algorithms." *University of Notre Dame Technical Report,* No. EE-701, January 1970.

**31.** Chevillat, P. and Costello, D. "A Multiple Stack Algorithm for Erasure Free Decoding of Convolutional Codes." *IEEE Transactions on Communications Technology* 25,12 (December 1977).

**32.** Ma, H. "The Multiple Stack Algorithm Implemented on a Zilog Z-80 Microcomputer." *IEEE Transactions on Communications Technology* COM-28 (November 1980): 1876–1887.

**33.** Elias, P. "Coding for Noisy Channels." *IRE Convention Record* (1955).

**34.** Epstein, M. "Algebraic Decoding for Binary Erasure Channel." *M.I.T. Research Laboratory of Electronics Technical Report,* No. 340, March 1958.

**35.** Chase, D. "A Class of Algorithms for Decoding Block Codes with Channel Measurement Information." *IEEE Transactions on Information Theory* IT-18,1 (January 1972): 170–82.

**36.** Dorsch, B. "A Decoding Algorithm for Binary Block Codes and *J*-ary Output Channels." *IEEE Transactions on Information Theory* 20 (May 1974): 391–94.

**37.** Weldon, E. "Decoding Binary Block Codes on *Q*-ary Output Channels." *IEEE Transactions on Information Theory* IT-17 (November 1971): 713–18.

**38.** Hartmann, C. and Rudolph, L. "An Optimum Symbol-by-Symbol Decoding Rule for Linear Codes." *IEEE Transactions on Information Theory* IT-22 (September 1976): 514–17.

**39.** Baumert, L. and McEliece, R. "Performance of Some Block Codes on a Gaussian Channel." *Proceedings of the International Telemetry Conference, Washington, D.C.* (1975): 189–95.

**40.** Ebert, P. and Tong, S. "Convolutional Reed-Solomon Codes." *Bell Systems Technical Journal* 48, 3 (March 1969).

**41.** Neuman, F. and Lumb, D. "Performance of Several Convolutional and Block Codes with Threshold Decoding." *NASA Technical Note,* TND-4402, March 1968.

**42.** Elias, P. "Error Free Coding." *IEEE Transactions on Information Theory* 1,1 (1965).

**43.** Forney, G. D. *Concatenated Codes.* Cambridge, Mass.: M.I.T. Press, 1966.

**44.** Gore, W. C. "Further Results on Product Codes." *IEEE Transactions on Information Theory* IT-16 (July 1970).

**45.** Pinsker, M. S. "On the Complexity of Decoding." *Problemy Peredachi Informatsii* 1,1 (1965).

**46.** Stiglitz, I. G. "Iterative Sequence Decoding." *IEEE Transactions on Information Theory* IT-15 (November 1969).

**47.** Odenwalder, J. P. "Optimal Decoding of Convolutional Codes." Ph.D. Dissertation, School of Engineering and Applied Sciences, University of California at Los Angeles, 1972.

**48.** Ramsey, J. S. "Cascaded Tree Codes." Ph.D. Dissertation, Department of Electrical Engineering, M.I.T., 1970.

**49.** Baumert, L. D. and McEliece, R. J. "A Golay-Viterbi Concatenated Coding Scheme for MJS'77." *JPL Technical Report,* 32–1526.

**50.** Cohn, D. L. and Levesque, A. H. "Concatenated Codes for Fading Channels." International Conference on Communications, 1976.

**51.** Wu, W. W. "New Convolutional Codes: Part III." *IEEE Transactions on Communications Technology* (September 1976).

**52.** Odenwalder, J. P.; Viterbi, A. J.; Jacobs, I. M.; and Heller, J. A. "Study of Information Transfer Optimization for Communication Satellites." *Final Report for*

*NASA Contract NAS2-6810,* prepared by Linkabit Corporation, NASA CR #114561, N73-20899.

**53.** Morakis, J.; Greene, E.; Miller, W.; Poland, W.; and Helgert, H. *Aerospace Data System Standards, Part 7: Coding Standard.* Goddard Space Flight Center, National Aeronautics and Space Administration, 1982.

**54.** Voulakis, D. "A Comparison of Known High-Rate Concatenated Codes." *International Journal of Electronics* 52,2 (1982): 177–87.

**55.** CCSDS Recommendation for Telemetry Channel Coding, prepared by the Consultative Committee for Space Data System, September 1983.

## 5.11  PROBLEMS

1. Wyner and Ash codes are optimal single error-correcting convolutional codes with code memory length $m$ and decoding constraint length $(m + 1)n_0$ where $n_0 = 2^m$ and code rate is $(n_0 - 1)/n_0$. Since there is no way to recognize an error as uncorrectable, we may characterize the reliability of the decoder by the frequency of errors in the decoded sequence. Define the average bit error probability in the $k$th block by

$$p_k = \frac{1}{n_0 - 1} \sum_{j=1}^{n_0 - 1} p_k^j$$

where $p_k^j$ is the bit error probability of the $j$th bit in the $k$th received block. Assume binary symmetric channel transition probability $p$; show that if $6mn_0p < 1$ and $k > 5m$, as assumed by Chen and Rutledge,

$$p_k < \frac{4mn_0^2 - 2.5mn_0 - 1.5\,n_0^2 + 2.5\,n_0 - 1}{n_0 - 1}\,p^2$$

$$+ \frac{(6mn_0p)^3}{6}.$$

2. In the error rate performance evaluation for convolutional codes with threshold decoding, the effective code constraint length $n_E$ is utilized as in (5.110). Prove that for any rate $R = k_0/n_0$ there exists a convolutional code with $J$ parity checks such that

$$n_E \le \frac{1}{2}\left[\frac{R}{1 - R}\,J^2 + k_0J + 2\right].$$

3. Without using the random coding bound argument, verify that both the error probabilities of sequential and maximum-likelihood decoding (a) have the same form, and (b) decrease exponentially with minimum free distance, and increase linearly with the number of minimum free-weight codewords, which are the number of paths in the subset of incorrect decoding with merging free distance.

4. For short constraint length $K$, rate $R = \dfrac{1}{n_0}$ convolutional code with Viterbi decoding, show that the minimum free distance can be approximated by the Heller's bound $K/2R$. Minimum free distance is defined as the unrestricted minimum distance between the set of codewords.

5. For systematic convolutional codes show that the undetectable error probability can be upper bounded by

$$P_{ud} < K2^{-K(1-R)E(1,Q)/R}$$

for rate $R < R_0$, where $K$ is the code constraint length. For a binary symmetric channel transition probability $10^{-3}$, what is $P_{ud}$ for a rate 1/2 and $K = 3$ convolutional code with maximum-likelihood Viterbi decoding?

6. In the evaluation of error performance for multiple stack algorithm (MSA), the decoder decodes the sequences erased by a conventional sequential decoder with some erasure-estimate error probability $P_{est} = P(\hat{x} \neq x/n_s > 1)$, which is the probability that the codeword estimate $\hat{x}$ differs from the transmitted codeword $x$, given that the first stack overflows. Where $n_s$ denotes the number of stacks formed before the end of the decoding tree is reached for the first time, establish the error probability for MSA in terms of $P_{est}$.

7. In weighted erasure decoding, investigate the significance of Reddy's reliability indicator, $R_\sigma = \max(0, d - 2F_\sigma)$.

8. By iterative extension of the algebraic analog, maximum-radius decoding (MRD) was suggested by Hartmann and Rudolph. Show that MRD is asymptotically optimum for cyclic codes in AWGN channels.

9. In the original proposed Aerospace Data System Standards of the NASA Goddard Space Flight Center, the concatenation of Viterbi and RS codes has been recommended for deep space applications. The following codes are included in the Coding Standards.
   (a) Calculate the first event probability for the rate 1/2, constraint length 7 Viterbi decodable code with generator polynomial

$$G_1 = 1\ 1\ 1\ 1\ 0\ 0\ 1$$

$$G_2 = 1\ 0\ 1\ 1\ 0\ 1\ 1.$$

   (b) Calculate the bit error rate of the $(31, 16, 17)$ RS code in $GF(2^5)$ modulo $1 + x^2 + x^5$.
   (c) Calculate the decoding performance when code (a) is used as the inner codec and code (b) is used as the outer codec.
   *Note:* Code (a) alone has been implemented in the Tracking Data Relay Satellite System (TDRSS).

10. In the shuttle program, the rate 1/3, constraint length $k = 7$ Viterbi decoder has been recommended for implementation. This code has the generator polynomial in binary form

$$G_1 = 1\ 1\ 1\ 1\ 0\ 0\ 1$$

$$G_2 = 1\ 0\ 1\ 1\ 0\ 1\ 1$$

$$G_3 = 1\ 1\ 0\ 0\ 1\ 0\ 1.$$

Evaluate the coding performance in a BSC with $p = 10^{-2}$ and $10^{-4}$.

11. The following Viterbi decoder with generator polynomial

$$G_1 = 1\ 0\ 1\ 1\ 0\ 1\ 1$$

$$G_2 = 1\ 1\ 1\ 1\ 0\ 0\ 1$$

$$G_3 = 1\ 1\ 1\ 0\ 1\ 0\ 1$$

has been suggested as a standard for the deep space planetary program. Evaluate this decoder's performance in terms of bit error rate versus low $E_b/N_0$, 2 to 6 dB in an AWGN channel.

12. In a BSC, evaluate the sequential decoders performance with the Fano algorithm. The rate 1/2 codes have the following generator polynomials:
(a) Systematic—$G_1 = 1000 \ldots \ldots \ldots 0$
    —$G_2 = 111, 001, 101, 100, 111, 011, 111, 000, 001, 011,$
             $001, 111, 100, 110, 101, 001$
(b) Nonsystematic—$G_1 = 111, 011, 011, 101, 011, 011, 110, 111, 110,$
                          $111, 011, 111, 011, 101, 101, 011$
    —$G_2 = 101, 011, 011, 101, 011, 011, 110, 111, 110,$
             $111, 011, 111, 011, 101, 101, 011.$

# Chapter 6

# ERROR-CORRECTING CODEC IMPLEMENTATION

## 6.1  INTRODUCTION

In Chapter 5 we presented the justification, variation, and performance aspects of channel coding techniques for satellite communication, and in this chapter we are concerned with how these techniques can be implemented either on the ground or in a satellite. In accordance with the emphasis of this book, this chapter contains both fundamental theory and practical examples for deep space and satellite application. The content of this chapter is intended to be selective, not exhaustive.

Sections 6.2 through 6.6 deal with binary multiple error-correction BCH codec design and its complexity. A new fast polynomial multiplication over $GF(2^m)$ needed for decoding is introduced in Section 6.3. The reason for thoroughly studying BCH codec design is that it represents a giant step toward other types of algebraic block codec designs. Section 6.7 discusses coset decoders, with the INTELSAT TDMA codec as an example of coset decoding design with double error-correcting BCH code. Section 6.8 describes the difference-set codec, with the particular example of (1057, 813, 33) encoder decoder designs. We single out this codec for satellite application because of its superior error-correction power, very high speed operation, and simplicity of implementation. Section 6.9 examines convolutional codes: Section 6.9.1 describes convolutional encoders, and Sections 6.9.2 through 6.9.4 present the three types of decoders for convolutional codecs, namely, threshold, sequential, and Viterbi (maximum likelihood). In each case fundamentals principles of decoding are given and codec examples are described.

Section 6.10 describes nonbinary character error-correcting Reed-Solomon block codes. Berlekamp's dual basis encoder is mentioned in Section 6.10.1 and Gore's fast Fourier transform decoder derivation is presented in Section 6.10.2. The (255, 223, 33) sixteen-character error-correcting Reed-Solomon codec with VLSI implementation is included as one of the examples.

When this chapter is completed the reader is not only expected to understand the fundamentals of the various channel coding techniques but also to be able to provide realistic recommendations as to which codec is best for a particular satellite system. Through the examples in this chapter, the reader should be able to draw his or her own conclusions as to the physical or economic feasibility of implementing a particular error-correcting codec.

## 6.2 BCH CODES

One of the most powerful classes of practical error-correcting codes is the BCH code [1, 2]. Detailed treatments of BCH decoding are available in the texts and literature [3, 4, 5, 6, 7, 8]. In the first part of this chapter, BCH codes are treated from a practical point of view, and only binary BCH codes, their implementation, and logic complexity estimation are discussed. Single and double error-correcting BCH codec implementation is relatively simple. Other techniques such as those described in Section 6.7 may also be used. In the following sections, the emphasis is on multiple error-correcting binary BCH codec designs. After a thorough understanding of the basic binary BCH codec implementation it is easier to extend to the designs of nonbinary BCH, Reed-Solomon codecs.

Although there is more than one way to define and describe BCH codes, the end result is the same and unique. Let $\alpha$ be an element of the finite field $GF(2^m)$, which has an order $n$, and all the nonzero elements can be expressed in terms of the powers of $\alpha$. An $E$ error-correcting BCH code may be defined as the set of all code polynomials $\{c(x)\}$ over $GF(2)$ of $n-1$ degree or less, such that

$$c(\alpha^i) = 0, \text{ for } i = 1, 2, \ldots, 2E \tag{6.1}$$

where

$$c(x) = \sum_{i=0}^{n-1} a_i x^i \tag{6.2}$$

and $a_i$ are symbols from $GF(2)$. $c(x)$ is a BCH codeword.

On the other hand, the set of codewords can be generated from the code generator polynomial $g(x)$, which relates to the code parity check polynomial $h(x)$ as $g(x) h(x) = x^n - 1$ and holds true for all cyclic codes. BCH codes can be defined in terms of $g(x)$ or $h(x)$. The codes can also be described by the code generator matrix $G$ or the code parity check matrix $H$. $g(x)$, $h(x)$, $G$, and $H$ are related as follows: The roots of $g(x)$ are the elements in the first column of $H$; and the inverse roots of $h(x)$ are the elements in $G$. For example, if $\alpha_1, \alpha_2, \alpha_3,$ $\ldots, \alpha_r$ are the roots of $g(x)$, then

$$g(\alpha_i) = 0, \qquad i = 1, 2, 3, \ldots, n - k = r. \tag{6.3}$$

These roots are the first column elements in $H$, as shown in (6.4). From finite field theory, $g(\beta) = 0$, if and only if $g(\beta^{2^i}) = 0$ for all $i$, where $\beta$ is an element in $GF(2^m)$. The other column elements in $H$ are simply the consecutive powers of the roots in the second column; that is,

$$H = \begin{bmatrix} \alpha^0 & \alpha_1 & \alpha_1^2 & \cdot & \cdot & \cdot & \alpha_1^{n-1} \\ \alpha^0 & \alpha_2 & \alpha_2^2 & \cdot & \cdot & \cdot & \alpha_2^{n-1} \\ \alpha^0 & \alpha_3 & \alpha_3^2 & \cdot & \cdot & \cdot & \alpha_3^{n-1} \\ \cdot & \cdot & \cdot & & & & \\ \cdot & \cdot & \cdot & & & & \\ \alpha^0 & \alpha_r & \alpha_r^2 & \cdot & \cdot & \cdot & \alpha_r^{n-1} \\ \text{0th} & \text{1st} & \text{2nd} & \cdot & \cdot & \cdot & (r-1)\text{th} \end{bmatrix}. \tag{6.4}$$

If $\beta_1$, $\beta_2$, $\beta_3$, . . . , $\beta_k$ are the roots of $h(x)$, then $h^*(x)$ has roots $\beta_1^{-1}$, $\beta_2^{-1}$, $\beta_3^{-1}$, . . . , $\beta_k^{-1}$. When the product of the polynomial vectors $\bar{g}(\bar{x}) \cdot \bar{h}^*(\bar{x}) = 0$, then the generator matrix contains the $\beta_i^{-1}$'s as its element.

$$G = \begin{bmatrix} \beta_1^0 & \beta_1^{-1} & (\beta_1^{-1})^2 & & & (\beta_1^{-1})^{n-1} \\ \beta_2^0 & \beta_1^{-1} & (\beta_2^{-1})^2 & \cdot & \cdot & (\beta_2^{-1})^{n-1} \\ \beta_3^0 & \beta_3^{-1} & (\beta_3^{-1})^2 & \cdot & \cdot & (\beta_3^{-l})^{n-1} \\ \cdot & \cdot & \cdot & \cdot\cdot\cdot & & \cdot \\ \cdot & \cdot & \cdot & & & \cdot \\ \cdot & \cdot & \cdot & & & \cdot \\ \beta_k^0 & \beta_k^{-1} & (\beta_k^{-1})^2 & & & (\beta_k^{-1})^{n-1} \end{bmatrix} \tag{6.5}$$

Both $\alpha_i$, $\beta_j$ may be expressed as the powers of the single elements $\alpha$, $\beta$ as $\alpha^i$, $\beta^j$, $h^*(x)$ is the corresponding inverse polynomial.

A valid codeword $\boldsymbol{\omega}$, $n$ digits long, must satisfy:

$$\boldsymbol{\omega} \cdot H^T = 0 \tag{6.6}$$

The first step in BCH error decoding is to recognize the nonzero pattern of $\boldsymbol{\omega} \cdot H$, the syndrome.

*Example 6.1:* As in Bose and Ray-Chaudhuri's original illustration, let $\alpha$ denote a root of the irreducible polynomial $x^4 + x + 1 = 0$ over GF($2^4$) [2]. Then $\alpha$ is a primitive element in GF($2^4$). The $2^4 - 1 = 15$ nonzero field elements are shown in Table 6.1.

The table is obtained by successive multiplication of $\alpha$ with the modulo of the polynomial $x^4 = 1 + x^2$. The parity check matrix is

**Table 6.1**
$2^4 - 1 = 15$ nonzero field elements.

| | | |
|---|---|---|
| $\alpha^0 = 1$ | | $= 1000$ |
| $\alpha^1 =$ | $\alpha$ | $= 0100$ |
| $\alpha^2 =$ | $\alpha^2$ | $= 0010$ |
| $\alpha^3 =$ | $\alpha^3 = 0001$ | |
| $\alpha^4 = 1 + \alpha$ | | $= 1100$ |
| $\alpha^5 =$ | $\alpha + \alpha^2$ | $= 0110$ |
| $\alpha^6 =$ | $\alpha^2 + \alpha^3 = 0011$ | |
| $\alpha^7 = 1 + \alpha$ | $+ \alpha^3 = 1101$ | |
| $\alpha^8 = 1$ | $+ \alpha^2$ | $= 1010$ |
| $\alpha^9 =$ | $\alpha$ | $+ \alpha^3 = 0101$ |
| $\alpha^{10} = 1 + \alpha + \alpha^2$ | | $= 1110$ |
| $\alpha^{11} =$ | $\alpha + \alpha^2 + \alpha^3 = 0111$ | |
| $\alpha^{12} = 1 + \alpha + \alpha^2 + \alpha^3 = 1111$ | | |
| $\alpha^{13} = 1$ | $+ \alpha^2 + \alpha^3 = 1011$ | |
| $\alpha^{14} = 1$ | $+ \alpha^3 = 1001$ | |
| $\alpha^{15} = 1$ | | |

$$H = \begin{bmatrix} \alpha^0 & \alpha & \alpha^2 & \alpha^3 & \alpha^4 & \alpha^5 & \alpha^6 & \alpha^7 & \alpha^8 & \alpha^9 & \alpha^{10} & \alpha^{11} & \alpha^{12} & \alpha^{13} & \alpha^{14} \\ \alpha^0 & \alpha^3 & \alpha^6 & \alpha^9 & \alpha^{12} & \alpha^{15} & \alpha^3 & \alpha^6 & \alpha^9 & \alpha^{12} & \alpha^0 & \alpha^3 & \alpha^6 & \alpha^9 & \alpha^{12} \\ \alpha^0 & \alpha^5 & \alpha^{10} & \alpha^{15} & \alpha^5 & \alpha^{10} & \alpha^{15} & \alpha^5 & \alpha^{10} & \alpha^0 & \alpha^5 & \alpha^{10} & \alpha^0 & \alpha^5 & \alpha^{10} \end{bmatrix}. \quad (6.7)$$

Substituting the corresponding 4-tupules of each field element $\alpha$ in $H$ and transposing, we have $H^T$ as shown in (6.8)

$$H^T = \begin{bmatrix} 1\,0\,0\,0 & 1\,0\,0\,0 & 1\,0\,0\,0 \\ 0\,1\,0\,0 & 0\,0\,0\,1 & 0\,1\,1\,0 \\ 0\,0\,1\,0 & 0\,0\,1\,1 & 1\,1\,1\,0 \\ 0\,0\,0\,1 & 0\,1\,0\,1 & 1\,0\,0\,0 \\ 1\,1\,0\,0 & 1\,1\,1\,1 & 0\,1\,1\,0 \\ 0\,1\,1\,0 & 1\,0\,0\,0 & 1\,1\,1\,0 \\ 0\,0\,1\,1 & 0\,0\,0\,1 & 1\,0\,0\,0 \\ 1\,1\,0\,1 & 0\,0\,1\,1 & 0\,1\,1\,0 \\ 1\,0\,1\,0 & 0\,1\,0\,1 & 1\,1\,1\,0 \\ 0\,1\,0\,1 & 1\,1\,1\,1 & 1\,0\,0\,0 \\ 1\,1\,1\,0 & 1\,0\,0\,0 & 0\,1\,1\,0 \\ 0\,1\,1\,1 & 0\,0\,0\,1 & 1\,1\,1\,0 \\ 1\,1\,1\,1 & 0\,0\,1\,1 & 1\,0\,0\,0 \\ 1\,0\,1\,1 & 0\,1\,0\,1 & 0\,1\,1\,0 \\ 1\,0\,0\,1 & 1\,1\,1\,1 & 1\,1\,1\,0 \end{bmatrix}. \quad (6.8)$$

Among the twelve columns in $H^T$, the last one is trivial and the ninth and tenth columns are repeats. After we delete columns 10 and 11, the rest of the columns are independent. The result is a parity check matrix of 10 parity checks and a block length of 15 BCH code. This code is designed in the data access channel of a domestic communications satellite system. We will use this example extensively to illustrate later discussions.

## 6.2.1   BCH Encoding

BCH codes are cyclic [4, 7]. Assume a cyclic code with a generator polynomial $g(x)$ of degree $n - k$; then the encoding process, due to an encoding sequence $I(x)$ of $k$ digits, is: multiply the encoding sequence by $x^{n-k}$; then divide the product by $g(x)$. This division gives a quotient $q(x)$ and a remainder $r(x)$, as

$$\frac{x^{n-k}I(x)}{g(x)} = q(x) + \frac{r(x)}{g(x)}. \tag{6.9}$$

Multiplying both sides of (6.9) by $g(x)$, we have

$$x^{n-k}I(x) = g(x)q(x) + r(x) \tag{6.10}$$

or

$$r(x) = x^{n-k}I(x) - g(x)q(x),$$

where the remainder polynomial $r(x)$ is the parity check polynomial. Its nonzero coefficients are the corresponding check digits which depend on the encoding sequence $I(x)$ and the code generating polynomial $g(x)$. This is obtained from the least common multiples of the minimum polynomials, specified by the roots, as described in the last section.

A BCH encoder can be implemented by a single shift register with length either $k$ or $n-k$. For message storage purposes, during encoding and decoding, $k$-stage implementation is better if $k > n-k$. For high rate codes, $n-k$ decreases; thus the $n-k$ stage encoder implementation can provide a significant saving. The number of modulo-2 adders needed depends on the type of encoding configuration; for total codec complexity evaluation we assume that a $k$-stage shift register configuration will be used.

*Example 6.2:*   The (31, 21, 5) BCH code, which was designed into the COM-SAT high-speed TDMA-1 and TDMA-2 experimental systems, has the generator polynomial $g(x) = 1 + x^3 + x^5 + x^6 + x^8 + x^9 + x^{10}$. If we assume a nonzero message sequence $101010 \ldots 01$, $I(x) = 1 + x^2 + x^4 + \ldots + x^{20}$; then the remainder polynomial, according to (6.11), is

$$r(x) = x^{31-21} I(x) - g(x) q(x),$$

where $q(x)$ is the quotient as a result of the division in (6.9). The parity check sequence from $r(x)$ is 1110001101. The $k$-stage implementation of the (31, 21, 5) encoder is shown in Figure 6.1.

### 6.2.2  BCH DECODING

Let the binary BCH codeword of length $n$ and minimum distance $d \geq 2E + 1$ be:

$$c(x) = \sum_{i=0}^{n-1} a_i x^i. \tag{6.11}$$

If we let the channel error sequence be

$$e(x) = \sum_{i=0}^{n-1} e_i x^i, \tag{6.12}$$

where $e_i = 0$, or 1, and $e_i = 1$ indicates an error in the $i$th bit, then the received codeword is:

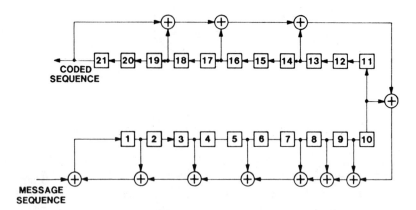

**Figure 6.1** BCH encoder of Example 6.2.

$$R(x) = \sum_{i=0}^{n-1} (c_i + e_i) x^i. \tag{6.13}$$

For $j = 0, 1, 2, \ldots, 2E-1$, the codeword $c(x)$ is a multiple of the minimal polynomial of the roots $\alpha^i$, and

$$c(\alpha^j) = 0; \tag{6.14}$$

thus the received codeword becomes

$$R(\alpha^i) = 0 + \sum_{i=0}^{n-1} e_i (\alpha^j)^i. \tag{6.15}$$

For an $E$-error-correcting BCH code, we consider the $E$ errors and let them be denoted as $x_0, x_1, \ldots, x_{E-1}$; then

$$R(\alpha^j) = \sum_{k=0}^{n-1} x_k^j = S_j, \tag{6.16}$$

where $S_j$ have been called power sums.

For implementation purposes, the first step in BCH decoding is to let $R(x)$ be divided by the corresponding minimum polynomials of $\alpha^i$; then the set of syndromes can be computed from the remainder as a result of the division, i.e., $R(x)/M_i(x)$. After the set of syndromes is obtained, the next step is to establish an error-locator polynomial $\sigma(x)$ with coefficients $\sigma_0, \sigma_1, \ldots, \sigma_E$. This is perhaps the most difficult part of the BCH decoding procedure. Basically, it calls for solving a set of Newton's identities:

$$S_1 - \sigma_1 = 0$$

$$S_3 - \sigma_1 S_2 + \sigma_2 S_1 - 3\sigma_3 = 0$$

$$S_5 - \sigma_1 S_4 + \sigma_2 S_3 - \sigma_3 S_2 + \sigma_4 S_4 - 5\sigma_5 = 0$$

$$\vdots \qquad\qquad \vdots$$

$$(6.17)$$

In the early years we had only Peterson's algorithm [3] which works as follows: Assume that there are $E$ or $E-1$ errors and one tries to solve the first $E$ equations. If the determinant of the set of equations is zero, then drop the last two equations and try again in order to find $\sigma_1$. Proceed in this manner until a solution is found. Peterson's algorithm requires a large number of computations even for fairly modest values of $E$.

To avoid the large number of computations for finding the coefficients of the error-locator polynomial, Berlekamp developed an iterative algorithm and Massey observed the synthesizing of minimum length shift register in order to avoid the direct computation of the set of identities [4, 8].

After the error-locator polynomial $\sigma(x)$ is found, the reciprocal roots of $\sigma(x)$ are the error locations that can be found by Chien's search [9]. To avoid a complex inversion step Burton provided an alternative solution [10]. The clarification of the above statements will be deferred until after the presentation of the key decoding mechanism.

The BCH decoder is a digital processor with computations carried out both in the finite field and ordinary arithmetic. As shown in Figure 6.2, the decoder consists of four basic functions:

1. Minimum polynomial division,
2. Power sum computation,

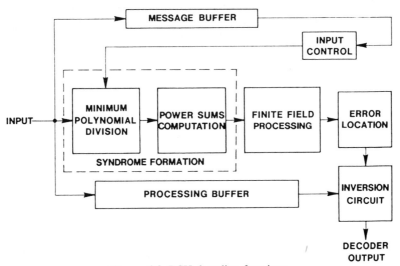

**Figure 6.2** BCH decoding functions.

3. Finite field processing, and
4. Error location.

The combination of items 1 and 2 is syndrome formation. For implementation, we separate the syndrome computation into minimum polynomial division and power sum computation. The incoming message buffer stores the next codeword while the present codeword is decoded in conjunction with the processing buffer. Each of the decoding functions and its circuit designs will be detailed in the following sections.

### 6.2.2.1  Minimum Polynomial Division Circuits

The first step in BCH decoding is syndrome formation; i.e., $S = R \cdot H^T$, where $S$ is the syndrome, $R$ is the received codeword, and $H^T$ is the transpose of the code parity check matrix. A partition of $S$ gives the set of power sums $S_j$ required in decoding. The operation $S = R \cdot H^T$ can be logically implemented if we first know the set of minimal polynomials. In this section we present a brief discussion of minimum polynomials and the design of the corresponding logic circuit. In general, the set of minimal polynomials is known once the roots, length, and minimum distance of a code are given. If not, it can be easily derived as follows.

Let $\alpha$ be a primitive element of GF($2^m$). Then $\alpha^{2^m-1} = 1$. $c(x)$ is a code vector if and only if $\alpha$, $\alpha^2$, $\alpha^3$, . . . , $\alpha^{2E}$ are roots of $c(x)$. If $M_i(x)$ denotes the minimum polynomial of the root $\alpha^i$, then $\alpha$, $\alpha^2$, $\alpha^4$, $\alpha^8$, . . . are the roots of $M_1(x)$, and $M_1(x) = M_2(x) = M_4(x) = M_8(x)$, . . . . Similarly, $\alpha^3$, $\alpha^6$, $\alpha^{12}$, $\alpha^{24}$, . . . are the roots of $M_3(x)$, and $M_3(x) = M_6(x)$ . . . . $\alpha^5$, $\alpha^{10}$, $\alpha^{20}$, $\alpha^{40}$, . . . are the roots of $M_5(x)$, and $M_5(x) = M_{10}(x) = M_{20}(x)$ . . . . $\alpha^7$, $\alpha^{14}$, $\alpha^{28}$, $\alpha^{56}$, . . . are the roots of $M_7(x)$, and $M_7(x) = M_{14}(x) = $ . . . .

Since $M_1(x)$ of degree $m$ generates GF($2^m$), which provides a set of $m$-tuple binary representations corresponding to the power of $\alpha$, to calculate $M_i(x)$, $i = 3, 5, 7, \ldots$, we substitute the roots $\alpha^i$ as

$$M_3(x) = (x - \alpha^3)(x - \alpha^6)(x - \alpha^{12}) \ldots$$
$$M_5(x) = (x - \alpha^5)(x - \alpha^{10})(x - \alpha^{20}) \ldots$$
$$M_7(x) = (x - \alpha^7)(x - \alpha^{14})(x - \alpha^{28}) \ldots \tag{6.18}$$

Then substitute the $\alpha$ powers in $M_i(x)$ in accordance with GF($2^m$). After collecting terms, substitute the binary representations of the $\alpha$'s. The modulo-2 summations of the $\alpha$'s are the coefficients of $x$. Since only binary codes are considered, the coefficients can be either 0 or 1. The above statements are best clarified by an example.

*Example 6.3:*  The (31, 21, 5) BCH code discussed in Example 6.2 has $M_1(x) = 1 + x^2 + x^5$, which generates GF($2^5$) and is shown in Table 6.2. The first

**Table 6.2**

Field of $1 + X^2 + X^5$.

| | 1 | $\alpha$ | $\alpha^2$ | $\alpha^3$ | $\alpha^4$ | | | | | |
|---|---|---|---|---|---|---|---|---|---|---|
| 0 | | | | | | 0 | 0 | 0 | 0 | 0 |
| $\alpha^0$ | 1 | | | | | 1 | 0 | 0 | 0 | 0 |
| $\alpha^1$ | | $\alpha$ | | | | 0 | 1 | 0 | 0 | 0 |
| $\alpha^2$ | | | $\alpha^2$ | | | 0 | 0 | 1 | 0 | 0 |
| $\alpha^3$ | | | | $\alpha^3$ | | 0 | 0 | 0 | 1 | 0 |
| $\alpha^4$ | | | | | $\alpha^4$ | 0 | 0 | 0 | 0 | 1 |
| $\alpha^5$ | 1 | | $+\alpha^2$ | | | 1 | 0 | 1 | 0 | 0 |
| $\alpha^6$ | | $\alpha$ | | $+\alpha^3$ | | 0 | 1 | 0 | 1 | 0 |
| $\alpha^7$ | | | $\alpha^2$ | | $+\alpha^4$ | 0 | 0 | 1 | 0 | 1 |
| $\alpha^8$ | 1 | | $+\alpha^2$ | $+\alpha^3$ | | 1 | 0 | 1 | 1 | 0 |
| $\alpha^9$ | | $\alpha$ | | $+\alpha^3$ | $+\alpha^4$ | 0 | 1 | 0 | 1 | 1 |
| $\alpha^{10}$ | 1 | | | | $+\alpha^4$ | 1 | 0 | 0 | 0 | 1 |
| $\alpha^{11}$ | 1 | $+\alpha$ | $+\alpha^2$ | | | 1 | 1 | 1 | 0 | 0 |
| $\alpha^{12}$ | | $\alpha$ | $+\alpha^2$ | $+\alpha^3$ | | 0 | 1 | 1 | 1 | 0 |
| $\alpha^{13}$ | | | $\alpha^2$ | $+\alpha^3$ | $+\alpha^4$ | 0 | 0 | 1 | 1 | 1 |
| $\alpha^{14}$ | 1 | | $+\alpha^2$ | $+\alpha^3$ | $+\alpha^4$ | 1 | 0 | 1 | 1 | 1 |
| $\alpha^{15}$ | 1 | $+\alpha$ | $+\alpha^2$ | $+\alpha^3$ | $+\alpha^4$ | 1 | 1 | 1 | 1 | 1 |
| $\alpha^{16}$ | 1 | $+\alpha$ | | $+\alpha^3$ | $+\alpha^4$ | 1 | 1 | 0 | 1 | 1 |
| $\alpha^{17}$ | 1 | $+\alpha$ | | | $+\alpha^4$ | 1 | 1 | 0 | 0 | 1 |
| $\alpha^{18}$ | 1 | $+\alpha$ | | | | 1 | 1 | 0 | 0 | 0 |
| $\alpha^{19}$ | | $\alpha$ | $+\alpha^2$ | | | 0 | 1 | 1 | 0 | 0 |
| $\alpha^{20}$ | | | $\alpha^2$ | $+\alpha^3$ | | 0 | 0 | 1 | 1 | 0 |
| $\alpha^{21}$ | | | | $\alpha^3$ | $+\alpha^4$ | 0 | 0 | 0 | 1 | 1 |
| $\alpha^{22}$ | 1 | | $+\alpha^2$ | | $+\alpha^4$ | 1 | 0 | 1 | 0 | 1 |
| $\alpha^{23}$ | 1 | $+\alpha$ | $+\alpha^2$ | $+\alpha^3$ | | 1 | 1 | 1 | 1 | 0 |
| $\alpha^{24}$ | | $\alpha$ | $+\alpha^2$ | $+\alpha^3$ | $+\alpha^4$ | 0 | 1 | 1 | 1 | 1 |
| $\alpha^{25}$ | 1 | | | $+\alpha^3$ | $+\alpha^4$ | 1 | 0 | 0 | 1 | 1 |
| $\alpha^{26}$ | 1 | $+\alpha$ | $+\alpha^2$ | | $+\alpha^4$ | 1 | 1 | 1 | 0 | 1 |
| $\alpha^{27}$ | 1 | $+\alpha$ | | $+\alpha^3$ | | 1 | 1 | 0 | 1 | 0 |
| $\alpha^{28}$ | | $\alpha$ | $+\alpha^2$ | | $+\alpha^4$ | 0 | 1 | 1 | 0 | 1 |
| $\alpha^{29}$ | 1 | | | $+\alpha^3$ | | 1 | 0 | 0 | 1 | 0 |
| $\alpha^{30}$ | | $\alpha$ | | | $+\alpha^4$ | 0 | 1 | 0 | 0 | 1 |
| $\alpha^{31}$ | 1 | | | | | 1 | 0 | 0 | 0 | 0 |

four powers of the primitive element $\alpha$ in GF($2^5$) must be present. $\alpha$, $\alpha^2$, $\alpha^4$, $\alpha^8$, $\alpha^{16}$ are the roots of $M_1(x) = M_2(x) = M_4(x)$. $\alpha^3$, $\alpha^6$, $\alpha^{12}$, $\alpha^{24}$, $\alpha^{48}$ are the roots of $M_3(x)$.

$$
\begin{aligned}
M_3(x) &= (x - \alpha^3)(x - \alpha^6)(x - \alpha^{12})(x - \alpha^{24})(x - \alpha^{48}) \\
&= (x - \alpha^3)[x - (\alpha + \alpha^3)][x - (\alpha + \alpha^2 + \alpha^3)] \\
&\quad [x - (\alpha + \alpha^2 + \alpha^3 + \alpha^4)][x - (1 + \alpha + \alpha^4)] \\
&= \alpha^0 + (\alpha^7 + \alpha^{14} + \alpha^{19} + \alpha^{25} + \alpha^{28})\, x + (\alpha + \alpha^2 + \alpha^4 + \alpha^8 + \alpha^{11} \\
&\quad + \alpha^{13} + \alpha^{16} + \alpha^{21} + \alpha^{22} + \alpha^{26})\, x^2 + (\alpha^5 + \alpha^9 + \alpha^{10} \\
&\quad + \alpha^{15} + \alpha^{18} + \alpha^{20} + \alpha^{23} + \alpha^{27} + \alpha^{29} + \alpha^{30})\, x^3 + (\alpha^3 + \alpha^6 + \alpha^{12} \\
&\quad + \alpha^{17} + \alpha^{24})\, x^4 + x^5 = 1 + x^2 + x^3 + x^4 + x^5.
\end{aligned}
$$

The last step is obtained by substituting the binary representation of the $\alpha$ powers under summation for each coefficient of $x$.

The generator polynomial $g(x)$ of a BCH code, as mentioned before, has the degree $n-k$. $g(x)$ also has $d-1$ nonrepeated roots.
For

$$d = n-k + 1 \tag{6.19}$$

$$g(x) = \text{L.C.M.} \prod_{i=1}^{d-1} M_i(x) \tag{6.20}$$

L.C.M. denotes the least common multiples. The division circuits are needed only for the factors of minimum polynomials that contribute to $g(x)$. The division circuit synthesis by feedback shift register is well known and will not be covered here [3, 4, 6, 8]. If $M_1(x)$, $M_3(x)$, $M_5(x)$, . . . , are the minimum polynomials of $g(x)$, then each minimal polynomial can be implemented by a feedback shift register. The feedback connections of the registers are according to the coefficients of the corresponding minimum polynomials. The inputs to the set of feedback shift registers are connected in common from a received codeword as shown in Figure 6.3, where each $M_i$ feedback shift register (FSR) is a division circuit. When a codeword is received all shift registers are clocked $n$ times simultaneously. At the $n$th clock time the remainders $r^{(i)}(x)$, $i = 1, 3, 5 . . .$, are contained in the registers. The remainders are needed to form the power sums.

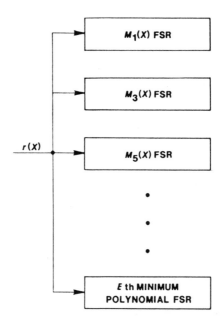

**Figure 6.3** Minimum polynomial division circuit.

*Example 6.4:* For the (31, 21, 5) code, discussed previously, the minimal polynomials of the code are $M_1(x) = 1 + x^2 + x^5$, and $M_3(x) = 1 + x^2 + x^3 + x^4 + x^5$. The division remainder circuit is shown in Figure 6.4.

The logic complexity of the minimum polynomial division circuit consists of $E$ shift registers. The total number of shift register stages is $n - k$. It would be a good approximation to say that each of the $E$ shift registers has $m$ stages. The number of modulo-2 adders is at most $Em$.

### 6.2.2.2  Power Sum Circuits

By software calculation the power sums $S_j$, $j = 1, 2, \ldots , 2E - 1$, and $S_{2j} = S_j^2$ can easily be obtained by multiplying the received code vector into the transpose of the code parity check matrix. Let us demonstrate this with a simple example. Using the (15,5,7) code parity check matrix as in Example 6.1 and, without loss of generality, assuming the all-zero sequence has been transmitted, if errors occur in $\alpha^3$ and $\alpha^{10}$ positions, then

$$S = rH^T = 1\ 1\ 1\ 1\ 1\ 1\ 0\ 1\ 1\ 1\ 0,$$

and we have $S_1 = 1\ 1\ 1\ 1 = \alpha^{12}$, $S_3 = 1\ 1\ 0\ 1 = \alpha^7$, $S_5 = 1\ 1\ 1\ 0 = \alpha^{10}$. The rest of the power sums are:

$$S_2 = S_1^2 = \alpha^{24} = \alpha^9 = 0\ 1\ 0\ 1$$

$$S_4 = S_2^2 = \alpha^{18} = \alpha^3 = 0\ 0\ 0\ 1$$

$$\text{and} \qquad S_6 = S_3^2 = \alpha^{14} \qquad = 0\ 1\ 1\ 1.$$

However, for hardware implementation, the power sums depend on the remainder polynomials $r^{(i)}(x)$ which were obtained in the last section. We calculate the set of power sums $S_j$ from $r^{(i)}(x)$ as in (6.21)

$$S_j = r^{(j)}(x^j) \bmod m_1(x) \tag{6.21}$$

$$j = 1, 2, \ldots , 2E - 1$$

with $S_{2j} = S_j^2$.

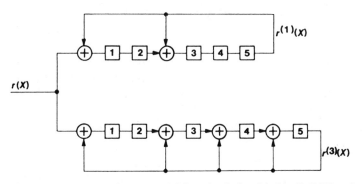

**Figure 6.4** Minimum polynomial division circuit for (31, 21, 5) BCH code.

The coefficient of each $x$ term in $r^{(j)}(x^j)$ determines the connections to the minimum polynomial division circuit. Note that $S_1$, $S_2$, $S_4$, $S_8$, correspond to $M_1(x)$, and $S_3$, $S_6$, $S_{12}$, $S_{24}$, . . . , correspond to $M_3(x)$, etc., in the connections from $S_j$ circuits to the $m_i(x)$ circuits.

*Example 6.5:* The minimal polynomials for the (15, 5, 7) BCH code are $m_1(x) = 1 + x + x^4$, $m_3(x) = 1 + x + x^2 + x^3 + x^4$, and $m_5(x) = 1 + x + x^2$. Then,

$$S_1 = r^{(1)}(x) = m_0 + m_1 x + m_2 x^2 + m_3 x$$

$$S_2 = r^{(2)}(x^2) = m_0 + m_1 x^2 + m_2 x^4 + m_3 x^6$$

$$= (m_0 + m_2) + m_2 x + (m_1 + m_3)x^2 + m_3 x^3$$

$$S_3 = r^{(3)}(x^3) = m_0 + m_1 x^3 + m_2 x^6 + m_3 x^9$$

$$= m_0 + m_3 x + m_2 x^2 + (m_1 + m_2 + m_3)x^3$$

$$S_4 = r^{(4)}(x^4) = (m_0 + m_1 + m_2 + m_3) + (m_1 + m_3)x$$

$$+ (m_2 + m_3)x^2 + m_3 x^3$$

$$S_5 = r^{(5)}(x^5) = m_0 + m_1 x^5 + m_2 x^{10} + m_3 x^{15}$$

$$= m_0 + m_1(x + x^2) + m_2(1 + x + x^2) + m_3 x^0$$

$$= (m_0 + m_2 + m_3) + (m_1 + m_2)x + (m_1 + m_2)x^2.$$

The power sum circuits for the (15, 5, 7) code are shown in Figure 6.5.

The power sum circuit consists of $2E - 1$ storage registers. Each of the storage registers has $m$ stages. The number of modulo-2 adders needed for the formation of the power sums is no more than $m^2(2E - 1)$, because each stage of the power sum storage register can have, at most, $m$ connections from an $m_i(x)$ shift register. There are $m$ stages in each $S_j$ storage register, and there are $2E - 1$ such registers.

### 6.2.2.3  Finite Field Processor

When the power sums $S_j$, $j = 1, 2, . . . , 2E - 1$ are determined as in Section 6.2.2.2, they are fed into the finite field processor which computes the coefficients, $\sigma_i$, $i = 1, 2, . . . , E$ of the error-locating polynomial $\sigma(x)$. For multiple error corrections (more than three) in binary BCH codes, the iterative algorithm developed by Berlekamp [4] is simplified for binary BCH codes in steps as follows:

Step 1: Let $1 + S = 1 + S_1 x + S_2 x^2 + \cdots + S_{2E-1}x^{2E-1}$ .

Step 2: Set $\sigma(0) = \tau(0) = D(0) = 1$ for $k = 0$. Where $k = 0, 1, . . . , 2E$, and $D(k)$ is the degree of $\sigma(k)$, $\tau(k)$ relates to $\sigma(k)$ as in Step 5.

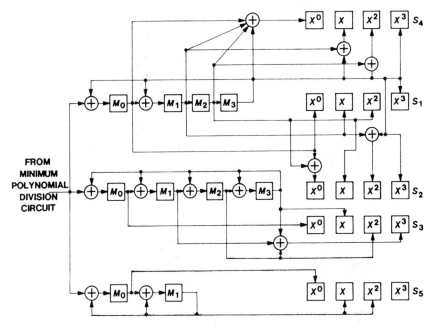

**Figure 6.5** Power sum circuit connections for (15, 5, 7) BCH decoder.

Step 3: If $S_{2k+1}$ is unknown, stop; otherwise,

Step 4: Define $\Delta(2k)$ as the coefficient of $x^{2k+1}$ in the product polynomial $(1 + S) \sigma(2k)$.

Step 5: Let $\sigma(2k + 2) = \sigma(2k) + \Delta(2k) \tau(2k)x$.

Step 6: Let $\tau(2k + 2) = x^2 \tau(2k)$, if $\Delta(2k) = 0$, or if $D(2k) > k + 1$. Otherwise, go to Step 7.

Step 7: $\tau(2k + 2) = \dfrac{x\sigma(2k)}{\Delta(2k)}$, if $\Delta(2k) \neq 0$ and $D(2k) \leq k + 1$.

Step 8: If a stop in Step 3 occurred, the last $\sigma(2k + 2)$ in Step 5 is the error-locator polynomial of the codeword.

The above algorithm is diagramed in Figure 6.6. As the algorithm indicates, the finite field processor must be able to add, subtract, multiply, and divide using the elements of finite field. Since in GF($2^m$) these field elements are expressed in terms of polynomials of, at most, degree $m-1$, the arithmetic operations mentioned above need to be performed on polynomials modulo the field generating polynomial. Addition and subtraction happen to be the same in

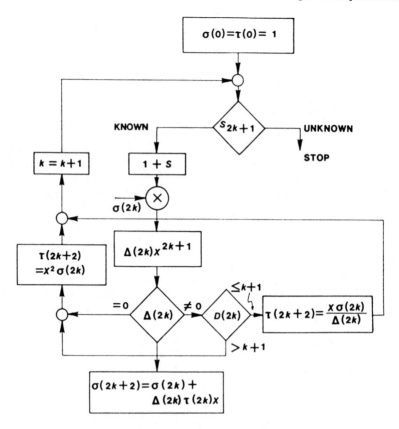

**Figure 6.6** The iterative algorithm for $\sigma_i$ calculation.

the binary field. The addition and subtraction of polynomials are relatively simple to implement. To implement polynomial multiplication, various methods will be discussed later.

*Example 6.6:* From Example 6.1, the field-generating polynomial of (15, 5, 7) BCH code is $1 + x + x^4$. The field, $GF(2^4)$, is shown in Table 6.1. Assume all the zero sequence was transmitted and $R = 0\ 1\ 0\ 0\ 1\ 0\ 1\ 0\ 0\ 0\ 0\ 0\ 0\ 0\ 0$ is received, i.e., $\alpha, \alpha^4, \alpha^6$ are in error. From (6.6) and (6.8)

$$S = R \cdot H^T = \underbrace{1\ 0\ 1\ 1}_{S_1}\ \ \underbrace{1\ 1\ 1\ 1}_{S_3}\ \ \underbrace{1\ 0\ 0\ 0}_{S_5}$$

The power sums $S_1, S_3, S_5$ can be obtained from the circuit described previously. With the aid of Table 6.1, $S_1 = 1\ 0\ 1\ 1 = \alpha^{13}$, $S_3 = 1\ 1\ 1\ 1 = \alpha^{12}$, and $S_5 = 1\ 0\ 0\ 0 = \alpha^0$. Since $S_{2j} = S_j^2$, it follows that $S_2 = S_1^2 = \alpha^{26} = \alpha^{11} = 0\ 1\ 1\ 1$, $S_4 = S_2^2 = \alpha^{22} = \alpha^7 = 1\ 1\ 0\ 1$, and $S_6 = S_3^2 = \alpha^{36} = \alpha^6 = 0\ 0\ 1\ 1$. After the power sums are obtained we proceed with the algorithm.

Step 1: $1 + S = 1 + S_1x + S_2x^2 + S_3x^3 + S_4x^4 + S_5x^5 + S_6x^6$

$$= 1 + \alpha^{13}x + \alpha^{11}x^2 + \alpha^{12}x^3 + \alpha^7x^4$$

$$+ \alpha^0x^5 + \alpha^6x^6.$$

Step 2: $\sigma(0) = \tau(0) = D(0) = 1.$

Step 3: $S_{2k+1}$ are known for $k=0,1,2.$

Step 4: Let $k=0$, $(1 + S)\sigma(0) = 1 + \alpha^{13}x$. Then $\Delta(0) = \alpha^{13}.$

Step 5: $\sigma(0 + 2) = \sigma(0) + \Delta(0)\tau(0)x = 1 + \alpha^{13}x.$

Step 6: $\Delta(0) \neq 0$, $D(0) \not> 1.$

Step 7: $\tau(0 + 2) = \tau(2) = \dfrac{x\sigma(0)}{\Delta(0)} = \dfrac{x}{\alpha^{13}} = \alpha^2x$, for $k = 0$, $\Delta(0) \neq 0$ and

$D(0) = 1 \le 0 + 1,$

Let $k = 1$,

$$(1 + S)\sigma(2) = (1 + \alpha^{13}x + \alpha^{11}x^2 + \alpha^{12}x^3 + \ldots)$$

$$(1 + \alpha^{13}x)$$

$$= \ldots + (\alpha^{13}\cdot\alpha^{11} + \alpha^{12})x^3 + \ldots$$

$$= \ldots + (\alpha^{24} + \alpha^{12})x^3 + \ldots$$

$$= \ldots + (\alpha^9 + \alpha^{12})x^3 + \ldots$$

$$= \ldots + \begin{bmatrix} 0 & 1 & 0 & 1 \\ 1 & 1 & 1 & 1 \end{bmatrix} x^3 + \ldots$$

$$= \ldots + [1\ 0\ 1\ 0]\, x^3 + \ldots$$

$$= \ldots + \alpha^8\, x^3 + \ldots$$

thus $\Delta(2) = \alpha^8$. Similarly,

$$\sigma(4) = \sigma(2) + \Delta(2)\tau(2)\, x$$

$$= 1 + \alpha^{13}x + \alpha^8 \cdot \alpha^2x \cdot x$$

$$= 1 + \alpha^{13}x + \alpha^{10}x^2.$$

Since $D(2) = \text{Deg}[\sigma(2)] = 1 < k + 1 = 2,$

$$\tau(4) = \frac{x\sigma(2)}{\Delta(2)} = \frac{x(1 + \alpha^{13}x)}{\alpha^8}$$

$$= \alpha^7x + \alpha^5x^2.$$

Let $k = 2$,

$$(1 + S)\sigma(4) = \ldots + (\alpha^7 \cdot \alpha^{13} + \alpha^{12} \cdot \alpha^{10})x^5 + \ldots$$

$$= \ldots + (\alpha^5 + \alpha^7)x^5 + \ldots$$

$$= \ldots + \alpha^{13} x^5 + \ldots$$

thus $\Delta(4) = \alpha^{13}$.

$$\sigma(6) = \sigma(4) + \Delta(4)\,\tau(4)x$$

$$= 1 + \alpha^{13}x + \alpha^{10}x^2 + \alpha^{13}(\alpha^7x + \alpha^5x^2)x$$

$$= 1 + \alpha^{13}x + \alpha^0 x^2 + \alpha^3 x^3.$$

As a verification, let $x = \alpha$ in $\sigma(6)$, then $\sigma(6) = 1 + \alpha^{14} + \alpha^2 + \alpha^6$. Referring to Table 6.1, we have $\sigma(6) = [1\ 0\ 0\ 0] \oplus [1\ 0\ 0\ 1] \oplus [0\ 0\ 1\ 0] \oplus [0\ 0\ 1\ 1] = [0\ 0\ 0\ 0]$.

The logic circuit that performs the above algorithmic computations is shown in Figure 6.7 with an explanation of notations indicated.

In general, the logic complexity of the finite field processor of an $E$ error-correcting BCH code is estimated as follows: The syndrome $1 + S$ storage requires $2E$ shift registers. Each shift register has $m$ stages. To obtain $\Delta(2k)$, $E$-polynomial multiplication circuits are required, each of which has, at most, $E + 1$ pairs of inputs. For checking whether $\Delta(2k)$ is zero, a set of AND gates is sufficient.

For each $k = 0, 1, 2, \ldots, E - 1$, the required checking corresponds to each $x^{2k+1}$. To determine the degree of $\sigma(2k)$ the nonzero detection of the $\sigma$-shift registers would show whether $D[2k] > k+1$ or $D[2k] \leq k+1$.

To obtain $\tau(2k+2)$, we first need to multiply $\sigma(2k)$ by $x$ and then multiply the polynomial by $\Delta^{-1}(2k)$. This operation requires $E$ polynomial multiplication circuits. The inversion of $\Delta(2k)$ to $\Delta^{-1}(2k)$ can be implemented by three shift registers, each of length $m + 1$, and an inverter gating circuit which operates the shift registers. This inversion step can be eliminated as described in [10]. This simplified scheme requires the addition of two $m$-stage shift registers. The five inverters may thus be eliminated.

The $\tau(2k)$ information requires $2E + 1$ shift registers each of length from $1, 2, \ldots, 2E + 1$ stages. There are $2E + 1$ storage shift registers for $\Delta(2k)x$ $\tau(2k)$. There are $E$ polynomial addition circuits to perform the $\sigma(2k) + \Delta(2k)x$ $\tau(2k)$ operation. For the $\tau(2k+2) = x^2\sigma(2k)$ calculation, a multiplication by $x^2$ circuit is required. The multiplication by $x$ or $x^2$ circuits is designed from a shift register of length $m$ or $m$ separate stages. Finally, there are four OR functions in the algorithm. This can be simply implemented by a set of OR gates.

From the above discussion, the number of logic complexities for the finite field processor with inversion can be approximated as:

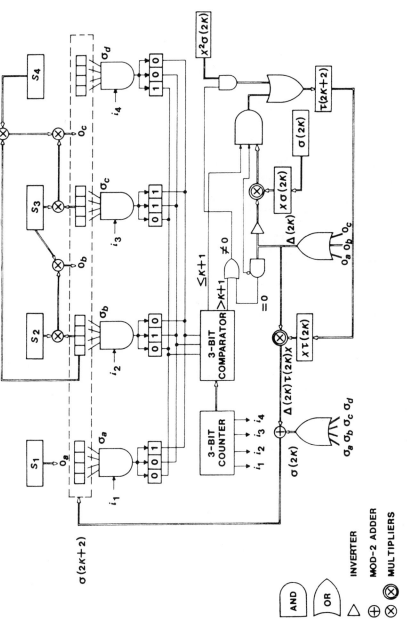

**Figure 6.7** Finite field processor of (15.5.7) BCH decoder.

| Logics | Number |
|--------|--------|
| Binary shift registers $m+1$ stages | $4(E+1)$ |
| Polynomial multiplication $m \times m$ | $2E$ |
| Summer ($m$-stage) | $E$ |
| $\otimes$ $x$ circuit $\Big\}$ $m$-stage | 2 |
| $\otimes$ $x^2$ circuit | 1 |
| AND gates | $E+15$ |
| OR gates | 4 |
| Inverter | $E$ |

Since both integrated circuits and microprocessors are function oriented, it does not make good sense to break down the logic complexity more than is presented. The polynomial multiplication and the multiplication of $\alpha$ power circuits will be treated after we discuss the error-location circuit.

### 6.2.2.4  Error-Location Circuit

Assume that $E$ errors have occurred in a codeword, and the error-locating polynomial is in general of the form

$$\sigma(x) = x^E + \sigma_1 x^{E-1} + \sigma_2 x^{E-2} + \ldots + \sigma_E. \tag{6.22}$$

The coefficients $\sigma_1, \sigma_2, \ldots, \sigma_E$ are sometimes called the elementary symmetric functions, which provide the error locations in terms of the power of $\alpha$. The purpose of this section is to present the Chien method for simpler implementation to obtain the $\sigma$'s without actually explicitly solving the equation $\sigma(x)$ [9].

This method is based on the Newton identity and the cyclic property of the BCH code. Let

$$\begin{aligned} &x^E + \sigma_1 x^{E-1} + \sigma_2 x^{E-2} + \ldots + \sigma_E \\ &= (x-\beta_0)(x-\beta_2) \ldots (x-\beta_E) \end{aligned} \tag{6.23}$$

so that

$$\sigma_1 = \sum_{j=1}^{E} \beta_j \tag{6.24}$$

$$\sigma_2 = \sum_{\substack{j,k=1 \\ j<k}}^{E} \beta_j \beta_k \tag{6.25}$$

$$\sigma_3 = \sum_{\substack{i,j,k=1 \\ i<j<k}}^{E} \beta_i \beta_j \beta_k. \tag{6.26}$$

If $\bar{\beta}_j = \alpha\beta_j$, $\bar{\beta}_j^{(\tau)} = \alpha^\tau\beta_j$, $j = 1, 2, \ldots, E$, are the set of transformed roots of $\bar{\sigma}(x) = x^E + \bar{\sigma}_1 x^{E-1} + \bar{\sigma}_2 x^{E-2} + \ldots + \bar{\sigma}_E = (x-\bar{\beta}_1)(x-\bar{\beta}_2) \ldots$

$(x - \bar{\beta}_E)$ and $\bar{\sigma}^{(\tau)}(x) = x^E + \bar{\sigma}_1^{(\tau)} x^{E-1} + \ldots + \bar{\sigma}_E^{(\tau)}$, with $\bar{\sigma}_k = \alpha^k \sigma_k$ and $\bar{\sigma}_k^{(\tau)} = \alpha^{k\tau} \sigma_k$, when $\alpha^0 = 1$ is a root of $\sigma(x)$, $x^k = 1$ for $k = 1, 2, \ldots, E$, we then have

$$\sigma(x = 1) = 1 + \sum_{k=1}^{E} \sigma_k = 0 \qquad (6.27)$$

or

$$\sum_{k=1}^{E} \sigma_k = 1.$$

In general, we have

$$\sum_{k=1}^{E} \bar{\sigma}_k^{(\tau)} = \sum_{k=1}^{E} \alpha^{k\tau} \sigma_k = 1 \qquad (6.28)$$

after $\tau_1, \tau_2, \ldots, \tau_t$ shifts, while $\tau_1, \tau_2, \ldots, \tau_E$ are the number of shifts from the initial conditions such that (6.28) is satisfied. For a $\sigma(x)$ of degree $E$, (6.28) is satisfied exactly $E$ times. The initial conditions are determined by $\bar{\sigma}_k$ or $\alpha^k \sigma_k$. Then the roots of $\sigma(x)$ are $\alpha^{n-\tau_1}, \alpha^{n-\tau_2}, \ldots, \alpha^{n-\tau_E}$, which are the error locations. The circuit, which detects the error locations, is shown in Figure 6.8.

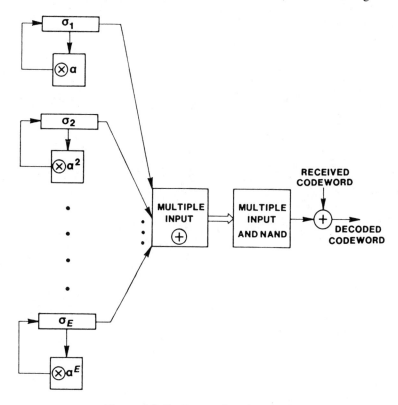

**Figure 6.8** Cyclic error location circuit.

The essential element in the cyclic error-location circuit is the set of $\alpha$ power multiplication units. This will be detailed in the next section.

*Example 6.7:*   Again, use Bose and Chaudhuri's original code, i.e., the BCH three-error-correcting (15, 5 ,7) code. The circuit for the error-locating calculation is shown in Figure 6.9. The design of the $\alpha$-power circuits will also be discussed in the next section.

The logic complexity of the error-locating circuit amounts to $E$ shift registers, each of which has $m$ stages. There are $m E$ input modulo-2 adders. For the set of multiplication by $\alpha$ power circuits, at most $(m-1)E$ modulo-2 adders is needed. The complexity of the error-locating circuit can be estimated as follows.

| Logics | Number |
|---|---|
| Shift registers (or equivalent of $m$ stages of flip-flops) | $E$ |
| Modulo-2 adders | $m+(m-1)E+1$ |
| $m$-input AND gate | 1 |
| Inverter | 1 |

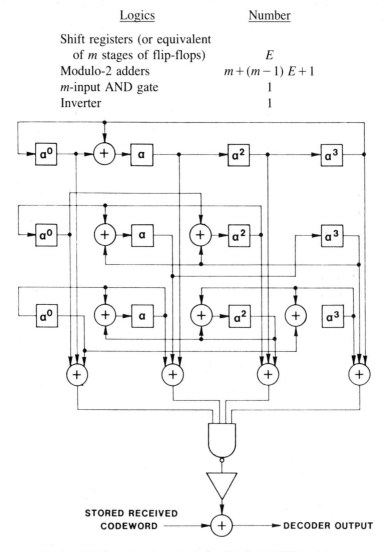

**Figure 6.9** Error-locating circuit for (15, 5, 7) BCH decoder.

### 6.2.2.5 Multiplication by Powers of α Circuit

We have seen in Section 6.2.2.4 that multiplication by powers of the primitive element α is essential in BCH decoding. In this section we discuss the fundamental theory and the design of such circuits.

Multiplication by the α function can be obtained by a feedback shift register according to a primitive polynomial $m(x) = \sum_{i=0}^{m} m_i x^i$. The feedback connections correspond to the presence or absence of the $m_i$ term in $m(x)$. There is feedback at $m_i$ if the $m_i$ term of $m(x)$ is nonzero. We note that multiplication by the α circuit is equivalent to a maximum length sequence shift register generator. Thus every maximum length sequence generator, which can be constructed from any primitive polynomial over a finite field, is a circuit capable of multiplication by the primitive element α.

For multiplications by higher powers of α, i.e., $\alpha^2, \alpha^3, \alpha^4, \ldots$, the circuits can be constructed from the following general procedure. Suppose a circuit of multiplying by $\alpha^\ell$ is needed, given a primitive integer which can vary from 0 to $2^m - 2$. Then

$$\alpha^\ell m(\alpha) = m_0 \alpha^\ell + m_1 \alpha^{\ell+1} + \ldots + m_{m-1} \alpha^{\ell+m-1}. \qquad (6.29)$$

Substitute $\alpha^{m-1}$ by $1 + m_1 \alpha^2 + m_2 \alpha^3 + \ldots + m_{m-2} \alpha^{m-2}$, $\alpha^m$ by $\alpha + m_1 \alpha^2 + m_2 \alpha^3 + \ldots + m_{m-2} \alpha^{m-1}$, and substitute the last $\alpha^{m-1}$ by $1 + m_1 \alpha + m_2 \alpha^2 + m_3 \alpha^3 + \ldots + m_{m-2} \alpha^{m-2}$ again. This procedure continues the same way the field table is constructed. Basically, it is just the ring of modulo the primitive polynomial $m(x)$ over GF($2^m$). After these substitutions, the polynomial (6.29) contains a collection of the coefficients of $m(x)$, with the degree of α's no more than $m - 1$. Putting the common α power term together, we have

$$\alpha^\ell \, m(\alpha) = \sum_j \oplus_i^j m_i \, \alpha^j, \qquad (6.30)$$

where $\oplus_i^j$ denotes the modulo-2 sum of the $m_i$'s in the term of $\alpha^j$. The algebraic sum of the $j$'s represents the polynomial $\alpha^\ell \, m(\alpha)$. The expression on the righthand side of (6.30) is used to synthesize the circuit that multiplies $\alpha^\ell$ in single clock time. For a primitive polynomial $m(x)$ of degree $m$, in general, $m$ flip-flops are needed. For each nonzero $\alpha^j$ term in (6.30), there is a modulo-2 adder at the input of the $m_i$th flip-flop. The inputs to this modulo-2 adder come from the coefficient of $\alpha^j$ in (6.30). If the contents of the flip-flops are set for ($\alpha^0, \alpha^1, \alpha^2, \ldots \alpha^{m-1}$), then the contents of the same flip-flops after one clock time are $\alpha^\ell(\alpha^0, \alpha^1, \alpha^2, \ldots, \alpha^{m-1})$. It is to be noted that unlike the circuit for multiplying by α, the higher power α multiplication circuits are, in general, not serially connected shift registers.

*Example 6.8:* For the simple primitive polynomial $m(x) = 1 + x + x^3$, which generates the field GF($2^3$), the multiply-by-α circuit is shown as Figure 6.10. The multiplication-by-$\alpha^2$ circuit is obtained as $\alpha^2(m_0 \alpha^0 + m_1 \alpha^1 + m_2 \alpha^2) = m_1 \alpha^0 + (m_1 + m_2)\alpha + (m_0 + \alpha_2)\alpha^2$. Then the circuit to perform $\alpha^2$ multiplication is shown in Figure 6.11.

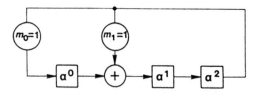

**Figure 6.10** Multiplication by $\alpha$ circuit for $1 + x + x^3$.

As an illustration, use $\alpha^3 = 1 + \alpha$; then the flip-flops set with $\alpha^0 = 1$, $\alpha^1 = 1$, $\alpha^2 = 0$. One clock time later, the flip-flops become $\alpha^0 = 1$, $\alpha^1 = 1$, and $\alpha^2 = 1$, which is $1 + \alpha + \alpha^2 = \alpha^5 = \alpha^2(\alpha^3)$.

For the multiplication by $\alpha^3$ we have, similarly,

$$\alpha^3(m_0\alpha^0 + m_1\alpha^1 + m_2\alpha^2) = (m_0 + m_2)\alpha^0 + (m_0 + m_1 + m_2)\alpha^1$$

$$+ (m_1 + m_2)\alpha^2.$$

The circuit for the multiplication by $\alpha^3$ is as shown in Figure 6.12.

If $\alpha^2 = 0\ 0\ 1$ is to be multiplied by $\alpha^3$, set $\alpha^0 = 0$, $\alpha^1 = 0$, and $\alpha^2 = 1$. One clock time later, the outputs become $1\ 1\ 1 = \alpha^5 = \alpha^3 \cdot \alpha^2 = \alpha^3(0\alpha^0 + 0\alpha^1 + \alpha^2)$.

*Example 6.9:*   From Example 6.7, where $m(x) = 1 + x + x^4$, the multiplication by $\alpha$ circuit is shown in Figure 6.13 with $m_0 = m_1 = 1$.

The multiplying by $\alpha^2$ circuit is obtained as

$$\alpha^2(m_0\alpha^0 + m_1\alpha + m_2\alpha^2 + m_3\alpha^3)$$

$$= m_0\alpha^2 + m_1\alpha^3 + m_2\alpha^4 + m_3\alpha^5$$

$$= m_0\alpha^2 + m_1\alpha^3 + m_2(1 + \alpha) + m_3(\alpha^2 + \alpha)$$

$$= m_2\alpha^0 + (m_2 + m_3)\alpha + (m_0 + m_3)\alpha^2 + m_1\alpha^3$$

which is shown in Figure 6.14.

Similarly, if an $\alpha^6$ multiplication circuit is desired,

$$\alpha^6(m_0 + m_1\alpha + m_2\alpha^2 + m_3\alpha^3) = (m_1 + m_2)\alpha^0$$

$$+ (m_1 + m_3)\alpha + (m_0 + m_2)\alpha^2$$

$$+ (m_0 + m_1 + m_3)\alpha^3.$$

The circuit in which the multiplication is done is shown in Figure 6.15.

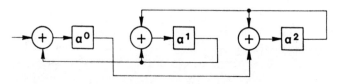

**Figure 6.11** Multiplication by $\alpha^2$ circuit for $1 + x + x^3$.

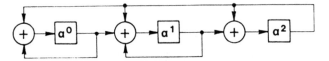

**Figure 6.12** Multiplication by $\alpha^3$ circuit for $1 + x + x^3$.

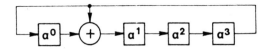

**Figure 6.13** Multiplication by $\alpha$ circuit for $1 + x + x^4$.

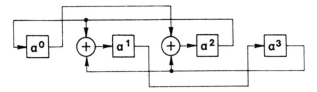

**Figure 6.14** Multiplication by $\alpha^2$ circuit for $1 + x + x^4$.

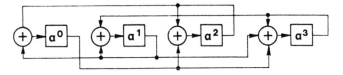

**Figure 6.15** Multiplication by $\alpha^6$ circuit for $1 + x + x^4$.

## 6.3 FAST POLYNOMIAL MULTIPLICATION CIRCUITS

Polynomial multiplication over a finite field is an essential and key operation in BCH and other algebraic decoding. For high data-rate transmission, the usefulness of an error-correcting codec is sometimes limited by its decoding speed. Since addition is fast, the decoding speed is essentially limited only by the mechanism of performing polynomial multiplication over a finite field.

In this section we present a scheme that simultaneously multiplies two polynomials of $m - 1$ degree over $GF(2^m)$ and modulos an irreducible polynomial that generates the field. The resultant polynomial can be obtained in one clock period by means of combination circuits. The synthesis of such a circuit is straightforward and the circuit itself is simple. It requires only logic AND and modulo-2 addition functions.

Let any two polynomials to be multiplied be $A(x) = A_0 + A_1x + A_2x^2 +$

$\ldots + A_{M_1-1}x$ and $B(x) = B_0 + B_1x + B_2x^2 + \ldots + B_{M_2-1}x^{M_2-1}$, where $A_i, B_j, (i = 0, 1, \ldots, M_1-1, j = 0, 1, \ldots, M_2-1)$ are elements from $GF(2^m)$, and $A(x), B(x)$ can be of degree less than $M_1, M_2$, respectively, which may be either greater or less than $m$. The purpose is to obtain the product

$$C(x) = A(x) \cdot B(x) \text{ [mod an irreducible polynomial over } GF(2^m)] \quad (6.31)$$

in one clock time. We consider the following two possible cases.

*Case 6.1:* $M_1 + M_2 < m + 2.$

Let
$$A(x) = A_0 + A_1x + \ldots + A_{M_1-1}x^{M_1-1}$$
$$B(x) = B_0 + B_1x + \ldots + B_{M_2-1}x^{M_2-1} \quad \text{and}$$
$$C(x) = A_0B_0 + (A_0B_1 + A_1B_0)x + \ldots +$$
$$A_{M_1-1}B_{M_2-1}x^{M_1+M_2-2}$$
$$= C_0 + C_1x + \ldots + C_px^p,$$

where $C_p$ denotes the coefficient corresponding to $x^{M_1+M_2-2}$. $C_0$ can be implemented by a 2-input AND gate with $A_0B_0$ as inputs. $C_1$ is obtained by two 2-input AND gates and a single modulo-2 adder. The pair of inputs to the AND gates are $A_0B_1$ and $A_1B_0$. The output of each AND gate is the input to the mod-2 adder. The design continues until all the coefficients of $C(x)$ are obtained.

*Case 6.2:* $M_1 + M_2 \geq m + 2.$

Without loss of generality, let us consider the multiplication of two polynomials $a(x), b(x)$ each of $M_1 = M_2 = m-1$ degree, and modulo an irreducible polynomial $m(x)$ of $m$ degree over $GF(2)$. The resultant polynomial is $c(x)$ which also has $m-1$ degree. The implementation to determine the coefficients of $c(x)$ is described in steps as follows:

Step 1: Let $c(x) = a(x) b(x)$, and the multiplication is performed as

$$a(x) = a_0 \quad + a_1x \quad + \ldots \quad + a_{m-1}x^{m-1}$$
$$b(x) = b_0 \quad + b_1x \quad + \ldots \quad + b_{m-1}x^{m-1}$$

---

$$a_0b_0 \quad + a_1b_0 x + \ldots \quad + a_{m-1}b_0 x^{m-1}$$
$$a_0b_1 x + a_1b_1 x^2 \quad + \ldots \quad + a_{m-1}b_1 x^m$$

$$\cdot \qquad\qquad \cdot \qquad\qquad \cdot$$

$$\cdot \qquad\qquad \cdot \qquad\qquad \cdot$$

$$\cdot \qquad\qquad \cdot \qquad\qquad \cdot$$

$$a_0 b_{m-1} x^{m-1} \qquad + a_1 b_{m-1} x^m + \ldots \qquad + a_{m-1} b_{m-1} x^{2(m-1)}$$

$$c(x) = a_0 b_0 + (a_1 b_0 + a_0 b_1) x + \ldots \ldots \ldots + a_{m-1} b_{m-1} x^{2(m-1)}.$$

Step 2:   Let $m(x) = m_0 + m_1 x + m_2 x^2 + \ldots + x^m$. Where $m_i = 0, 1$ for $i = 0, 1, \ldots, m$. For $m(x) = 0$,

$$x^m = m_0 + m_1 x + m_2 x^2 + \ldots + m_{m-1} x^{m-1}$$

$$x^{m+1} = m_0 x + m_1 x^2 + \ldots + m_{m-1} x^m$$

$$x^{m+2} = m_0 x^2 + m_1 x^3 + \ldots + m_{m-1} x^{m+1}$$

$$x^{2(m-1)} = m_0 x^{m-2} + m_1 x^{m-1} + \ldots + m_{m-1} x^{2m-1}. \qquad (6.32)$$

On the righthand side of (6.32), substitute the nonzero $x^m$, $x^{m+1}, x^{m+2}, \ldots \ldots \ldots, x^{2(m-1)}$ iteratively so that each equality has $m-1$ degree or less.

Substituting $x^m, x^{m+1}, x^{m+2}, \ldots \ldots x^{2(m-1)}$ of $c(x)$ in Step 1 with the corresponding polynomials obtained above, the same $x$ power terms are collected. The result is $c(x)$ with $m-1$ degree or less. Each term $x^k$ has a sum of the product in the form of $a_i b_j$ $i, j = 0, 1, \ldots$, $m-1$. At this point the coefficients of $c(x)$ are determined. $c_k$ corresponds to $x^k$, $k = 0, 1, \ldots, m-1$.

Step 3:   Form an $(m \times m)$ 2-input logic AND gates as shown in Figure 6.16. $a_i b_j$ denote the two inputs $a_i$, $b_j$ to the AND gate. The outputs from the set of AND gates go to a set of $m$ modulo-2 adders. The precise connections from the outputs of the AND gates to the modulo-2 adders are in accordance with the sum of the product $a_i b_j$ which makes $c_k$ as in Step 2. The outputs of the modulo-2 adders are the set of coefficients of $c(x)$.

*Example 6.10:*   Let $c(x) = a(x) \cdot b(x)$ in GF($2^3$), $m = 3$.

Step 1:

$$a(x) = a_0 \quad + a_1 x + a_2 x^2$$

$$b(x) = b_0 \quad + b_1 x + b_2 x^2$$

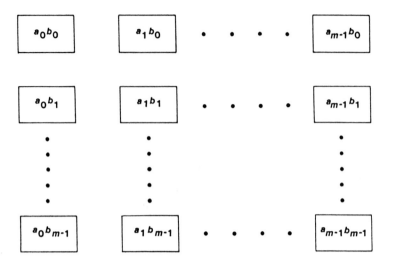

**Figure 6.16** $M \times m$ array of 2-input AND gates.

$$a_0 b_0 + a_1 b_0 x + a_2 b_0 x^2$$

$$a_0 b_1 x + a_1 b_1 x^2 + a_2 b_1 x^3$$

$$a_0 b_2 x^2 + a_1 b_2 x^3 + a_2 b_2 x^4$$

$$c(x) = a_0 b_0 + (a_1 b_0 + a_0 b_1) x + (a_0 b_2 + a_1 b_1 + a_2 b_0) + x^2$$

$$+ (a_1 b_2 + a_2 b_1) x^3 + a_2 b_2 x^4 .$$

Step 2: Let $m(x) = 1 + x + x^3$

$$x^3 = 1 + x$$

$$x^4 = x + x^2.$$

If we substitute $x^3$, $x^4$ in $c(x)$ of Step 1, we have

$$c(x) = a_0 b_0 + (a_1 b_0 + a_0 b_1) x + (a_0 b_2 + a_1 b_1 + a_2 b_0) x^2$$

$$+ (a_1 b_2 + a_2 b_1) (1 + x) + a_2 b_2 (x + x^2)$$

$$= a_0 b_0 + a_1 b_2 + a_2 b_1 + (a_0 b_1 + a_1 b_0 + a_1 b_2$$

$$+ a_2 b_1 + a_2 b_2) x + (a_0 b_2 + a_1 b_1 + a_2 b_0 + a_2 b_2) x^2 .$$

Thus

$$c_0 = a_0 b_0 + a_1 b_2 + a_2 b_1$$

$$c_1 = a_0 b_1 + a_1 b_0 + a_1 b_2 + a_2 b_1 + a_2 b_2$$

$$c_2 = a_0 b_2 + a_1 b_1 + a_2 b_0 + a_2 b_2 .$$

Step 3: The complete circuit connection is also shown in Figure 6.17. As a verification, any product of the powers of the primitive element $\alpha$ in $GF(2^3)$, i.e.,

$$
\begin{array}{c c c c}
 & x^0 & x^1 & x^2 \\
\alpha^0 & 1 & 0 & 0 \\
\alpha^1 & 0 & 1 & 0 \\
\alpha^2 & 0 & 0 & 1 \\
\alpha^3 & 1 & 1 & 0 \\
\alpha^4 & 0 & 1 & 1 \\
\alpha^5 & 1 & 1 & 1 \\
\alpha^6 & 1 & 0 & 1 \; ,
\end{array}
$$

can be achieved in one clock period. If $a(x) = \alpha^1$, $a_0 = 0$, $a_1 = 1$, $a_2 = 0$ and $b(x) = \alpha^4$, $b_0 = 0$, $b_1 = b_2 = 1$, then $c(x) = a(x) \cdot b(x) = \alpha^1 \cdot \alpha^4 = \alpha^5$ with $c_0 = c_1 = c_2 = 1$.

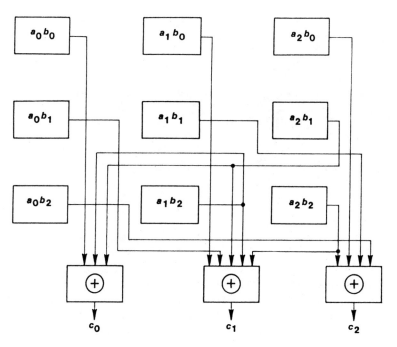

**Figure 6.17** Fast polynomial multiplication circuit in $GF(2^3)$.

*Example 6.11:*

Step 1: For GF($2^5$), $m = 5$,

$$c(x) = a(x) \cdot b(x) = (a_0 + a_1 x + a_2 x^2 + a_3 x^3 + a_4 x^4)$$

$$(b_0 + b_1 x + b_2 x^2 + b_3 x^3 + b_4 x^4)$$

$$= a_0 b_0 + (a_1 b_0 + a_0 b_1) x + (a_2 b_0 + a_1 b_1 + a_0 b_2) x^2$$

$$+ (a_3 b_0 + a_2 b_1 + a_1 b_2 + a_0 b_3) x^3 + (a_4 b_0 +$$

$$a_3 b_1 + a_2 b_2 + a_1 b_3 + a_0 b_4) x^4 + (a_4 b_1 +$$

$$a_3 b_2 + a_2 b_3 + a_1 b_4) x^5 + (a_4 b_2 + a_3 b_3 + a_2 b_4) x^6$$

$$+ (a_4 b_3 + a_3 b_4) x^7 + a_4 b_4 x^8 .$$

Step 2: Let $m(x) = 1 + x^2 + x^5$

$$x^5 = 1 + x^2$$
$$x^6 = x + x^3$$
$$x^7 = x^2 + x^4$$
$$x^8 = x^3 + x^5 = x^3 + 1 + x^2.$$

Substituting $x^5, \ldots, x^8$ in $c(x)$ of Step 1, we have, after collecting terms,

$$c(x) = a_0 b_0 + a_4 b_4 + (a_1 b_0 + a_0 b_1 + a_4 b_2 + a_3 b_3 + a_2 b_4) x$$

$$+ (2_2 b_0 + a_1 b_1 + a_0 b_2 + a_4 b_3 + a_3 b_4 + a_4 b_4) x^2$$

$$+ (a_3 b_0 + a_2 b_1 + a_1 b_2 + a_0 b_3 + a_4 b_2 + a_3 b_3 + a_2 b_4$$

$$+ a_4 b_4) x^3 + (a_4 b_0 + a_3 b_1 + a_2 b_2 + a_1 b_3 + a_0 b_4$$

$$+ a_4 b_3 + a_3 b_4) x^4$$

$$= c_0 + c_1 x + c_2 x^2 + c_3 x^3 + c_4 x^4 .$$

The circuit connection of Step 3 is omitted here.

The number of $AB$ products as the coefficients of $x$ term with degrees less than $m$ is

$$1 + 2 + 3 + \ldots + m = \frac{m(m+1)}{2} .$$

The additional products to be accounted for in the final coefficients of the $x$'s depend on the number of terms in the irreducible polynomial $m(x)$ that generates GF($2^m$). If we assume $m(x)$ has $t$ terms, then the number of additional products that we need to account for is

$$t[(m-1) + (m-2) + \ldots + m-(m-1)] ,$$

where $m-1$ is the number of products belonging to $x^m$, $m-2$ is the number of products belonging to $x^{m+1}, \ldots,$ and $m-(m-1)$ is the number of products belonging to $x^{2(m-1)}$.

The logic complexity of fast polynomial multiplication is estimated as follows. It requires, at most, $m^2$ 2-input AND gates to compute the $a_ib_j$ product. It requires $m$ multiple input modulo-2 addition circuits for the summation. Each modulo-2 addition circuit has a maximum of $2m$ inputs, because this is the worst situation in which every $x$ power term of $m(x)$ is nonzero.

## 6.4 SEQUENTIAL POLYNOMIAL MULTIPLICATION CIRCUIT

Another polynomial multiplication method, as described by Peterson and Berlekamp, is as follows [3, 4]. Let the coefficients of the multiplicand polynomial $u(x) = M_0 + u_1x \ldots u_{m-1}x^{m-1}$ be stored in a feedback shift register $u$ which provides $\alpha u(x)$ upon each shift. The coefficients of the multiplier polynomial $v(x) = v_0 + v_1x + \ldots + v_{m-1}x^{m-1}$ are stored in another shift register V. To perform the multiplication, there is another shift register Z, as shown in Figure 6.18, which is initially set to zero. The contents of the U shift register are added to the Z shift register if $v_0 = 1$; if $v_0 = 0$, $z$ is left unchanged. The V register is then shifted to the right, the U register is multiplied by $\alpha$ by a single

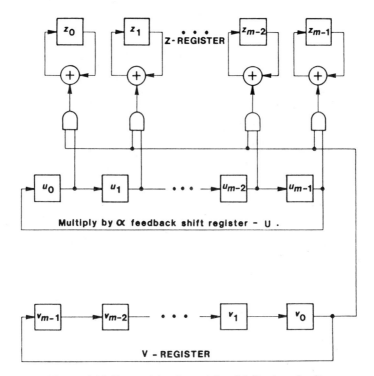

**Figure 6.18** Sequential polynomial multiplication circuits.

shifting, and the process is repeated. After $m$ such steps, i.e., when $v_{m-1}$ reaches $v_0$, the multiplication is completed. Z contains the product of the contents of the U and V registers.

*Example 6.12:*  Let $\alpha$ be a primitive root of the irreducible polynomial $1 + x + x^3$. Then multiplication by $\alpha$ is accomplished by the feedback shift register U shown in Figure 6.19. Let us multiply $\alpha^2 = 0\ 0\ 1$ by $\alpha^3 = 1\ 1\ 0$. We load $u_0 = 0$, $u_1 = 0$, $u_2 = 1$ in the U register and $v_2 = 0$, $v_1 = 1$, $v_0 = 1$ in the V register. The first shift produces $z_0 = z_1 = z_2 = 1$. The Z register contents are unchanged by the third shift because $v_0 = 0$ at this time. Thus the product is $1\ 1\ 1\ = \alpha^5 = \alpha^2 \cdot \alpha^3$.

This method of multiplication requires, in general, $m$ times the clock rate to complete the two-polynomial multiplication.

## 6.5   MAXIMUM LENGTH SEQUENCE POLYNOMIAL MULTIPLICATION

A scheme described by Bartee and Schneider for the multiplication of polynomials is described in this section [11]. This scheme works only for a related field table and does not directly provide the result of polynomial multiplication over GF($2^m$) as required in general decoding. The method is simplified into steps as follows.

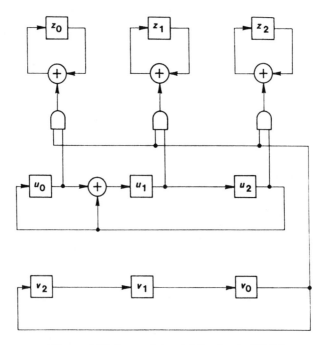

**Figure 6.19**  Sequential multiplication in GF(2)$^3$.

Step 1:  Given a maximum length sequence, $m_0\, m_1\, m_2\, m_3 \ldots m_{L-1}$, where $L = 2^m - 1$, form the following shifts

$$\gamma^0 = m_0 m_1 \ldots m_{m-1}$$

$$\gamma^1 = m_1 m_2 \ldots m_m$$

$$\gamma^2 = m_2 m_3 \ldots m_{m+1}$$

$$\cdot \qquad \cdot$$

$$\cdot \qquad \cdot$$

$$\cdot \qquad \cdot$$

$$\gamma^{L-1} = m_{L-1}\, m_L \ldots m_{L+m-2} \tag{6.33}$$

Step 2:  From the $\gamma$-table of Step 1, let $\omega_1 = \gamma^0$, and identify all single nonzero $m$-tuples of $\gamma^i$. Since there can be a nonzero value in each position, there are $m$ such $\gamma^i$ including $\gamma^0$. Let $\omega_i$ denote the $m$-tuple whose $i$th digit is nonzero, i.e.,

$$\omega_i = \ldots 0\ 0\ 0\ 1\ 0\ 0\ 0 \ldots = \gamma^k.$$

$$\uparrow \qquad\qquad \uparrow \qquad\qquad \uparrow \tag{6.34}$$

$$\text{1st} \qquad\quad \text{$i$th} \qquad\quad \text{$m$th}$$

There can be only $m$ such $\omega_i$; i.e., $\omega_1, \omega_2, \ldots, \omega_m$. Construct an $m \times m$ square matrix as

$$M = [\omega_i \omega_j]$$

$$i, j = 1, 2, \ldots, m \tag{6.35}$$

and substituting $\gamma^k$, $\gamma^q$ for $\omega_i$, $\omega_j$. The elements of matrix $M$ are now in terms of $\gamma^{k+q}$. Next, substituting all the corresponding binary representations of $\gamma^{k+q}$ into $M$, which has just been converted into a binary matrix, retain the $m$-tuples as in $\gamma^{k+q}$.

Step 3:  Split the binary $M$ matrix into $m$ binary submatrices of size $m \times m$. The submatrix $M_1$ is composed of all the first columns of the $m$-tuples in $M$. $M_2$ is composed of all the second columns of the $m$-tuples in $M$. The columns of $M_m$ are derived from the $m$th columns of the $m$-tuples in $M$.

Step 4:  Let $a = (a_1, a_2, \ldots, a_m) = a_1 + a_2 x + \ldots a_m x^{m-1}$, $b = (b_1, b_2, \ldots, b_m) = b_1 + b_2 x + \ldots + b_m x^{m-1}$ and $c = a \cdot b = (c_1, c_2, \ldots c_m)$ modulo the minimal polynomial $m_1(x)$. Then

$$c_1 = a\, M_1\, b$$

$$c_2 = a\, M_2\, b$$

$$\cdot \qquad \cdot$$

$$\cdot \qquad \cdot$$

$$\cdot \qquad \cdot$$

$$c_m = a\, M_m\, b \qquad\qquad (6.36)$$

where $M_1, M_2, \ldots, M_m$ were produced from Step 3. After the matrix multiplications, the $c$'s can be expressed in terms of sum of the products of $a_i b_j$, $i, j = 1, 2, \ldots, m$.

*Example 6.13:*

Step 1: The maximum length sequence generated by $1 + x + x^3$ is $m_0 m_1 m_2 \ldots m_7 = 1\ 0\ 0\ 1\ 0\ 1\ 1\ 1\ 0\ 0\ 1\ 0\ 1\ 1 \ldots$. Then

$$\gamma^0 = 1\ 0\ 0$$
$$\gamma^1 = 0\ 0\ 1$$
$$\gamma^2 = 0\ 1\ 0$$
$$\gamma^3 = 1\ 0\ 1$$
$$\gamma^4 = 0\ 1\ 1$$
$$\gamma^5 = 1\ 1\ 1$$
$$\gamma^6 = 1\ 1\ 0$$

Step 2: Let $\omega_1 = \gamma^0$, $\omega_2 = \gamma^2$, and $\omega_3 = \gamma$; since $m = 3$, these are all the $\omega$'s. Next construct the $M$ matrix as

$$M = \begin{bmatrix} \omega_1\omega_1 & \omega_2\omega_1 & \omega_3\omega_1 \\ \omega_1\omega_2 & \omega_2\omega_2 & \omega_3\omega_2 \\ \omega_1\omega_3 & \omega_2\omega_3 & \omega_3\omega_3 \end{bmatrix}$$

$$= \begin{bmatrix} \gamma^0\gamma^0 & \gamma^2\gamma^0 & \gamma\,\gamma^0 \\ \gamma^0\gamma^2 & \gamma^2\gamma^2 & \gamma\,\gamma^2 \\ \gamma^0\gamma & \gamma^2\gamma & \gamma\,\gamma \end{bmatrix}$$

$$= \begin{bmatrix} \gamma^0 & \gamma^2 & \gamma \\ \gamma^2 & \gamma^4 & \gamma^3 \\ \gamma & \gamma^3 & \gamma^2 \end{bmatrix}$$

$$= \begin{bmatrix} 1\ 0\ 0 & 0\ 1\ 0 & 0\ 0\ 1 \\ 0\ 1\ 0 & 0\ 1\ 1 & 1\ 0\ 1 \\ 0\ 0\ 1 & 1\ 0\ 1 & 0\ 1\ 0 \end{bmatrix}$$

Step 3: From Step 2, the last expression of $M$ matrix with binary elements, the submatrices are determined

$$M_0 = \begin{bmatrix} 1 & 0 & 0 \\ 0 & 0 & 1 \\ 0 & 1 & 0 \end{bmatrix}$$

$$M_1 = \begin{bmatrix} 0 & 1 & 0 \\ 1 & 1 & 0 \\ 0 & 0 & 1 \end{bmatrix}$$

$$M_2 = \begin{bmatrix} 0 & 0 & 1 \\ 0 & 1 & 1 \\ 1 & 1 & 0 \end{bmatrix}.$$

Step 4:  Let $a = (a_0, a_1, a_2)$ represent any 3-tuple of $\gamma^a$ and $b = (b_0, b_1, b_2)$ represent any 3-tuple of $\gamma^b$, then $c = a \cdot b = (a_0, a_1, a_2) \cdot (b_0, b_1, b_2) = (c_0, c_1, c_2)$
and

$$c_0 = aM_0b^T = (a_0, a_1, a_2) \begin{bmatrix} 1 & 0 & 0 \\ 0 & 0 & 1 \\ 0 & 1 & 0 \end{bmatrix} \begin{bmatrix} b_0 \\ b_1 \\ b_2 \end{bmatrix}$$

$$= a_0b_0 + a_2b_1 + a_1b_2$$

$$c_1 = aM_1b^T = (a_0, a_1, a_2) \begin{bmatrix} 0 & 1 & 0 \\ 1 & 1 & 0 \\ 0 & 0 & 1 \end{bmatrix} \begin{bmatrix} b_0 \\ b_1 \\ b_2 \end{bmatrix}$$

$$= a_1b_0 + (a_0 + a_1)b_1 + a_2b_2$$

$$c_2 = aM_2b^T = (a_0, a_1, a_2) \begin{bmatrix} 0 & 0 & 1 \\ 0 & 1 & 1 \\ 1 & 1 & 0 \end{bmatrix} \begin{bmatrix} b_0 \\ b_1 \\ b_2 \end{bmatrix}$$

$$= a_2b_0 + (a_1 + a_2)b_1 + (a_0 + a_1)b_2 .$$

The circuit can be implemented by AND logics and modulo-2 adders as described in the two previous sections. Any product $\gamma^k\gamma^q$ from the $\gamma$-table will result in the binary representation of $\gamma^{k+q}$. If $\gamma^2$ (0 1 0) is multiplied by $\gamma^3$ (1 0 1), then $\gamma^5$ (1 1 1) is the solution.

*Example 6.14:*   From the maximum length sequence generated by $1 + x + x^4$, the $\gamma$-table is

$$\gamma^0 = 1\ 0\ 0\ 0$$
$$\gamma^1 = 0\ 0\ 0\ 1$$
$$\gamma^2 = 0\ 0\ 1\ 0$$
$$\gamma^3 = 0\ 1\ 0\ 0$$
$$\gamma^4 = 1\ 0\ 0\ 1$$
$$\gamma^5 = 0\ 0\ 1\ 1$$

$$\gamma^6 = 0\ 1\ 1\ 0$$
$$\gamma^7 = 1\ 1\ 0\ 1$$
$$\gamma^8 = 1\ 0\ 1\ 0$$
$$\gamma^9 = 0\ 1\ 0\ 1$$
$$\gamma^{10} = 1\ 0\ 1\ 1$$
$$\gamma^{11} = 0\ 1\ 1\ 1$$
$$\gamma^{12} = 1\ 1\ 1\ 1$$
$$\gamma^{13} = 1\ 1\ 1\ 0$$
$$\gamma^{14} = 1\ 1\ 0\ 0.$$

Thus Step 1 is completed.

Step 2:  Let

$$\omega_1 = \gamma^0 = 1\ 0\ 0\ 0$$

$$\omega_2 = \gamma^3 = 0\ 1\ 0\ 0$$

$$\omega_3 = \gamma^2 = 0\ 0\ 1\ 0$$

$$\omega_4 = \gamma^1 = 0\ 0\ 0\ 1$$

$$M = \begin{bmatrix} \omega_1\omega_1 & \omega_2\omega_1 & \omega_3\omega_1 & \omega_4\omega_1 \\ \omega_1\omega_2 & \omega_2\omega_2 & \omega_3\omega_2 & \omega_4\omega_2 \\ \omega_1\omega_3 & \omega_2\omega_3 & \omega_3\omega_3 & \omega_4\omega_3 \\ \omega_1\omega_4 & \omega_2\omega_4 & \omega_3\omega_4 & \omega_4\omega_4 \end{bmatrix}$$

$$= \begin{bmatrix} \gamma^0\gamma^0 & \gamma^3\gamma^0 & \gamma^2\gamma^0 & \gamma^1\gamma^0 \\ \gamma^0\gamma^3 & \gamma^3\gamma^3 & \gamma^2\gamma^3 & \gamma^1\gamma^3 \\ \gamma^0\gamma^2 & \gamma^3\gamma^2 & \gamma^2\gamma^2 & \gamma^1\gamma^2 \\ \gamma^0\gamma^1 & \gamma^3\gamma^1 & \gamma^2\gamma^1 & \gamma^1\gamma^1 \end{bmatrix}$$

$$= \begin{bmatrix} \gamma^0 & \gamma^3 & \gamma^2 & \gamma^1 \\ \gamma^3 & \gamma^6 & \gamma^5 & \gamma^4 \\ \gamma^2 & \gamma^5 & \gamma^4 & \gamma^3 \\ \gamma^1 & \gamma^4 & \gamma^3 & \gamma^2 \end{bmatrix}$$

$$= \begin{bmatrix} 1\ 0\ 0\ 0 & 0\ 1\ 0\ 0 & 0\ 0\ 1\ 0 & 0\ 0\ 0\ 1 \\ 0\ 1\ 0\ 0 & 0\ 1\ 1\ 0 & 0\ 0\ 1\ 1 & 1\ 0\ 0\ 1 \\ 0\ 0\ 1\ 0 & 0\ 0\ 1\ 1 & 1\ 0\ 0\ 1 & 0\ 1\ 0\ 0 \\ 0\ 0\ 0\ 1 & 1\ 0\ 0\ 1 & 0\ 1\ 0\ 0 & 0\ 0\ 1\ 0 \end{bmatrix}$$

Step 3:

$$M_1 = \begin{bmatrix} 1 & 0 & 0 & 0 \\ 0 & 0 & 0 & 1 \\ 0 & 0 & 1 & 0 \\ 0 & 1 & 0 & 0 \end{bmatrix}$$

$$M_2 = \begin{bmatrix} 0 & 1 & 0 & 0 \\ 1 & 1 & 0 & 0 \\ 0 & 0 & 0 & 1 \\ 0 & 0 & 1 & 0 \end{bmatrix}$$

$$M_3 = \begin{bmatrix} 0 & 0 & 1 & 0 \\ 0 & 1 & 1 & 0 \\ 1 & 1 & 0 & 0 \\ 0 & 0 & 0 & 1 \end{bmatrix}$$

$$M_4 = \begin{bmatrix} 0 & 0 & 0 & 1 \\ 0 & 0 & 1 & 1 \\ 0 & 1 & 1 & 0 \\ 1 & 1 & 0 & 0 \end{bmatrix}$$

Step 4: Let $a = (a_1, a_2, a_3, a_4)$ and $b = (b_1, b_2, b_3, b_4)$, then $c = a \cdot b$ modulo $1 + x + x^4$.

$$c_1 = aM_1b = a_1b_1 + a_4b_2 + a_3b_3 + a_2b_4$$
$$c_2 = aM_2b = a_2b_1 + (a_1 + a_2)b_2 + a_4b_3 + a_3b_4$$
$$c_3 = aM_3b = a_3b_1 + (a_2 + a_3)b_2 + (a_1 + a_2)b_3 + a_4b_4$$
$$a_4 = aM_4b = a_4b_1 + (a_3 + a_4)b_2 + (a_2 + a_3)b_3 + (a_1 + a_2)b_4$$

The circuit connections in order to obtain the coefficients of $c(x)$ are shown in Figure 6.20 (a), (b), (c), and (d). As a verification, let $a = (0\ 1\ 1\ 0) = x + x^3$, $b = (1\ 1\ 0\ 1) = 1 + x + x^3$. From the $\gamma$-table, note that $a = \gamma^6$ and $b = \gamma^7$, and $c_1 = 1$, $c_2 = 1$, $c_3 = 1$, $c_4 = 0$, which correspond to $\gamma^{13} = c = a \cdot b = \gamma^6 \cdot \gamma^7 = \gamma^{13}$.

*Example 6.15:*   From the maximum length sequence generated by the primitive polynomial $1 + x^2 + x^5$, the $\gamma$-table is

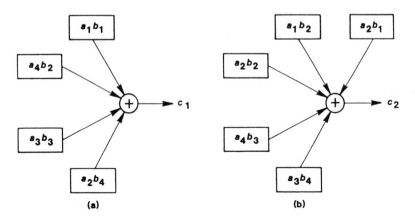

**(a)**                                    **(b)**

**Figure 6.20** Circuits for the calculation of the coefficients of $c(x)$.

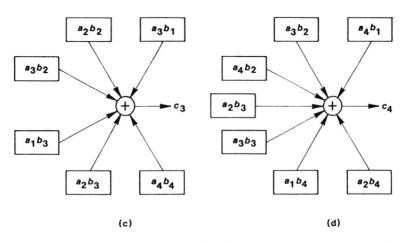

(c)                                                                        (d)

**Figure 6.20 (continued)** Circuits for the calculation of the coefficients of $c(x)$.

| | |
|---|---|
| $\gamma^0$ | 1 0 0 0 0 |
| $\gamma^1$ | 0 0 0 0 1 |
| $\gamma^2$ | 0 0 0 1 0 |
| $\gamma^3$ | 0 0 1 0 0 |
| $\gamma^4$ | 0 1 0 0 1 |
| $\gamma^5$ | 1 0 0 1 0 |
| $\gamma^6$ | 0 0 1 0 1 |
| $\gamma^7$ | 0 1 0 1 1 |
| $\gamma^8$ | 1 0 1 1 0 |
| $\gamma^9$ | 0 1 1 0 0 |
| $\gamma^{10}$ | 1 1 0 0 1 |
| $\gamma^{11}$ | 1 0 0 1 1 |
| $\gamma^{12}$ | 0 0 1 1 1 |
| $\gamma^{13}$ | 0 1 1 1 1 |
| $\gamma^{14}$ | 1 1 1 1 1 |
| $\gamma^{15}$ | 1 1 1 1 0 |
| $\gamma^{16}$ | 1 1 1 0 0 |
| $\gamma^{17}$ | 1 1 0 0 0 |
| $\gamma^{18}$ | 1 0 0 0 1 |
| $\gamma^{19}$ | 0 0 0 1 1 |
| $\gamma^{20}$ | 0 0 1 1 0 |
| $\gamma^{21}$ | 0 1 1 0 1 |
| $\gamma^{22}$ | 1 1 0 1 1 |
| $\gamma^{23}$ | 1 0 1 1 1 |
| $\gamma^{24}$ | 0 1 1 1 0 |
| $\gamma^{25}$ | 1 1 1 0 1 |
| $\gamma^{26}$ | 1 1 0 1 0 |
| $\gamma^{27}$ | 1 0 1 0 1 |

$$\begin{array}{ll} \gamma^{28} & 0\ 1\ 0\ 1\ 0 \\ \gamma^{29} & 1\ 0\ 1\ 0\ 0 \\ \gamma^{30} & 0\ 1\ 0\ 0\ 0 \end{array}$$

Let

$$\begin{aligned} \omega_1 &= \gamma^0 = 1\ 0\ 0\ 0\ 0 \\ \omega_2 &= \gamma^4 = 0\ 1\ 0\ 0\ 0 \\ \omega_3 &= \gamma^3 = 0\ 0\ 1\ 0\ 0 \\ \omega_4 &= \gamma^2 = 0\ 0\ 0\ 1\ 0 \\ \omega_5 &= \gamma^1 = 0\ 0\ 0\ 0\ 1 \end{aligned}$$

$$M = \begin{bmatrix} \omega_1\omega_1 & \omega_2\omega_1 & \omega_3\omega_1 & \omega_4\omega_1 & \omega_5\omega_1 \\ \omega_1\omega_2 & \omega_2\omega_2 & \omega_3\omega_2 & \omega_4\omega_2 & \omega_5\omega_2 \\ \omega_1\omega_3 & \omega_2\omega_3 & \omega_3\omega_3 & \omega_4\omega_3 & \omega_5\omega_3 \\ \omega_1\omega_4 & \omega_2\omega_4 & \omega_3\omega_4 & \omega_4\omega_4 & \omega_5\omega_4 \\ \omega_1\omega_5 & \omega_2\omega_5 & \omega_3\omega_5 & \omega_4\omega_5 & \omega_5\omega_5 \end{bmatrix}$$

$$M = \begin{bmatrix} \gamma^0 & \gamma^4 & \gamma^3 & \gamma^2 & \gamma \\ \gamma^4 & \gamma^8 & \gamma^7 & \gamma^6 & \gamma^5 \\ \gamma^3 & \gamma^7 & \gamma^6 & \gamma^5 & \gamma^4 \\ \gamma^2 & \gamma^6 & \gamma^5 & \gamma^4 & \gamma^3 \\ \gamma^1 & \gamma^5 & \gamma^4 & \gamma^3 & \gamma^2 \end{bmatrix}$$

Substituting the binary representation of the $\gamma$'s in the above matrix $M$, we have:

$$M = \begin{bmatrix} 1\ 0\ 0\ 0\ 0 & 0\ 1\ 0\ 0\ 1 & 0\ 0\ 1\ 0\ 0 & 0\ 0\ 0\ 1\ 0 & 0\ 0\ 0\ 0\ 1 \\ 0\ 1\ 0\ 0\ 0 & 1\ 0\ 1\ 1\ 0 & 0\ 1\ 0\ 1\ 1 & 0\ 0\ 1\ 0\ 1 & 1\ 0\ 0\ 1\ 0 \\ 0\ 0\ 1\ 0\ 0 & 0\ 1\ 0\ 1\ 1 & 0\ 0\ 1\ 0\ 1 & 1\ 0\ 0\ 1\ 0 & 0\ 1\ 0\ 0\ 1 \\ 0\ 0\ 0\ 1\ 0 & 0\ 0\ 1\ 0\ 1 & 1\ 0\ 0\ 1\ 0 & 0\ 1\ 0\ 0\ 1 & 0\ 0\ 1\ 0\ 0 \\ 0\ 0\ 0\ 0\ 1 & 1\ 0\ 0\ 1\ 0 & 0\ 1\ 0\ 0\ 1 & 0\ 0\ 1\ 0\ 0 & 0\ 0\ 0\ 1\ 0 \end{bmatrix}$$

The corresponding submatrices and the $c$'s can be obtained as follows:

$$M_1 = \begin{bmatrix} 1\ 0\ 0\ 0\ 0 \\ 0\ 1\ 0\ 0\ 1 \\ 0\ 0\ 0\ 1\ 0 \\ 0\ 0\ 1\ 0\ 0 \\ 0\ 1\ 0\ 0\ 0 \end{bmatrix}$$

$$\begin{aligned} c_1 = a\,M_1 b &= a_1 b_1 + (a_2 + a_5) b_2 + a_4 b_3 + a_3 b_4 + a_2 b_5 \\ &= a_1 b_1 + a_2 b_2 + a_2 b_5 + a_3 b_4 + a_4 b_3 + a_5 b_2 \end{aligned}$$

$$M_2 = \begin{bmatrix} 0\ 1\ 0\ 0\ 0 \\ 1\ 0\ 1\ 0\ 0 \\ 0\ 1\ 0\ 0\ 1 \\ 0\ 0\ 0\ 1\ 0 \\ 0\ 0\ 1\ 0\ 0 \end{bmatrix}$$

$$c_2 = a M_2 b = a_2 b_1 + (a_1 + a_3) b_2 + (a_2 + a_5) b_3 + a_4 b_4 + a_3 b_5$$

$$= a_1 b_2 + a_2 b_1 + a_2 b_3 + a_3 b_2 + a_3 b_5 + a_4 b_4 + a_5 b_3$$

$$M_3 = \begin{bmatrix} 0 & 0 & 1 & 0 & 0 \\ 0 & 1 & 0 & 1 & 0 \\ 1 & 0 & 1 & 0 & 0 \\ 0 & 1 & 0 & 0 & 1 \\ 0 & 0 & 0 & 1 & 0 \end{bmatrix}$$

$$c_3 = a M_3 b = a_3 b_1 + (a_2 + a_4) b_2 + (a_1 + a_3) b_3 + (a_2 + a_5) b_4$$
$$+ a_4 b_5$$

$$= a_1 b_3 + a_2 b_2 + a_2 b_4 + a_3 b_1 + a_3 b_3 + a_4 b_5 + a_5 b_4$$
$$+ a_4 b_2$$

$$M_4 = \begin{bmatrix} 0 & 0 & 0 & 1 & 0 \\ 0 & 1 & 1 & 0 & 1 \\ 0 & 1 & 0 & 1 & 0 \\ 1 & 0 & 1 & 0 & 0 \\ 0 & 1 & 0 & 0 & 1 \end{bmatrix}$$

$$c_4 = a M_4 b = a_4 b_1 + (a_2 + a_3 + a_5) b_2 + (a_2 + a_4) b_3$$
$$+ (a_1 + a_3) b_4 + (a_2 + a_5) b_5$$

$$= a_1 b_4 + a_2 b_3 + a_2 b_5 + a_2 b_2 + a_3 b_2 + a_3 b_4$$
$$+ a_4 b_3 + a_4 b_1 + a_5 b_2 + a_5 b_5$$

$$M_5 = \begin{bmatrix} 0 & 1 & 0 & 0 & 1 \\ 0 & 0 & 1 & 1 & 0 \\ 0 & 1 & 1 & 0 & 1 \\ 0 & 1 & 0 & 1 & 0 \\ 1 & 0 & 1 & 0 & 0 \end{bmatrix}$$

$$c_5 = a M_5 b = a_5 b_1 + (a_1 + a_3 + a_4) b_2 + (a_2 + a_3 + a_5) b_3$$
$$+ (a_2 + a_4) b_4 + (a_1 + a_3) b_5$$

$$= a_1 b_2 + a_1 b_5 + a_2 b_3 + a_2 b_4 + a_3 b_2 + a_3 b_3$$
$$+ a_3 b_5 + a_4 b_2 + a_4 b_4 + a_5 b_1 + a_5 b_3$$

According to the expressions for $c_1, c_2, \ldots, c_5$, the logic circuits can be designed like the previous two examples with the $m^2 = 5^2$ 2-input AND gates omitted.

After describing the two algorithms and presenting some examples, we shall point out the resemblance and difference between the polynomial multiplications of $\alpha$ and $\gamma$. First, the total number of $\alpha$ and $\gamma$ is the same. The $\gamma$'s can be obtained from the $\alpha$'s, as follows. Given a primitive polynomial of degree $m$, the $\alpha$-table of $GF(2^m)$ can be generated in terms of the powers of a primitive root $\alpha$. In this field table, the binary representation of the $m$-tuples are denoted from left to right as $x^0$, $x^1$, $x^2$, ..., $x^{m-1}$. The binary symbols in the 0th column, i.e., under $x^0$, are the maximum length sequence combination generated by the primitive polynomial. This relation is best illustrated by a simple example.

*Example 6.16:*   From the $\alpha$-table of Example 6.10, the 0th column under $x^0$ in $GF(2^3)$ is 1 0 0 1 0 1 1 . . . which is just the maximum length sequence used in Example 6.13 to produce the $\gamma$-table. In this example the $\gamma$'s are related to the $\alpha$'s as

$$\gamma^0 = 1\ 0\ 0 = \alpha^0$$

$$\gamma^1 = 0\ 0\ 1 = \alpha^2$$

$$\gamma^2 = 0\ 1\ 0 = \alpha^1$$

$$\gamma^3 = 1\ 0\ 1 = \alpha^6$$

$$\gamma^4 = 0\ 1\ 1 = \alpha^4$$

$$\gamma^5 = 1\ 1\ 1 = \alpha^5$$

$$\gamma^6 = 1\ 1\ 0 = \alpha^3$$

In general, the $\gamma$-table can be obtained as demonstrated once the $\alpha$-table is given. However, it is important to recognize that multiplication in the $\gamma$-table is not equivalent to the multiplication in the $\alpha$-table. For instance, $\gamma^3(1\ 0\ 1) \cdot \gamma^2(0\ 1\ 0) = \gamma^5(1\ 1\ 1)$. But $\alpha^6(1\ 0\ 1) \cdot \alpha(0\ 1\ 0) \neq \alpha^5(1\ 1\ 1)$. That is to say, polynomial multiplication among the $\gamma$'s cannot be used for the multiplication of the $\alpha$'s. In algebraic decoding, such as BCH, it is the polynomial multiplication of the $\alpha$'s that is needed, not the $\gamma$'s.

## 6.6   BCH CODEC COMPLEXITY

By combining the encoder and the estimated complexities of the three sections of a decoder, described in the previous sections, the total complexity of a BCH codec can be obtained, as shown in Table 6.3. Since we used the conservative estimation in each case, the total estimation is thus an upper bound. It should be mentioned that the logic complexity list does not include such necessary logics as clock, clock driver, timing controls and counters, etc. But the essential elements of the codec itself are reasonably estimated. Based on the total logic estimation of Table 6.3, Figure 6.21 gives the number of shift register stages versus the number of errors that are capable of being corrected by the code. With code length as parameters, the register stages increase with code length

**Table 6.3**

The logic complexity of BCH codecs.

| | Shift registers of length: ( ) | Modulo-2 adders | AND gates | OR gates | Inverter | Flip-flops |
|---|---|---|---|---|---|---|
| Encoding | $1:(k$ or $n-k)$ | $k$ or $n-k$ | 4 | 3 | — | 2 |
| Syndrome calculation | $3E-1:(m)$ | $Em+m^2(2E-1)$ | — | — | — | — |
| Field processor | $4(E+1):(m+1)$ | $m$ multi-input | $m^2+E+15$ | 4 | $E$ | $3m+E$ |
| Error locator | $E:(m)$ | $m+(m-1)E+1$ | $1$ ($m$ input) | — | 1 | — |
| Total | $8E+4$ | $m^2(2E-1)+2m$ $(E+1)-E+k+1$ | $m^2+E+20$ | 7 | $E+1$ | $3m+E+2$ |

when the number of correctable errors are fixed. Figure 6.22 gives the maximum number of modulo-2 adders needed. Figure 6.23 combines the number of AND, and OR gates and flip-flops.

## 6.7   COSET DECODERS

A linear $(n, k, d_m)$ group code has $2^k$ binary codewords $c_j$, $j = 1, 2, \ldots, 2^k$. The cosets of a code can be defined over GF(2) as the collection of group elements $g_i$, $i = 0, 1, \ldots, 2^{n-k} - 1$, where $i$ ranges from weights less than $d_m$. That is, $g_i$ are a set of sequences of length $n$, which are not codewords. The $g_i$'s are referred to as coset leaders. The cosets can be formed by the addition of $g_i + c_j$. Each collection $i$ in $g_i + c_j$, with the numeration of all $j$, is a coset. By numeration of all $i$ and $j$ for $g_i + c_j$, an array can be formed for decoding. Note that the $c$'s in this section are different from the last section.

*Example 6.17:*   The codewords of the (5, 3, 2) group code are $c_1 = (10011)$, $c_2 = (01010)$, $c_3 = (11001)$, $c_4 = (00101)$, $c_5 = (10110)$, $c_6 = (01111)$, and $c_7 = (11100)$. The minimum weight of the code is 2, which comes from $c_2$ and $c_4$ for the only two nonzero word components. Because the code is linear, the minimum distance of the code is equal to its minimum weight. The coset leaders are $g_1 = (00000)$, $g_2 = (10000)$, $g_3 = (01000)$, $g_4 = (00100)$, $g_5 = (00010)$, and $g_6 = (00001)$, which are all the possible sequences of length of $n = 5$ with distinctive single weight locations. The first coset is formed from $g_1 + c_1$, $g_1 + c_2$, $\ldots$, $g_1 + c_6$. The second coset is formed from $g_2 + c_1$, $g_2 + c_2$, $\ldots$, $g_2 + c_6$; and the last coset is formed from $g_6 + c_1$, $g_6 + c_2$, $\ldots$, $g_6 + c_6$. The coset array is shown in Table 6.4. Since there are repeated elements between cosets of $g_3$ and $g_5$, $g_4$ and $g_6$, the result is reduced to Table 6.5.

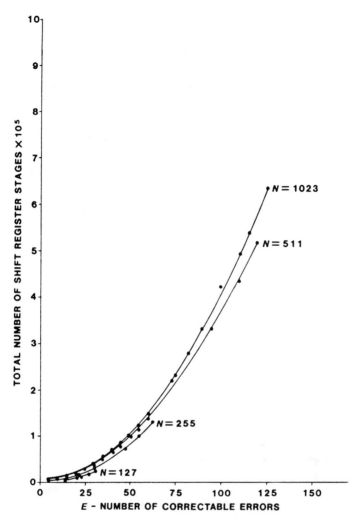

**Figure 6.21** Number of shift register stages vs. number of correctable errors.

The decoding strategy is implemented by first computing the syndrome and finding the leader of the coset that contains this syndrome, then subtracting this coset leader from the received sequence. The result is the decoded codeword.

*Example 6.18:*   The parity check matrix of the simple code in Example 6.17 is

$$H = \begin{bmatrix} 1 & 1 & 0 & 1 & 0 \\ 1 & 0 & 1 & 0 & 1 \end{bmatrix}.$$

Assume that the transmitted codeword is $c_6 = 01111$ and an error is made

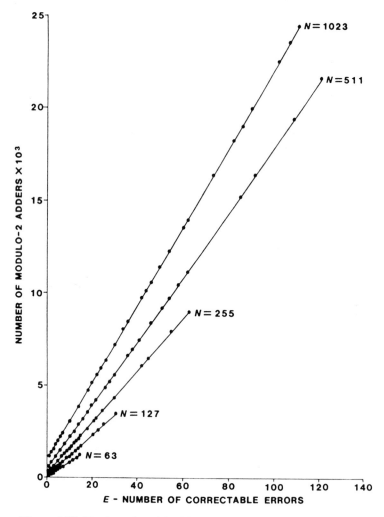

**Figure 6.22** Number of mod-2 adders vs. number of correctable errors.

in the last digit. Thus, with the received codeword $R = 01110$, the syndrome is formed as

$$S = RH^T = [0\ 1\ 1\ 1\ 0] \begin{bmatrix} 1 & 1 \\ 1 & 0 \\ 0 & 1 \\ 1 & 0 \\ 0 & 1 \end{bmatrix} = [0\ 1].$$

The coset leader that contains the syndrome pattern $[01]$ is $g_6 = (00001)$. Subtracting $g_6$ from $R$ we obtain $01111$ which is the correctly decoded word $c_6$.

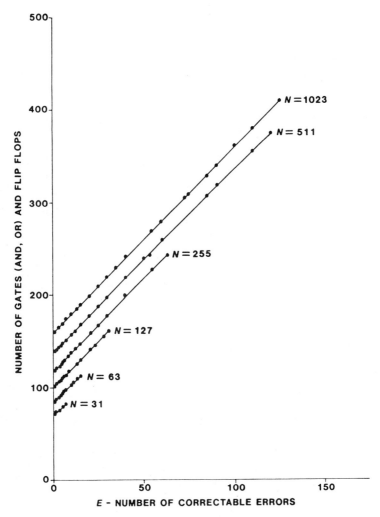

**Figure 6.23** Number of AND and OR gates vs. number of correctable errors.

### 6.7.1 Coset Decoding of BCH Codes

Section 6.7 demonstrates the ideal of simple coset decoding strategy which was known from early coding theory development [3, 4, 12]. For a code to decode successfully with coset leaders, the number of codewords of certain weights in the code, i.e., the weight enumeration, must be known. The weight enumeration problem has been comprehensively treated by Berlekamp [4]. The weight enumeration of binary BCH codes has been investigated by Kasami, Sidelnikov, and Berlekamp [13, 14]. A complete coset decoding algorithm for a primitive double error-correcting BCH code of length $n = 2^{m-1}$ can be accomplished

**Table 6.4**
Coset list for Example 6.17

| $g_i$ \ $c_j$ | $c_1$ 1 0 0 1 1 | $c_2$ 0 1 0 1 0 | $c_3$ 1 1 0 0 1 | $c_4$ 0 0 1 0 1 | $c_5$ 1 0 1 1 0 | $c_6$ 0 1 1 1 1 |
|---|---|---|---|---|---|---|
| $g_1 = 0\,0\,0\,0\,0$ | 1 0 0 1 1 | 0 1 0 1 0 | 1 1 0 0 1 | 0 0 1 0 1 | 1 0 1 1 0 | 0 1 1 1 1 |
| $g_2 = 1\,0\,0\,0\,0$ | 0 0 0 1 1 | 1 1 0 1 0 | 0 1 0 0 1 | 1 0 1 0 1 | 0 0 1 1 0 | 1 1 1 1 1 |
| $g_3 = 0\,1\,0\,0\,0$ | (1 1 0 1 1) | 0 0 0 1 0 | 1 0 0 0 1 | 0 1 1 0 1 | 1 1 1 1 0 | 0 0 1 1 1 |
| $g_4 = 0\,0\,1\,0\,0$ | 1 0 1 1 1 | <0 1 1 1 0> | 1 1 1 0 1 | 0 0 0 0 1 | 1 0 0 1 0 | 0 1 0 1 1 |
| $g_5 = 0\,0\,0\,1\,0$ | 1 0 0 0 1 | 0 1 0 0 0 | (1 1 0 1 1) | 0 0 1 1 1 | 1 0 1 0 0 | 0 1 1 0 1 |
| $g_6 = 0\,0\,0\,0\,1$ | 1 0 0 1 0 | 0 1 0 1 1 | 1 1 0 0 0 | 0 0 1 0 0 | 1 0 1 1 1 | <0 1 1 1 0> |

through the code weight enumerators, i.e., how many codewords of weight $w_c$ are there in the coset of weight $w_s$. For $w_c > 5$, $w_s$ depends on the coset leaders, which are all the $\binom{n}{1}$ and $\binom{n}{2}$ combinations. This forms the foundation for the following applications.

### 6.7.2   The INTELSAT TDMA Error-Correcting Codec

Based on the simplicity of the coset decoding technique, a tradeoff between operating speed and storage for the cosets can be compromised. With high rate and reasonable coding gain, the (128, 112, 6) double error-correcting and triple error-detecting BCH code was implemented, evaluated, and recommended for the 120-Mbps INTELSAT system by KDD, NEC, and COMSAT [15, 16]. This section concerns only the implementation aspects of this codec. The two error-correcting (128, 112, 6) BCH code is a modification from the primitive (127, 113, 5) BCH code. The modified BCH code has a generator polynomial

$$g(x) = x^{15} + x^{14} + x^{13} + x^{12} + x^{11} + x^{10}$$
$$+ x^7 + x^2 + x + 1 .$$

**Table 6.5**
Coset list for the (5, 3, 2) code.

| $g_i$ \ $c_j$ | $c_1$ | $c_2$ | $c_3$ | $c_4$ | $c_5$ | $c_6$ |
|---|---|---|---|---|---|---|
| $g_1$ | $g_1 + c_1$ | $g_1 + c_2$ | $g_1 + c_3$ | $g_1 + c_4$ | $g_1 + c_5$ | $g_1 + c_6$ |
| $g_2$ | $g_2 + c_1$ | $g_2 + c_2$ | $g_2 + c_3$ | $g_2 + c_4$ | $g_2 + c_5$ | $g_2 + c_6$ |
| $g_3$ | $g_3 + c_1$ | $g_3 + c_2$ | $g_3 + c_3$ | $g_3 + c_4$ | $g_3 + c_5$ | $g_3 + c_6$ |
| $g_6$ | $g_6 + c_1$ | $g_6 + c_2$ | $g_6 + c_3$ | $g_6 + c_4$ | $g_6 + c_5$ | $g_6 + c_6$ |

The corresponding encoder is shown in Figure 6.24. The input message control is gated for every 64-symbol or 128-bit block. The encoder shift register is pulsed for only 4 nanoseconds for each shift for the 120 Mbps operation of the codec. For encoding when the number of messages for the last block in a TDMA burst is less than 112 bits, it is necessary to provide dummy zeros.

The decoding process of the (128, 112, 6) BCH code consists of two parts, and each part is responsible for its respective parity checking according to its code generator subpolynomial. That is:

$$g_1(x) = x + 1$$

$$g_2(x) = x^{14} + x^{12} + x^{10} + x^6 + x^5$$

$$+ x_4 + x^3 + x^2 + 1,$$

where $g(x) = g_1(x) g_2(x)$. The decoding function, shown in Figure 6.25, consists of data storage, error-detection, and error-correction functions. The data storage function is provided by a 208-bit buffer. Error detection is accomplished by parity checking and modulo-2 addition (see Chapter 5). In this case three errors per block of 128 bits can be detected. If the code minimum distance is $d_m$, the number of errors correctable is $E$, and the number of errors detectable is $d$ in a block of this type code, then $d_m \geq d + E + 1$. For this case equality holds. The error-correction function consists of syndrome formation and error location. Syndrome formation is accomplished by 8-bit parallel processing. The reason for using 8-bit parallel processing is that the TDMA frame format is formed from an increment of 4 symbols or 8 bits. The error-location function is accomplished by the $16 \times 16$ Kbit table of read-only-memory (ROM). The purpose of the gate is to stop any additional erroneous error correction once three errors have been detected.

The (128, 112, 6) coset decoder is shown in Figures 6.26 and 6.27. For the 4-PSK modem (Chapter 4, Volume I), the received P and Q message sequences

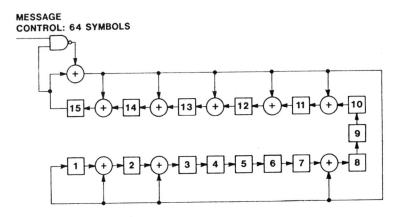

**Figure 6.24** INTELSAT TDMA BCH encoder, shift register pulse duration 4 ns.

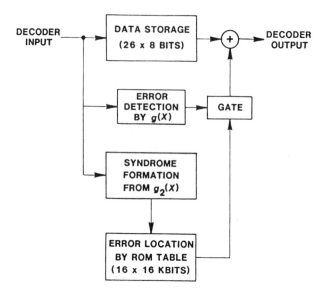

**Figure 6.25** Decoding function of the (128, 112, 6) BCH code.

fed into the decoder are converted to eight streams of parallel signals, which are fed to the data storage and the syndrome generator. The syndrome generator operates at a speed of ⅛ times that of serial processing and the syndrome is generated every time one block of data is received and transferred to the address buffer. One bit of the syndrome stored in the address buffer corresponds to a parity check, and the remaining 14 bits of the syndrome data are used to assign the address of the ROM table. In the EPROM (Intel 2716) the error locations in the received block are written, and the content of the ROM corresponding to the syndrome is read out and stored in the comparator buffer (L). For the memory capacity of the ROM, 16 PROMs (2K × 8 bits/each) are required since the number of addresses assigned by the syndrome is $2^{14} = 16k$ and two error locations are written into each address by 7 bits. The three higher bits of the syndrome are used to select ROM through 3–8 decoder as shown in Figure 6.26. In the comparator circuit, the binary timing signal synchronized with data supplied from the block counter and the error location signal from the above-mentioned comparator buffer are compared with each other. When the two signals coincide with each other, an error correction pulse is fed to the exclusive OR circuit in the data storage output. The inverted output of the parity check counter and the data of $O_7$ of ROM are logically summed at the OR circuit, and when the output of OR circuit is "1" error correction takes place.

This codec employs CML-IC (current mode logic-integrated circuit), Schottky-TTL, TTL (transistor-transistor logic), and EPROM for achieving high-speed operation. Major IC's in use are the CML-IC MCI600 and CML-IC MC100000 series (Motorola), the CML-IC up B10000 series (NEC), and the EPROM Intel

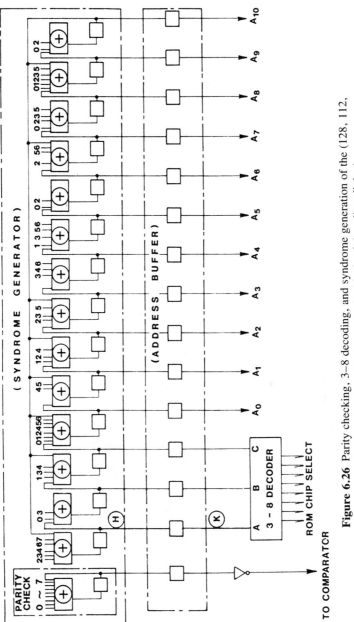

**Figure 6.26** Parity checking, 3–8 decoding, and syndrome generation of the (128, 112, 6) coset decoder. (0–7 refer to the block of decoding digits.)

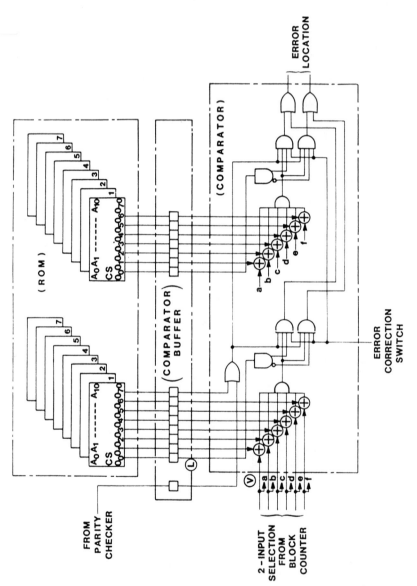

**Figure 6.27** Error location of the (128, 112, 6) coset decoder.

2716 (2k × 8bit erasable PROM). The high-speed error-correcting codec, as an integral part of the INTELSAT TDMA system, has a 1985 operational date.

## 6.8   THE (1057, 813, 33) DIFFERENCE-SET CODEC DESIGN

In Sections 5.5.2.4 and 5.5.3 we discussed threshold decodable block codes and block code performances. In particular, Figures 5.14 and 5.15 show the performance of the (1057, 813, 33) difference-set code, which has an outstanding coding gain.

The superior coding performance was recognized for satellite application by Wu and Golding and, independently, for deep space application by Dorsch and Dolainsky [17, 18]. The (1057, 813, 33) codec was designed and computer-simulated for a number of experimental satellite systems [19]. In this section, we outline the design and describe the complexity of this codec. Detailed designs and simulation and testing programs for this codec are available (along with other codec simulation programs) from the author.

The difference set elements and generator polynomial of the code are given by Weldon [20] as

$$\{D\} = \{d_0, d_1, d_2, \ldots \ldots, d_{k-1}\}$$

$$= \{ \quad 0, \quad 1, \quad 3, \quad 7, \quad 15, \quad 31, \quad 54, \quad 63, \quad 109, \quad 127,$$

$$138, \quad 219, \quad 255, \quad 277, \quad 298, \quad 338, \quad 348, \quad 439$$

$$452, \quad 511, \quad 528, \quad 555, \quad 597, \quad 677, \quad 697, \quad 702$$

$$754, \quad 792, \quad 879, \quad 905, \quad 924, \quad 990, \quad 1023,\}$$

and the exponents of the code generator polynomial are:

|      |      |      |      |      |      |      |      |      |        |
|------|------|------|------|------|------|------|------|------|--------|
| 0,   | 1,   | 3,   | 4,   | 5,   | 11,  | 14,  | 17,  | 18,  | 22, 23 |
| 26,  | 27,  | 28,  | 32,  | 33,  | 35,  | 37,  | 39,  | 41,  | 43     |
| 45,  | 47,  | 48,  | 51,  | 52,  | 55,  | 59,  | 62,  | 68,  | 70     |
| 71,  | 72,  | 74,  | 75,  | 76,  | 79,  | 81,  | 83,  | 88,  | 95     |
| 96,  | 98,  | 101, | 103, | 105, | 106, | 108, | 111, | 114  |        |
| 115, | 116, | 120, | 121, | 122, | 123, | 124, | 126, | 129  |        |
| 131, | 132, | 135, | 137, | 138, | 141, | 142, | 146, | 147  |        |
| 149, | 150, | 151, | 153, | 154, | 155, | 158, | 160, | 161  |        |
| 164, | 165, | 166, | 167, | 169, | 174, | 175, | 176, | 177  |        |
| 178, | 179, | 180, | 181, | 182, | 183, | 184, | 186, | 188  |        |
| 189, | 191, | 193, | 194, | 195, | 198, | 199, | 200, | 201  |        |
| 202, | 203, | 208, | 209, | 210, | 211, | 212, | 214, | 216  |        |
| 222, | 224, | 226, | 228, | 232, | 234, | 236, | 242, | 244  |        |

That is, $g(x) = 1 + x + x^3 + x^4 + x^5 + x^{11} + \ldots + x^{242} + x^{244}$. The encoder can be implemented according to $g(x)$ as described for BCH encoders in Section 6.2.1. The first and last parts of the encoder are shown in Figures 6.28 and 6.29.

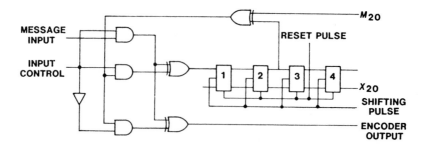

**Figure 6.28** First part of the (1057, 813, 33) difference-set encoder.

The function of the decoder is shown in Figure 6.30 where the syndrome is a 244-stage shift register. The parity checking connections before the threshold circuits are determined by the composite parity check polynomial $W^i(D)$ in terms of the difference-set elements

$$
\begin{aligned}
W^i(D) = & D^{d_i - d_{i-1} - 1} + D^{d_i - d^{i-2} - 1} + \dots \\
& + D^{d_i - d_0 - 1} + D^{d_i - 1} + \dots \\
& + D^{n-1 - d_l + d_i} + D^{n-1 - d_{l-1} + d_i} + \dots \\
& + \dots + D^{n-1},
\end{aligned} \tag{6.37}
$$

where $W^i(D)$ is enumerated exhaustively for all the difference-set elements $d_i$, and $d_l \neq d_i$ is another element in the set.

*Example 6.19:*  For the (7, 3, 4) difference-set code, the generator polynomial is $g(x) = 1 + x^2 + x^3 + x^4$, and the code is derived from the difference-set elements $d_0 = 0$, $d_1 = 2$, and $d_2 = 3$. Then for $i = 0, 1, 2$, we have

$$
\begin{aligned}
W^2(D) &= D^{d_2 - d_1 - 1} + D^{d_2 - d_0 - 1} + D^{n-1} \\
&= 1 + D^2 + D^6 \\
W^1(D) &= D^{d_1 - d_0 - 1} + D^{n-1 + d_1 - d_2} + D^{n-1} \\
&= D + D^5 + D^6 \\
W^0(D) &= D^{n-1 + d_0 + d_2} + D^{n-1 + d_0 + d_1} \\
&\quad + D^{n-1} \\
&= D^3 + D^4 + D^6.
\end{aligned}
$$

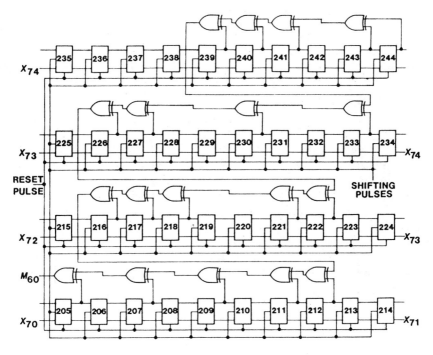

**Figure 6.29** Last part of the (1057, 813, 33) encoder.

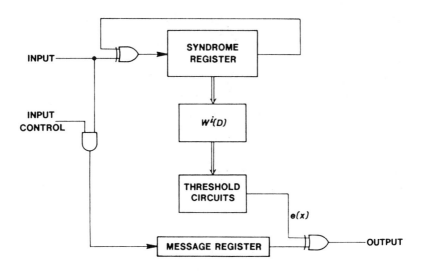

**Figure 6.30** Threshold decoding difference-set codes.

For the (1057, 813, 33) code, the set of $W^i(D)$ is computed and the nonzero terms are listed in Table 6.6, from which the parity check connections from the syndrome shift register can be determined. The nonzero terms in $W^i(D)$ are the tap positions for the syndrome shift register of the $i$th polynomial, and there are 33 such polynomials. Physically, let the collection of the nonzero terms in $W^i(D)$ be the inputs to the $J(k)$ box, where $k = 0, 1, 2, \ldots, 32$. The first part of the syndrome register connection is shown in Figure 6.31, where $J(k)_r$ denotes the $r$th input for the $J(k)$ box. All $J(k)$ boxes are implemented with exclusive-OR gates and logic delays. The output of $J(k)$ is denoted as $T(k)$. Thus, from Table 6.6, the complete modulo-2 implementation of all the $J(k)$ box for the decoder is shown from Figures 6.32 to 6.40. We note that for some $k$'s the $J(k)$ box has the identical configuration. We also note that the inputs to the $J(k)$ box are only numbered, and the actual connections should be in accordance with Table 6.6.

**Table 6.6**
Syndrome connections for the (1057, 813, 33) difference-set decoder.

| $k$ | $i$ | Powers of Nonzero $D$ in $W^i(D)$ |
|---|---|---|
| 0 | 0 | 33, 66, 132, 151, 177 |
| 1 | 1 | 0, 34, 67, 133, 152, 178 |
| 2 | 3 | 1, 2, 36, 69, 135, 180 |
| 3 | 7 | 3, 5, 6, 40, 73, 139, 158, 184 |
| 4 | 15 | 7, 11, 13, 14, 48, 81, 147, 166, 192 |
| 5 | 31 | 15, 23, 27, 29, 30, 64, 97, 163, 182, 208 |
| 6 | 54 | 22, 38, 46, 50, 52, 53, 87, 120, 186, 205, 231 |
| 7 | 63 | 8, 31, 47, 55, 59, 61, 62, 96, 129, 195, 214, 240 |
| 8 | 109 | 45, 54, 77, 93, 101, 105, 107, 108, 142, 175, 241 |
| 9 | 127 | 17, 63, 72, 95, 111, 119, 123, 125, 126, 160, 193 |
| 10 | 138 | 10, 28, 74, 83, 106, 122, 130, 134, 136, 137, 171, 204 |
| 11 | 219 | 80, 91, 109, 155, 164, 187, 203, 211, 215, 217, 218 |
| 12 | 255 | 35, 116, 127, 145, 191, 200, 223, 239 |
| 13 | 277 | 21, 57, 138, 149, 167, 213, 222 |
| 14 | 298 | 20, 42, 78, 159, 170, 188, 234, 243 |
| 15 | 338 | 39, 60, 82, 118, 199, 210, 228 |
| 16 | 348 | 9, 49, 70, 128, 209, 220, 238 |
| 17 | 439 | 90, 100, 140, 161, 183, 219 |
| 18 | 452 | 12, 103, 113, 153, 174, 196, 232 |
| 19 | 511 | 58, 71, 162, 172, 212, 233 |
| 20 | 528 | 16, 75, 88, 179, 189, 229 |
| 21 | 555 | 26, 43, 102, 115, 206, 216 |
| 22 | 597 | 41, 68, 85, 144, 157 |
| 23 | 677 | 79, 121, 148, 165, 224, 237 |
| 24 | 697 | 19, 99, 141, 168, 185, 244 |
| 25 | 702 | 4, 24, 104, 146, 173, 190 |
| 26 | 754 | 51, 56, 76, 156, 198, 225, 242 |
| 27 | 792 | 37, 89, 94, 114, 194, 236 |
| 28 | 879 | 86, 124, 176, 181, 201 |
| 29 | 905 | 25, 112, 150, 202, 207, 227 |
| 30 | 924 | 18, 44, 131, 169, 221, 226 |
| 31 | 990 | 65, 84, 110, 197, 235 |
| 32 | 1023 | 32, 98, 117, 143, 230 |

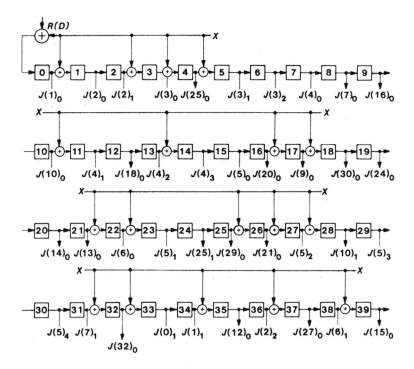

**Figure 6.31** First part of the (1057, 813, 33) decoder syndrome connections.

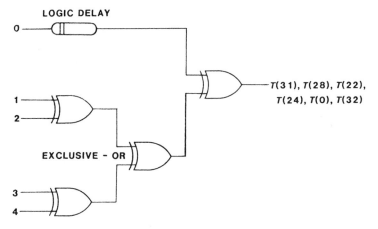

**Figure 6.32** Logic diagrams for circuits $J(31)$, $J(28)$, $J(22)$, $J(24)$, $J(0)$, $J(32)$.

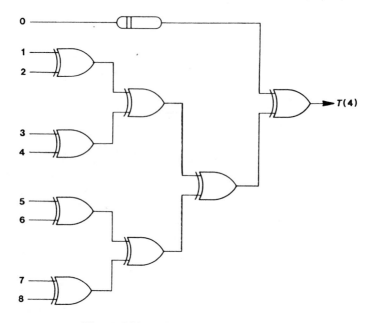

**Figure 6.33** Logic diagram for circuit $J(4)$.

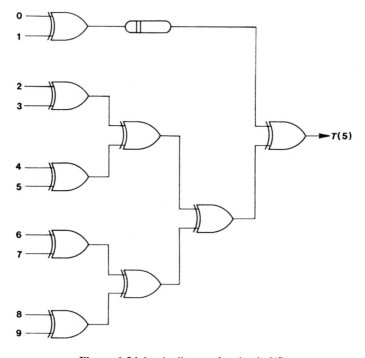

**Figure 6.34** Logic diagram for circuit $J(5)$.

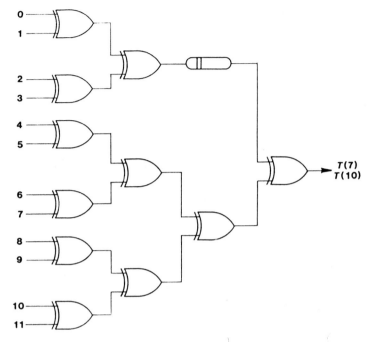

**Figure 6.35** Logic diagrams for circuits $J(7)$, $J(10)$.

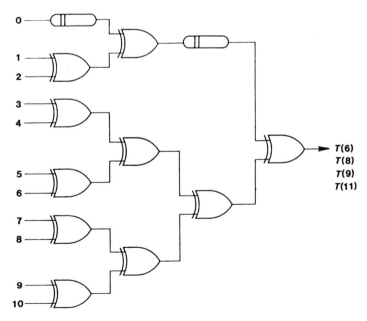

**Figure 6.36** Logic diagrams for circuits $J(6)$, $J(8)$, $J(9)$, $J(11)$.

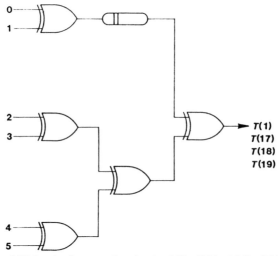

**Figure 6.37** Logic diagrams for circuits $J(1)$, $J(17)$, $J(18)$, $J(19)$.

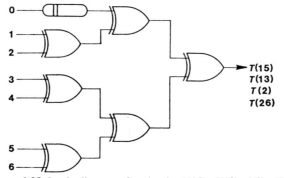

**Figure 6.38** Logic diagrams for circuits $J(15)$, $J(13)$, $J(2)$, $J(26)$.

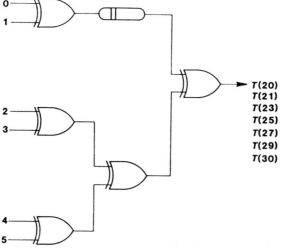

**Figure 6.39** Logic diagrams for circuits $J(20)$, $J(21)$, $J(23)$, $J(25)$, $J(27)$, $J(29)$, $J(30)$.

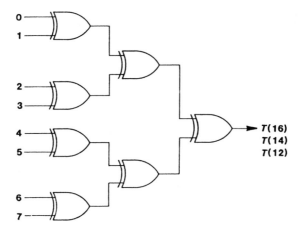

**Figure 6.40** Logic diagrams for circuits $J(16)$, $J(14)$, $J(12)$.

Next, the 33 outputs, $T(k)$s of $J(k)$ boxes, are fed as inputs to the threshold circuits. The function of these threshold circuits is

$$e(x) = \begin{cases} 1, \text{ if and only if there are} \\ \quad 17 \text{ or more nonzero } J(k)\text{s;} \\ 0, \text{ otherwise,} \end{cases} \tag{6.38}$$

where $e(x)$ is the output of the threshold circuits, and $e(x) = 1$ indicates error presence at that bit time. Looking inside the threshold circuits, we show how

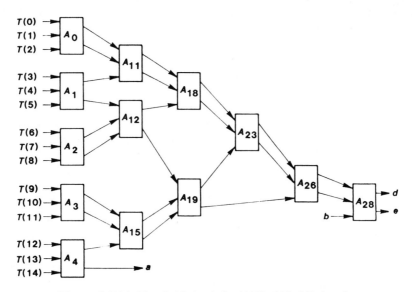

**Figure 6.41(a)** Threshold circuit for (1057, 813, 33) decoder.

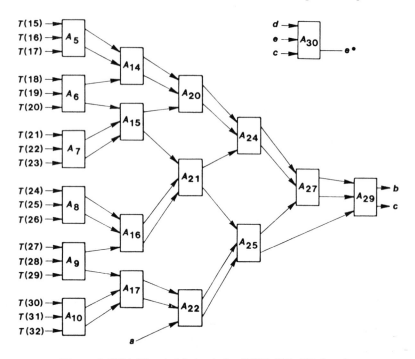

**Figure 6.41(b)** Threshold circuit for (1057, 813, 33) decoder.

they can be implemented with 3-input logic adders. With 31 such adders connected as tree-like structures, as shown in Figure 6.41, the threshold circuits are completely revealed.

With Figure 6.30, the operation of the codec is described as follows:

1. Reset both syndrome and message shift registers to zero.
2. With input control, shift 813 information bits into the registers.
3. Disable the input control gate and shift $1057 - 813 = 244$ bits for the syndrome register. During this time the message register remains stationary.
4. Shift 813 bits for both syndrome and message registers. Error correction takes place through the exclusive-OR gate as each bit shifts out from the message register.
5. The procedure is repeated for the next block.

From the above discussion, the complete codec can be implemented with two 244-stage shift registers each for the encoder and for syndrome formation, an 813-stage shift register for message storage, 202 modulo-2 adders, 31 3-input adders, and less than ten 2-input AND gates. With present-day multiple-stage, multiple-logic-function ICs, the size of the codec can be reduced to a few cards.

## 6.9   CONVOLUTIONAL CODES

As mentioned in the last chapter, convolutional codes can be decoded by means of threshold, sequential, and Viterbi (maximum likelihood) decoding techniques. Convolutional codes can be represented by either generator polynomials or corresponding matrices. The codes can be described by code trees, and the codecs can be implemented by feed-forward shift registers. Block codes, on the other hand, are implemented with feedback shift registers, as we have seen in the previous sections.

Convolutional codes are powerful alternatives for error correction in satellite communication. Depending on the specific system application, some classes of codes (or specific codes) are more useful than others. Some criteria for code selection have been outlined in Chapter 5. In these next four sections, we shall present the practical aspects of convolutional codes particularly as they relate to satellite applications.

### 6.9.1   Convolutional Code Encoder

Although convolutional codes have traditionally been distinguished from block codes for blockless encoding and decoding processes, we wish to point out here that this distinction can be quite misleading, and that in general convolutional encoding and decoding also process in blocks. The difference from block codes is that in convolutional codes the blocks are processed for every digit and each block contains a smaller number of digits.

Figure 6.42(a) describes the foundation of block encoding in which a block of $k$ digits of information is coded into a sequence of $n$ digits, while a systematic convolutional encoder as shown in Figure 6.42(b) takes a block of $k_o$ digits of information and codes it into a block of $n_o$ digits, usually $k_o < k$ and $n_o < n$.

For a nonsystematic convolutional encoder as shown in Figure 6.42(c), a successive block of $k_o$ information digits produces a block of $n_o$ digits at the encoder output. The encoding process of convolutional codes is performed for every bit timing, while for block coding a complete block is required in decoding. To say convolutional coding is continuous means that the collection of blocks in convolutional codes is continuously decoded for every bit time. Another key difference between the two types of encoding process is the fact that most block encoding processes are systematic; that is, the information and the parity checking parts are separated, while convolutional encoding can be either systematic or nonsystematic. For the latter the identity of the information sequence is no longer retained.

For binary convolutional codes the encoders consist of two basic elements: shift registers and modulo-2 adders. Sample convolutional encoders are shown from Figures 6.43 to 6.48. Figure 6.43 shows a five-stage, rate ½, systematic convolutional encoder with generator polynomial $G(D) = 1 + D^3 + D^4 + D^5$ or $g_1 = 100111 = 10^2 1^3$. The corresponding tree for this code is shown in Figure 6.49, where $t_i(j)$, $i = 0, 1, \ldots, 5; j = 1, 2$, denote the encoder output. Figures 6.44 to 6.46 are nonsystematic convolutional encoders for rate ½, ⅔,

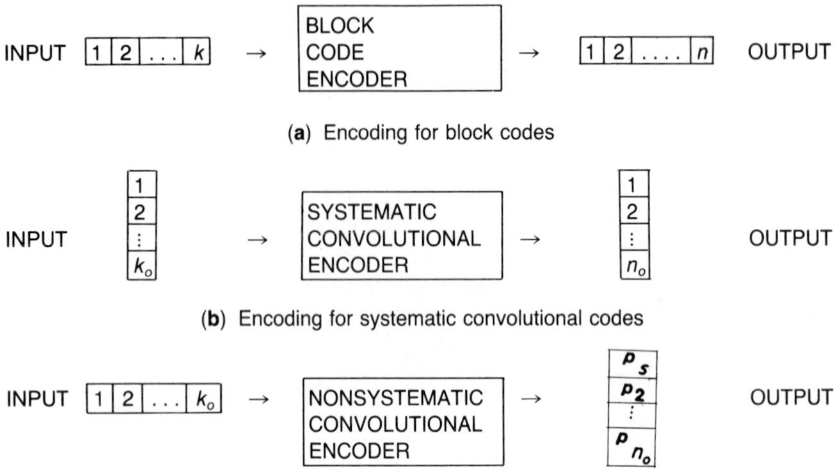

INPUT $\boxed{1\,|\,2\,|\,\ldots\,|\,k}$ → BLOCK CODE ENCODER → $\boxed{1\,|\,2\,|\,\ldots.\,|\,n}$ OUTPUT

**(a)** Encoding for block codes

INPUT $\boxed{\begin{array}{c}1\\2\\\vdots\\k_o\end{array}}$ → SYSTEMATIC CONVOLUTIONAL ENCODER → $\boxed{\begin{array}{c}1\\2\\\vdots\\n_o\end{array}}$ OUTPUT

**(b)** Encoding for systematic convolutional codes

INPUT $\boxed{1\,|\,2\,|\,\ldots\,|\,k_o}$ → NONSYSTEMATIC CONVOLUTIONAL ENCODER → $\boxed{\begin{array}{c}p_s\\p_2\\\vdots\\p_{n_o}\end{array}}$ OUTPUT

**(c)** Encoding for nonsystematic convolutional codes

**Figure 6.42** A comparison of encoding processes.

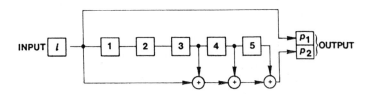

**Figure 6.43** Rate $\frac{1}{2}$, $m = 1$, systematic conventional encoder.

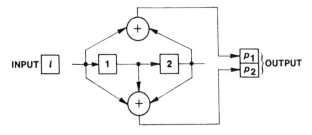

**Figure 6.44** Rate $\frac{1}{2}$, $m = 2$, nonsystematic convolutional encoder.

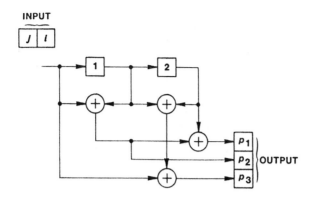

**Figure 6.45** Rate ⅔, $m = 2$, nonsystematic convolutional encoder.

and ¾ with $m = 2$ and $m = 5$. Figures 6.47 and 6.48 show systematic convolutional encoders of rate ¾ and ⅞. Viterbi decoders use nonsystematic encoders. Systematic convolutional encoders are used for both threshold decoding and sequential decoding; however, most sequential decoders are limited to low-rate codes.

The generator polynomials for Figures 6.43 to 6.48 are shown in Table 6.7.

### 6.9.2   Threshold Decoders

Threshold decoding of convolutional codes began with Massey [21]. It has been demonstrated with operational satellite systems that threshold decoders are the least complex technique for high rate, high speed, and moderate coding gains [22]. Other advantages of threshold decoders are as follows:

1. When threshold decoders are implemented with self-orthogonal codes, they are free from error propagation at the output of the decoder. Thus they are ideal for code concatenation.

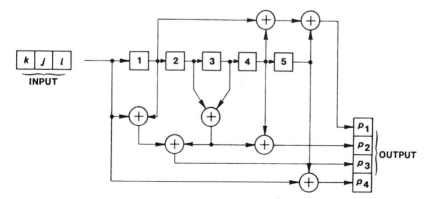

**Figure 6.46** Rate ¾, $m = 5$, nonsystematic convolutional encoder.

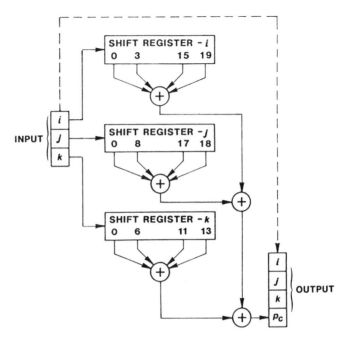

**Figure 6.47** Rate ¾, $m(\text{max}) = 19$, systematic convolutional encoder.

2. The decoders are capable of correcting more errors than the code minimum distance guaranteed.
3. A large number of convolutional self-orthogonal codes for threshold decoding is available.

Threshold decoding also depends on syndrome formations. A convolutional code can be decoded by examining the syndrome sequence

$$S(D) = R(D) \cdot H \tag{6.39}$$

where $R(D)$ is a received message sequence, and $H$ is the parity check matrix of the code. For codes with rate $k_o/(k_o + 1)$,

$$H = [\Delta j : j = 1, 2, \ldots, k_o]. \tag{6.40}$$

Each $\Delta j$ is an $m \times m$ matrix whose elements are in accordance with the subgenerator polynomial $g^{(j)}(D)$. $S_j(D)$ depends only on $g^{(j)}(D)$ and the corresponding error sequence, $E_j(D)$. Figure 6.50 shows the formation of a syndrome sequence for a systematic convolutional code for a binary, additive, stationary noise sequence, $F_j(D) = E_j(D)$. For simplicity of analysis we let $k_o = 1$, then $\Delta j = \Delta$, $G^{(j)}(D) = G(D)$, $E_j(D) = E(D)$, $F_j(D) = F(D)$, and $S_j(D) = S(D)$. The block diagram represents the codec structure of rate ½ code because there is only a single code generator polynomial $G(D)$. When the input message $I(D)$ is coded by the encoder with the generator polynomial $G(D)$, the parity check

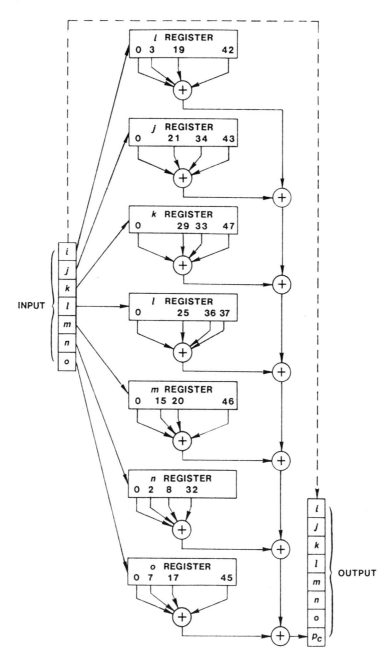

**Figure 6.48** Rate ⅞, $m(\text{max}) = 47$, systematic convolutional encoder.

polynomial, or parity check sequence $P(D)$ is the convolution of $I(D)$ and $G(D)$; i.e., $I(D)* G(D)$. Because the decoding scheme is systematic; i.e., the first $k_o$ (in this case I) digits are transmitted as $I(D)$, and $I(D)$ encounters channel error

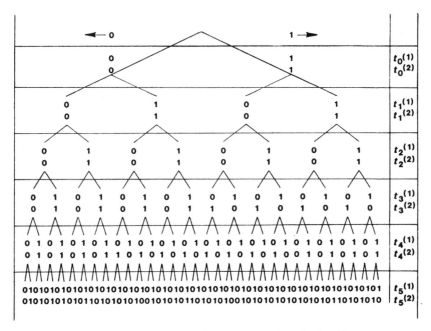

**Figure 6.49** Tree structure for $G^{(1)}(D) = 1 + D^3 + D^4D^5$, $G^{(0)}(D) = 1$.

sequence $E(D)$. At the receiving end the input to the encoder replica is $I(D) + E(D)$, and the output of the encoder replica becomes $G(D) * [I(D) + E(D)]$. On the other hand, if the channel error sequence to the coded parity check $P(D)$ is $F(D)$, then $P(D) + F(D)$ is one of the two inputs forming the syndrome sequence $S(D)$. That is, with all the additional modulo-2 for binary transmission,

$$S(D) = G(D)* [I(D) + E(D)]$$
$$+ I(D) * G(D) + F(D). \qquad (6.41)$$

**Table 6.7**
Convolutional encoders.

| Figures | Code rate | Generator polynomial |
|---------|-----------|----------------------|
| 6.43 | ½ | $10^2 1^3$ |
| 6.44 | ½ | $101, 1^3$ |
| 6.45 | ⅔ | $1^3, 1^2 0, 1^3$ |
| 6.46 | ¾ | $010^2 1^2, 1^5 0, 1^4 0^2$ |
| 6.47 | ¾ | $10^2 10^{11} 10^3 1, 10^7 10^8 1^2,$ |
|  |  | $10^5 10^4 101$ |
| 6.48 | ⅞ | $10^2 10^{15} 10^{23} 1, 10^{20} 10^{12} 10^8 1,$ |
|  |  | $10^{28} 10^3 10^{13} 1, 10^{24} 10^{10} 1^2$ |
|  |  | $10^{14} 10^4 10^{25} 1$ |
|  |  | $1010^5 10^{25} 1$ |
|  |  | $10^6 10^9 10^{27} 1$ |

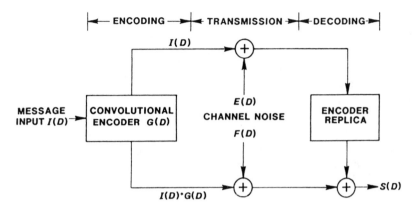

**Figure 6.50** Syndrome formation of systematic convolutional codes.

In the case $E(D) = F(D)$, then from the above equation

$$S(D) = E(D) * [1 + G(D)]. \tag{6.42}$$

We observe that the syndrome sequence $S(D)$ depends on only the code expressed by $G(D)$ and the channel errors expressed by $E(D)$. Physically, $S(D)$ is the contents in the syndrome shift register, as shown in Figure 6.51.

For a rate $k_o/(k_o + 1)$ code, $k_o$ identical threshold circuits are needed. Each threshold circuit connects to the corresponding syndrome locations. The output of each threshold circuit is responsible for the error corrections of that information digit position within the block. However, a single threshold circuit may be shared in order to simplify decoder complexity. A combination gating circuit is necessary for the use of a single threshold circuit. The combination circuit selects the set of $J$ syndrome connections that corresponds to a particular information digit.

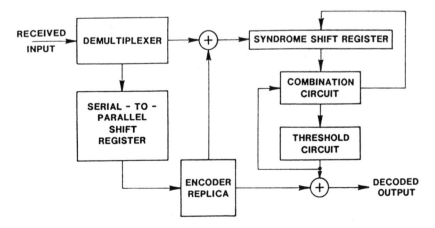

**Figure 6.51** Threshold decoding of convolutional codes.

For the rate $k_o/(k_o + 1)$ code, the combination circuit operates $k_o$ times per block. As long as the appropriate set of $J$ connections (specified by the coefficients of the subgenerator) corresponding to the information is selected, the order of the selection is not important. However, it may be easier to compute the thresholds in accordance with the order of the information digits in the block. A simple example will follow.

*Example 6.20:* For the rate ¾ systematic convolution encoders in Table 6.7 the syndrome shift register connections are shown in Figure 6.52. The combination circuit samples the three sets of four connections sequentially within each unit time. The threshold circuit performs the simple logic functions listed in Table 6.8, where $A$, $B$, $C$, and $D$ denote the syndrome connections and thus are inputs to the threshold circuit.

The threshold function can be implemented by two 3-input standard binary adders. From the first three inputs ($A$, $B$, and $C$), the sum ($S$) and carry ($CA$) are obtained. With $S$, $CA$, and the input, $D$, as the inputs to the second adder, the weight, $W$, is determined. The output of the threshold provides a 1 for $W \geq 3$ to indicate that an error has occurred. Not shown in Figure 6.52 are the proper feedbacks and the correction shift registers, which are part of the decoder for this example.

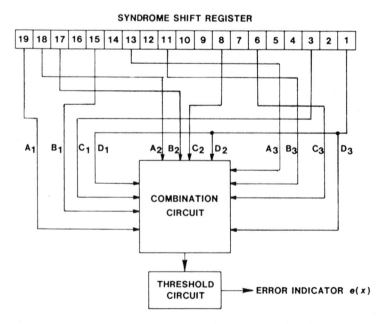

**Figure 6.52** Syndrome connections for rate ¾ systematic convolutional decoder.

**Table 6.8**

Threshold function.

| A | B | C | S | CA | D | W | e(x) |
|---|---|---|---|----|---|---|------|
|   |   |   |   |    | 1 | 4 | 1 |
| 1 | 1 | 1 | 1 | 1  | 0 | 3 | 1 |
|   |   |   |   |    | 1 | 3 | 1 |
| 1 | 1 | 0 | 0 | 1  | 0 | 2 | 0 |
|   |   |   |   |    | 1 | 3 | 1 |
| 1 | 0 | 1 | 0 | 1  | 0 | 2 | 0 |
|   |   |   |   |    | 1 | 2 | 0 |
| 1 | 0 | 0 | 1 | 0  | 0 | 1 | 0 |
|   |   |   |   |    | 1 | 3 | 1 |
| 0 | 1 | 1 | 0 | 1  | 0 | 2 | 0 |
|   |   |   |   |    | 1 | 2 | 0 |
| 0 | 1 | 0 | 1 | 0  | 0 | 1 | 0 |
|   |   |   |   |    | 1 | 2 | 0 |
| 0 | 0 | 1 | 1 | 0  | 0 | 1 | 0 |
|   |   |   |   |    | 1 | 1 | 0 |
| 0 | 0 | 0 | 0 | 0  | 0 | 0 | 0 |

### 6.9.3   Sequential Decoders

The fundamentals of sequential decoding were presented in Section 5.6.3, which includes the rules for decoding, error performance, and algorithmic comparison between Fano and Stack algorithms. In particular the potential superiority in operating speed of the multiple stack algorithm was put forward. In this section, the practical aspects of sequential decoding are examined.

Recall that the basic idea of sequential decoding is a probabilistic procedure to decode a code sequence by tracing the most likely path through the code tree. This tree tracing technique is done by moving ahead one tree node at a time. At each node the decoder evaluates a branch metric for each branch leaving that node. The branch metric is a probabilistic quantity that relates the coded and received symbols for a chosen branch. Corresponding with the high correlation to the coded sequence, the decoder chooses the branch with the largest value of metric. This branch metric is then added to a stored path metric, which is the running sum of the branch metric presently being traced in the code tree.

The path metric has a running threshold. As long as the value of the path metric is above the threshold, the decoder is presumably in the correct path, and the threshold is incrementally increased through each free node. When the path metric is below the threshold at a particular node, the decoder recognizes this as a wrong path and traces the path backward a step and lowers the threshold level to search for the maximum value of the path metric again. When the value of path metric increases continuously, the decoder traces a path in the tree structure as the decoded sequence. For more background information on sequential decoding, the reader is referred to the references [6, 23–33]. For digital satellite application an excellent chapter is written on sequential decoding by Haccoun [39].

Practical sequential decoders have long been recognized as one of the most powerful channel coding techniques, and have been designed, simulated, and evaluated for more than 20 years. Most of the early work was done at the M.I.T. Lincoln Laboratory [24–27]. Sequential decoding has been recommended for deep space applications by Jacobs, and by Layland and Lushbaugh, and for satellite communication by Forney and Langelier [28–30]. The modularized sequential decoder described by Forney operated at 5.0 Mbps with 5.0 dB coding gain, and the codec was built by the Codex Corporation [31, 32]. The code used was a rate $\frac{1}{2}$, $m = 48$ convolutional code, and the decoder can convert an error rate from $4.5 \times 10^{-2}$ to $10^{-5}$. Other decoding performances in terms of $E_b/N_0$ are shown in Figures 5.31 to 5.33. A number of implementation shortcuts have been pioneered in this codec. Soon after the 5.0 Mbps sequential decoder by the Codex Corporation, the Linkabit Corporation built a 40.0 Mbps, rate $\frac{1}{2}$, constraint length 41 sequential decoder [33]. Because of its uniqueness in implementation, we shall describe this decoder further.

*Example 6.21:* The Linkabit code has the generator polynomial $1^3 0^2 1^2 0 1^2 0^2 1^3 0 1^5 0^5 1 0 1^2 0^2 1^5 0^2 1^2$. As shown in Figure 6.53, the syndrome sequences are formed through the encoder replica-1. Then the syndrome sequences are written into sequential locations in the random access memory (RAM-1) which has read/write capabilities. The contents of the addressed sequence are first read out from the RAM before the new syndrome sequence is written in.

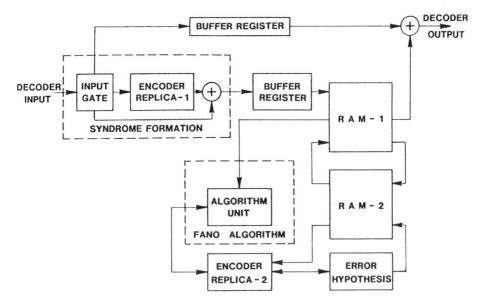

**Figure 6.53** High-speed sequential decoding.

The heart of the decoder is the algorithmic logic unit which performs the functions of Fano sequential decoding algorithm. This logic unit is affected by the conditions of the memory units in accordance with the path-finding phenomenon in the code tree. Thus, the algorithm logic unit controls the status of the decoder through the decoding tree. For each computation the algorithm logic unit performs the following decision functions: (a) move forward, backward, or parallel, (b) checking error at the node, (c) increase or decrease threshold, and (d) change amount of decoding metric.

The function of RAM-1 is to store the syndrome sequence while it is awaiting information from RAM-2. The total throughput requirement of RAM-1 for the 40 Mbps decoder operation is 140 Mbps, which includes a computation rate of about $10^8$ Fano algorithm computations per second. RAM-2 functions as follows: when a syndrome sequence is read from RAM-1 it stores in RAM-2 for error validation. Then this sequence is returned to RAM-1 ready for error correction as instructed by an error metric indicator.

If a bit error is hypothesized at a node, the decoder backs up and the error sequence is returned, and the original syndrome is reconstituted in the encoder replica-2 by shifting the error sequence and by exclusive-OR the contents of the encoder with the error indication at every node at which an error was hypothesized. The error sequence is shifted out and returned to the RAM-2 buffer when the tree path proceeds forward.

Encoder replica-2 has the capability of shifting either right or left. If it shifts right it arrives at the hypothesized error sequence, which is stored in the backup buffer. When the decoder is backing up along the tree, the contents of encoder replica-2 are shifted out from the lefthand side. The buffer registers provide proper delays to accommodate processing time in the other parts of the decoder.

Although the complexity of a sequential decoder in general is linearly proportional to the encoder memory $m$, the amount of memory and buffer requirements for a high data rate operation is large in comparison with the other two types of decoders for convolutional codes.

### 6.9.3.1  Multiple Stack Algorithm Decoder

Section 5.6.3.4 of Chapter 5 described the decoding function and the performance of the multiple stack algorithm. The end result of this approach is the tradeoff between computational speed and storage complexity, which primarily determines the complexity of the decoder. Ma evaluated the complexity aspect of the multiple stack algorithm (MSA) decoder and simulated the algorithm on a Zilog Z-80 microcomputer [34]. Storage comparison is made for both MSA and VA (Viterbi algorithm) in terms of code constraint length. As can be seen in Figure 6.54 the storage requirement of a conventional VA increases exponentially with constraint length, while the storage requirement of the MSA increases linearly with code constraint length. Note that the storage requirement crosses over beween the two algorithms at about the constraint length of 12.

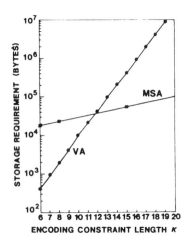

**Figure 6.54** Storage requirements of the MSA and Viterbi algorithm [34].

### 6.9.4   The Viterbi Decoder

The performance aspects of Viterbi decoding were described in Section 5.6.2, where the decoding performance calculation, error probability derivation, and a sample of decoder evaluations were presented. In this section we examine the implementation aspects of Viterbi decoding of convolutional codes and give specific examples.

Viterbi decoders minimize the decoding error probabilities in a memoryless AWGN biphase modulation channel by minimizing the inner product between a quantized received sequence $r$ and the transmitted sequence $y$. This inner product, which is a measure, is called a metric. Obviously, this metric is a function of the quantization level.

*Example 6.22:*   A decoder with 3-bit quantization has $2^3 = 8$ levels, 000, 001, 010, 011, 100, 101, 110, and 111. If we let the binary 0 and 1 be represented by 000 and 111, respectively, and assuming the received sequences are 001, 011, 010, 110 corresponding to the transmitted coded sequence 0101, then the decoding metric is:

$$(001)\ (000) = 001 \to 1$$

$$(011)\ (111) = 100 \to 4$$

$$(010)\ (000) = 010 \to 2$$

$$(110)\ (111) = 001 \to 1.$$

The metric for the received sequences is $1 + 4 + 2 + 1 = 8$. We note that the inner product is performed as modulo-2 addition for the case of 0, 1 binary symbols.

Thus the maximum likelihood decoding for the memoryless AWGN channel is obtained by performing the metric calculation between the received sequence and the code sequence, and choosing the path that exhibits the smallest value of the metrics, or the largest value of the correlation. The objective of the decoder is to find this minimum metric path. Because a path consists of a number of branches, branch metrics need to be calculated first. At a particular time during the decoding process, the stored metrics for each state, which has been the maximum of the correlation (or minimum of the metrics) of all the paths leading to this state, are added by the branch metrics for the branches departing from this state. Next, comparisons are made among all pairs of the branches entering each state, and then the maximum likelihood path is selected and stored. The decoder is now ready to repeat the same procedure as described above for the next state.

For an $m$-stage shift register encoder, the constraint length of the code is $m + 1$. There exist $2^m$ number of paths corresponding to the information sequence. These paths are called the survivor sequence as a result of the higher correlation comparisons with the received sequence. The smaller correlated path entering each state, which can never become a high correlated path, is eliminated. From an implementation standpoint the procedure can be summarized in the following steps:

Step 1: If we correlate the $m + 1$ bits of received sequence with each of the two possible branches of each of the $2^m$ states, the result generates a branch metric, which is the number of bit positions in which the received branch differs from the branch with which it is being correlated. There are $2^{m-1}$ such correlations to be performed. The branch metric for the two paths leaving each state are added to the previous branch metrics.

Step 2: Branch metrics for the two paths terminating in each of the next $2^m$ states are compared, and the path that has the lowest metric is retained. The survivor sequences for the paths terminating in each of the $2^m$ states and the branch metrics are then stored.

Step 3: Repeat steps 2 and 3.

Step 4: Make the bit decision after a number of constraint length, usually $5(m + 1)$, digits is sufficient.

### 6.9.4.1  Decoding Mechanization

The basic functions of the Viterbi decoding algorithm, shown in Figure 6.55, consist of branch metric calculation, state or path decision, decision memory, and output selection. Not included in the figure are the functions of input arrangement and synchronization, which are not considered here as essential in understanding the basic features of decoder implementation.

The branch metric calculation computes the metrics for all the branches at each bit time. To compute these metrics, it needs to sum the bit metrics that

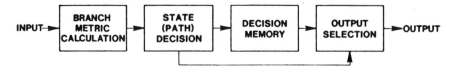

Figure 6.55 Basic functions of the Viterbi decoding algorithm.

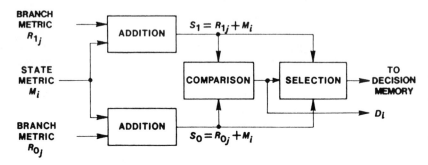

Figure 6.56 Function of state decision.

make up the branch. This can be easily accomplished by modulo-2 adder and logic summers, as in Example 6.22.

### 6.9.4.2 The State Decision Unit

The key element in a Viterbi decoder is the state decision unit that performs the basic calculations of the decoder.

The state (or path) decision circuit determines from which previous states the present state has arrived. These decisions are accomplished by the addition, comparison, and selection circuits, as shown in Figure 6.56. These decisions are made from the branch metrics from state to state. The branch metrics are first added to a pair of state matrices, then the two sums of path metrics are compared, and the smaller value of the two sums is selected. The decision is stored in the path memory unit and this smaller sum now becomes the new state metric.

*Example 6.23:* The first Viterbi hardware decoders were built by Lushbaugh at the Jet Propulsion Laboratory based on the simulation result by Heller [35, 36]. Lushbaugh described Viterbi decoders with $m = 2, 3,$ and 4 at code rate ½ or ⅓ and operated at 1.0 megabit data rate. The state decision circuit was arranged into a 6" × 6" circuit board. The logic diagram of the state decision functions (Figures 6.57(a) and (b)), has a 10-bit shift register memory for state metric and 32 bits for the survivor bit stream associated with that state. To increase the speed of operation, two 10-bit registers are used to store the metrics. The metrics are shifted at 10 Mbps with 1.0 μs per node. The result of the comparison is

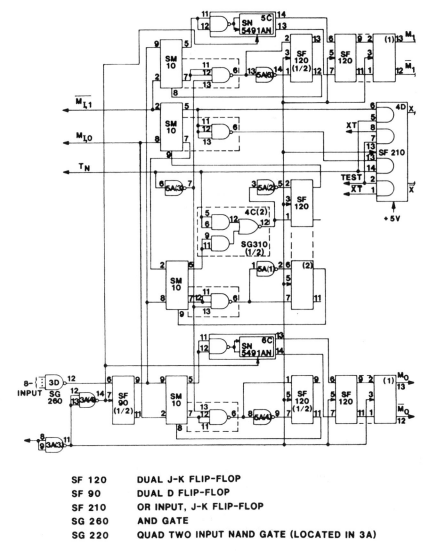

| SF 120 | DUAL J-K FLIP-FLOP |
| SF 90 | DUAL D FLIP-FLOP |
| SF 210 | OR INPUT, J-K FLIP-FLOP |
| SG 260 | AND GATE |
| SG 220 | QUAD TWO INPUT NAND GATE (LOCATED IN 3A) |

**Figure 6.57(a)** Logic diagram of Viterbi decoder main board.

stored in the $x$ flip-flop, which is the most recent metric information of the surviving path. The detailed connections of the state decision unit to the other parts of the decoder are omitted, as the purpose of this example is to show the relative simplicity of the heart of the low-rate Viterbi decoder.

*Example 6.24:* The state decision function for $m = 3$, rate $\frac{1}{2}$ convolutional code was implemented by the Linkabit Corporation early in 1971 [33]. The code generators are 1101 and 1111. For high-speed operation using emitter-coupled

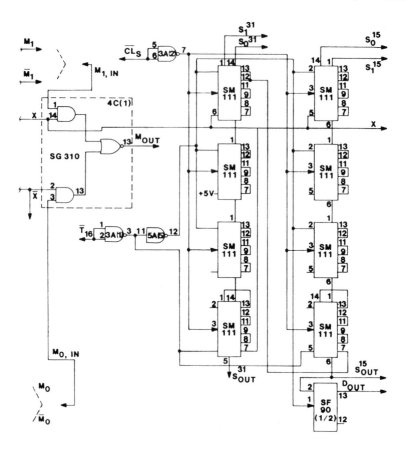

SG 310        TWO INPUT AND/NOR GATE
SG 380        HEX INVERTER (LOCATED IN 5A)
SM 10         FULL ADDER; 3 INPUT NAND GATE
SM 111        4-BIT SHIFT REGISTER
SN 5491AN     8-BIT SHIFT REGISTER

**Figure 6.57(b)** Logic diagram of Viterbi decoder main board (continued).

logic (ECL), the circuit is shown in Figure 6.58. $M_{xj}$ denotes the metric associated with state $x_j$, and $R_{ij}$ is the branch metric, which equals the log likelihood of the received digits $r_1$, $r_2$ (because of the rate ½ code) for that branch, given that $i$ and $j$ are transmitted. In other words, $R_{ij} = \log P(r_1 \mid i \text{ transmitted}) + \log P(r_2 \mid j \text{ transmitted})$.

The complementary values of $M_{xj}$ and $R_{ij}$ shown in Figure 6.57 are for the purpose of logic implementation. Figures 6.58(a) and 6.58(b) show the adders that add the state metrics to the branch metrics. Figure 6.58(c) shows the circuit for the comparison and selection functions. The adders are the carry-save type

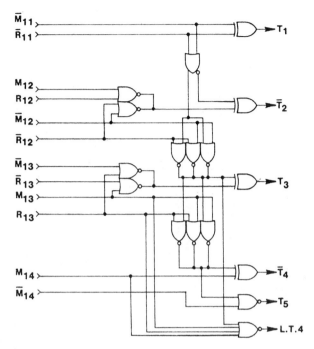

**Figure 6.58(a)** State decision circuit for $m = 3$, rate ½ Viterbi decoder.

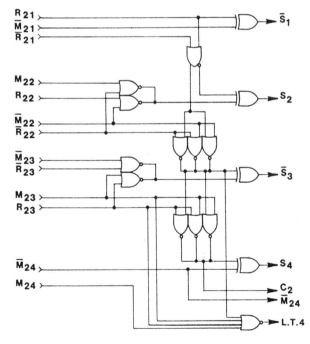

**Figure 6.58(b)** State decision circuit for $m = 3$, rate ½ Viterbi decoder (continued).

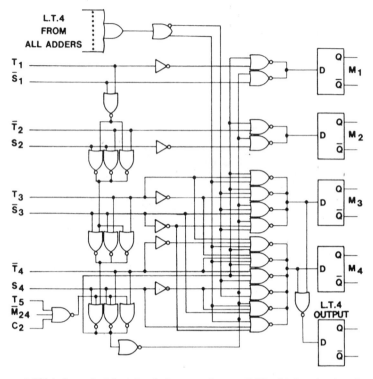

**Figure 6.58(c)** State decision circuit for $m = 3$, rate $\frac{1}{2}$ Viterbi decoder (continued).

in which ripple carries are used between the stages. A detection function of the result less than 4 (L.T.4) is part of the state decision normalization function. If none of the computations of the state decision circuit produces a result less than 4, then 4 must be subtracted from the results. This normalization serves to ease the selection process.

After the state decision function, the set of binary decisions from the state decision circuit are fed as input to the decision memory selection circuit, which is used to determine the new values of the state metrics.

### 6.9.4.3   The Output Selection Unit

From a $5(m + 1)$-bit shift register memory, the last bit is shifted out as the output of the decoder, which makes a decision either to discard or keep this bit as part of the decoded word at that particular state. The output selection circuit selects as decoder output the bit from the state with the smallest value of metrics from the state decision circuit. The output selection circuit can be implemented in a number of ways; however, a general block diagram of the output selection function will suffice for our purposes. Figure 6.59 shows the output selection circuit consisting of an $X$-box, a $D$-box, a path selection unit, a threshold unit, and a multiple OR gate. The selection process is based on the examination of

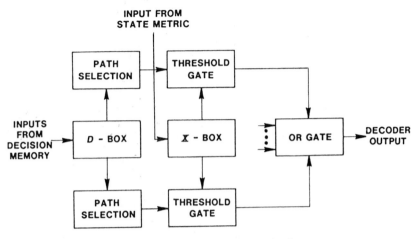

**Figure 6.59** Functions of output selection.

each state metric $x_j$, which is used to gate the output at state $j$ from the output shift register memory. The contents of the $D$-box are decision values from the state decision memory unit with the addition of proper bit delays for tracing the decoding path through the trellis. The path selection unit consists of a set of gates or switches which are commanded by the $D$-box. The threshold unit determines the selection either from the $X$-box or from the path selection box. This selection is performed with a predetermined state metric value as threshold. The number of inputs toward the final OR gate depends on the levels of quantization. For 3-bit quantization there are eight chains; i.e., path selection and gating merge as inputs to the OR gate.

Heller and Jacobs suggested Viterbi decoding for satellite and space channels; Spilker and others also described Viterbi decoding for satellite applications; and Morakis, Greene, Miller, Poland, and Helgert applied Viterbi decoding to part of NASA's Aerospace Data and Coding Standard [37–40]. Schweikert and Hagenauer applied Viterbi decoders to maritime satellites; Bernstein, Heggestad, Mui, and Richer, and Yasuda, Hirata, Nakamura, and Otani, to variable rate applications [41–43].

### 6.9.4.4 Decoder Complexity

The complexity of a conventional Viterbi decoder depends on the code constraint length, the code rate, and the quantization level. As the constraint length increases by one, the number of states doubles, which leads to the doubling of the number of state decision units, which in turn requires more than doubling the decision memory unit. For a $k_o/n_o$ code $2^{n_o}$ branch metrics need to be computed in the input part of the decoder: thus this part grows exponentially with $n_o$. The number of information symbols $k_o$ affects the complexity of a decoder in the following manner: the number of states is an exponentially decreasing function of $k_o$, which leads to the decreasing number of state metric buffer requirements, but the state decisions increase exponentially with $k_o$. For high rate decoders, the increasing of state decisions has more effect on the complexity of a decoder. As indicated in the previous sections, the size of the branch and state metrics, as well as the output selection unit, is a (linear) function of the quantization level.

Orndorff, Chou, Krcmarik, Colesworthy, Doak, and Koralek have described a VLSI integrated circuit version of a Viterbi decoder [44]. Clark, Davis, Cain, and Geist of the Harris Corporation have advanced a number of hardware-saving implementation schemes of Viterbi decoders [45–47]. An interesting approach to building a Viterbi decoder using analog components is described in the following example.

*Example 6.25:* Applying microwave technology to state metric calculations, Acampora and Gilmore of Bell Telephone Laboratories demonstrated a 50.0 Mbps data rate, or 100 Mpbs decoder output speed, for the rate ½. $m = 2$ code with code generators $g_1 = 101$ and $g_2 = 111$ [48]. The analog implementation of the Viterbi decoder was intended for high data rate communication satellite channels.

Maximum likelihood decoding is achieved through recursive application of selection metrics transferring back to the input sample and hold circuit for that state. Addition, comparison, and selection functions of the state decision can all be implemented by the operational amplifier principle. The result is not only that the decoder speed is increased but the component complexity is decreased compared with its digital implementation. The disadvantage of the analog approach is that the decoder produces errors due to amplifier drift, capacitor leakage, and nonlinear effects of the analog devices. The errors contribute to the performance degradation of the decoder.

Yasuda and others at the KDD Laboratories evaluated Viterbi decoder complexity when the punctured convolutional coding scheme (with bit deletion) is used [43]. With a punctured code the number of computations in the state decision (addition, comparison, and selection function) can be reduced from $2^{n_o - 1}$ comparisons for the conventional Viterbi decoder to binary comparisons at each state. For example, with the same constraint length $K = 7$ code the total number of estimated standard logics required for the rate ½ codec is 50,000. For the rate ¾ punctured codec it is 68,000 and for the rate ¾ nonpunctured codec it is 138,000. A total of 78,000 logics are required for the state decision function of the ¾ nonpunctured codec in comparison with 18,000 for the same function of rate ¾ punctured codec. Thus the total hardware saving is about ½ when the punctured code is used and there is negligible performance degradation.

Because of the significant hardware saving at each decision state of the decoding process, high rate punctured convolutional codes can be decoded with the Viterbi algorithm at low code rate complexity.

The first work to link Viterbi and sequential decoding algorithms as a generalized probabilistic decoding algorithm for convolutional codes was by Haccoun and Ferguson [49]. In their approach, decoding tree paths in an ordered stack for conventional sequential decoding are extended. More than one path can be extended simultaneously. Trellis type path remerging is used as in the standard Viterbi algorithm to delete redundant paths from the stack. The result is the reduction of computational variations at the expanse of increasing the number of decoding computations. However, the idea has generated an impact in research in this area.

## 6.10   NONBINARY CHARACTER ERROR-CORRECTING BLOCK CODES

For burst error correction or for code concatenation as described in Chapter 5, nonbinary character error-correcting (NBCEC) codecs are a very powerful means for effective error control. Section 5.5.2.1 presents such code properties and Sections 5.8–5.8.4 bring forth the superiority in performance when NBCEC codes are used in either random or burst error satellite channels. For block NBCEC codes, this class of codes are Reed-Solomon codes, which have been treated in coding texts, in particular by Berlekamp [4]. A convenient way to become familiar with Reed-Solomon codes is through BCH code structures as follows:

The binary BCH codes, described in Chapter 5 and earlier sections of this chapter, can be generalized to nonbinary BCH codes in a relatively simple manner by recognizing that a q-ary nonbinary BCH code has a code length of $n = q^s - 1$ and is capable of correcting up to $E$ errors in code block characters. In other words, letting $\alpha$ be a primitive element in $GF(q^s)$, the generator polynomial $g(x)$ of such nonbinary BCH code has $\alpha, \alpha^2, \ldots, \alpha^{2E}$ consecutive powers of $\alpha$ as its roots. In terms of its minimal polynomials $M_i(x)$,

$$g(x) = \text{L.C.M.} \prod_{i=1}^{2E} M_i(x)$$

L.C.M. denotes the least common multiples, where the degree of each $M_i(x)$ is less than $s$. Thus the degree of $g(x)$ is at most $2sE$. When $s = 1$, we have the Reed-Solomon codes. From a practical standpoint, we are most interested in Reed-Solomon codes with $q = 2^m$, where $m$ is any positive integer, the number of binary digits in a character.

### 6.10.1   Berlekamp's Reed-Solomon Code Encoding

The fundamentals of conventional encoding and decoding techniques of RS codes are very similar to binary BCH encoding and decoding techniques described in Sections 6.2.1 to 6.2.2. As shown in (6.9) to (6.11), the RS encoding process is identical to that of a binary BCH code. In fact, all cyclic block codewords can be generated by the division of its code generator polynomial and the use of its remainder as parity checks for decoding. However, the conventional representation of finite field elements on a basis consisting of powers of the primitive element $\alpha$ complicates the codec hardware implementation. Berlekamp [50] introduces dual basis ($\beta$) of $\alpha$-basis through trace functions (which were defined and used in the generation of Bent function sequences for frame synchronization described in Section 5.5.2 of Volume I, Chapter 5).

The new $\beta$-basis provides bit-serial field multiplications and thus simplifies codec implementation. The amount of hardware savings can be revealed through NASA's Space Telescope example cited by Berlekamp. The specific RS code applied is the (255, 239, 17) eight-error correcting code operated over GF($2^8$) with eight level interleaving. The encoder-interleavers consist of 27 standard ICs (integrated circuits) and operate at above 1.0 Mbps. The conventional approach of implementing the same encoder-interleavers requires over 80 ICs.

Liu [51] discusses RS encoder designs with VLSI (very large-scale integrated) circuits. The conceptual design of the (255, 223, 33) RS encoder requires four identical interconnected VLSI chips of 235 × 235 mil CMOS/bulk, radiation-hardened with a 7 $\mu m$ standard cell approach on all logic. The logic structure of the encoder chip is shown in Figure 6.60. Where $g_j$, $j = 0, 1, \ldots, 2E$, are the coefficients of the code generator polynomial. All distinct coefficients except 1 are stored in the read-only memory on the chip. In general, an $E \times m$ bit size table is required. The output of this table becomes the input to the set of multiplexers. The number of multiplexers depends on the number of chips, in this case 4. The multiplexing ratio is the number of character errors correctable by the code divided by the number of chips to 1, and in this case it is (16/4) = 4 to 1. The outputs of the multiplexers, after selection, are then fed as the inputs of the multipliers M1, M2, M3, and M4. The outputs of each multiplier are fed into an 8-bit shift register with parallel in and serial out. The output of this 8-bit shift register is added with the corresponding 40-bit shift register ($S1$ to $S8$). The addition is modulo-2. NASA/Goddard Space Flight Center has assessed the feasibility of a single encoder chip design, which has been substantiated by researchers at the University of Idaho under NASA/CCSDS sponsorship [52–54].

### 6.10.2   Reed-Solomon Code Decoding

The functions of an RS decoder are shown in Figure 6.61. Similar to the BCH decoding functions of Figure 6.2, the RS decoder also first performs syndrome formation from a received codeword. At the same time the received codeword is also stored in the message buffer register. The finite field processing function of the decoder computes the error location and error evaluation polynomials. The next function is a Chien searcher, which checks each position of the code-word to see if the position, as it is shifted out, is in error. The error magnitude function operates in parallel with the Chien searcher; if an error is determined by the searcher, then the searcher output allows the error magnitude to perform error correction.

A comparison of BCH decoding (Figure 6.2) and RS decoding (Figure 6.61) reveals that the basic difference between the two decoding functions lies in the additional computation of character error magnitude required within Reed-

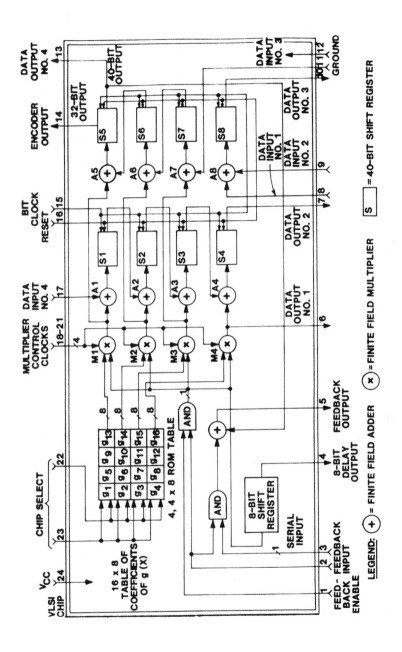

**Figure 6.60** VLSI RS encoder chip logic structure [51].

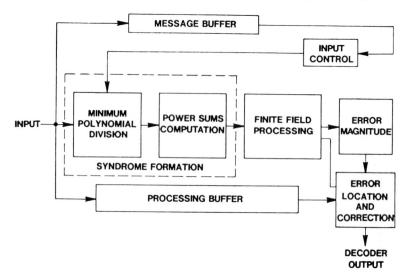

**Figure 6.61** Reed-Solomon code decoding functions.

Solomon decoding where one needs not only to locate which character is in error but also to identify which bits in the character are in error. At this point the best way to recognize such a difference is through a simple example of RS decoding procedure.

*Example 6.26:* We consider the RS code over $GF(2^4)$ which is generated by the primitive polynomial $1 + x + x^4$. The elements of this field are shown in Table 6.1. The code has length $n = 2^4 - 1 = 15$, 5 information characters, and code minimum distance $7(E = 3)$. The generator polynomial of the code is

$$g(x) = \prod_{i=1}^{2E} (x + \alpha^i)$$

$$= \alpha^6 + \alpha^9 x + \alpha^6 x^2 + \alpha^4 x^3 + \alpha^{14} x^4$$

$$+ \alpha^{10} x^5 + x^6.$$

Assume the all-zero message sequence has been transmitted and that the errors occur at the location $X_1 = \alpha^3$ and $X_2 = \alpha^{10}$. Then the received codeword is

$$\bar{r} = (0\ 0\ 0\ 1\ 0\ 0\ 0\ 0\ 0\ 0\ 1\ 0\ 0\ 0\ 0)$$

where the nonzero positions in $\bar{r}$ indicate character errors. The corresponding received codeword polynomial is $r(x) = x^3 + x^{10}$. The syndrome components are:

$$S_1 = r(\alpha) = \alpha^3 + \alpha^{10}$$

$$S_2 = r(\alpha^2) = \alpha^6 + \alpha^{20} = \alpha^6 + \alpha^5$$

$$S_3 = r(\alpha^3) = \alpha^9 + \alpha^0$$

$$S_4 = r(\alpha^4) = \alpha^{12} + \alpha^{40} = \alpha^{12} + \alpha^{10}$$

$$S_5 = r(\alpha^5) = \alpha^{15} + \alpha^{20} = \alpha^0 + \alpha^5$$

$$S_6 = r(\alpha^6) = \alpha^{18} + \alpha^{60} = \alpha^3 + \alpha^0.$$

Using Table 6.1 and modulo-2 additions, we have $S_1 = (0\ 0\ 0\ 1) +$ $(1\ 1\ 1\ 0) = (1\ 1\ 1\ 1) = \alpha^{12}$. Thus we have: $S_1 = \alpha^{12}$, $S_2 = \alpha^9$, $S_3 = \alpha^7$, $S_4 = \alpha^3$, $S_5 = \alpha^{10}$, and $S_6 = \alpha^{14}$. As in three-error-correcting binary BCH decoding, the error-locating polynomials of the RS decoder are:

$$S_1\sigma_3 + S_2\sigma_2 + S_3\sigma_1 = S_4$$

$$S_2\sigma_3 + S_3\sigma_2 + S_4\sigma_1 = S_5$$

$$S_3\sigma_3 + S_4\sigma_2 + S_5\sigma_1 = S_6$$

However, for two errors $\alpha^3$ and $\alpha^{10}$ we have only

$$\alpha^9\sigma_2 + \alpha^7\sigma_1 = \alpha^3$$

$$\alpha^7\sigma_2 + \alpha^3\sigma_1 = \alpha^{10}$$

let

$$\Delta_x = \begin{vmatrix} \alpha^9 & \alpha^7 \\ \alpha^7 & \alpha^3 \end{vmatrix} = \alpha^{12} + \alpha^{14}$$

$$= \alpha^5$$

$$\sigma_1 = \frac{\begin{vmatrix} \alpha^9 & \alpha^3 \\ \alpha^7 & \alpha^{10} \end{vmatrix}}{\Delta_x} = \frac{\alpha^4 + \alpha^{10}}{\alpha^5}$$

$$= \frac{\alpha^{17}}{\alpha^5} = \alpha^{12}$$

$$\sigma_2 = \frac{\begin{vmatrix} \alpha^3 & \alpha^7 \\ \alpha^{10} & \alpha^3 \end{vmatrix}}{\Delta_x} = \frac{\alpha^6 + \alpha^2}{\alpha^5}$$

$$= \frac{\alpha^3}{\alpha^5} = \alpha^{13}.$$

Hence the error-locating polynomial is

$$\sigma(x) = x^2 + \sigma_1 x + \sigma_2$$

$$= x^2 + \alpha^{12}x + \alpha^{13}.$$

For checking let $x = \alpha^3$ and $x = \alpha^{10}$.

$$\sigma(\alpha^3) = \alpha^6 + \alpha^{15} + \alpha^{13}$$

$$= (0\ 0\ 1\ 1) + (1\ 0\ 0\ 0) + (1\ 0\ 1\ 1)$$

$$= (0\ 0\ 0\ 0)$$

$$\sigma(\alpha^{10}) = \alpha^{20} + \alpha^{22} + \alpha^{13}$$

$$= \alpha^5 + \alpha^7 + \alpha^{13}$$

$$= (0\ 1\ 1\ 0) + (1\ 1\ 0\ 1) + (1\ 0\ 1\ 1)$$

$$= (0\ 0\ 0\ 0).$$

To compute the error magnitudes $Y_i$'s we take the first two ($j = 2$ because $E = 2$) syndrome equations from

$$S_j = \sum_{i=1}^{E} Y_i X_i^j$$

$$S_1 = Y_1 X_1 + Y_2 X_2 = \alpha^{12}$$

$$S_2 = Y_1 X_1^2 + Y_2 X_2^2 = \alpha^9$$

Substituting the error locations $X_1 = \alpha^3$ and $X_2 = \alpha^{10}$ in $S_1$ and $S_2$ of the above two equations we have

$$Y_1 \alpha^3 + Y_2 \alpha^{10} = \alpha^{12}$$

$$Y_1 \alpha^6 + Y_2 \alpha^5 = \alpha^9.$$

Let

$$\Delta_y = \begin{vmatrix} \alpha^3 & \alpha^{10} \\ \alpha^6 & \alpha^5 \end{vmatrix} = \alpha^8 + \alpha = \alpha^{10}$$

$$Y_1 = \frac{\begin{vmatrix} \alpha^{12} & \alpha^{10} \\ \alpha^9 & \alpha^5 \end{vmatrix}}{\Delta_y} = \frac{\alpha^2 + \alpha^4}{\alpha^{10}} = 1$$

$$Y_2 = \frac{\begin{vmatrix} \alpha^3 & \alpha^{12} \\ \alpha^6 & \alpha^9 \end{vmatrix}}{\Delta y} = \frac{\alpha^{12} + \alpha^3}{\alpha^{10}} = 1,$$

where $Y_1$ and $Y_2$ are the error magnitudes, which are needed in all Reed-Solomon code decoding.

### 6.10.2.1  Gore's Decoding Method

When Reed-Solomon codes are encoded by means of the Chinese Remainder Theorem (see Chapter 2), it is shown by Mandelbaum that the Chien's search and the calculations of error values in decoding are no longer needed, and they can be replaced by a polynomial division and added calculation in determining the syndrome [55]. This is an improvement in reduction of RS codec complexity. However, the most significant improvement in RS decoding is due to Gore [56]. Gore is the first to suggest the use of the Fast Fourier transform (FFT) in order to reduce the number of multiplications and additions required in an RS decoder. The following analysis is based on part of his original work.

The basic decoding procedure for the RS codes is dependent upon the fact that there is a linear relation between the coefficients of the error locator polynomial $\sigma(z) = 1 + \sigma_1 z + \ldots + \sigma_E z^E$ and the power sums $S_j$, $j = 1, 2, \ldots,$ $2E - 1$. This relation is developed as follows. Let

$$\sigma(z) = \prod_{i=1}^{E} (1 - X_i z) \tag{6.43}$$

where $E$ is the number of errors and $X_i$ are the error locators. Further let:

$$S_j = \sum_{i=1}^{E} Y_i X_i^j$$

$$= r(\alpha^j), \quad 1 \le j \le 2E - 1 \tag{6.44}$$

where $\alpha$ is a primitive element of GF($2^m$), in which the RS code is operated, $Y_i$ are the error magnitudes, and $r(x) = r_0 + r_1 x + \ldots + r_{n-1} x^{n-1}$ is the received codeword. Then, since $\sigma(X_i^{-1}) = 0$, we have

$$1 + X_i^{-1} \sigma_1 + \ldots + X_i^{-E} \sigma_E = 0.$$

Multiplying by $Y_i X_i^{j+E}$ we get

$$Y_i X_i^{j+E} + Y_i X_i^{j+E-1} \sigma_1 + \ldots + Y_i X_i^j \sigma_E = 0.$$

If these equations are summed over $i$ we get

$$S_{j+E} + S_{j+E-1} \sigma_1 + \ldots + S_j \sigma_E = 0. \tag{6.45}$$

By letting $1 \le j \le 2E - 1$ in (6.45) we obtain the set of $E$ linear equations which can be used to solve for the $E$ values of $\sigma_i$. The solving of these equations was simplified by the invention of the Berlekamp algorithm, and the utility of this algorithm was extended by Massey's observation that solving equations (6.45) for the $\sigma_i$'s is the problem of synthesizing a minimum length shift register to produce the given sequence $\{S_j, S_{j+1}, S_{j+2}, \ldots\}$

In what follows we let

$$c(x) = c_0 + c_1 x + \ldots + c_{n-1} x^{n-1} = \text{the transmitted codeword}$$

$$e(x) = e_0 + e_1 x + \ldots + e_{n-1} x^{n-1} = \text{the error pattern}$$

$$r(x) = c(x) + e(x) \qquad\qquad = \text{the received codeword}.$$

For any vector (polynomial) $f(x) = f_0 + f_1 x + \ldots + f_{n-1} x^{n-1}$ we define a new polynomial $F(y) = F_0 + F_1 y + \ldots + F_{n-1} y^{n-1}$ with

$$F_j = f(\alpha^j) \tag{6.46}$$

We will call $F(y)$ the transform of $f(x)$ and let

$$F(y) = T\{f(x)\} \tag{6.47}$$

Equations (6.46) and (6.47) can be used to decode nonsystematic RS codes in general. For decoding systematic RS codes, as in most practical cases, we proceed as follows: It is well known that there exists an inverse transform [1]:

$$f(x) = T^{-1}\{F(y)\} \tag{6.48}$$

with

$$f_i = -F(\alpha^{-i}). \tag{6.49}$$

If we assume that $e(x)$ has $E$ nonzero terms with coefficients $Y_1, Y_2, \ldots, Y_E$ in positions $i_1, i_2, \ldots i_E$ and that $X_S = \alpha^i S$, then

$$e(\alpha^j) = \sum_{i=1}^{E} Y_i X_i^j = S_j \qquad \text{for all } j. \tag{6.50}$$

Because $c(x)$ is a codeword, $c(\alpha^j) = 0$ for $1 \le j \le 2E$.

Further:  $\quad r(\alpha^j) \quad = c(\alpha^j) + e(\alpha^j)$

or  $\qquad T\{r(x)\} = T\{c(x)\} + T\{e(x)\}$

$$R(y) \quad = C(y) + E(y) \tag{6.51}$$

Gore's decoding procedure is to compute the transform of the received vector. The first $2E$ coefficients of $R(y)$ are identical to those of $E(y)$. Using (6.50) and (6.45) we can find $E(y)$ by synthesizing the shift register specified by (6.45) (the Berlekamp-Massey algorithm) and then exercising the shift register for $n - 2E$ additional shifts. Having found $E(y)$ and computed $R(y)$, we take their difference to find $C(y)$. The corrected codeword is found by taking the inverse transform of $C(y)$. Such a decoder is shown in Figure 6.62.

The first block of Figure 6.62 computes a linear transform of the received word. The first $n - k$ components of this transform are the syndrome components. These first $n - k$ components are the input to the second block, while all $n$ components of the transform are stored in a buffer register. The second block is a Berlekamp-Massey algorithm computer which implements the Berlekamp algorithm in such a way that it synthesizes a shift register as shown by Massey [57]. Once this algorithm is concluded, the shift register that is synthesized is exercised to produce the transform of the error pattern. This transform is subtracted from the transform of the received word in the buffer register; and the difference is the input to the third block. The third block computes the inverse of the transform performed in the first block to give the corrected received word. It has been demonstrated that in many cases the first and third blocks in Figure 6.62 are no more complex, and may in fact be less complex, than the syndrome computation of the conventional RS decoder. Further, in some cases the third block of Figure 6.62 may be omitted from the decoder altogether.

We next investigate the complexity of the transformer. Considering our polynomials as row vectors, and

$$rT = R \tag{6.52}$$

the transform of $r(x)$ to produce $R(y)$; then $T$ is the $n \times n$ matrix:

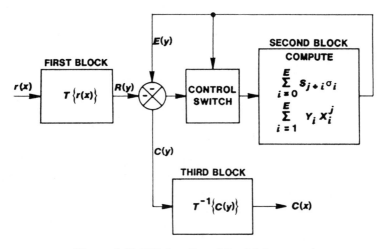

**Figure 6.62** FFT decoding of Reed-Solomon codes.

$$
T = \begin{bmatrix}
1 & 1 & 1 & \dots 1 \\
1 & \alpha^1 & \alpha^2 & \dots \alpha^{(n-1)} \\
1 & \alpha^2 & \alpha^4 & \dots \alpha^{2(n-1)} \\
\cdot & \cdot & \cdot & \cdot \\
\cdot & & & \\
\cdot & & & \\
1 & \alpha^{(n-1)} & \alpha^{2(n-1)} & \dots \alpha^{(n-1)(n-1)}
\end{bmatrix} \tag{6.53}
$$

This matrix is formally equivalent to the matrix $A$ (where $j = \sqrt{-1}$ ):

$$
A = \begin{bmatrix}
1 & 1 & 1 & \dots 1 \\
1 & \varepsilon^{j\frac{2\pi}{n}} & \varepsilon^{j\frac{2\pi}{n}2} & \dots \varepsilon^{j\frac{2\pi}{n}(n-1)} \\
1 & \varepsilon^{j\frac{2\pi}{n}2} & \varepsilon^{j\frac{2\pi}{n}4} & \dots \varepsilon^{j\frac{2\pi}{n}2(n-1)} \\
\cdot & \cdot & & \\
\cdot & & & \\
\cdot & & & \\
1 & \varepsilon^{j\frac{2\pi}{n}(n-1)} & \varepsilon^{j\frac{2\pi}{n}(n-1)} & \dots \varepsilon^{j\frac{2\pi}{n}(n-1)(n-1)}
\end{bmatrix} \tag{6.54}
$$

with the basic corresponding elements of the $T$, $A$ matrices as:

$$
\alpha \leftrightarrow \varepsilon^{j\frac{2\pi}{n}} . \tag{6.55}
$$

Matrix $A$ is recognized as the discrete Fourier transform matrix; and it is well known then when $n$ is composite the transformation can be made with fewer multiplications and additions than 6.52 would indicate. If $n = n_1 n_2 \dots , n_S$ then the fast Fourier transform [58] requires $n(n_1 + n_2 + \dots + n_S)$ multiplications and additions while a direct evaluation of (6.52) requires $n^2$ multiplications and additions. It should be noted that to compute the $r$ syndrome components $S_1, S_2, \dots S_{n-k}$ of a conventional RS decoder requires $nr$ multiplications and additions. Thus, when $n_1 + n_2 + \ldots + n_S \leq n - k$, the entire transformation can be performed with fewer multiplications and additions than the number required in a conventional decoder in order to compute the syndrome.

The transform decoding method of RS codes has been further investigated at the University of Southern California, Jet Propulsion Laboratory, IBM, MITRE Corporation, GTE, Sylvania Electric and C.C.E.T.T. of France, by Reed, Truong, Miller, and Benjauthrit; Blahut; Michelson; Liu; Botrel and Harrari and others [59–63]. Redinbo applied FFT to BCH decoding [64]. Liu [63] at JPL provides the logic structures of VLSI decoder chips and suggests RS decoder designs based on the VLSI chips. His 8-stage VLSI syndrome generator chip logic structure is shown in Figure 6.63. Two such chips can be connected to a shift register to form an 8-character error-correcting syndrome generator. With a 4-stage VLSI linear feedback shift register synthesis (Massey's method to compute the elementary symmetric functions $\sigma_i$ from the set of $S_j$), the transform function $T\{r(x)\}$ for the (31, 15, 17) RS code can be connected as shown in Figure 6.64. The inverse transform function $T^{-1}\{c(y)\}$ can be accomplished through the connection of four 8-stage VLSI syndrome generator chips (Figure 6.63), as shown in Figure 6.65. As noted in these figures, registers B, C, and S are part of the LFSR logic structure.

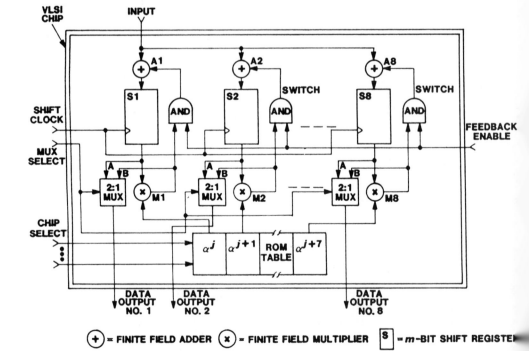

Figure 6.63 8-stage VLSI syndrome generator chip logic structure.

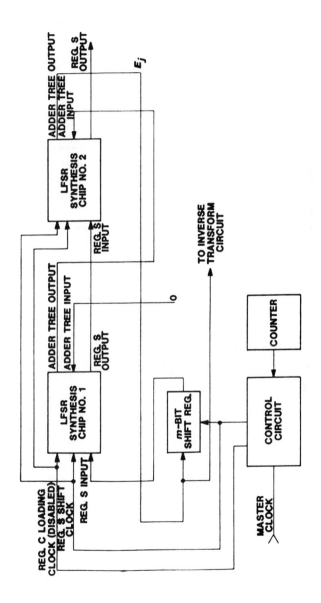

**Figure 6.64** Chip implementation of the transform function of the (31, 15, 17) Reed-Solomon code by Liu [63].

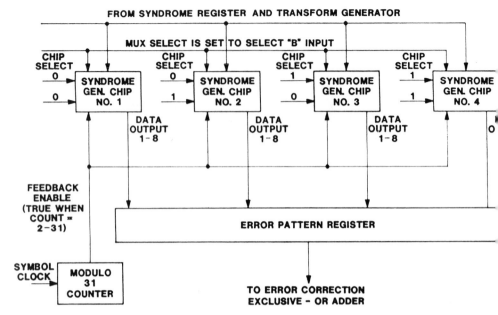

**Figure 6.65** Implementation of inverse transform for (31, 15, 17) Reed-Solomon decoder [63].

### 6.10.2.2 Decoding Technology

The RS decoding technology has been advanced in two fronts, through clever implementation and product improvement. Cyclotomics under the direction of Dr. E. Berlekamp manufactures exclusively RS codecs, with speeds from 64.0 kbps to 120.0 Mbps, applied to deep space, military, and commercial satellite communication [65, 66].

The complexity of an RS codec depends on a number of factors such as the coding rate, the number of bits per character, code length, signaling speed, and error-correcting capability of the code. RS codec hardware complexity issues are extensively discussed in [65]. For any fixed value of block length $n$, an $E$-character-error-correcting RS code actually requires a smaller Galois field than the corresponding $E$-error-correcting binary BCH codes of the same length. Not only can RS decoders operate faster than BCH decoders, but they also require less hardware than comparable BCH decoders. Thus RS codecs are not necessarily more complex than comparable BCH codecs. But, as we have noted, an understanding of the BCH codec implementation is a giant step toward RS codec designs.

In Section 5.8.4 of Chapter 5 we describe the NASA coding standard [later evolved as Consultative Committee For Space Data System (CCSDS)] and present a coding performance example in Figure 5.68, illustrating a concatentation of the rate ½, constraint length 7 Viterbi codec and the (255, 223, 33) RS codec, which is intended for a satellite to ground station communication link. Since Viterbi codec complexity has been discussed in the previous sections, we need only explore the design and implementation aspects of the RS codec, which we consider as the second front in RS decoder implementation technology.

The generator polynomial of the RS code over GF($2^8$) selected by CCSDS is

$$G(x) = 1 + x + x^2 + x^7 + x^8$$

with $m = 8$ bits per RS code character. The decoder can correct up to 16 character errors in a code block of 255 characters. When the code is expressed in terms of GF($2^8$) and a primitive element $\beta$, in GF($2^8$), the code generator polynomial is

$$g(x) = \prod_{j=112}^{143} (x - \beta^j) .$$

The RS encoder has been built as shown in Figure 6.60, where all lines are 8-bit parallel buses and addition and multiplication are performed in GF($2^8$) modulo $g(x)$. The decoder implementation is shown in Figure 6.66 [52]. The decoder is mapped into six VLSI circuit boards. Each VLSI is equivalent to a thousand 16 pin LSI chips. Board 1 contains the syndrome generator, syndrome interface, and serial parallel converter; Board 2 and Board 3 contain the logics of performing the required finite field processing; Board 4 and Board 5 contain the functions of error location and error magnitude; and Board 6 is the decoder output section where error correction takes place. Board 1 contains a $1k \times 8$ RAM and Board 6 contains a $2k \times 8$ RAM. The RAM's are used for message storage. As mentioned previously, the encoder requires a single such VLSI circuit

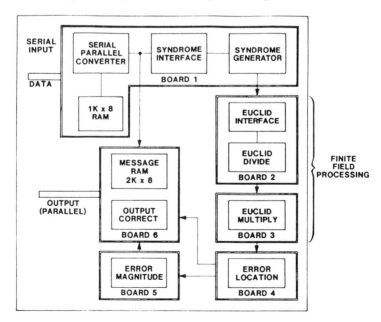

**Figure 6.66** Decoder implementation of the University of Idaho codec.

board. When designed with $T^2L$ logic elements the encoder can be operated at 10.0 Mbps.

## 6.11  CONCLUDING REMARKS

From a practical viewpoint this chapter discusses basic coding and decoding techniques for both block and convolutional codecs. Actual code implementations for either satellite or space application are illustrated. When combining the information of a codec design and complexity from this chapter and other considerations such as the code performance from Chapter 5, the reader should be able to recommend a basic error-correcting codec for a particular satellite communication system.

## 6.12  REFERENCES

1. Hocquenghem, A. "Codes correcteurs d'erreurs" Chiffres, Paris, vol. 2 (1959).
2. Bose, R. C., and Ray-Chaudhuri, D. K. "On a Class of Error Correcting Binary Group Codes," *Information Control*, vol. 3 (1960).
3. Peterson, W. W. *Error Correcting Codes*. Cambridge, Mass.: M.I.T. Press, 1961.
4. Berlekamp, E .R. *Algebraic Coding Theory*. New York: McGraw-Hill, 1968. Revised Edition, Aegean Park Press, 1984.
5. Mattson, H. F., and Solomon, G. "A New Treatment of BCH Codes." *Journal of the Society of Industrial Applied Mathematics* 9 (1961): 654–669.
6. Gallager, R. G. *Information Theory and Reliable Communications*. New York: John Wiley & Sons, 1968.
7. Massey, J. L. "Step by Step Decoding of the BCH Codes." *IEEE Transactions on Information Theory*, IT-11 (October 1965).
8. Massey, J. L. "Shift-Register Synthesis and BCH Decoding." *IEE Transactions on Information Theory*, IT-15 (January 1969).
9. Chien, R. T. "Cyclic Decoding Procedures for BCH Codes." *IEEE Transactions on Information Theory*, IT-10 (October 1964).
10. Burton, O. H. "Inversionless Decoding of Binary BCH Codes." *IEEE Transactions on Information Theory*, IT-17 (July 1971).
11. Bartee, T. C., and Schneider, D. I. "Computation with Finite Fields." *Information and Control*, vol. 6 (1983).
12. Slepian, D. "A Class of Binary Signalling Alphabets." *Bell Systems Technical Journal 35* (1956): 203-234.
13. Kasami, T. "Weight Distributions of Bose-Chaudhuri-Hocquenghem Codes." In *Combinatorial Mathematics and Its Applications*. R. Bose and T. Dowling, eds. Chapel Hill, N.C.: University of North Carolina Press, 1968.
14. Sidelnikov, V. "Weight Spectrum of Binary Bose-Chaudhuri-Hocquenghem Codes." *Problems of Information Transmission*, vol. 7, no. 1 (1971): 11–17.
15. Muratani, T.; Saitoh, H.; Koga, K.; Mizuno, T.; Yasuda, Y.; and Snyder, J. "Application of FEC Coding to the INTELSAT TDMA System." *Proceedings of 4th International Conference on Digital Satellite Communications* (October 1978): 108–115.
16. "High-Speed Error Correction Codec." Final Implementation Report, by KDD/NEC for INTELSAT Contract IS-838. January 1979.

17. Wu, W., and Golding, L. "Error Coding Performance for Digital Communication Systems." COMSAT Technical Memorandum, Supplement to CL-46-71, September 1971.

18. Dorsch, B., and Dolainsky, F. "Performance of Difference-Set Cyclic Codes." *Proceedings of the NATO Conference* (1974): 109–115.

19. Wu, W., and Hu, M. "Computer Simulation Results of Error Correcting Codecs." COMSAT Technical Report, April 1971. Also "User's Manual for Coding Systems." COMSAT Scientific Computer Applications Publication 73-4-4, 1973.

20. Weldon, E., Jr. "Difference-Set Cyclic Codes." *Bell Systems Technical Journal* 45 (September 1966): 1045–1055.

21. Massey, J. *Threshold Decoding.* Cambridge, Mass.: M.I.T. Press, 1963.

22. Wu, W. "New Convolutional Codes—Part I." *IEEE Transactions on Communications Technology* COM-23 (September 1975): 942–956.

23. Lin, S., and Costello, D. *Error Control Coding: Fundamentals and Applications.* Englewood Cliffs, N.J.: Prentice-Hall, 1983.

24. Bluestein, G., and Jordan, K. "An Investigation of the Fano Sequential Decoding Algorithm by Computer Simulation." Report 62G-5, Lincoln Laboratory, M.I.T., July 1963.

25. Savage, J. "The Computation Problem with Sequential Decoding." Technical Report 371, Lincoln Laboratory, M.I.T., February 1965.

26. Niessen, C. "An Experimental Facility for Sequential Decoding." Technical Report 396, Lincoln Laboratory, M.I.T., September 1965.

27. Lebow, I., and McHugh, P. "A Sequential Decoding Technique and Its Realization in the Lincoln Experimental Terminal." *IEEE Transactions on Communications Technology*, COM-15, no. 4 (August 1967): 477–491.

28. Jacobs, I. "Sequential Decoding for Efficient Communication for Deep Space." *IEEE Transactions on Communications Technology*, COM-15, no. 4 (August 1967): 492–501.

29. Layland, J., and Lushbaugh, W. "A Flexible High-Speed Sequential Decoder for Deep Space Channels." *IEEE Transactions on Communications Technology*, COM-19, no. 5 (October 1971): 813–820.

30. Forney, G., Jr., and Langelier, R. "A High Speed Sequential Decoder for Satellite Communications." *ICC* (1969): 33.9–33.17.

31. Forney, G., and Bower, E. "A High-Speed Sequential Decoder: Prototype Design and Test." *IEEE Transactions on Communications Technology*, COM-19, no. 5 (October 1971): 821–825.

32. "Final Report on a High-Speed Sequential Decoder Study." Contract DAAB07-68-C-0093, Codex Corp. April 1968.

33. Gilhousen, K.; Heller, J.; Jacobs, I.; and Viterbi, A. "Coding System Study for High Data Rate Telemetry Links." Contract Report No. NAS2-6024, Linkabit Corporation, January 1971.

34. Ma, H. "The Multiple Stack Algorithm Implemented on a Zilog Z-80 Microcomputer." *IEEE Transactions on Communications Technology*, COM-28 (November 1980): 1876–1887.

35. Lushbaugh, W. "Information Systems: Hardware Version of an Optimal Convolutional Decoder." Jet Propulsion Laboratory Technical Report 32–1526, Vol. II.

36. Heller, J. "Sequential Decoding: Short Constraint Length Convolutional Codes," Space Summary 37–54, Vol. III, pp. 171–177, Jet Propulsion Laboratory, Dec. 1968.

37. Heller, J. and Jacobs, I. "Viterbi Decoding for Satellite and Space Communication." *IEEE Transactions on Communications Technology*, COM-19 (October 1971).

38. Spilker, J. *Digital Communications by Satellite*. Englewood Cliffs, N.J.: Prentice-Hall, 1977.

39. Bharagava, V.; Haccoun, D.; Matyas, R.; and Nuspl, P. *Digital Communications by Satellite*. New York: John Wiley & Sons, 1981.

40. Morakis, J.; Greene, E.; Miller, W.; Poland, W.; and Helgert, H. *Aerospace Data System Standards, Part 7: Coding Standard*. NASA Report, 1982.

41. Schweikert, R., and Hagenauer, J. "Channel Modeling and Multipath Compensation with Forward Error Correction For Small Satellite Ship Earth Stations." 6th ICDSC (September 1983): XII-32–XII-38.

42. Bernstein, S.; Heggestad, H.; Mui, S.; and Richer, I. "Variable-Rate Viterbi Decoding in the Presence of RFI." *NTC Record* (1977): 36.6-1–36.6-5.

43. Yasuda, Y.; Hirata, Y.; Nakamura, K.; and Otani, S. "Development of Variable-Rate Viterbi Decoder and Its Performance Characteristics." 6th ICDSC (September 1983): XII-24–XII-31.

44. Orndorff, R.; Chou, P.; Kremarik, J.; Colesworthy, R.; Doak, T.; and Koralek, R. "Viterbi Decoder VLSI Integrated Circuit for Bit Error Correction." NTC 1981, E1.7.1–E1.7.4.

45. Clark, G. "Implementation of Maximum Likelihood Decoders for Convolutional Codes." *International Telemetry Conference Record*, 1971.

46. Clark, G., and Davis, R. "Two Recent Applications of Error-Correction Coding to Communication System Design." *IEEE Transactions on Communications Technology*, COM-19, no. 5 (October 1971): 856–863.

47. Cain, J.; Clark, G.; and Geist, J. "Punctured Convolutional Codes of Rate $(n - 1)/n$ and Simplified Maximum Likelihood Decoding." *IEEE Transactions on Information Theory*, vol. IT-25 (January 1979): 97–100.

48. Acampora, A., and Gilmore, R. "Analog Viterbi Decoding for High School Digital Satellite Channels." *IEEE Transactions on Communications Technolgy*, COM-26 (October 1978): 1463–1470.

49. Haccoun, D., and Ferguson, M. "Generalized Stack Algorithms for Decoding Convolutional Codes." *IEEE Transactions on Information Theory*, IT-21 (November 1975): 638–651.

50. Berlekamp, E. R. "Bit-Serial Reed-Solomon Encoders." *IEEE Transactions on Information Theory*, IT-28, no. 6, (November 1982): 869–874.

51. Liu, K. "Architecture for VLSI Design of Reed-Solomon Encoders." *IEEE Transactions on Computers*, C-31, no. 2 (February 1982): 170–175.

52. Maki, G.; Mankin, R.; Owlsey, P.; Kim, G.; and K. Winters. *NASA Design Review*, Electrical Engineering Department, University of Idaho, September 1983.

53. Morakis, J.; H. Helgert; W. Poland; and W. Miller. *Aerospace Data System Standard (Draft) Part 7: Coding Standard*. NASA/GSFC (January 1982).

54. "Consultative Committee for Space Data Systems (CCSDS) Recommendation for Telemetry Channel Coding." *Blue Book* (February 1984).

55. Mandelbaum, D. "On Decoding Reed-Solomon Codes," *IEEE Transactions on Information Theory*, IT-17, no. 6 (November 1971): 707–712.

56. Gore, W. "Transmitting Binary Symbols with Reed-Solomon Codes." Report No. 73-5, The Johns Hopkins University Electrical Engineering Department, April 1973.

57. Massey, J. L. "Shift-Register Synthesis and BCH Decoding." *IEEE Transactions*

*on Information Theory*, IT-15, no. 1 (January 1969): 122–127.

**58.** Cooley, J. W., and Tukey, J. W. "An Algorithm for the Machine Calculation of Complex Fourier Series." *Mathematical Computation*, vol. 19 (April 1965): 297–301.

**59.** Reed, I. S.; Truong, I. K.; Miller, R. L. and Benjauthrit, B. "Further Results on Fast Transforms for Decoding Reed-Solomon Codes over GF($2^m$) for $m$ = 4, 5, 6, 8." *Deep Space Network Progress Report 42-50*, pp. 132–154, Jet Propulsion Laboratory, Pasadena, Calif., January 1979.

**60.** Blahut, R. "Transform Techniques for Error Control Codes." *IBM Journal of Research and Development*, vol. 23 (May 1979): 299–315.

**61.** Michelson, A. "A Fast Transform in Some Galois Fields and an Application to Decoding Reed-Solomon Codes." *Proceedings of IEEE International Symposium on Information Theory*. Ronneby, Sweden (June 1976).

**62.** Botrel, J., and Harrari, S. "The Efficiency of the Decoding Algorithms for Reed-Solomon Codes and Their Generalization." *Proceedings of International Symposium on Information Theory*, Grignano, Italy (June 1979).

**63.** Liu, K. "Architecture for VLSI Design of Reed-Solomon Decoders." *IEEE Transactions on Computers*, C-33, no. 2 (February 1984): 178–189.

**64.** Redinbo, R. "Finite Field Arithmetic in an Array Processor with Applications to BCH Decoding." *Proceedings of International Symposium on Information Theory*, Grignano, Italy (June 1979).

**65.** Berlekamp, E. "The Technology of Error-Correcting Codes." *Proceedings of the IEEE*, vol. 68, no. 5 (May 1980): 564–593.

**66.** *Economic Feasibility of Reed-Solomon Decoding*. Berkeley, Calif.: Cyclotomics, Inc., 1983.

## 6.13 PROBLEMS

1. The basic mechanism for the realization of all error-correcting codes and decoders is a shift register, which consists of multiple stages called flip-flops. A flip-flop is a two-state (0, 1) multivibrator.
   (a) Describe how a multivibrator works.
   (b) Investigate all the input and output of each stage when two flip-flops are connected in cascade with common shifting pulses.
   (c) Draw a state transition diagram for the two-stage shift register as described in (b).

2. For (31, 21, 5) BCH code encoding and decoding, if the test sequence is a 21-bit 1010 . . . . 1 pattern, what output will assure you that the codec is operated correctly?

3. Derive the generator polynomials for (15, 7, 5), (31, 26, 3), and (63, 51, 5) BCH codes.

4. Design the $k$-configuration and the $n - k$-configuration types of encoder for the (15, 7, 5) BCH code.

5. Design the power sum circuits for the (15, 7, 5) BCH decoder.

6. Design the finite field processor and the error-location circuit for the (15, 7, 5) decoder.

7. Design the coset decoder for the (15, 7, 5) BCH code and compare the codec logic complexity with the results from Problems 5 and 6.

8. For the (7, 3, 4) difference-set code with generator polynomial $g(x) = 1 + X^2 + X^3 + X^4$ and the set elements of 0, 2, 3, . . . , design the complete codec with shift registers, and AND, and OR logic gates.

9. List all the cosets and coset leaders for decoding the (7, 3, 4) Hamming code and compare the complexity of the coset decoder with the threshold decoder of Problem 8.

10. With a 4-bit quantization in Viterbi decoding, what is the metric for a received sequence 10101?

11. For a two-stage ($m = 2$), rate ½ nonsystematic convolutional code with generators $g_1 = 101$, $g_2 = 111$, design the Viterbi decoder (down to the logic level) by sketching
    (a) the branch metric calculation unit,
    (b) the state (path) decision unit, and
    (c) the output selection unit.

12. Design in principle the state decision unit for the Viterbi decoder with the rate ¾ convolutional encoder as shown in Figure 6.46.

13. Conceptually design a rate ½, constraint length 41 sequential decoder with multiple stack algorithm. Compare the estimated decoder complexity with Example 6.21 in which the sequential decoder uses a syndrome.

14. Provide a relative comparison table for the codes described in this chapter in terms of coding gain, coding rate, data rate, code constraint length, burst error sensitivity, computational effort, decoding delay, storage requirement, and codec complexity. In some items assumptions need to be made, for example, for the same coding gain, in order to compare other items.

15. Provide the decoder design using fast Fourier transform of the (15, 5, 7) Reed-Solomon code, as in Example 6.26.

# Chapter 7

# INTEGRATED SERVICES DIGITAL SATELLITE NETWORKS AND PROTOCOLS

## 7.1 INTEGRATED SERVICES DIGITAL SATELLITE NETWORK (ISDSN)

The integrated services digital satellite network (ISDSN) may be defined as a combination of the integrated service digital network (ISDN) and the integrated digital satellite network (IDSN), as shown in Figure 7.1. These two networks have evolved almost independently. The ISDN has basically evolved from tele-

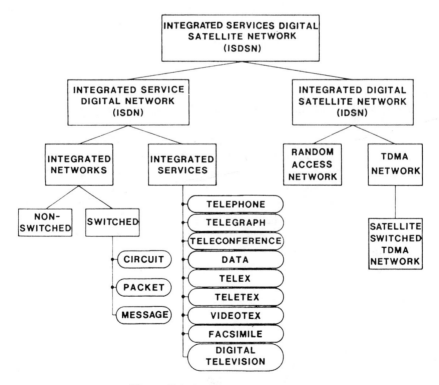

**Figure 7.1** An ISDSN Architecture.

phone networks with an emphasis on multiplexing and switching arrangements, while the IDSN (or just DSN) has evolved for conservation of satellite resources such as power and bandwidth with an emphasis on multiple access and modulation techniques. Both networks, however, share the common goal of serving a large number of users and offering a variety of services.

The networks are discussed separately in this chapter because network distinctions do exist. The first part of the chapter is devoted to issues concerned with the ISDN. The second part of the chapter is devoted to the details of the network aspects of the first international digital satellite network as an exmaple of the IDSN. The last part of the chapter highlights the theoretical foundations of network protocols and flow controls.

From an access viewpoint, the IDSN can be implemented by a TDMA network, a satellite switched network, a random access network, or any network combination of the access schemes described in Chapter 3 of Volume I. In presenting the IDSN, Figure 7.1 does not show the alternatives of modulation methods, synchronization techniques, error detection and correction, encryption and decryption and message security, and network optimization procedures. These topics are discussed in detail elsewhere in this book and it is not a difficult task to apply them to IDSN once the objectives, the types of services, and the user's demands are determined.

The types of integrated services covered by ISDN are not described in our discussion of IDSN, the basic principle being that any service provided by ISDN should also be provided by IDSN. As for the integrated networks of ISDN, there are basically switched and nonswitched, synchronous, asynchronous, plesiosynchronous, and public and private networks. Circuit-switched, packet-switched, and message-switched networks already existed for switched networks.

Services are characterized by user-to-user compatibility, a subscribers' listing, standardized testing and maintenance procedures, and charging policies. These services include telephone, telegraph, data, teletex, videotex, and facsimile transmission. Teletex is a new telegraph service that offers a combination of local facilities with transmission functions for communication with remote stations through the public switched network. With a transmission speed higher than the 50 baud ordinary telegraph, teletex provides a faster exchange of correspondence. Videotex refers to systems that transmit messages to a user's visual display in the form of either text or graphics. A good introductory reference on videotex is the book by Woolfe [1].

The ISDN may be considered from two viewpoints, that of the user and that of the network, or provider. The user is more concerned with integrated service, while the network designer and provider must consider different types of messages and their effect on the function of a common facility to switch and transmit all services. Both should cooperate to achieve a truly efficient, economic, and longer-lasting integrated network or a set of networks for international as well as regional digital communication, including satellites.

When ISDN interfaces with IDSN, the following factors are to be considered for ISDSN:

- Universality: An ISDSN should have a global appeal and be able to accommodate a wide variety of applications. The network should handle small and large amounts of traffic, as well as serve small and large numbers of users. The ISDSN should be capable of facilitating communication between and among regional or national networks. A universal ISDSN also implies easy and economic accessibility to user networks.

- Flexibility: An important feature must be flexible design. The ISDSN should be simple to change and easily expandable so that the network can have long-lasting value even with changes in traffic pattern and user number.

- Compatibility: One of the challenges of developing a universal ISDSN is to make various existing and planned digital networks compatible, so that they can operate under a truly integrated network architecture. For future networks the task should be easier as they can be designed in accordance with the ISDSN guidelines. Compatibility requires the following considerations.

- Familiarity: The characteristics of the existing digital networks that may be integrated or interconnected need to be identified. There are the network services of TWX, Telex, Telenet, Tymnet, Dataphone, SBS, ARPA-net in the U.S.; DATAPAC in Canada; Euronet in Europe; Videotex and Teletex in France and in the U.K.; and DDX in operational with circuit switching in Japan. In West Germany an all-digital telephone network has been proposed, to be called DTN, for voice, data, text, and FAX communications. A satellite data communication experimental network called INTERNET has been participated in by the U.S., the U.K., Norway, and West Germany. Familiarity with these and other existing, or developing networks is essential for the purposes of network integration.

- Reliability: Network failures due to excessive noise, interference, congestion, or network components must be minimized. Reliability should be an important factor in every phase of the ISDSN design.

- Efficiency: As in any large-scale system design, network operational efficiency should be emphasized. It is not enough for subsystems to be efficient. All elements of the overall network must be integrated to perform with comparable efficiency.

- Connectivity: Based on existing network characteristics, the best interface mechanism must be found in terms of bit rates, line codes, framing structure, and signal synchronization, to ensure compatibility, efficiency, minimum delay, and reliability. The objectives of simplicity and cost effectiveness must be kept in mind.

- Maintainability: In ISDSN design consideration should be included for minimum service outage and operational failure as well as minimal ease in identifying maintenance cost.

- Multiplicity: Because of the existence of different traffic demands due to geographical and national needs multiple compatible terrestrial ISDNs should be considered.

- Simplicity: The basic design principle for a complicated network, the user-to-network interfaces, and network-to-network interfaces must be simple. In many system designs, simplicity can enhance reliability and maintainability, and as well minimize the total cost of the network installation.

The above general, desirable factors for network design can be changed to precise specifications once types of services, number of users, traffic patterns, performance criteria, and cost factors are determined. Very often these user factors logically determine a network architecture.

The possible network elements of an ISDSN shown in Figure 7.2 consist of user networks, such as telephone and data networks; local networks, such as regional or national data networks; transit networks, such as domestic satellite networks; satellite network, such as the global space segment of the satellite communication network and its terrestrial interfaces.

Network standards should be developed separately in accordance with the elemental networks of an ISDSN. With a thorough understanding of the separated networks, interface guidelines can then be developed in order to facilitate interface designs.

## 7.2 INTEGRATED SERVICE DIGITAL NETWORK (ISDN)

The principles of ISDN as recommended by the International Telegraph and Telephone Consultative Committee Study Group XVIII (see next section for CCITT) are stated as follows:

- The main feature of an ISDN is to support the voice and nonvoice services in the same network where all signals are eventually transmitted in digital form. A key element of service integration for an ISDN is to provide a limited set of multipurpose user/network interface arrangements as well as a limited set of multipurpose ISDN services.
- ISDNs support a variety of applications including both switched or non-switched connections. Switched connections in an ISDN include both circuit-switched and packet-switched connections and their concatenations. New services introduced into an ISDN should be arranged to be compatible with 64 kbit/s switched digital connections.
- An ISDN will contain intelligence for the purpose of providing service features, maintenance, and network management functions. This intelligence may not be sufficient for some new services and may have to be supplemented by either additional intelligence within the network or possibly compatible intelligence in the user terminals.
- A layered protocol structure should be used for the specification of the access to an ISDN. Access from a user to ISDN resources may vary depending upon

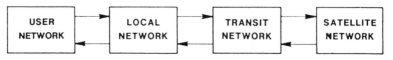

**Figure 7.2** Network elements of an ISDSN.

the services required and also upon the status of the implementation of national ISDNs. The reason for using a layered protocol will be mentioned later.

- It is recognized that ISDNs may be implemented in a variety of configurations according to specific national situations. This is a good far-reaching goal.

For ISDN realization the approaches are recommended as follows:

- ISDNs will be based on, and evolved from, digital telephone networks by progressively incorporating additional functions and network features including those of any other dedicated networks such as circuit switching and packet switching for data so as to provide for existing and new services.
- The transition from an existing network to a comprehensive ISDN may require a time period extended over one or more decades. During the transition period arrangements must be developed for interworking of services on ISDNs and on other networks.
- The evolution towards digital end-to-end connectivity will be obtained via plant and equipment used in the existing networks, such as digital transmission, time-division multiplex and/or space-division multiplex switching.
- In the early stages of the evolution of an ISDN, some interim user/network arrangements may need to be adopted in certain countries to facilitate early penetration of digital service capabilities. Some of the interim arrangements as hybrid access arrangements are recommended by the CCITT. Other arrangements corresponding to national variants may comply partially or fully with CCITT recommendations on ISDN.

The ISDN functions have been narrowed down to local connection-related functions such as user/network signaling, narrow band (64 kbit/s) circuit switching and nonswitched functions, common signaling channel, broadband-switched and nonswitched functions, and packet-switched functions. These functions need not be provided by separate networks but may be combined appropriately for a particular implementation.

Higher-layer functions implemented within (or associated with) an ISDN may be accessed by means of any of the above-mentioned functional entities. Functional entities could be implemented totally or partly within an ISDN as with dedicated networks. Both cases may provide all the ISDN services. Where packet services are provided, the local center will have to perform some functions related to packet handling, and may provide packet switching of data circuits for voice transmission.

Circuit switching and common channel signaling functions will be performed by ISDN local centers. Packet switching may be located in the exchanges of dedicated public networks for data transmission according to CCITT X series recommendations.

Described in the guidelines for design, both ISDSN and ISDN should have a universal appeal for most terrestrial networks. An ISDN should be able to interface and communicate with a number of standardized network structures. The integrated network itself should be designed with both flexibility for future expansion and simplicity to minimize cost and enhance ease of maintenance. Maximum freedom should be given to national or regional networks with a minimal set of standards for international connections. Although existing network parameters should be adopted whenever possible, the main emphasis of the integrated network architecture should be on long-lasting design. A good summary of ISDN as related to international standardization and the impact of new technology is provided by Hass [2].

## 7.2.1   CCITT and CCIR

The International Telegraph and Telephone Consultative Committee (CCITT) has concerned itself with the recommendations and standardizations of international telecommunication issues. The International Radio Consultative Committee (CCIR) has been concerned with the recommendations and the standardizations of spectrum utilization, transmission media, and satellite services. The work of these two committees has considerable influence on system planners, network designers, and equipment manufacturers. Operators are affected to a lesser degree. For this reason we briefly describe here the origin of the groups and organizations, their responsibilities and functions, and how ISDN issues are related to them.

CCITT came into existence in 1956 as a result of the merger of two organizations, CCIT (the Technical Consultative Committee for Telephone) and CCIF (Technical Consultative Committee for Telegraphy). CCIT and CCIF had been established in 1924 and 1925 respectively, and were a part of the organization called International Telecommunication Union (ITU), which includes CCIR (the Technical Consultative Committee for Radio), established in 1927. Although ITU evolved from work in the 1860s in western Europe, the organization became international only in 1934. The United States and Canada did not actually ratify the telephone and telegraph regulation until 1975. In its early days the functions of ITU were much simpler and were limited to maintaining, regulating, and coordinating international telecommunication capability. Unfortunately, standardization efforts were not as successful as expected; this was due to regional, economic, or political reasons. But the situation is changing, as we have witnessed during the last decade.

As part of the permanent structure of the ITU, CCITT has as its purpose to standardize, promote, and ensure the operation of international telecommunication systems. This is to be accomplished by establishing recommended standards for performance, interconnection, and maintenance of international networks for telephone, telegraph, and data communication. Certain tariff principles are also established as part of the CCITT activities. In 1985 there are 160 administrations (or governments), more than 50 recognized private operating agencies (representing governments in global telecommunication issues), 32 international organizations, and 137 scientific and industrial organizations belonging to CCITT.

Within the structure of CCITT there are a number of study groups that provide the work classificaiton or specilization for the purpose of developing the recommendations. The number of these study groups is not fixed but depends on the amount of work being done for a given study period. For example, during the study period of 1981 to 1984 there were 18 study groups. Corresponding group descriptions are shown in Table 7.1. The main responsibility of ISDN falls in Study Group XVIII, Digital Networks. Specific questions related to the various aspects of ISDN issues are listed in Table 7.2.

Among the many efforts put forward to define and establish ISDN the activities of CCITT Study Group XVIII have led the way in coordinating, reviewing, discussing and recommending suggestions from other groups and from worldwide organizations. The responsibilities associated with data communication networks fall upon Study Group VII, which has produced the well-known X-series recommendations from X.3 to X.75. These have paved the way for standardization of interfaces for public data networks and protocols for international packet-switched data networks. The results and continuing efforts of Study Group VII will provide significant building blocks for ISDN and in turn for ISDSN. Study groups I, II, and XVII also have participated and contributed to the advancement of ISDN realization. CCIR has concerned itself with frequency allocations, propagation phenomena, and broadcasting of sound and television services. In particular, CCIR Study Group IV, Fixed Satellite Service, has concerned itself

**Table 7.1**
Titles designated for the CCITT study groups .

| Group Number | Activity |
| --- | --- |
| I | Definition and operational aspects of telegraph and user services (facsimile, teletex, videotex, etc.) |
| II | Telephone operation and quality of service |
| III | General tariff principles |
| IV | Transmission maintenance of international lines, circuits and chains of circuits; maintenance of automatic and semi-automatic networks |
| V | Protection against dangers and disturbances of electromagnetic origin |
| VI | Protection and specifications of cable sheaths and poles |
| VII | Data communication networks |
| VIII and (XIV) | Terminal equipment for services (facsimile, teletex, videotex, etc.) |
| IX (and X) | Telegraph networks and terminal equipment |
| XI | Telephone switching and signaling |
| XII | Telephone transmission performance and local telephone networks |
| XV | Transmission systems |
| XVI | Telephone circuits |
| XVII | Data communication over the telephone network |
| XVIII | Digital networks |

with satellite orbiting and frequency sharing characteristics, methods of modulation and multiple access, antennas and earth stations. Recently, CCIR Study Group IV also became involved with digital baseband processing techniques. The classification of CCIR study group activity is shown in Table 7.3.

**Table 7.2**

Questions concerning ISDN under consideration by CCITT Study Group XVIII.

| Question Number | Activity |
|---|---|
| A/XVIII | General network aspects of an Integrated Services Digital Network (ISDN) |
| B/XVIII | General network performance aspects of Integrated Digital Networks |
| C/XVIII | Signalling for the ISDN |
| D/XVIII | Switching for the ISDN |
| E/XVIII | Availability for the ISDN |
| F/XVIII | Interfaces in digital networks |
| G/XVIII | Definition for digital networks |
| H/XVIII | Maintenance philosophy of the digital network |
| I/XVIII | Implementation of maintenance philosophy |
| J/XVIII | Synchronization in digital networks |
| K/XVIII | Encoding of speech and voice-band signals using methods other than PCM |
| L/XVIII | Characteristics for digital sections |
| M/XVIII | Characteristics for digital line systems on metallic-pair cables |
| N/XVIII | Characteristics for digital line systems on optical fibre cables |
| O/XVIII | Performance characteristics of PCM channels at audio frequencies |
| P/XVIII | Characteristics of PCM multiplexing equipment and other terminal equipments for voice frequencies |
| Q/XVIII | Characteristics of digital multiplex equipment and multiplexing arrangements for telephony and other signals |
| R/XVIII | Interworking between digital systems based on different standards |
| S/XVIII | Encoding/decoding of FDM groups, supergroups, etc. |
| T/XVIII | Interference from external sources to transmission systems and equipment |
| U/XVIII | Performance of transmultiplexers |
| V/XVIII | Characteristics of transmultiplexing equipments |
| X/XVIII | Digital routing of data links |
| Y/XVIII | Digital speech interpolation systems |
| Z/XVIII | Network aspects of existing and new levels in the digital hierarchy |
| AA/XVIII | User/network interface |

**Table 7.3**

CCIR Study Group activities.

| Group Number | Activity |
|---|---|
| I | Spectrum utilization and monitoring |
| II | Space research and radioastronomy |
| III | Fixed service at frequencies below about 30 MHz |
| IV-1 | Fixed-satellite service |
| IV/IX-2 | Frequency sharing and coordination between systems in the fixed-satellite service and radio-relay systems |
| V | Propagation in non-ionized media |
| VI | Propagation in ionized media |
| VII | Standard frequencies and time signals |
| VIII | Mobile services |
| IX-1 | Fixed service using radio-relay systems |
| X-1 | Broadcsting service (sound) |
| X/XI-2 | Broadcasting-satellite service (sound and television) |
| XI-1 | Broadcasting service (television) |
| XII | Transmission of sound broadcasting and television signals over long distances |

Because CCIR study groups are concerned with fixed and broadcasting satellite services their work is expected to have an impact on the standardization of ISDSN.

## 7.2.2 Transmission Characteristics

Based on present recommendations, the basic transmission characteristics of the ISDN consist of a set of basic transmission rates (higher transmission rates are integer multiples of lower rates), maximum transmission rates at the network interfaces, user-network access strategies, and a signaling frame structure. These issues are addressed in this section, while other network transmission characteristics such as performance and protocol will be discussed separately in Sections 7.2.3 to 7.2.5.

### 7.2.2.1 Channel Types

A channel is defined here as a specific portion of the information-carrying capacity of a network interface. A channel is specified by a specific transmission rate. With common characteristics, channels are classified by CCITT into six types depending on their designated usage and information-carrying mechanism. These channel types, denoted as A, B, C, D, E, and H, are described as follows:

The A-channel is reserved for conventional analog telephone channel communication.

The B-channel is a 64 kbps user information channel carrying either voice or data in digital form. In principle a user has complete freedom of usage of the B-channel for user information transmission provided that proper protocol has been established at the beginning of the transmission. The B-channel does not carry signaling information for circuit switching by ISDN. It may carry user information not recognized by CCITT.

The B-channel can be used to provide access to a variety of communication modes within the ISDN such as circuit- and package-switching modes. For circuit switching the ISDN can provide either a transparent end-to-end 64 kbit/s connection or a connection specifically suited to a particular service, such as a telephone, in which case the connection is not transparent. The B-channel will carry X.25 protocols for packet switching. Both circuit switching and packet switching modes of communication can provide temporary connections.

The C-channel, which may have different bit rates, is primarily intended for telemetry and packet-switched data. The C-channel uses layered protocols to establish communication, and may be mixed with signaling information for an analog channel.

The bit rate of the D-channel is 64 kbit/s, which is the basic transmission rate for digitized voice at present.

The D-channel, which may also have different bit rates, is used primarily for signaling information on circuit-switched option. Layered protocols and X.25 have been recommended for the D-channel. It has been suggested that the D-channel also include other information such as packet-switched data and the mixed analog channel capability of the C-channel. The application of D-channel protocols for packet-switching is presently under study by CCITT.

The E-channel is a 64 kbit/s channel intended to transmit signaling information for circuit-switched ISDN. In conjunction with the CCITT Signaling System No. 7 (to be described in Section 7.2.5), the E-channel may be used for multiplexing and providing an arrangement for multiple access interfaces. The network control procedures for the E-channel are compatible with the third layer protocols of the D-channel. The protocol layers will be discussed in Section 7.2.4.

The H-channel, for a positive integer $n > 1$, has a transmission rate of $384n$ kbit/s. This channel is intended for a variety of user information functions with bit rates greater than the 64 kbit/s of the B-channel. The H-channel is not expected to handle signaling information for circuit switching, which will instead be handled by a D-channel. The creation of the H-channel provides for wider services with much higher bit rates than the B-channel.

The B-channel is defined at 64 kbit/s, but CCITT also recognizes channels with speeds less than 64 kbit/s, such as 8, 16, and 32 kbit/s. A simple method that may be used for a network to transmit at lower speeds involves grouping the lower speeds of 8, 16, or 32 kbit/s, then multiplexing them to 64 kbit/s to form a B-channel.

*Example 7.1 [3]:*   The multiplexing methods adapting sub-B-channel rates of 8, 16, and 32 kbit/s to 64 kbit/s can be arranged as shown in Table 7.4. ISDN

**Table 7.4**
Examples of rate adaptation and multiplexing.

| Subchannel Rate (kbit/s) | Subchannel Information | Representation at 64 kbit/s |
|---|---|---|
| 32 | 01 01 01 . . . | 0 0 1 1 0 0 1 1 0 0 1 1 . . . |
| 16 | 01 0 . . . | 0 0 0 0 1 1 1 1 0 0 0 0 . . . |
| 8 | 01 . . . | 0 0 0 0 0 0 0 0 1 1 1 1 1 1 1 1 |
| 32 | a b c d e . . . | a m b n c o d p e q . . . |
| 32 | m n o p q . . . | |
| 16 | a b c . . . | aa mm bb nn cc oo . . . |
| 16 | m n o . . . | |
| 16 | a b c . . . | |
| 16 | m n o . . . | a t m u b v n w c x o y . . . |
| 32 | t u v w x y . . . | |
| 64 (2 × 16) | aa mm bb nn cc oo | a t m u b v n w c x o y . . . |
| 32 | t u v w x y . . . | |

is not expected to experience any problem with sub-B-channel rates.

Although we have described all the desirable transmission rates (called channel types by CCITT), we can summarize them into a single quantity as

$$R_T = 2^i \, 3^j \, 5^k, \text{ bit/s.} \qquad (7.1)$$

All the desirable transmission rates can be derived from (7.1) by simply specifying the exponents $i$, $j$, and $k$. For example, B-channel is obtained by letting $i = 9, j = 0, k = 3$. Kilobits can always be multiplied by $2^3 \times 5^3$. The coefficient 384 of the H-channel can be obtained by letting $i = 7, j = 1$, and $k = 0$. As the needs and technologies change the rates can also be changed and specified accordingly.

### 7.2.2.2  *Channel Structures*

Channel structures refer to the combinations of different types of channels. A channel structure defines the maximum digital information carrying capacity in terms of bit rate across an ISDN interface. At present the basic channel structure is composed of two B-channels and one D-channel, that is, 2B + D. The bit rate of the D-channel in this channel structure is 16 kbit/s. The B-channels may be used independently, i.e., in different connections at the same time. With the basic channel structure recommended by CCITT, two B-channels and one D-channel are always present at the ISDN user network physical interface. One or both B-channels, however, may not be supported by the network, and therefore the only basic access capabilities possible are 2B + D, B + D, or D.

The C-channel structure consists of one C-channel. The C-channel is associated with a conventional analog channel in a hybrid access arrangement. Depending on the type of hybrid access arrangement, the bit rate of the C-channel is 8 kbit/s or 16 kbit/s. The intermediate channel structures are to be determined. The primary rate multiplex channel structures are composed of B-channels and one D-channel. Since the bit rate of the D-channel is 64 kbit/s, at 1544 kbit/s primary rate, the channel structure is recommended as 23B + D. At 2048 kbit/s primary rate, the channel structure is recommended as 30B + D. With the primary rate multiplex channel structure, the designated number of B-channels is always present at the ISDN user/network physical interface. A medium rate is defined as mB + D with 2 < m < 24 or 30.

Alternatively, instead of a D-channel, an E-channel may be combined with a B-channel for more than one primary rate multiplex access interface. At 1544 kbit/s and 2048 kbit/s primary rates the corresponding channel structures recommended by CCITT are 23B + E and 30B + E, respectively.

*Example 7.2:*   Station or terminal equipment and network are denoted by TE and NT, respectively. Figure 7.3 illustrates a typical access arrangement and for this particular configuration it is not necessary to apply the same channel structure at reference points S and T. For example, basic channel structures may be used for interfaces located at reference point S. Either basic or primary rate multiplex or other channel structures may be used at interfaces located at reference point T.

Figure 7.4 illustrates a possible configuration for a variety of hybrid access arrangements. A hybrid access arrangement consists of a digital channel structure used in conjunction with an analog channel. A physical interface is shown at reference points S and T, where the C-channel structure or the basic channel structure may be used. In addition to the analog channel, the hybrid access arrangement includes one of the following digital access capabilities: C, D, B + D, or 2B + D. This example is proposed from reference [3].

The high-speed channel structure at present is composed of one H-channel and one D-channel, H + D. The bit rate in this channel structure equals that of the corresponding multiplex channel structure. The bit rate of the D-channel in this channel structure is $x$ kbit/s, the value of $x$ to be determined.

With the high-speed channel structure, one H-channel and one D-channel are always present at the ISDN user/network physical interface. The high-speed channel capacity is H + D.

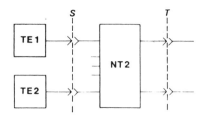

**Figure 7.3** Example of the reference configurations for CCITT ISDN user/network interfaces applied to a physical configuration employing multiple connections.

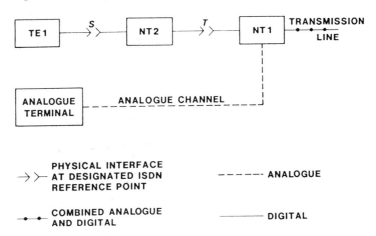

**Figure 7.4** Reference configuration for hybrid access arrangements for CCITT ISDN user/network interfaces.

### 7.2.2.3  Network Access

Close to the subject of network interface is network access. At this time CCITT supports two versions, the all digital 2B + D basic access and the ability to provide an analog channel A, together with the basic access having capabilities of D, B + D, or 2B + D options. The version that includes one or more of these options is called a hybrid access.

### 7.2.2.4  A Signaling Frame Structure

When the channel structure is 2B + D, the signaling frame structure is recommended, for example, as shown in Figure 7.5, where each frame consists of 48 bits. Since 2B + D has a data rate of $2 \times 64 + 64 = 192$ kbit/s, the 48-bit frame spans a duration of $48/192 = 250$ microseconds. The frame structures of Figure 7.5 are shown at interface reference points S and T. The top frame is from NT to TE and the bottom frame is from TE to NT. As shown, a time delay of two bits exists between received and transmitted frames. For transmission from NT to TE, the frame transmitted by NT contains an echo channel that is used to retransmit the D-channel bits received from the stations as E-channel bits. The D-channel (echo) is used for contention resolution. The last $L$ bits in the frame are used for channel transition balancing. An auxiliary framing $(F_A)$ bit is provided to enhance either the message security or reliability. For transmission from TE to NT, each frame consists of a framing signal, $F_A$/s (s for spare), and B1 and B2 channels. Multiframe can be obtained in either direction by counting the number of $F_A$ or $F_A$/s.

The various frame formats can be precisely structured when the architecture of an ISDN is determined.

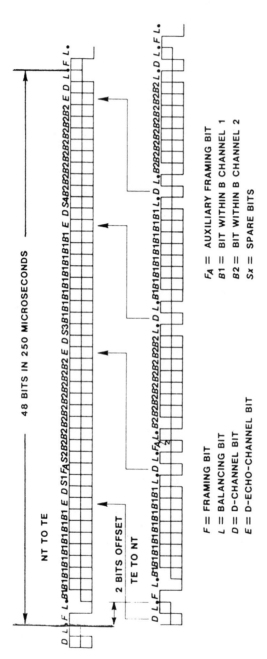

**Figure 7.5** Frame structure at reference points S and T [3].

### 7.2.3  Network Performance

The performance of a network may be evaluated in terms of network availability or throughput, transmission efficiency or message delay, reliability or network failure, and network degradation or error performance. Factors contributing to the performance of a network depend on specific network topology, traffic conditions, transmission mechanisms, buffer size, etc. Because we do not have a general configuration for the ISDSN at present, we cannot realistically address the above performance standards except to consider the criteria for network degradation, which may be discussed without a specific network architecture in mind. Therefore in this section we examine only what has been proposed as a possible standard of error performance for ISDN, analyzing these ideas and discussing their shortcomings. At the end of this section we propose a different approach, which seems to make better sense technically.

#### 7.2.3.1  Error Performance

For some time both CCIR and CCITT study groups have advocated error free second (EFS) as the error performance standard for all digital transmissions. Recently, the study groups have recognized the inadequacy of using EFS modifications and additions based on EFS.

The idea of error free second was originated by Mahoney [4], defining the percentage of EFS as

$$P(\text{EFS}) = (1 - P_e)^B \times 100 \qquad (7.2)$$

where $B$ is any data rate, $P_e$ is the bit error rate. Many contributions from CCITT, CCIR, and other organizations of the ITU have supported the use of EFS as a performance measure for ISDN. However, its usefulness as a performance measure is very much limited not only for digital satellite communication purposes but also for other digital communication channel description. Basically, the definition of EFS, as described in Annex 5 of CCIR Doc. 4/117, Doc. 9/80, Doc. CMTT/107, is restricted to a 25,000 Km hypothetical reference digital path of 64 kbps (see Section 7.2.3.2). As has already been noticed in Doc. 4/225, the validity of EFS was questioned. In particular, if EFS was to be adopted as the international standard, it would not provide enough information for possible applications of forward-error correction in ISDN and it does not bring forth the transmission channel error statistics that are required either to select an error-correcting code or to combine FEC with ARQ in digital networks.

There are other concerns with regard to the concept of EFS:

**Accuracy.**  EFS represents only one statistical parameter of a transmission channel. Thus EFS is not sufficient to reflect the true channel error condition because it cannot give the burst errors information for a digital transmission channel.

*Example 7.3:*  Assume that there are $n$ digits per second on the average with $e$ errors per $n$ digits. Assume a 3-second interval is under observation. The first interval consists of a burst of $e$ errors. The second interval consists of evenly spaced errors within $n$ digits. The third interval consists of two bursts each with $e/2$ errors. The average bit error rate is $P_e = e/n$. As long as the ratio $e/n$ has the same value, P(EFS) yields the same result. Clearly, each 1-second interval has $e$ bit errors. Each interval meets the same bit error of rate $P_e = e/n$. Each 1-second interval thus results in the same value of P(EFS), assuming the signaling rate $B$ does not change. But the three distinct error patterns, as described above have not been reflected by P(EFS). Thus, we may conclude that any definition, concept, or criterion derived or based upon an averaged bit error rate cannot describe a channel with burst errors. What is needed is error distribution statistics as a function of time in order to truly represent channel error behavior.

***Speed.***    As shown in (7.2) P(EFS) is exponentially related to the data rate $B$. With the definition of EFS, every time a system is designed for a different speed a new value of P(EFS) needs to be specified. But for the purpose of establishing a criterion, it is not desirable to introduce another independent performance factor.

***Timing.***    The long- or short-term mean bit error ratios, clustering index, block error rates, or their combinations as described in the previous CCITT contributions are used in channel statistics. Unfortunately, for time-invariant channels, as most channels are, including satellite channels, the mean bit error rate is independent of the time. From sampling theory what really counts is the confidence interval of an established bit error rate.

***Detectability.***    The concept of EFS was primarily suggested for ARQ. Even for ARQ the percentage of P(EFS) cannot guarantee the channel reliability as intended, i.e., there exist error patterns that can satisfy P(EFS), but with undetectable error blocks.

*Example 7.4:*    Assume a 3-bit all-zero message sequence is encoded by a (7,3,3) single error-correcting code for ARQ transmission. The result is a 7-bit all-zero coded sequence. The operation in ARQ detection is a simple parity check matrix multiplication. If the channel has an error pattern of 0110011, the detector performs as

$$[0110011] \cdot \begin{bmatrix} 0 & 0 & 1 \\ 0 & 1 & 0 \\ 0 & 1 & 1 \\ 1 & 0 & 0 \\ 1 & 0 & 1 \\ 1 & 1 & 0 \\ 1 & 1 & 1 \end{bmatrix} = [000]$$

which indicates no error. In fact, four errors exist but the percentage of P(EFS) cannot evaluate or determine such a channel condition.

Summarizing, the concept of EFS alone has a limited scope in general network applications. It is inaccurate in transmission channel description and provides insufficient and unnecessary parameters to the real objective as a universal error performance criterion.

A summary of bit and block error probabilities, error free seconds, and error free 100-ms intervals is provided by the U.S. contribution to the CCITT Study Group XVIII on the effect of clustering on the selection of error probability parameters. This document (see [3]) suggests two parameters, i.e., bit error rate and a time-based error probability such as EFS or EF100ms. Unfortunately, neither block error probability or EFS nor the combination of bit error rate and EFS solves the basic problem, the reasons being that block error rate depends on block length and error free second depends on bit error rate, and the conditional bit error probability cannot reflect how often such error clusters occur (see Problem 3).

### 7.2.3.2 Hypothetical Reference Connections

The need to support performance objectives such as error rate and path delay between various parts of a network leads to the idea of hypothetical reference connections (HRXs), where transmission channel performances are expressed in partitioned hypothetical distances.

HRX is based on a near worst international connection that could be experienced between two network users. The idea was originally used on terrestrial facilities and it is under consideration for ISDN including satellites. The error allocation of HRX is distance dependent. For uniform switching centers of simple telephone networks HRX has been used as an error performance standard with justification. For future general networks, however, with integrated services including satellites the recommendation of HRX as the performance measure is not clearly justifiable. It is important to point out that laws of addition are needed for digital circuits which allow one to translate standards specified for hypothetical reference circuits to those for any given actual circuit. These exist for analog circuits, but not for digital circuits.

An example of HRX from COM XVIII–No. R11 [3] is shown in Figure 7.6, where the distances are partitioned according to national and international boundaries. As shown, the percentages of degradation are allocated on the basis of the corresponding hypothetical distances in kilometers.

As another example of the suggested use of HRXs for ISDN performance measures by Study Group XVIII, Contribution 164 (March 1983) lists a number of national networks, international transit centers, and three types of satellite systems, as shown in Figure 7.7.

Figure 7.7 does not show the number of satellite hops in any connection. The figure only indicates satellites as part of the HRX. $B_1$, $B_2$ and $F_1$, $F_2$ are the local exchanges; $D_1$, $D_2$ and $E_1$, $E_2$ are international satellite exchanges and $C_1$, $C_2$ are national satellite exchanges.

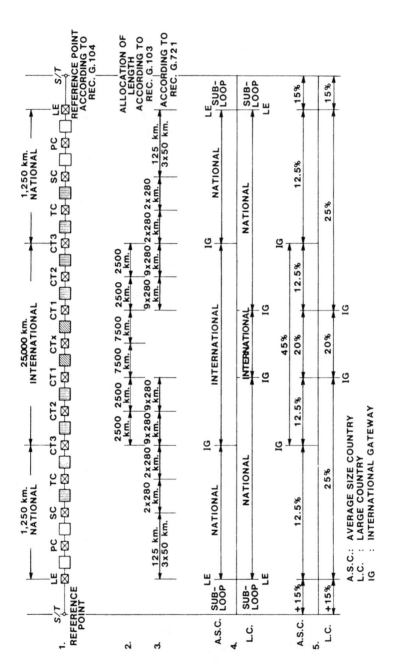

**Figure 7.6** Percentage of transmission channel degradation based on hypothetical distance.

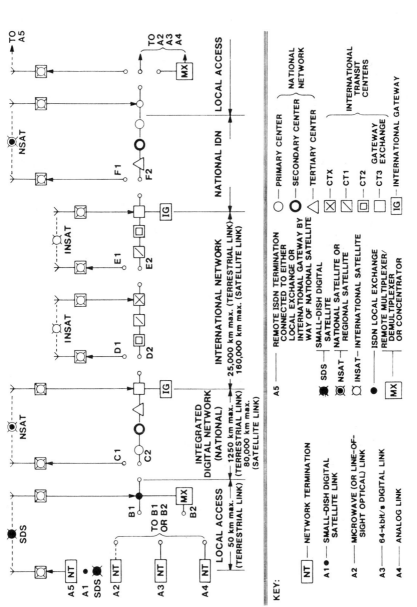

**Figure 7.7** Hypothetical distance measures including satellite networks.

Based on the above discussions we cannot endorse or recommend using either physical or hypothetical distance as a transmission channel performance measure. The distance measurement does not and cannot reflect the true and more complex transmission channel conditions. It will be regrettable if CCITT adopts HRX as the error performance standard for ISDN, which in turn is part of the ISDSN. We will next show that a simpler, more natural and accurate method may be used for the error performance criteria of ISDN and ISDSN.

### 7.2.3.3    Error Burst Distribution

In general, an error sequence $E(D)$ of length $n$ may be represented in polynomial form in $D$-domain (delay operator).

$$E(D) = e_0 + e_1 D + e_2 D^2 + \ldots + e_{n-1} D^{n-1}. \tag{7.3}$$

For binary transmission the error position, $e_i = 1$ or $0$ only, for $i = 0, 1, \ldots,$ $n - 1$. $e_j = 1$ means that the $j$th bit is in error, and $e_k = 0$ means that the $k$th bit is not in error. A sequence of consecutive 1's between two 0's is referred to as error length, which is the number of consecutive errors between two error free digits. Let this number be denoted as $N_e$, where $N_e$ is a random variable. The sum of all the $N_e$'s equals the total number of errors in the sequence. For simplicity, the error sequence can be represented by the corresponding binary sequence form.

*Example 7.5:*  Let $n = 11$; assume

$$E(D) = 1 + D + D^3 + D^6 + D^8 + D^9 + D^{10}$$

$$= 1\,1\,0\,1\,0\,0\,1\,0\,1\,1\,1$$

$$= 1^2\,0\,1\,0^2\,1\,0\,1^3.$$

We then have $N_1 = 2, N_2 = N_3 = 1$. The subscript of $N$ equals the exponent in the corresponding binary error sequence. The total number of errors is

$$2N_1 + N_2 + N_3 = 2 \times 1 + 2 + 3 = 7.$$

Extending this example to a general case, we can define the error burst probability distribution function (EBD) as

$$\Pr(N_e \leq e) = \sum_{i=1}^{e} P(e=1) \tag{7.4}$$

which is determined by menas of measurement of transmission channel statistics. This function may be plotted out with $\Pr(N_e = e)$ versus $e$. All error information of a channel can be obtained from such a curve. When a channel is burstless, a special case of the EBD reduces to a single average value, which is the bit error rate. Thus, with EBD we do not need to specify bit error rate.

*Example 7.6:* Let $n = 20$, and assume

$$E(D) = 1\ 1\ 0\ 1\ 0\ 1\ 1\ 1\ 0\ 1\ 1\ 1\ 1\ 0\ 1\ 0\ 1\ 1\ 0\ 1,\ \text{then}$$

$$P(e=1) = 3/20,\ P(e=2) = 2/20,\ P(e=3) = 1/20,\ \text{and}$$

$$P(e=4) = 1/20.\ \text{Pr(EBD)} = (3+2+1+1)/20 = 7/20.$$

The corresponding EBD plot is shown in Figure 7.8. The idea of EBD is similar to the run length statistics used in Chapter 5 to describe the performance of error control decoders.

From the above general analysis and examples we see the merit of the error burst distribution (EBD) as a possible transmission channel measure reference. From the EBD we can obtain the bit error rate (BER). Though BER information is contained in EBD, EBD information cannot be obtained from BER. Any other digital error information including EFS can be identified from EBD.

With EBD, channels for binary transmission can be described more accurately than EFS. EBD can be applied in general and is independent of time, distance, speed, and other factors. EBD can be applied to ARQ, FEC, hybrid ARQ/FEC, or any other scheme to be developed in the future. EBD can be used for satellite as well as terrestrial network. If EBD is used as a reference measure of error performance, it will not only have wider impact in terms of applications but longer-lasting values.

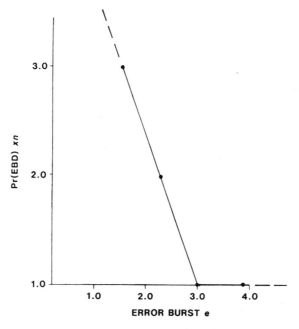

**Figure 7.8** EBD plt for $E(D) = 1^2 0101^3 01^4 010101^2 01$.

## 7.2.4 ISO's OSI Protocols

The International Organization for Standardization (ISO), representing more than 80 countries, exists to develop standards to facilitate the international exchange of products, information systems, and services. ISO is responsible and has established guidelines and standards for data communication processors. One of the reference models developed by ISO for a data communication standard is the basic reference model for Open System Interconnection (OSI) [5]. The openness of the system refers to the set of network systems, segments, or entities that communicate under the OSI recommendations.

As it is stated in the draft recommendation, the purpose of OSI is to provide a common base for the coordination of standards for system interconnections. A number of contributions from CCIR and CCITT have recommended OSI for ISDN. Thus, we will first discuss the OSI recommendations, next with examples see how OSI may be applied to ISDN, and finally point out a disadvantage of such a direct adaptation to digital satellite communication.

A protocol is a set of necessary procedures or a format that establishes the timing, instructs the processors, and recognizes the messages. The protocol of OSI consists of several specific layers which represent the component processes. The collection of these component processes defines the overall data communication protocol. Each layer has a set of elements associated with the users in that layer. The elements in the same layer are referred to as peers. It takes at least two layers in order to establish the most simple communication. The seven layers of OSI are described as follows:

The first layer is called the physical layer. It provides mechanical, electrical, functional, and algorithmic properties to activate, maintain, and deactivate transmission interface connections. The second layer is the data link layer. It provides functional and systematic procedures to establish, maintain, and release data link connections among different segments in the network. The main objective of the data link layer is to provide data transmission services, with error detection or correction from the physical layer to the network layer. This second layer also maps data units into data link protocols for transmission.

The basic function of the third layer, the network layer, is to transfer all data from the transport layer. The network layer establishes, maintains, and terminates connections in accordance with the user's instructions. The fourth layer, the transport layer, exists to provide transfer data between session-to-session messages among the network segments. In addition, the protocols in this layer have end-to-end significance. Connections in the transport layer represent a two-way simultaneous data path between a pair of transport addresses. The transport layer contains establishment, transfer, and termination phases.

The fifth layer, the session layer, manages, organizes, and synchronizes data exchanges among the network segments. In this layer services are provided to establish a session connection between two network (presentation) segments. The sixth layer, the presentation layer, presents information to communicating (application) segments and resolves syntax problems. The seventh layer, the application layer, selects appropriate services to be supplied from the lower layers based on requests from network users.

In actuality, the first three layers of the CCITT recommendation X.25 are similar to the first three layers of OSI, i.e., the physical, data link, and network layers. The data link layer is supported by the already established high-level data link control (HDLC) standards and the presentation layer relates to CCITT Recommendations X.3, X.28, and X.29 [6]. The other layers' standards have not yet been established.

A summary of the services provided and layer functions of the OSI reference model is presented in Table 7.5. For a more detailed description of the items in the seven layers refer to Appendix A in the book by Meijer and Peeters[7], or to Appendix A in the paper on the differences of OSI and SNA (Systems Network Architecture) by Rutledge [8].

**Table 7.5**
Summary of the OSI reference model.

1. Physical Layer

| Services | Functions |
|---|---|
| Physical connections | Management of interface with transmission |
| Physical service data units | medium |
| Data circuit identification | Interface control |
| Sequencing | CCITT X.21, V.24, V.35 |
| Fault condition notification | |
| Quality of service parameters | |

2. Data Link Layer

| Services | Functions |
|---|---|
| Data link service data units | Data link connection activation and |
| Data link connection endpoint identifiers | deactivation synchronization |
| Sequencing | Sequence control, data link multiplexing |
| Error notification | Error detection |
| Quality of service parameters | Error recovery |
| | Flow control |
| | Identification and parameter exchange with |
| | peer elements |
| | Control of data circuit interconnection |
| | Layer management |
| | Data to data link protocol mapping |

3. Network Layer

| Services | Functions |
|---|---|
| Provide data transport between layer 4 | Network addressing and identification |
| elements | CCITT X.25 |
| Network addresses | Recommendation |
| Network connections | Routing and relaying |
| Network connection endpoint identifiers | Network connections and multiplexing |

## Table 7.5
### (continued)

Network service data unit transfer          Segmenting and blocking
Quality of service parameters               Error detection
Error notification                          Error recovery
Sequencing                                  Sequencing
Flow control                                Flow control
Expedited network service data unit transfer Expedited data transfer
Release services                            Service selection
                                            Layer management

### 4. Transport Layer

| Services | Functions |
|---|---|
| Identification | Establishment, maintenance, and release of |
| — Transport addresses | connections but still maintain a 2-way path |
| — Transport connections | between a pair of layer 4 users (or layer 5 |
| — Transport connection endpoint identifiers | elements) |
| Establishment services | Data transfer between pairs of layer 5 |
| — Transport connection establishment | elements |
| — Class of service selection | — Sequencing |
| Data transfer services | — Blocking |
| — Transport service data unit | — Concatenation |
| — Expedited transport service data unit | — Flow control |
| Transport connection release | — Error detection |
|  | — Error recovery |
|  | — Expedited data transfer |
|  | — Transport service data unit delimiting |
|  | — Transport connection identification |
|  | Layer management |

### 5. Session Layer

| Services | Functions |
|---|---|
| Session connection establishment and release | Control of data flow between the element |
| Normal data exchange | pairs of layer 6 |
| Quarantine service | Mapping of session connection onto transport |
| Expedited data exchange | connection |
| Interaction management | Expedited data transfer |
| Session connection synchronization | Session connection recovery and release |
|  | Layer management |

### 6. Presentation Layer

| Services | Functions |
|---|---|
| Data transformation | Resolving requirement of information |
| Data formatting | presentation |
| Syntax selection | Session establishment and termination |
| Presentation connections | requests |
|  | Presentation image negotiation and |
|  | renegotiation |
|  | Data transformation and formatting |
|  | Addressing and multiplexing |
|  | Layer management |

**Table 7.5**

(continued)

7. Application Layer

| Services | Functions |
|---|---|
| User identification | Systems management |
| Services to application processes | Application management |
| Order establishment | Activation |
| Data control procedures | Error control |
| Information transfer | System access |

### 7.2.5 Protocol Reference Model—The Applications of OSI to ISDN

The ISDN protocol reference model is the basis for a set of ISDN protocol configurations that describe the interchange of information between the user and the network elements, between network elements of the same network, and between different networks. This model is based on the layered structures and signaling system (SS) No. 7 for both circuit switching and packet switching with type B and D channels. The layered structure is based on the reference model of the Open Systems Interconnection (OSI) described in Section 7.2.4. Type B and D channels are defined in Section 7.2.2. We now describe in the following what SS No. 7 is.

Signaling System (SS) No. 7 evolves from Signaling System No. 1, developed for the first telephone signaling system in the 1930's. It is the latest common channel signaling protocol switching system under development by CCITT Study Groups VI, VII, and XVIII. Although Signaling System No. 7 is specified primarily for 64 kbit/s operation, it can also be used as other bit rates, in particular at 4.8 kbit/s. The protocol structure of SS No. 7 is layered and has a message capacity of 58 bytes. For the fourth layer the message capacity is 256 bytes. The number of circuits that can be signaled in SS No. 7 is 40,000.

In general with layered structures a possible ISDN protocol reference model may be constructed by taking into consideration three types of information in the network. First there are the messages, including digitized voice, data, and other services, which can be handled by ISDN as described in Section 7.1. Let the message information be denoted by U and a seven-layer structure be used to describe the various types of U messages. Second, let all control information, such as access control, and network usage control be denoted by the letter S. The in-band signaling may be identified by X.21, X.25, and the out-of-band signaling may be identified with the common channel Signaling System No. 7. The S information is represented by seven OSI layers. Third, all the local station management aspects associated with the transfer of user information and the signaling and control information, designed by M, are responsibile for the management functions associated with U and S. Such functions include traffic control and network failure monitoring activities. Most interactions between U and S need to go through M, the exceptions being in the higher layers, as suggested in [3].

From the three types of information defined above we can structure a fundamental protocol building block as shown in Figure 7.9. In some applications, such as the basic access, layer 1 can be shared by U and S in the fundamental building block, while for PABX, private automatic branch exchange, U and S may be accessed separately. Information transfer between one fundamental building block and another takes place over the physical media attached to U and S. Interactions at the top face of the building block provide system (or block) management, user and signaling functions; the application processes where the layer protocols are associated with U and S as shown in Figure 7.10 are not included.

With the fundamental protocol building blocks just described we can develop a general ISDN protocol reference model as shown in Figure 7.11, where TE, S/T denote station or terminal equipment and signaling transfer, respectively. S/T refers to the user/network interface point. For the reference model within the network segment different media connections may be used between two inner building blocks. For network elements that perform information exchange and transfer functions U is represented by the lower three levels. For network elements that perform end-system functions, such as message processing, all the seven layers are used.

*Example 7.7:* The ISDN protocol reference model, based on the multilayer protocol building blocks, can be applied to the circuit-switched connection protocols as shown in Figure 7.12. In this application all the seven-layer protocols are used for all S and U information of the station equipment, and only layer 1 is used for all U information within the network. For packet-switched connection through the D-channel the ISDN protocol model is shown in Figure 7.13, where all U and S information outside the network is represented by seven layers, but U information within the network utilizes one and three layers as shown. The protocol reference model for packet-switched connection through the B-channel model is shown in Figure 7.14, which illustrates the difference in the number of layers within and outside the network. We can easily observe that the signal

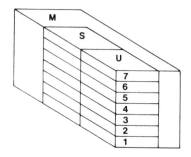

**Figure 7.9** A basic protocol building block (CCITT); M: management, S: control, U: message.

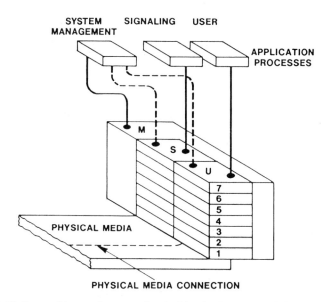

**Figure 7.10** External interactions associated with a fundamental building block (CCITT).

for packet-switched connection through the B-channel is almost one fundamental protocol building block less than through the D-channel. For mixed circuit-switched and packet-switched connections of U information the network aspect of ISDN protocol reference configuration is shown in Figure 7.15, where $U_c$ and $U_p$ denote the respective circuit-switched and packet-switched connections of the U information building blocks.

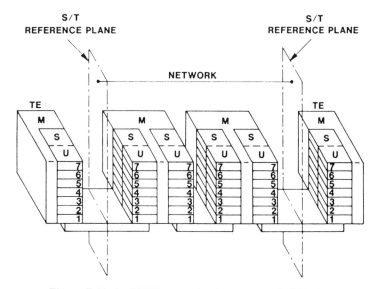

**Figure 7.11** An ISDN protocol reference model (CCITT).

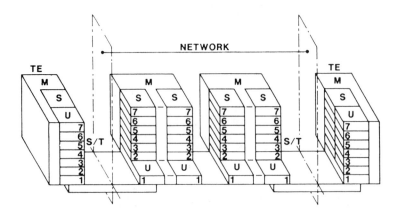

**Figure 7.12** CCITT ISDN protocol reference configuration for circuit-switched connection.

In the following, we will provide two more examples to show how the various ISDN protocols may be configured utilizing the layered structures. Examples 7.7 to 7.9 are based on the recommendations from Reference [3].

*Example 7.8:* COMXVIII-No. R-8 of CCITT suggested a network configuration and protocol for circuit switching as shown in Figure 7.16, which includes provision for the conversion of signaling information in the D-channel to the ISDN user, in this case, Signaling System No. 7. The suggested mapping of the signaling information at level 3 of the D-channel to the user part of Signaling System No. 7 is shown in Figure 7.17.

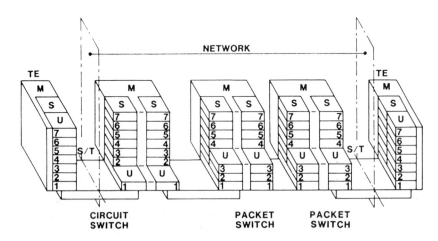

**Figure 7.13** CCITT ISDN protocol reference configuration for packet-switched connection through D channel.

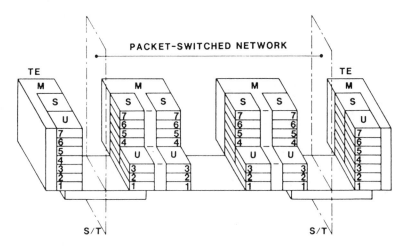

**Figure 7.14** CCITT ISDN protocol reference model for packet-switched connection through B channel.

In Figure 7.17, UP on the top (level 4) of SS No. 7 denotes the user's part, and the MTP in the lower levels denotes the message transfer part. Level 4 in SS No. 7 corresponds to layers 4 to 7 of the OSI, thus accommodating both telephone and data users.

The three levels of MTP closely track the first three layers of the OSI Reference Model. These three levels are the signaling data link, the signaling link function, and the signaling network functions. Level 1 accommodates a variety of data transmission facilities and variable transmission rates. Similar to X.25, level 2 provides error and sequence control, failure detection, and link recovery. Level 3 of SS No. 7 provides message handling and routing and network management. Level 3 also defines but does not connect a network protocol.

The network configurations and protocols for packet switching using both B and D-channels are shown in Figures 7.18 and 7.19, respectively. In each figure

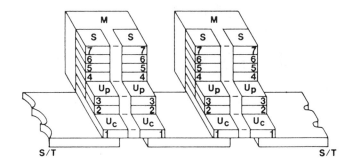

**Figure 7.15** CCITT ISDN protocol reference configuration; multi-media application: circuit-switched connection type Uc packet-switched connection type Up.

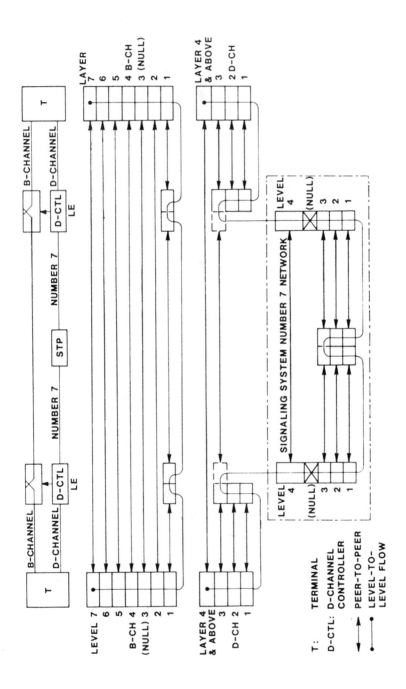

**Figure 7.16** Network configuration and protocols for ciruit switching (CCITT).

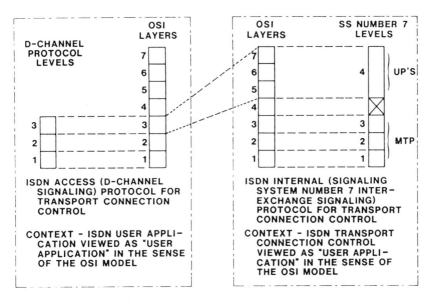

**Figure 7.17** Illustration of correspondence between D-Channel signaling protocol levels, Signaling System No. 7 signaling levels, and OSI layers for the case of transport connection control signaling (CCITT). The figure is not applicable to the possible case of transport of user data via the D-Channel and Signaling System No. 7 signaling network.

the terminals, controllers, switching facilities, and level-to-level flows are indicated.

For the original system interconnections of the OSI the systems are referred to as computers, peripherals, and stations. When it applies to the diversified services for digital satellite communications the OSI may become too rigid to be efficient for use in the space segment, for example, in on-board processing.

*Example 7.9:* For various services that can be supported by ISDN, the protocol models may be arranged from the OSI fundamental layers, as shown in CCITT COMXVIII-No. R15. A 64 kbit/s circuit-switched transparent service protocol model is shown in Figure 7.20(a). For this type of service a user may choose any set of higher layer (layer 4 to 7) protocols for communication because ISDN does not provide compatibility at higher layers (4–7) between users. The protocol model for access and interaction with information storage and processing facilities is shown in Figure 7.20(b). In this case higher layers of the three B-channels are used, and the connection is provided between the two S/T reference points, or between an S/T reference point and a high layer function element. For telecommunication services the protocol models of the 64 kbit/s circuit-switched teletex and 64 kbit/s circuit-switched videotex are shown in Figure 7.21(a) and (b), respectively. The difference lies in the layer level of TE1 and the videotex

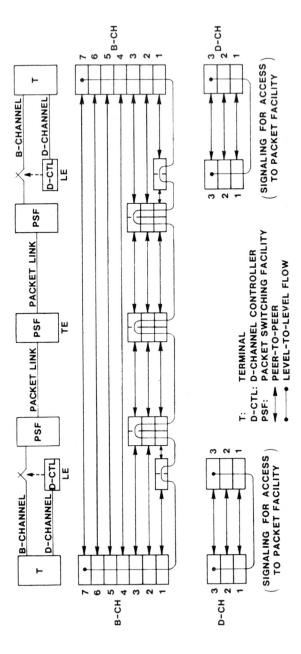

**Figure 7.18** Network configuration and protocols for packet switching using B-Channel with circuit-switched access (CCITT).

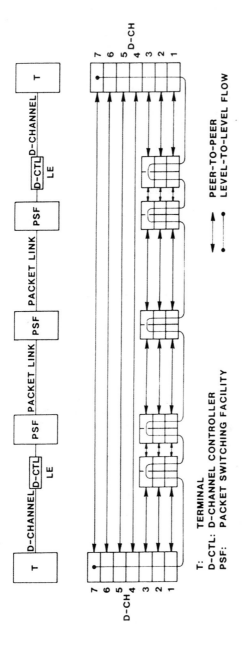

**Figure 7.19** Network configuration and protocols for packet switching using D-Channel (CCITT).

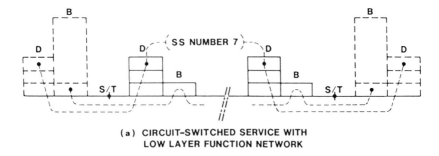

(a) CIRCUIT–SWITCHED SERVICE WITH
LOW LAYER FUNCTION NETWORK

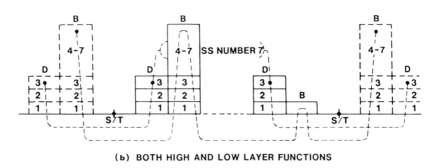

(b) BOTH HIGH AND LOW LAYER FUNCTIONS

**Figure 7.20** Protocol models for 64 kbit/s circuit-switched CCITT ISDN.

center. We note here that for teletex transmission packet-switched connections are feasible. We also note that all the protocol models in this example have assumed signaling through Signaling System No. 7.

## 7.3   THE INTELSAT TDMA NETWORK AND PROTOCOL

In Section 7.1 we defined the integrated service digital satellite network, outlined the architecture of such a network, provided the criteria and guidelines for the realization of such a network, and hinted at the difficulties and challenges in designing such a network. In Section 7.2 we supported the international standardization of such a network, described the most recent trend of thought with regard to the elements of such a network, advocated some of the concepts, critically disagreed with others, and gave the reasons why. In the next seven sections (Sections 7.3 to 7.9) we present in detail the inner workings of the INTELSAT TDMA network and protocol procedures.

Under the subject of multiple access, the INTELSAT 120.0 Mbps TDMA system was discussed in Chapter 3 of Volume I. For an introduction and description of the system the reader is referred to Section 3.12 of that volume.

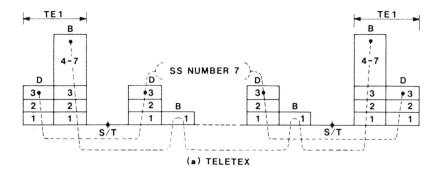

(a) TELETEX

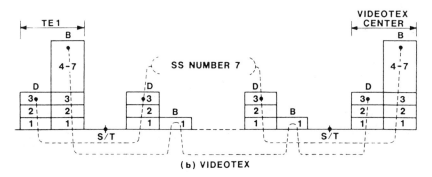

(b) VIDEOTEX

**Figure 7.21** Protocol models for 64 kbit/s circuit-switched CCITT ISDN (communication services).

### 7.3.1 Network Elements

The INTELSAT TDMA network consists of four reference stations, owned and operated by INTELSAT, two in each hemisphere covered by Zone beam, and up to 116 traffic stations owned and operated by the participating signatories. The exact number of traffic stations depends on a particular network connection. Three examples will be given in Section 7.9, one of which is the actual IN-TELSAT major path-2, Atlantic Ocean region eight-traffic station TDMA network operation.

The four reference stations can provide reference bursts for all TDMA Hemi and Zone beam transponders. Thus the reference stations operate in pairs, one pair in each coverage area. Within a coverage area, one reference station will be in a master primary mode and the other in a secondary mode. In the opposite coverage area, one reference station will be in a primary mode and the other in a secondary mode. The status of the reference stations is to be initially determined manually and thereafter changed in accordance with the reference station change-over procedures and the reference station role matrix, described in Section 7.3.4.3.

The network elements are the satellite and its connections, the reference stations, and the set of traffic stations. The reference stations are designed to start up the network, initiate acquisition and synchronization procedures, provide network management, coordinate status changes, monitor failures, and maintain network operations. All these functions are performed through the proper protocols, to be described below.

All the protocol information is transmitted through the service channel (SC), which conveys 32-bit messages in every multiframe available from both the reference burst and the traffic burst, and the control delay channel (CDC), which also conveys 32-bit-per-multiframe messages, but available only from the reference burst. The CDC contains a 2-bit control code or 2-bit reference station status code. The representative codes are listed in Table 7.6. The contents of the table will be described later.

### 7.3.2   Network Connections

The following four basic configurations of beam connections are represented in the INTELSAT TDMA network:

1. The first type of beam connection is shown in Figure 7.22, in which the reference stations can receive their own messages. This configuration corresponds to the global beam connectivity. Because each reference station is capable of receiving any TDMA burst from any station, various network procedures are much simplified.
2. The second type of connection considered is shown in Figure 7.23, in which the pair of reference stations in the same hemisphere (either East or West) can receive messages from the reference stations of the other hemisphere, but the pair of reference stations cannot receive its own messages. This type of connection is also referred to as hemispheric crossing only.
3. The third type of connection is shown in Figure 7.24, in which in addition to the hemispheric crossing connection described in Figure 7.23 there can be a loop-back connection in either hemisphere so that bursts from its own hemispheric stations can be monitored.

**Table 7.6**

Status and control codes.

| Word | Status Code | Control Code |
| --- | --- | --- |
| 00 | Inoperative Reference Station Code | Do Not Transmit (DNTX) Code |
| 01 | Secondary Reference Burst (SRB) Code | Initial Acquisition Phase 1 (IAP1) Code |
| 10 | Primary Reference Burst (PRB) Code | Initial Acquisition Phase 2 (IAP2) Code |
| 11 | Master Primary Reference Burst (PRB) Code | Synchronization (SYNC) Code |

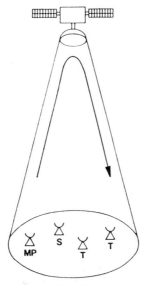

MP: MASTER PRIMARY REFERENCE STATION (OR MPR)
S: SECONDARY REFERENCE STATION (OR SR)
T: TRAFFIC STATION

**Figure 7.22** Network connection (1).

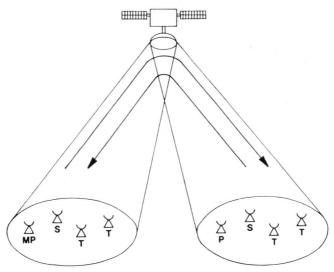

MP: MASTER PRIMARY REFERENCE STATION
P: PRIMARY REFERENCE STATION
S: SECONDARY REFERENCE STATION
T: TRAFFIC STATION

**Figure 7.23** Network connection (2).

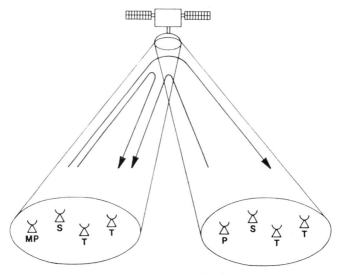

MP:  MASTER PRIMARY REFERENCE STATION
P:   PRIMARY REFERENCE STATION
S:   SECONDARY REFERENCE STATION
T:   TRAFFIC STATION

**Figure 7.24** Network connection (3).

4. The fourth type of network connection is shown in Figure 7.25, in which reference stations in one hemisphere can receive their own messages as well as those transmitted by the two reference stations in the other hemisphere. This connection is the complete configuration including both hemispheric crossings and loop-backs.

Each basic connection may consist of a number of variations. For example, three possible network connections shown in Figure 7.26 can all be reduced to the second basic type of configuration. All four basic configurations are summarized in Figure 7.27 in terms of the reference stations, denoted as P for primary, MP for master primary, S for secondary, and T for traffic stations, and O for loop-back configuration in the E for East and W for West hemispheres. The numbers 1,2,3,4 in Figure 7.27 represent the types of stations and network beam coverages. Most of the descriptions are based on [18].

Since the network applies to stations in the East and West zones as well as East and West hemispheres, the above basic connectivities apply to zone beam also. The network beam connections can thus be represented by a simple 4 × 4 beam connection matrix, as shown in Figure 7.28, where a "1" indicates there is a connection and a "0" indicates otherwise. Because two channels are used for the TDMA operation, each channel exhibits such a matrix connection. Thus for all the possible crossing and loop-back connectivities and zone configurations, eight transponders are needed with each transponder-per-beam connection.

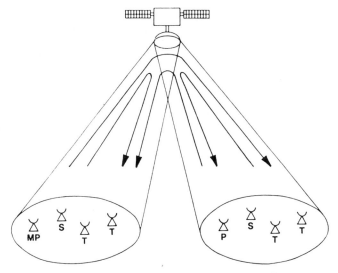

MP:   MASTER PRIMARY REFERENCE STATION
P:    PRIMARY REFERENCE STATION
S:    SECONDARY REFERENCE STATION
T:    TRAFFIC STATION

**Figure 7.25** Network connection (4).

In terms of the set of network stations, the beam coverage may be listed in discrete form similar to that shown in Figure 7.28 in which "0" denotes no coverage and "1" denotes that the station is covered by the corresponding beam. If there are $B$ beam coverages and $S$ stations in the network, the size of the beam coverage matrix is $B \times S$. For the 4 beam, 16 stations and 2-channel TDMA operation the matrix is shown in Figure 7.29. Such binary matrix representation can easily keep track of the beam connectivity and stations used as input format for the network simulation.

The relation between receive (multi)frame timing and reception time of unique words of RB1 and RB2 is shown in Figure 7.30. $T1_i$ and $T2_i$ represent the offset time required in order to derive the start of receive frame (SORF) from the reception time of the unique word (UW) of RB1 and RB2 respectively in transponder $i$. In the first frame of the multiframe, the start of receive frame is also referred to as the start of receive multiframe (SORMF).

The hierarchy of the frame structure is shown in Figure 3.30 of Volume I, where F denotes frame, MF denotes multiframe, CF denotes control frame, SF denotes superframe, RB denotes reference burst, UW denotes unique word, and SOTF denotes the start of transmit frame. Each reference station can transmit up to four reference bursts per frame.

As mentioned, for each CDC the stations are addressed sequentially in a 32 MF (or a CF) cycle. Each MF of the CF is dedicated to the distribution of information (a CDC message) to one of the controlled stations. This MF is

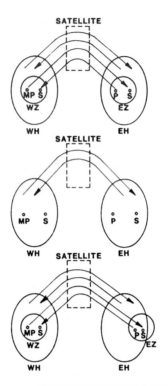

SATELLITE
MP
S
WZ
WH
P
S
EZ
EH
SATELLITE
MP
S
WH
P
S
EH
SATELLITE
MP
S
WZ
WH
P
S
EZ
EH
WH: WEST HEMI
EH: EAST HEMI
WZ: WEST ZONE
EZ: EAST ZONE
MP: MASTER PRIMARY REFERENCE STATION
P: PRIMARY REFERENCE STATION
S: SECONDARY REFERENCE STATION

**Figure 7.26** Network configurations reduced to connection (2).

| | $P_W$ | $S_W$ | $T_W$ | $O_W$ |
|---|---|---|---|---|
| MP | 2 | 2 | 2 | 1 |
| $S_E$ | 2 | 2 | 2 | 1 |
| $T_E$ | 2 | 2 | 2 | 1 |
| $O_E$ | 1 | 1 | 1 | |

**Figure 7.27** Summary of beam connectivities.

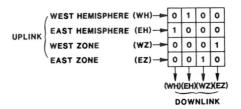

**Figure 7.28** Beam connection matrix.

referred to as the control multiframe (CMF) for the controlled station, and the CDC message includes control code and transmit delay $D_n$. Each controlled station derives its start of transmit multiframe (SOTMF) by appending the transmit delay to the SORMF. The derivation of the SOTMF is shown in Figure 7.31. When a transmit delay is given to a controlled station, it is not applied immediately but to the SORMF at $t_k + 3T_M$ for the first time. Here, $t_k$ is the SORMF of CMF in which delay $D_n(k)$ is given to the controlled station.

In the first transmit multiframe generated according to a new value of $D_n$, the controlled station transmits back the value of the used $D_n$ through the service channel (SC) of its burst. The receive multiframe in which the controlling reference station receives the returned $D_n$ from a controlled station is referred to as the measurement multiframe (MMF), because the reference station measures the burst position error of the controlled station in this multiframe; $t_s$ is the amount of time determined by the satellite and the earth stations. The detailed protocols with respect to CMF and MMF will be considered later.

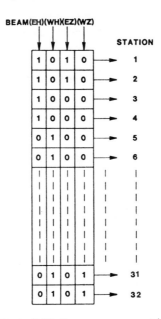

**Figure 7.29** Beam coverage matrix.

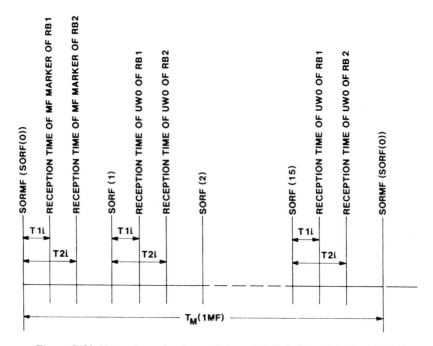

**Figure 7.30** Network receive frame timing. (UWO is first of the four UWs.)

### 7.3.3 Network Timing

The network operation can be established by the transmission of reference bursts from the reference stations, which are designated as master primary, primary, and two secondary stations. Either the master primary or the primary reference station is paired with a secondary reference station in one hemispheric or zone coverage, as described in Section 7.3.1.

The master primary reference station determines the overall network timing. The basic timing unit of the network is a frame of 2.0 ms, and every 16 frames

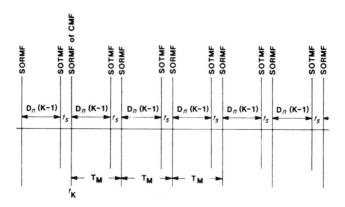

**Figure 7.31** Network timing.

constitutes a multiframe, which the controlling reference station uses to address all other stations in the network. Each station is to be addressed during one multiframe. Every 32 continuous multiframes constitutes a control frame during which all reference and traffic stations are provided with control and delay information through the control and delay code, CDC. Every 16 control frames constitutes a superframe, which is primarily used for coordination of the network operation (burst time) plan changes and for calculation of satellite coordinates.

### 7.3.4 Network Coordination

The basic functions of network coordination consist of network start-up, reference station role replacement, and operational plan change procedures. These functions are the responsibilities of the reference stations in the network and are not performed by the traffic stations.

#### 7.3.4.1 Reference Station Start-Up

The reference station start-up procedure is used to start the operation of the TDMA system. The procedure depends on the network connectivities. The network start-up begins when the master primary reference station first transmits its reference burst(s). At this point the MP starts the acquisition support procedure for the primary reference station. When the primary is brought into synchronization, both primary and master primary reference stations start acquisition of the secondary reference stations and of the ranging traffic terminals. When these stations are brought into synchronization, phase 1 of the start-up is completed. In this phase, the acquisition delays are also obtained using the satellite position prediction method. The burst position measurements, performed to maintain synchronization of the stations participating in phase 1 of the network start-up, also provide ranging information. This information is used for real-time satellite position determination.

When the MP has obtained accurate knowledge of the satellite position, phase 2 of the network start-up begins. In this phase, all remaining traffic terminals are acquired and brought into synchronization. If a satellite position prediction method is used instead of the real-time satellite position determination, phase 2 will not be executed and all traffic terminals will be brought into synchronization in phase 1.

There are two methods of start-up. Method 1 is used when the master primary reference station cannot receive its own burst. Method 2 is used when the master primary reference station can receive its own burst. Under normal operating conditions the network start-up procedure is the responsibility of both the master primary reference station in one coverage and the primary reference station in the opposite coverage. Because the roles of MP and P reference stations are different, the start-up procedures are therefore different for these stations. Since there exist two start-up methods, there are four possible start-up procedure variations, but all variations begin with a command from the operation center when the acquisition and synchronization procedure is unsuccessfully terminated (we obviously don't need start-up if synchronization is already successful).

**7.3.4.1.1  Start-Up without Loop-Back**  Depending on whether the master primary reference station can receive its own TDMA burst, the protocol of the two methods of start-up are shown in Figures 7.32 to 7.36, respectively. In Method 1 (without loop-back), shown in Figure 7.32, the information needed at MP is: which stations are to become P and S reference stations, which stations are designated as ranging stations in the network, and the nominal distance between the satellite and the stations. MP acquires P by prohibiting transmissions from all other stations by setting all control codes to DO NOT TRANSMIT (DNTX) except the primary reference station P. The initial protocol from MP to P is to set the control code in the MP reference burst to SYNC and provide the transmit delay $D_n$ for the station. P is acquired with the establishment of SORMF. MP then starts the terminal acquistion and synchronization support (TASS) procedures (Section 7.5) for the purpose of acquiring the secondary reference station S. If the inoperative (I) reference burst is received from the station to become S, it is considered to be acquired by MP, and the message BECOME SECONDARY is sent to the station in accordance with the inoperative to secondary promotion procedure (Section 7.3.4.3).

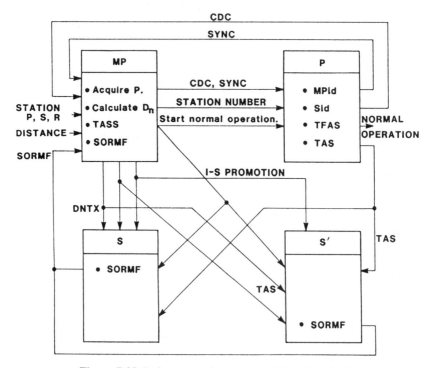

**Figure 7.32** Reference station start-up without loop-back.

Through CDC and SYNC, P identifies MP and S ($MP_{id}$, $S_{id}$). P performs transmit frame and acquisition and synchronization (TFAS, Section 7.6.3) and terminal acquisition support (TAS, Section 7.5.1). This part of the start-up procedure for MP, which cannot receive its own burst, is shown in Figure 7.33(a).

After the TASS the next procedure is the inoperative (I) to secondary (S) station promotion, as shown in Figure 7.33(b). The I to S procedure performed by the controlling reference station will be shown in Figure 7.41. If a satellite ranging method (determination or prediction) has been selected, then the TASS procedure is applied to the principle ranging station except RS. If the TASS-

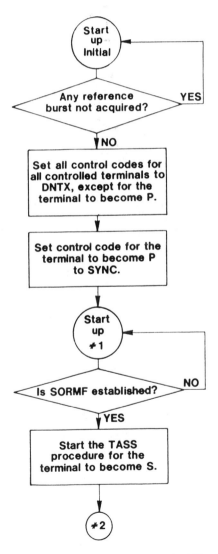

**Figure 7.33(a)** Flow chart of reference station start-up by MP without loop-back.

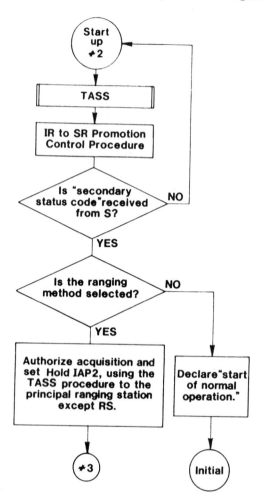

**Figure 7.33(b)** Flow chart of reference station start-up by MP without loop-back (continued). RS = reference station.

SYNC condition is not achieved after 32 control frames, then the procedure is applied to the back-up ranging station. If the real-time satellite position data is available, the start-up procedure terminates with the declaration of start of normal operation, as shown in Figure 7.33(c).

As part of the start-up procedure the setting and resetting of Hold IAP2 is also shown in Figures 7.33(a) and 7.33(b). The declaration of the "start of normal operation" as a result of the start-up procedure is also shown clearly. The flow chart of real-time satellite position determination is shown in Figures 7.34(a) to 7.34(c). For background information on satellite position determination the reader is referred to Section 5.11 of Volume I.

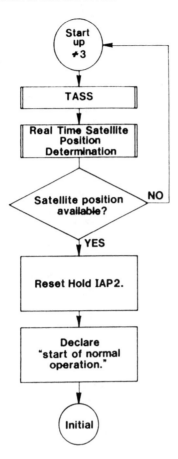

**Figure 7.33(c)**  Flow chart of reference station start-up by MP without loop-back (continued).

Satellite position prediction may also be accomplished using an algorithm which predicts the satellite's geocentric position at a given time. The algorithm for determining the satellite's geocentric position involves the use of a 24th-order Chebyshev polynominal expansion. Such a high-order polynomial provides ± 1.0 meter accuracy. Each of three Cartesian position components is expanded separately. The expansions are obtained from a fit to 25 sets of position components spaced at two-hour intervals. Though the data interval is 48 hours, the polynominal expansions are valid only for the central 24-hour period.

The method 1 procedure performed by P is shown in Figures 7.35(a) to 7.35(c), in which P identifies the station numbers of the MP, the station to become S, and the principal and back-up ranging stations. The initial protocol of P to MP is the same as the initial protocol from MP to P; that is, all control codes are

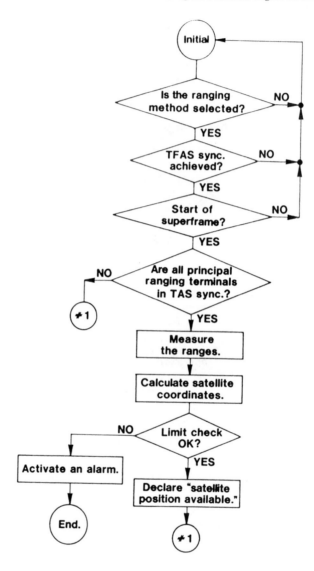

**Figure 7.34(a)** Flow chart of real-time satellite position determination.

set to DNTX except for MP, for which the control code is set to SYNC. After the establishment of SORMF and the reception of the SYNC code from MP, P starts the transmit frame. Next the P reference station starts the TASS procedure for the other reference station to become secondary. With steps similar to the start-up procedure performed by MP described earlier, the functions of I-S promotion, and ranging, real-time satellite position determination are also performed by P as indicated.

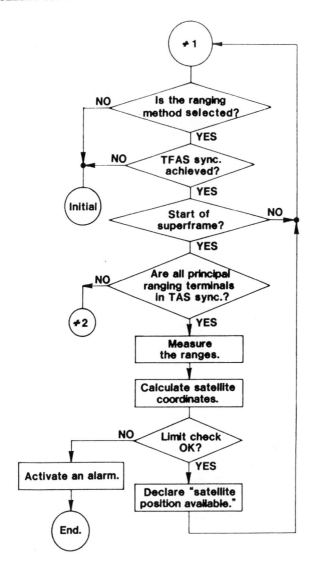

**Figure 7.34(b)** Flow chart of real-time satellite position determination (continued).

*7.3.4.1.2  Start-up with Loop-Back*  For the Method 2 (with loop-back) start-up procedure the MP needs to know its own station number, the numbers of the stations that will become P and S reference stations, and the CDC and SYNC messages from partner S, as shown in Figure 7.36. Other steps in the start-up procedure are as follows:

- Set all control codes to DNTX, except for the station to become MP.
- Start the TASS procedure for the station to become P and two S's when

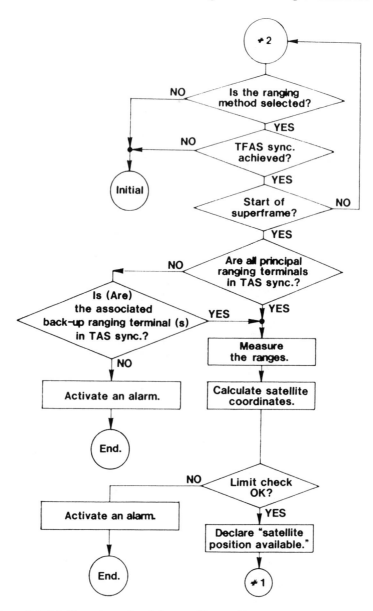

**Figure 7.34(c)** Flow chart of real-time satellite position determination (continued).

SORMF is established in the MP.

- Send BECOME SECONDARY message to the station to become S in accordance with the INOPERATIVE TO SECONDARY PROMOTION procedure.
- Terminate the procedure and enter the normal operation when the MP station

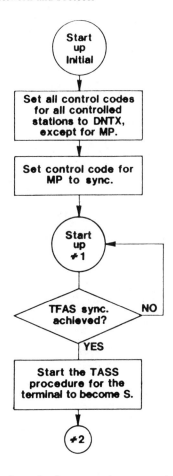

**Figure 7.35(a)** Flow chart of reference station start-up by P without loop-back.

receives SYNC code from S station in the same beam coverage and PRE-DICTION ALGORITHM READY or SATELLITE POSITION AVAILABLE is declared, as in earlier cases.

The flow chart of the MP start-up procedure with loop-back is shown in Figure 7.37(a) to (c).

For the Method 2 start-up the inputs required by P are:

- TFAS-SYNC achieved or not.
- CDC message from MP.
- MP terminal number.
- Ranging terminal (R) number.
- TASS-SYNC achieved/TASS-SYNC not achieved.

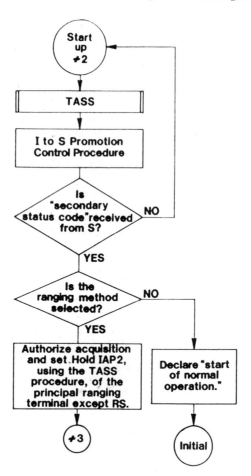

**Figure 7.35(b)** Flow chart of reference station start-up by P without loop-back (continued).

The procedure to be carried out by P, shown in Figure 7.38(a) and 7.38(b), is simpler than the function that needs to be performed by the same primary reference station without loop-back (Figure 7.35).

The start-up procedures for the set of reference stations as described in the last two sections depend on the network connectivities. Table 7.7 summarizes the responsibilities of the MP and P stations toward the seconary stations. In the table, the secondary station in the same zone as P is denoted as S', and the connectivities refer to the configurations are described in Section 7.3.2. A 1 for both MP and P and a 1 for S' indicate (as in Figure 7.24) the secondary station in the same zone as P.

As it turns out for the start-up of secondary stations, there are only two alternatives for the four network connectivities. For Figures 7.22 and 7.25 con-

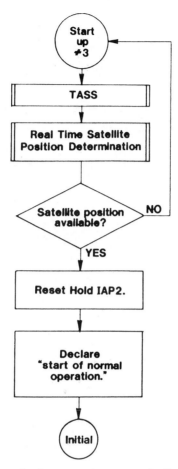

**Figure 7.35(c)**   Flow chart of reference station start-up by P without loop-back (continued).

nections, MP starts all other reference stations, and for Figures 7.23 and 7.24 connections, MP starts P and S′, and P starts S.

### 7.3.4.2   Traffic Rearrangement

The procedure for traffic rearrangement among the traffic stations is initiated and controlled only by the MP reference station with the cooperation of the P reference station. This procedure is carried out as follows:

- In the SC message of the reference burst, MP transmits an INITIATE TRAFFIC COUNTDOWN sequence to P.
- If P received the countdown sequence from MP, P starts a local countdown. If four or more countdown sequences are received, P declares INITIATE TRAFFIC COUNTDOWN. After waiting for three multiframes from SORMF

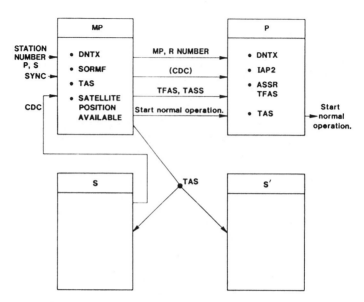

**Figure 7.36** Reference station start-up with loop-back.

plus the amount of delay of P, P starts the NOTIFICATION OF TIME PLAN CHANGE. If fewer than four sequences are received, the procedure is terminated.

• After the INITIATE TRAFFIC COUNTDOWN, the MP waits for 12 multiframes and starts the NOTIFICATION OF TIME PLAN CHANGE.

Automatic traffic rearrangement among traffic stations is controlled by MP and P. The procedures are shown separately in Figures 7.39, 7.40(a), and 7.40(b). Secondary stations do not participate in this procedure, which is separately described for MP and P as follows: For MP the procedure starts by sending a destination-directed INITIATE TRAFFIC COUNTDOWN sequence to P. After waiting 12 $T_M$ from the SI (significant instant) of the INITIATE TRAFFIC COUNTDOWN, which is defined as the SOTMF of the multiframe including the countdown code 00000000, a NOTIFICATION OF TIME PLAN CHANGE countdown sequence is next sent over 32 multiframes. COUNTDOWN COMPLETE is then declared at countdown code 00000000 and the procedure is terminated.

The automatic traffic rearrangement procedure performed by P begins with the reception of the INITIATE TRAFFIC COUNTDOWN sequence. If less than four countdown codes are received, no action is taken and the procedure is terminated. If four or more countdown codes are received, INITIATE TRAFFIC COUNTDOWN is declared at the SI of the message. After waiting 3 $T_M + D_p$, where $D_p$ is the delay for P, from the SI of the INITIATE TRAFFIC COUNTDOWN, the NOTIFICATION OF TIME PLAN CHANGE countdown sequence

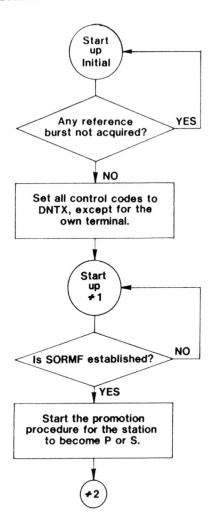

**Figure 7.37(a)** Flow chart of reference station start-up by MP with loop-back.

is next sent over 32 multiframes. Last COUNTDOWN COMPLETE is declared at countdown code 00000000.

Thus both automatic traffic rearrangement procedures provide NOTIFICA-TION OF TIME PLAN CHANGE and DECLARE COUNTDOWN COM-PLETE. In the case of MP, additional initial traffic countdown is provided through the SC message.

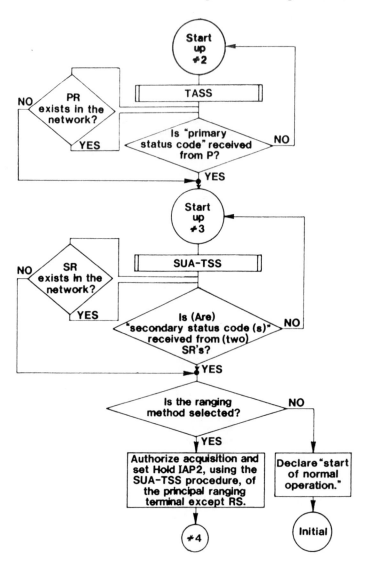

**Figure 7.37(b)**  Flow chart of reference station start-up by MP with loop-back (continued).

### 7.3.4.3  Reference Station Role Change

Because of a station failure, or for other reasons, the MP, P, and S roles of set reference stations can be changed. There are two types of necessary role change that can be applied to a reference station: the inoperative to secondary and the secondary to primary promotions. The significance of station change (or pro-

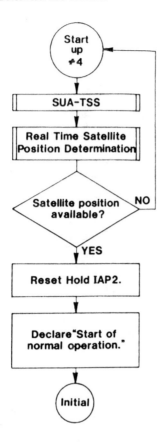

**Figure 7.37(c)** Flow chart of reference station start-up by MP with loop-back (continued).

motion) was mentioned in Section 3.12.4 of Volume I. The protocols required to establish the inoperative to secondary promotion are described as follows:

- Start the procedure when TAS-SYNC ACHIEVED is declared on the inoperative reference burst.
- If there exists a partner reference burst in BQ burst qualified status of the inoperative reference burst, examine the control frame synchronization, superframe synchronization, and nominal identity of the reference bursts.
- Transmit BECOME SECONDARY message to the inoperative terminal, if control frame synchronization and superframe synchronization have been delcared between the inoperative reference burst and its partner reference burst and if the manual control is enabled.
- If there does not exist a partner reference burst in BQ status, send BECOME SECONDARY message when manual control is enabled.

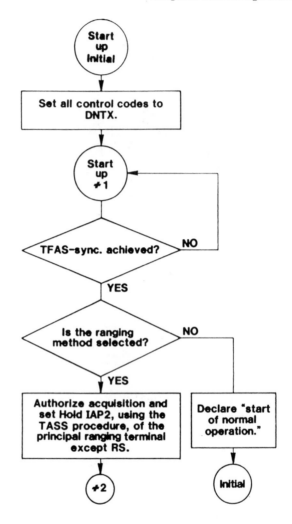

**Figure 7.38(a)** Flow chart of reference station start-up by P with loop-back.

- Terminate the procedure if TAS-SYNC NOT ACHIEVED condition is declared on the inoperative reference burst or if the promotion to secondary is completed.

The flow chart of the inoperative to secondary procedure performed by a controlling reference station is shown in Figure 7.41.

The inputs to the promotion procedure performed by the inoperative reference (I) terminal are: Superframe in synchronization or not; Control frame in synchronization or not; and BECOME SECONDARY message. The procedure simply assumes a secondary role from the first SOTMF of a control frame just

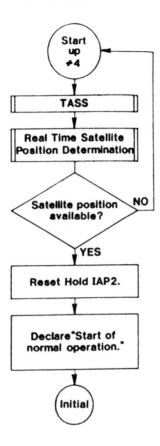

**Figure 7.38(b)** Flow chart of reference station start-up by P with loop-back (continued).

following the reception of the BECOME SECONDARY message, if superframe and control frame synchronization procedures have been completed. The output of this procedure is the promotion to the secondary role. The flow chart of the inoperative to secondary promotion procedure performed by the inoperative (I) reference station is shown in Figure 7.42.

With the FAILED or NOT FAILED status of reference bursts, replacement matrix, received status code, and role matrix as inputs the secondary to primary (or master primary) promotion steps performed by the failure detection terminal are:

- Start the procedure if all reference bursts from PR are declared BURST FAILED.
- Wait 128 $T_M$.
- Terminate the procedure if one or more RB's from the PR station are in BURST NOT FAILED condition.

**Table 7.7**

Secondary station start-up.

| Reference Station / Connectivity | MP | P | S | S' |
|---|---|---|---|---|
| Figure 7.22 | 1 | 1 | 1 | 1 |
| Figure 7.23 | 1 | 1 | 0 | 1 |
|  | 0 | 1 | 1 | 0 |
| Figure 7.24 | 1 | 1 | 0 | 1 |
|  | 0 | 1 | 1 | 0 |
| Figure 7.25 | 1 | 1 | 1 | 1 |

- Otherwise, determine the replacement station number for the failed station, using the replacement matrix.
- Declare the failed station a partner station if the station's own number is the replacement station number.
- If the failed station is a partner station, determine the next role using the role matrix.
- If the failed station is not a partner station, send a BECOME PRIMARY message to the replacement station.
- Repeat transmission of the BECOME PRIMARY message at intervals of approximately one control frame.
- Stop the transmission of the message and terminate the procedure if the replacement station has become primary.
- Raise a prompt maintenance alarm if 32 control frames have elapsed since the first message was sent.

The flow chart of the procedure is presented in Figure 7.43. The outputs of this procedure are the BECOME PRIMARY message, prompt maintenance alarm, and promotion to primary.

With the BECOME PRIMARY message and role matrix (Table 3.5, Volume I, pp. 146–147) inputs the procedure performed by the secondary reference terminal is:

- Determine the next role using the role matrix, if the secondary reference station receives the BECOME PRIMARY message.
- Assume the new role from the first SOTMF of the following control frame.
- Neglect the BECOME PRIMARY message if the next role in the role matrix is invalid.

The flow chart of the secondary to primary promotion procedure by the secondary reference station is shown in Figure 7.44.

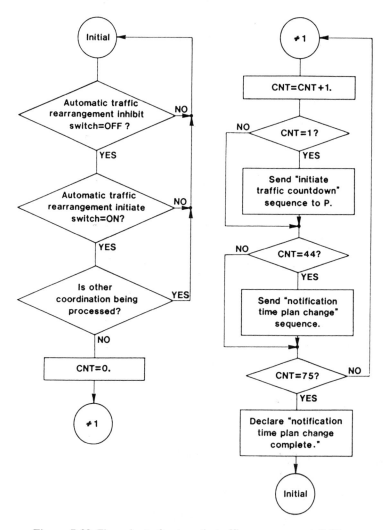

**Figure 7.39** Flow chart of automatic traffic rearrangement (MP).

### 7.3.4.4 Burst Time Plan Change

The basic operation of a TDMA network depends on its burst, or frame, structure in which both the protocol and message information are contained. Upon synchronization, the network operates in accordance with such a burst format; for efficient network operation this burst format (or burst time plan) needs to be changed from time to time to follow changes in traffic pattern. The protocols required to change the burst length as well as burst location by the set of reference

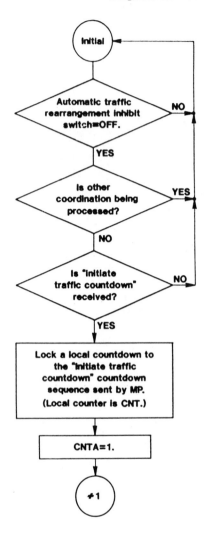

**Figure 7.40(a)** Automatic traffic rearrangement (P).

stations are described below. The protocols performed by the MP as shown in Figures 7.45(a) to 7.45(d) are:

- Send the START OF PLAN CHANGE countdown sequence to P; also send DNTX to controlled stations that are receiving the IAP1 code. Then wait for $12\ T_M$ from SI (significant instant) of START OF PLAN CHANGE, which is defined as the SORMF of the last multiframe of the countdown sequence.
- Send REQUEST FOR READY TO CHANGE message to each controlled traffic station, S and I, except the ones sent DNTX.

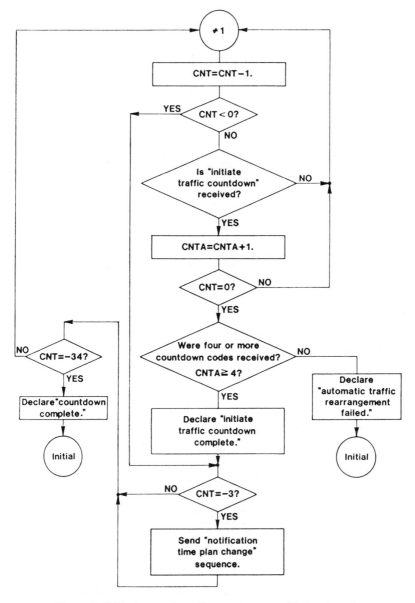

**Figure 7.40(b)**  Automatic traffic rearrangement (P) (continued).

- Declare READY TO CHANGE if READY TO CHANGE messages are received from all the stations to which the MP sent REQUEST FOR READY TO CHANGE messages within 96 $T_M$ from the SI of START OF PLAN CHANGE. Otherwise, declare BTP CHANGE FAILED and terminate BTP change procedure.

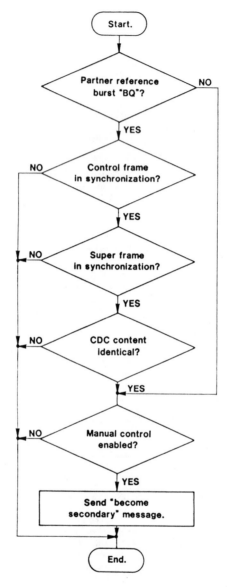

**Figure 7.41** Inoperative to secondary procedure performed by controlling reference station.

- Declare READY TO INITIATE COUNTDOWN if READY TO CHANGE message is received from PR during the 32 multiframe interval, commencing 128 multiframes after the multiframe marker of the first multiframe of RE-

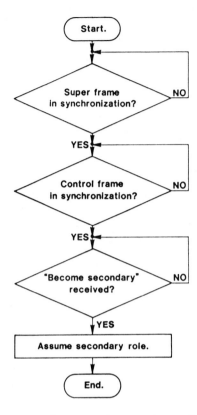

**Figure 7.42** Inoperative to secondary promotion procedure performed by the inoperative reference station.

QUEST FOR READY TO CHANGE message transmission, and if MP itself has declared READY TO CHANGE. Otherwise, declare BTP CHANGE FAILED and terminate BTP change procedure.

- Send INITIATE COUNTDOWN countdown sequence to PR from 415 $T_m + D_n$ after the next superframe instant following the declaration of READY TO INITIATE COUNTDOWN.
- Wait 12 $T_M$ from SI of INITIATE COUNTDOWN, which is defined as the multiframe marker of the last multiframe of the countdown sequence.
- Send NOTIFICATION OF TIME PLAN CHANGE countdown sequence.
- Lock a local countdown to the NOTIFICATION OF TIME PLAN CHANGE countdown sequence sent by P.
- If four or more countdown codes are received, declare COUNTDOWN COMPLETE at countdown code 00000000.

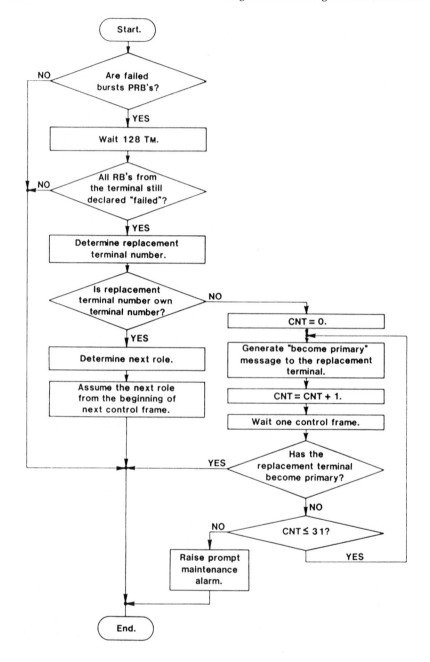

**Figure 7.43** Secondary to primary promotion procedure executed by failure detecting station.

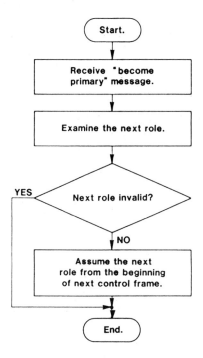

**Figure 7.44** Secondary to primary promotion procedure by the secondary reference station.

- If the COUNTDOWN COMPLETE is declared, implement the new transmit time plan at the SOTMF at $t_o + 3T_M + D_n$, where $t_o$ is the SORMF of the multiframe including the countdown code 00000000, and implement the new receive time plan at the SORMF at $t_o + 12T_M$.

The protocols performed by P as shown in Figures 7.46(a) to 7.46(f) are:

- Lock a local countdown to received START OF PLAN CHANGE countdown.
- If four or more countdown codes are received and the manual BTP interlock is off, START OF PLAN CHANGE is declared. Otherwise, no action is taken.
- Send DNTX to controlled stations that are receiving IAP1 code.
- Wait $3\, T_M + D_p$ from SI of START OF PLAN CHANGE, which is defined as the SORMF of multiframe including countdown code 00000000 of START OF PLAN CHANGE countdown sequence.
- Send REQUEST FOR READY TO CHANGE message to each controlled traffic station, S and I, except the ones sent DNTX.
- Declare READY TO CHANGE if READY TO CHANGE messages are received from all the stations to which the PR sent REQUEST FOR READY

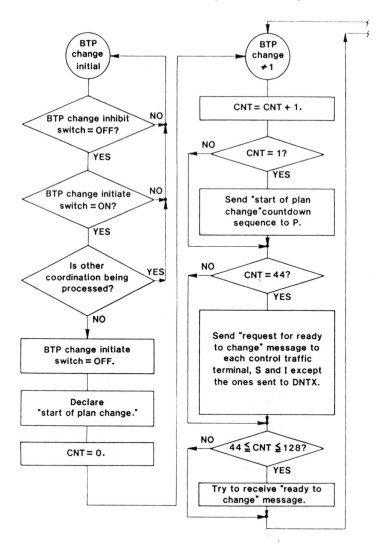

**Figure 7.45(a)** BTP change by the master primary station.

TO CHANGE messages within 96 $T_M$ from the SI of START OF PLAN CHANGE. Otherwise, declare BTP CHANGE FAILED and terminate BTP change procedure.

- Send READY TO CHANGE message to MPR for one control frame from multiframe commencing 128 multiframes after the SI of start of plan change.

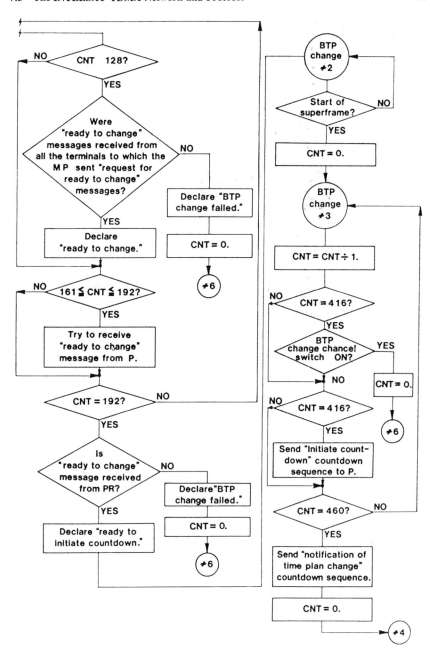

**Figure 7.45(b)** BTP change by the master primary station (continued).

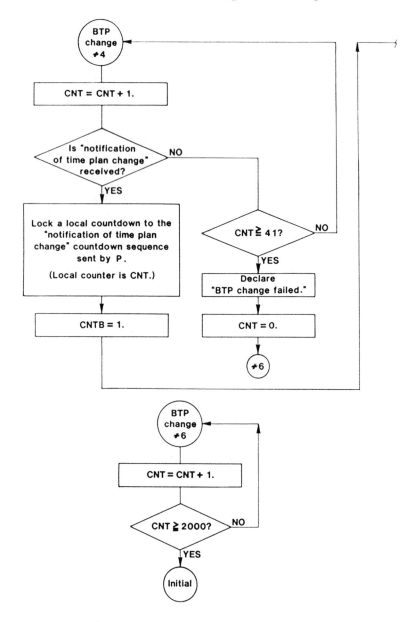

**Figure 7.45(c)** BTP change by the master primary station (continued).

- Lock a local countdown to the received INITIATE COUNTDOWN countdown sequence.
- Declare INITIATE COUNTDOWN if four or more countdown codes are received. Otherwise, no action is taken.

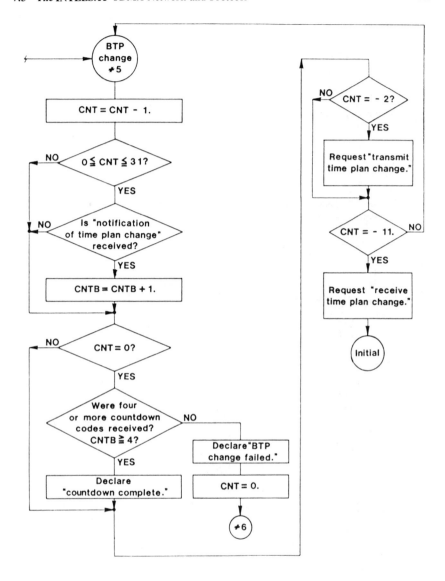

**Figure 7.45(d)** BTP change by the master primary station (continued).

- Wait $3\,T_m\,7D_p$ from SI of INITIATE COUNTDOWN, which is defined as the SORMF of the last multiframe of INITIATE COUNTDOWN sequence. This SORMF instant is derived from the local countdown sequence.
- Send NOTIFICATION OF TIME PLAN CHANGE countdown sequence.

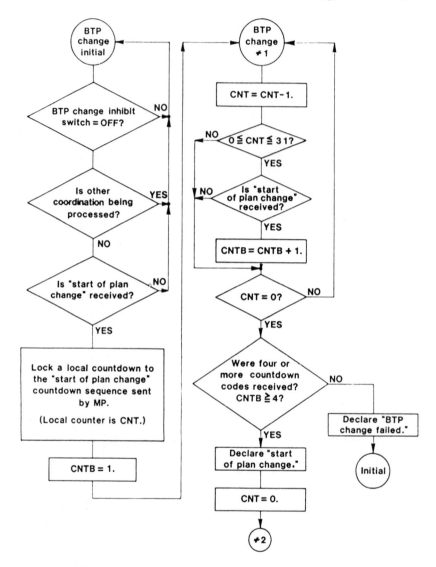

**Figure 7.46(a)** BTP change by the primary reference station.

- If INITIATE COUNTDOWN is not declared within 1920 multiframes from the SI of start of plan change, the burst time plan change procedure terminates and BTP CHANGE FAILED is declared.
- Lock a local countdown to the NOTIFICATION OF TIME PLAN CHANGE countdown sequence sent by MPR.
- If four or more countdown codes are received, declare COUNTDOWN COMPLETE at countdown code 00000000.

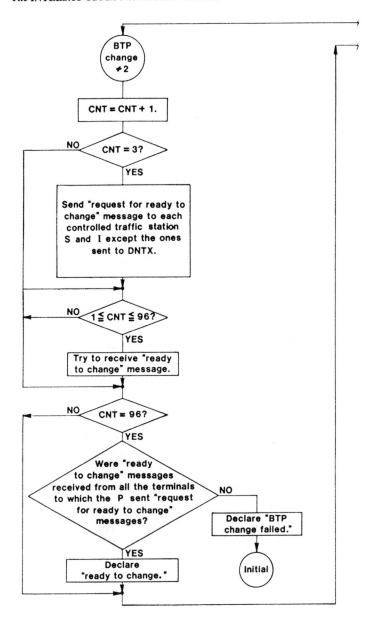

**Figure 7.46(b)** BTP change by the primary reference station (continued).

- If COUNTDOWN COMPLETE is declared, implement the new transmit time plan at the SOTMF at $t_o + 3 T_M + D_n$, where $t_o$ is the SORMF of the multiframe including the countdown code 00000000, and implement the new received time plan at the SORMF at $t + 12 T_M$.

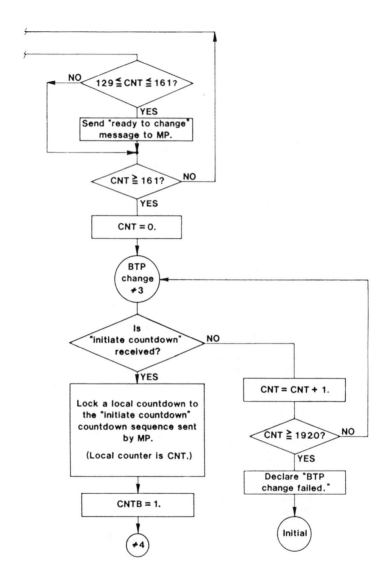

**Figure 7.46(c)** BTP change by the primary reference station (continued).

The protocols performed by S and I stations for the BTP change are:

- Activate PLAN CHANGE ENABLE facility when REQUEST FOR READY TO CHANGE message is received, containing a BTP number identical to the number stored in the background memory, unless manually inhibited.

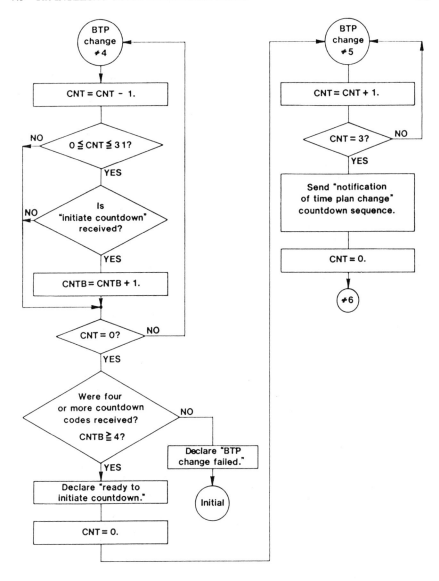

**Figure 7.46(d)** BTP change by the primary reference station (continued).

- Start sending READY TO CHANGE message according to the priority table of SC message.
- Stop sending READY TO CHANGE message when NOTIFICATION OF TIME PLAN CHANGE countdown sequence is declared COUNTDOWN COMPLETE or 1920 multiframes have elapsed since the first reception of the REQUEST FOR READY message.

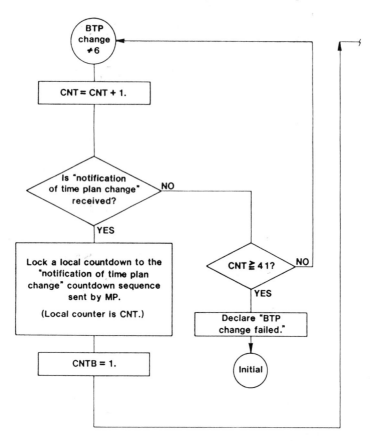

**Figure 7.46(e)** BTP change by the primary reference station (continued).

- Lock a local countdown to received START OF PLAN CHANGE countdown sequence.
- If four or more countdown codes are received, declare COUNTDOWN COMPLETE at countdown code 00000000.
- If the COUNTDOWN COMPLETE is declared, implement the new transmit time plan at the SOTMF at $t_o + 3 T_M + D_n$, where $t$ is the SORMF of the multiframe including the countdown code 00000000, and implement the new receive time plan at the SORMF at $t + 12 T_M$.

The flow chart of BTP change by S and I is shown in Figures 7.47(a) and 7.47(b).

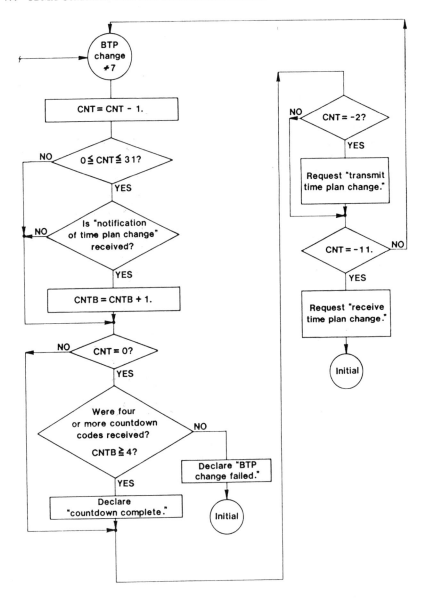

**Figure 7.46(f)** BTP change by the primary reference station (continued).

## 7.4 TDMA CONDENSED TIME PLAN TRANSMISSION PROTOCOLS

The purpose of this section is to highlight the protocols and procedures to be used for the transmission of condensed time plans (CTPs) to and from TDMA traffic stations and the INTELSAT Operation Center TDMA Facility (IOCTF)

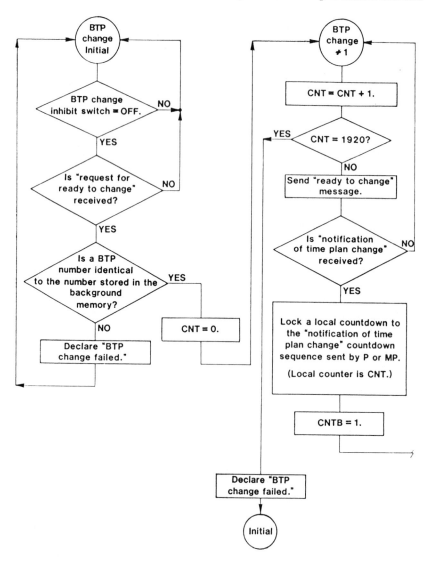

**Figure 7.47(a)** BTP change by secondary and inoperative stations.

in the INTELSAT TDMA network. The recommendations are based on references [9,10,14]; a condensed description can be found in [11].

Traffic stations are controlled by reference stations in accordance with the current burst time plan (BTP). Each station will receive a version of the BTP that details all those elements of the BTP that will affect a particular station. This reduced version is called a master time plan (MTP) and will be produced in printed form and delivered to the station well before its anticipated implementation. To obviate disruption to the network following a BTP change, various

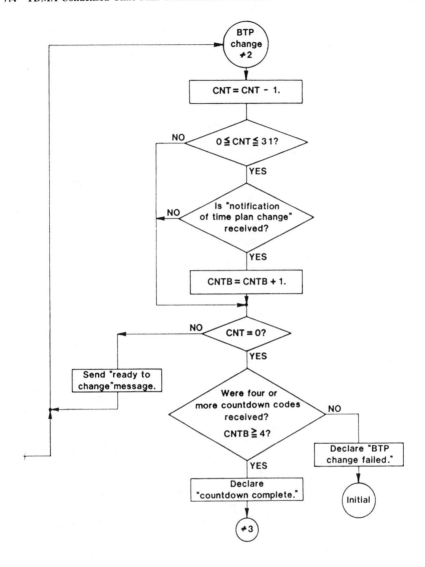

**Figure 7.47(b)** BTP change by secondary and inoperative stations (continued).

critical parameters contained in the MTP are assembled into a condensed time plan (CTP), which is sent from the IOCTF directly to the station's memory. To ensure that the station has correctly received the CTP the station is required to send the CTP back to the IOCTF for bit-by-bit comparison. The retransmitted CTP from the reference station is called a RCTP.

The Working Group assisting INTELSAT in the drafting of the TDMA specification indicated that the transmission circuit and link protocols should be based on CCITT X.25 packet switching protocol levels 1 and 2. The data line is full duplex with a maximum bit rate of 600 bits/s and maximum information field length of 1024 bits. The rest of this section concerns the CTP transmission through the X.25 protocol.

The CCITT X.25 recommendation provides the interface between the data station equipment (DTE) and the data circuit terminating equipment (DCE) for stations operating in primarily the packet mode on public data networks [6,16]. For interfacing, X.25 consists of the following three levels:

Level 1—The physical, electrical, functional, and procedural characteristics to establish, maintain, and disconnect the physical link between the DTE and the DCE.

Level 2—The link access procedure for data interchange across the link between the DTE and the DCE.

Level 3—The packet format and control procedures for the exchange of packets containing control information and user data between the DET and the DCE.

The first two levels can be applied to a network without packet-switched data. This is the case for the INTELSAT TDMA system, in which the IOCTF reference station is considered a DCE and the controlled stations are considered DTEs.

The advantages of the X.25 interface are: easy access, fault detection, error recovery, flow control, network maintenance, and message distribution and tracking. In addition, the X.25 protocol has been recognized internationally as providing cost-effective services through gateways between X.25 networks such as ARPANET (U.S.), TRANSPAC (France), and DATEX-P (Federal Republic of Germany).

The first level of the X.25 recommendation is actually based on the previous recommendaiton of X.21, which provides detailed protocol arrangements for data or telephone call connection and disconnection between DTE and DCE through circuit-switched services [16]. This documentation also gives diagrammatic examples of interface signaling sequences and the establishment of a data transfer phase for point-to-point operation. Level 2 of the X.25 protocol recommends HDLC (High-Level Data Link Control) with a balanced link access protocol (BLAP) as the standard for link level protocol. HDLC is specified by the International Organization for Standardization (ISO). The link access procedure (LAP) uses 48 bits plus the information format of a frame. Two 8-bit flag sequences are used for signifying the beginning and end of the frame, 8 bits each are allocated to addressing and control respectively, and 16 bits are reserved for parity checking of error detection in the frame.

For more convenient processing, CTP data is transmitted in blocks of 125 bytes in the information portion of the X.25 frame format (125 bytes is the maximum length of information allowed in a frame). If the amount of CTP information is less than 125 bytes, then dummy bits are required.

### 7.4.1   Circuit Establishment

The station CTP communications software program monitors and controls the circuit establishment procedure. The circuit is a voice grade channel which will support low-speed full duplex synchronous data transmission. The circuit will be established through a circuit-switched network.

The IOCTF initiates circuit establishment by automatically dialing a call on the circuit-switched network provided by the INTELSAT Engineering Service Circuit. The IOCTF will then wait to detect a carrier. When the IOCTF detects a carrier, it will begin to transmit acknowledgment.

If a carrier is not detected by the IOCTF, the IOCTF will not transmit the carrier to the station. If the IOCTF does not receive a carrier detect signal within approximately 56 seconds, the IOCTF will go on-hook for 3 seconds before attempting to retry the circuit establishment procedure. Two further attempts to establish the circuit are made before the IOCTF operator is informed.

When the station detects a carrier from the IOCTF, then the circuit is considered to be established. This condition will initiate the HDLC link establishment procedure. This may be done automatically by the station CTP communication software or manually by the operator.

### 7.4.2   Link Establishment

Circuit establishment initiates link establishment procedures within HDLC. Once IOCTF detects a carrier, the IOCTF will attempt to establish the link by periodically transmitting HDLC SABM (set asynchronous balanced mode) commands. The IOCTF then expects the station to transmit a UA (unnumbered acknowledge) response. The link establishment procedure at the station is either manually or automatically initiated once the terminal detects the carrier from the IOCTF.

The procedure for establishing a station link within the HDLC supports either an active or a passive link establishment. For passive link establishment, the station waits for receipt of the IOCTF SABM command. Upon receipt of the SABM, the station sends a UA response to the IOCTF. Directly after transmitting the UA, the station initiates the CTP data transfer. For active link establishment, the station immediately sends periodic SABM messges to the IOCTF. The station ignores receipt of all messages from the IOCTF except the SABM and the UA messages. Upon receipt of a UA, the station stops sending SABM and immediately initiates the CTP data transfer. Upon receipt of a SABM from the IOCTF, the station discontinues sending SABM, sends a UA message, and initiates the CTP data transfer.

Once the link is established, the station can automatically initiate CTP data transfer. Only one CTP data transfer (transmit CTP or transmit RCTP) occurs per link establishment/disestablishment session. The link must be disestablished and reestablished before another CTP data transfer may be attempted.

## 7.4.3 Data Transfer

There are two types of data transfers. CTP transmission refers to CTP data transmission from the IOCTF to the station. RCTP transmission refers to RCTP data transmission from the station to the IOCTF.

When the link is established, only one data transfer (CTP or RCTP) may occur. More than one data transfer may occur per circuit establishment. In order to transmit CTP and RCTP data during a single circuit establishment, the link is disestablished after the first transmission, then reestablished prior to the second transmission.

Automatic retransmission error recovery required by events including bit errors, line outages, protocol errors, timeouts, flow control conditions, etc., will be handled by the CCITT X.25 Link Level (Level 2) HDLC software.

If at any time after link establishment, and before orderly link disestablishment, the X.25 Level 2 software (HDLC) resets, the IOCTF operator will be immediately notified. It is recommended that the station operator be similarly notified of this event. This condition indicates that the error recovery procedures within HDLC failed to recover; under these circumstances, IOTCF will disestablish the circuit. A further attempt at the CTP transfer will be coordinated by the operators at both sites.

A normal CTP data transfer will proceed uninterrupted, except for HDLC frame level retransmissions, once begun, until all CTP data is transferred.

When all CTP data is transferred (transmitted by the sender), the sender must wait approximately 20 seconds before automatically disestablishing the link. This time period allows the HDLC software to retransmit the end of the CTP data, if required, before the link is disestablished.

The receiving software makes a final determination of successful CTP reception by applying a check-sum test on each of the received CTP data blocks and a block count test. The success or failure of this test is communicated to the operator.

## 7.4.4 Link and Circuit Disestablishment

After each data transfer is complete, the link is automatically disestablished by the sender. The sender is the IOCTF for CTP transmission and the station for RCTP transmission. The link disestablishment procedure consists of transmitting the DISC command approximately 20 seconds after transmission of the last information frame of the CTP. The DISC command is retransmitted every 5 seconds (system parameter T1) if not acknowledged by a UA response.

The receiver responds to a DISC command by transmitting a UA response and thereafter ignoring all the frames from the sender except an SABM command. The link is now disconnected. The station CTP communications software always initiates circuit disestablishment upon receiving a UA response to a DISC command.

The circuit is disestablished by instructing the circuit facilities, by signaling the ON HOOK condition to the circuit-switching equipment. Circuit disconnection by the station can be caused either by the reception of a UA response to a DISC command or by a failure to detect a carrier.

## 7.5 NETWORK CONTROL PROTOCOLS

The main protocol for network control is the controlling mechanism of the reference stations, to assist all the stations in the network into synchronization. The procedure to accomplish the synchronization objective is through terminal acquisition and synchronization support (TASS), which in turn consists of two parts, the terminal acquisition support (TAS) and the terminal synchronization support (TSS), to be described next.

### 7.5.1 Terminal Acquisition Support

For controlled terminal, or station, acquisition there are two options. The terminal may either have a dedicated acquisition window in the frame or it may share a common acquisition window. For the dedicated window acquisition, two or more traffic terminals can be acquired simultaneously; this type of acquisition is called parallel acquisition. For a common window acquisition, only one terminal at a time can use the window; thus it is called sequential acquisition. To perform sequential acquisition an acquisition cycle of 16 control frames is used. The cycle is divided into four intervals, which are referred to as acquisition cycle intervals (ACI), each interval being associated with one of the sequential acquisition terminals. During each of the four intervals the TAS procedure allows only the authorized terminal to acquire.

The terminals to be controlled by the sequential TAS procedure in the superframe are selected at the start of the superframe. From each ACI the terminal number assigned to ACI is obtained. Parallel and sequential acquisition procedures for all the controlled terminals are then performed by TASS in accordance with the acquisition type for the controlled terminals. A block diagram of the TASS flow by MP and P is shown in Figures 7.48(a) and 7.48(b).

### 7.5.2 Sequential Terminal Acquisition Support (TAS)

With authorized stations and assigned ACI the sequential TAS procedure starts with the reference station transmitting the IAP1 code and the calculated acquisition delay value of the station to be acquired in its CDC message. After the IAP1 code is transmitted through the CMF1, the unique word acquisition, or specifically the acquisition phase UW detection (APD) procedure, takes place as shown in Figures 7.49(a) and 7.49(b). Corresponding to the first control frame a measurement multiframe is defined (see Section 7.3.3) for purposes of measuring TDMA burst position. As the first measurement multiframe (MMF1) burst is received, burst lost conditions are checked. The difference between burst received and burst lost is detected through the unique word combinations. If both conditions are satisfied under normal acquisition, the IAP2 is selected as a control code and transmitted through the CMF2. Otherwise either DNTX is

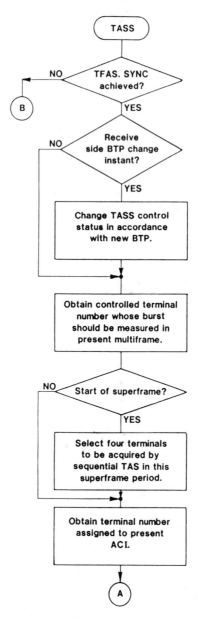

**Figure 7.48(a)** Flow diagram of TASS by MP and P.

transmitted for burst received but lost or IAP1 is transmitted for burst not received and not lost.

During the second measurement multiframe (MMF2), the SC messages in the burst are decoded and the measurements of the burst position are attempted. If the SC messages are unsuccessfully decoded, invalid delay is declared. Otherwise

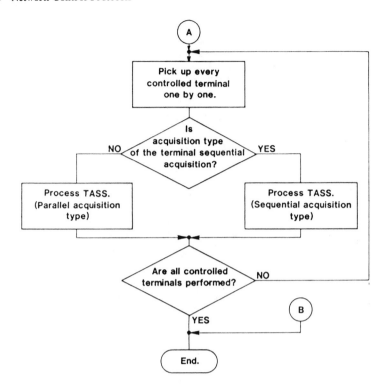

**Figure 7.48(b)** Flow diagram of TASS by MP and P (continued).

a valid delay condition is declared. During and at the end of the second mea-surement multiframe the position and length of the burst can also be measured. Before the transmission of CMF3 the control code is selected again. If the control code was IAP1 in CMF2 the control code is selected in the same manner as in CMF2. DNTX is selected if DNTX was the control code in CMF2. If IAP2 was transmitted and the delay is valid and received burst is not lost, then the SYNC is selected in the control code for CMF3 transmission. When the SYNC code is successfully selected, TAS SYNCHRONIZATION ACHIEVED is declared at the start of MMF3 and the TAS procedure is terminated. For each abnormal operating condition the control code is set with DNTX (do not transmit) as shown. For each measurement multiframe the acquisition delay is calculated.

In Figure 7.49 $D_A(M)$ denotes the acquisition delay value for station M. The more detailed protocol steps of the sequential TASS status and processing pro-cedure are shown in Figures 7.50(a) and 7.50(b). To keep track of the control and measurement multiframe numbers for this part of the protocol Figure 7.51 is provided for clarification.

If the selected code in Control Multiframe 3 (CMF3) was IAP2, the burst position is measured and the decoded SC message from the terminal in Mea-

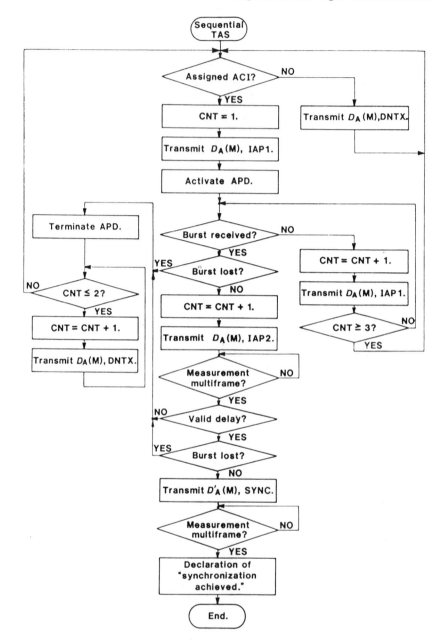

**Figure 7.49(a)** Flow diagram of sequential TAS procedure.

surement Multiframe 3 (MMF3) examined. Control code SYNC is transmitted if the received burst is not lost and has a valid delay. Otherwise, DNTX is transmitted in control Multiframe 4, if the control code in Control Multiframe 3 was IAP2. If the selected code in Control Multiframe 4 (CMF4) was SYNC, a

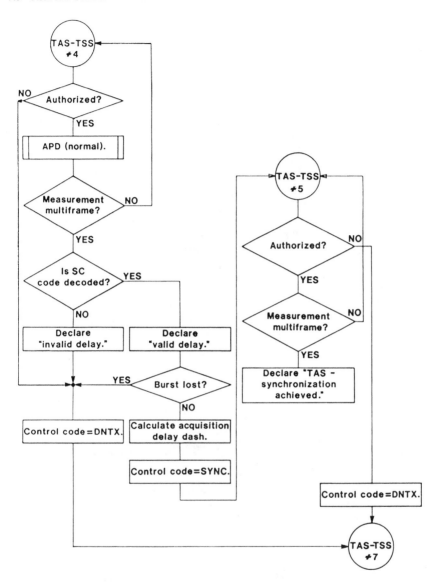

**Figure 7.49(b)** Flow diagram of sequential TAS procedure (continued).

TAS-SYNC ACHIEVED is delcared at the start of Measurement Multiframe 4 (MMF4) and the TAS procedure is terminated. If the selected code was DNTX, no action is taken in Measurement Multiframe 4. The process is repeated in the next assigned ACI. Until then the DNTX code is transmitted in every control multiframe. In Figure 7.51 the delay information transmission for each control multiframe has been omitted.

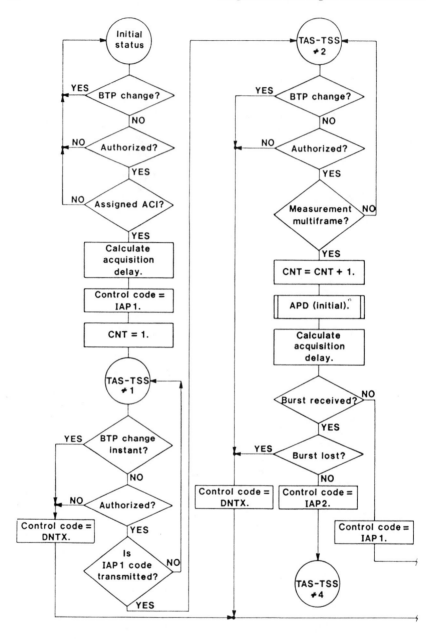

**Figure 7.50(a)** Status and processing flow chart of sequential TAS-TSS subunit.

### 7.5.3 Parallel Terminal Acquisition Support (TAS)

Similar to the sequential TAS procedure, the parallel TAS procedure also starts with no BTP change and protocol takes place only with authorized stations in

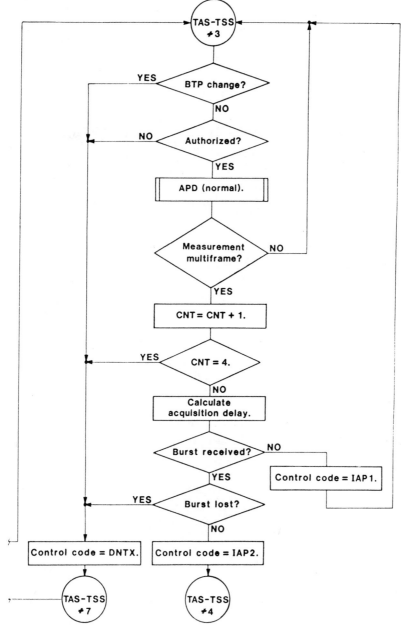

**Figure 7.50(b)** Status and processing flow chart of sequential TAS-TSS subunit (continued).

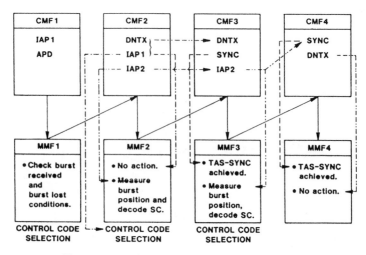

**Figure 7.51** Protocols between CMFs and MMFs.

the network. The parallel TAS procedure differs from the sequential TAS procedure in that it has no ACI. The procedure is outlined as follows:

1. The reference station transmits the IAP1 code and the value of acquisition delay $D_A(M)$ in CMF1 to the terminal to be acquired.
2. Activate APD procedure for the controlled terminal.
3. Terminal control code selected in accordance with Table 7.8 and $D_A(M)$ in CMF2.
4. If the selected code is IAP1, then repeat from step 3.
5. If the selected code is DNTX, then repeat from step 1.

**Table 7.8**

Control code selection

(Previous code was IAP1).

|  | Burst* Received | Burst Not* Received |
|---|---|---|
| Received* Burst Lost | DNTX | X |
| Received* Burst Not Lost | IAP2 | IAP1 |

X    = This condition cannot occur.

*    = Burst condition (from APD procedure) to be sampled not earlier than one multiframe prior to transmission of the control code.

IAP1 = INITIAL ACQUISITION PHASE 1 Code.

IAP2 = INITIAL ACQUISITION PHASE 2 Code.

DNTX = DO NOT TRANSMIT Code.

6. If the selected code is IAP2, measure the burst position and examine the decoded SC message from the terminal in the corresponding measurement multiframe.
7. If the SC message is not lost, declare VALID DELAY; otherwise, declare INVALID DELAY.
8. Transmit control code selected in accordance with Table 7.9 for the following control multiframe: if the selected code is DNTX and the delay is $D_A(M)$, then repeat from step 3; otherwise, the delay is $D'_A(M)$, which is a function of the decoded delay in the previous measurement multiframe.
9. If the control code selected is SYNC or IAP2, TAS-SYNC is achieved at the start of following measurement multiframe, and the TAS procedure terminates.

The logic of the parallel TAS procedure is shown in Figure 7.52 and the detailed protocol flow diagram is shown in Figures 7.53(a) to 7.53(c).

### 7.5.4   Terminal Synchronization Support (TSS)

The TSS procedure is devised for purposes of a reference station to support the synchronization function of the controlled stations in the network. Specifically, the procedure supports the transmit frame synchronization (TFS) procedure performed by the controlled stations.

The TSS procedure begins with the declaration of synchronization achieved from either sequential or parallel TAS. Then the protocol proceeds as follows:

- In the received measurement multiframe, decode the SC message and measure the burst position. Declare VALID DELAY if the SC message is not lost; otherwise, declare INVALID DELAY.
- Declare VALID MEASUREMENT if four or more valid unique words are detected; otherwise declare INVALID MEASUREMENT.
- In control multiframe, transmit control code is SYNC if four consecutive unique words are detected. Transmit control code DNTX if unique words are missed in four consecutive frames and if after that 1536 frames or more elapse without the unique word detection in four consective frames.
- Determine the value of transmit delay in accordance with Table 7.10.

**Table 7.9**
Control code selection.

| | Valid Delay | | Invalid Delay |
|---|---|---|---|
| | Hold IAP2 | Not Hold IAP2 | |
| Received Burst Lost | DNTX | | DNTX |
| Received Burst Not Lost | IAP2 | SYNC | DNTX |

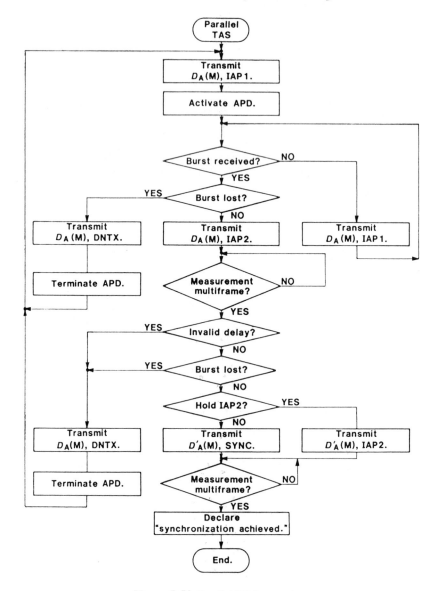

**Figure 7.52** Parallel TAS procedure.

- If the selected control code is DNTX, terminate the TSS procedure after the transmission of the code and declare TAS-SYNC NOT ACHIEVED.

The flow diagram of the TSS procedure is shown in Figure 7.54.

## 7.6  PROTOCOLS FOR ACQUISITION AND SYNCHRONIZATION

An indispensable function in a TDMA system is acquisition and synchronization. This functional complexity increases in the INTELSAT TDMA network due to

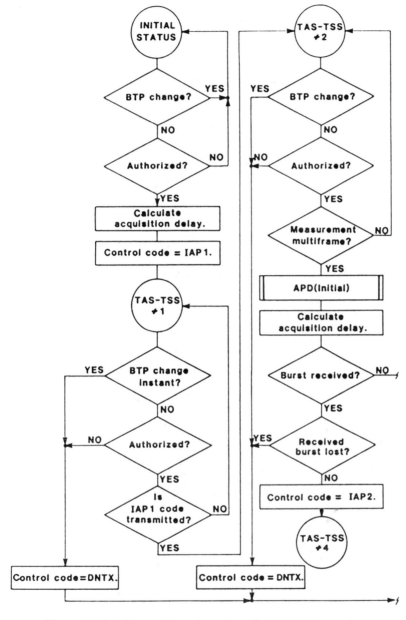

**Figure 7.53(a)** Protocol flow chart of parallel TAS-TSS procedure.

the provision of cross hemispheric connectivities as described in Section 7.3.2. The INTELSAT network specifies Acquisition and Steady State Reception (ASSR), Receive Frame Synchronization (RFS), and Transmit Frame Acquisition and Synchronization procedures. These three procedures apply to both the reference

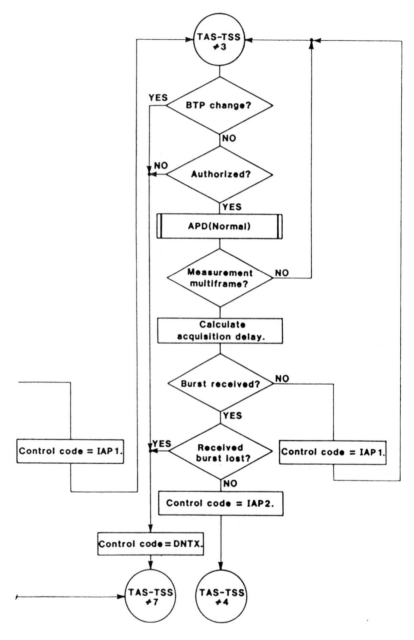

**Figure 7.53(b)** Protocol flow chart of parallel TAS-TSS procedure (continued).

stations and the traffic stations in the network, except for the set of reference stations where additional control frame synchronization and superframe synchronization procedures are required.

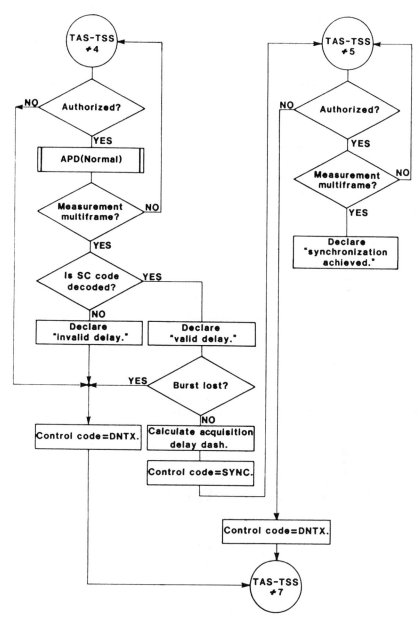

**Figure 7.53(c)** Protocol flow chart of parallel TAS-TSS procedure (continued).

## 7.6.1 Acquisition and Steady State Reception (ASSR)

Given a series of received reference bursts, the ASSR procedure performs the following main functions:

- finds locations of reference bursts and maintains the locations
- establishes timing for each received reference bursts and maintains the timing
- identifies status of the received reference bursts

The ASSR procedure consists of five subprocedures, namely, SMA (Search Mode Acquistion), GMA (Gated Mode Acquisition), SSR (Steady State Reception), BTS (Burst Time Synchronization), and RBID (Reference Burst Identification).

### 7.6.1.1   Search Mode Acquisition (SMA)

The SMA subprocedure is used in the initial acquisition of reference bursts. At the initial stage, both reference bursts are assumed to be in the BNA (burst not acquired) condition.

For the multiframe mode operation, the SMA procedure takes the UW detection status (Correct or Miss) and the UW arrival timing as inputs and proceeds as follows:

- Reset the counter. If the counter reaches 1 second in the steps to follow before BA (burst acquired) declaration, terminate the procedure and maintain BNA condition.
- Take in a set of data for a multiframe from the input data. Inspect only the UW of a reference burst in the first frame of the multiframe.
- If its detection status is a miss, neglect the set of data for the multiframe, maintain BNA condition and go back to above step.
- If the UW detection status is correct, declare the BA and terminate the SMA subprocedure.
- UW detection timing refers to (multiframe marker) arrival time with the correct status.

Thus the outputs from the SMA subprocedure are the declaration of BA or BNA, and UW detection timing, which is essential to subsequent operations.

**Table 7.10**
Selection of transmit delay for TSS procedure.

| Control Code Selection | Delay and Measurement Status | | |
|---|---|---|---|
| | Valid Delay | Invalid Delay | |
| | | Valid Measurement | Invalid Measurement |
| SYNC | $D_d - e$ | $D_{pm} - e$ | $D_{pm}$ |
| DNTX | $D_{pm}$ | $D_{pm}$ | $D_{pm}$ |

$D_d$ = delay decoded in previous measurement multiframe (returned delay)
$D_{pm}$ = delay transmitted in previous control multiframe in case the present SC is lost
$e$ = burst position error measured in previous measurement multiframe

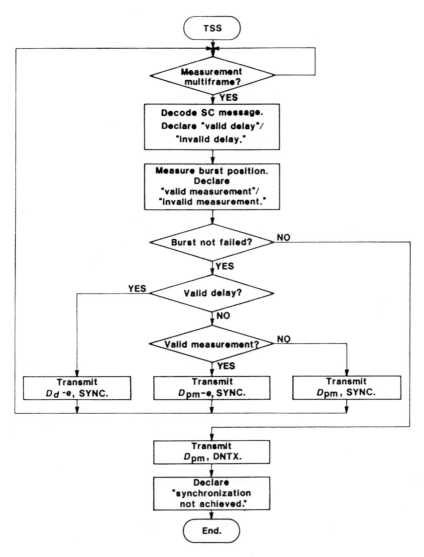

**Figure 7.54** Flow chart of the TSS procedure.

### 7.6.1.2 Gated Mode Acquisition (GMA)

The GMA subprocedure is applied only to the acquisition of a reference burst when the other reference bursts (RB1 or RB2) are in the BQ (burst qualified) condition. This condition exists only when the ASSR procedure of a burst is successfully completed and the reference burst is identified. The inputs required for the multiframe mode operation of GMA are: UW detection status (Correct or Miss), UW arrival timing, BQ/BNQ condition for the reference burst already acquired, detection timing of the reference burst already acquired, and the burst time plan.

The difference in the input conditions of GMA and those of SMA can be clarified by referring to Table 7.11.

The procedure for GMA is as follows:

- Input data for a multiframe is taken in for processing a block at a time from the data file.
- Maintain BNA until the successful termination of GMA.
- If the reference burst already acquired at the beginning becomes BNQ during GMA, terminate GMA unsuccessfully and resume SMA.
- With regard to the reference burst to be acquired, inspect the UW detection status and the arrival timing. If the UW status is correct and the arrival timing is in the UW detection window, declare BA as the objective reference burst and terminate the GMA normally maintaining the UW detection timing; otherwise, try another block of input data for the next multiframe.

The outputs of this procedure are declaration of BNA or BA, and UW detection timing.

### 7.6.1.3   Steady State Reception (SSR)

The SSR subprocedure is applied to any of the reference bursts in the BA condition. The SSR maintains reception of the reference bursts, declares the BA or BNA condition, and the UW detection timing for the BTS procedure. The SSR terminates only when BNA is declared. The SSR takes the UW detection status and arrival timing and performs the following in the multiframe mode.

- One block of multiframe data at a time is taken in for processing.
- Inspect the UW detection status and the arrival timing of the reference burst in the first frame of a multiframe repeatedly for a series of multiframes.
- If the UW detection status is a miss or the arrival timing is out of the UW detection window, initiate the multiframe counter.
- If a correct status of the multiframe marker is detected in the UW detection window, reset the counter to zero.

**Table 7.11**
Input conditions of GMA.

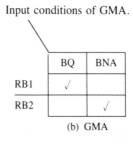

|      | BA | BNA |
|------|----|-----|
| RB1  |    | √   |
| RB2  |    | √   |

(a)  SMA

|      | BQ | BNA |
|------|----|-----|
| RB1  | √  |     |
| RB2  |    | √   |

(b)  GMA

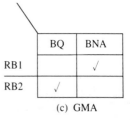

|      | BQ | BNA |
|------|----|-----|
| RB1  |    | √   |
| RB2  | √  |     |

(c)  GMA

- If the counter reaches 96 multiframes or 1536 frames equivalently, declare the BNA condition and terminate the SSR procedure to initiate the SMA or the GMA as appropriate.
- Since only the first frame of a multiframe is considered, the UW detection window is centered 32 msec, following the previous UW detection timing. Its width is 65 symbols. If the previous UW is not detected, the center of the UW detection window is 32 msec, following the center of the previous detection window.
- UW detection timing is set equal to the UW arrival timing for correct detection. Use the false detection timing instead for UW detection timing, in case of a false UW detection, and record the UW detection status.

SSR provides BA/BNA, UW detection timing and status as its outputs.

### 7.6.1.4 Burst Time Synchronization (BTS)

Based on the output of the SSR, the BTS subprocedure provides either burst time acquired (BTA), or burst time not acquired (BTNA) condition, which is used to start the TFAS procedure (to be described). The BTS checks the validity of the UW detected by the SSR and substitutes it for UW and multiframe marker timing when it is a missed.

Again for a multiframe mode, the BTS procedure inputs are: the BNA/BA condition declared by the SMA, GMA, or SSR procedure; UW detection status (Correct or Miss); UW detection timing, and the multiframe number in which BA or BNA was declared.

The procedure is as follows:

- Inspect the UW detection status of the first frame for a series of multiframes.
- Initially declare the BTNA and reset the multiframe counter.
- Increment the counter by one, only if the UW status is correct for the first frame or a multiframe; otherwise, reset the counter.
- Declare the BTA condition if the counter reaches 2.
- Terminate the BTS procedure if the BTA is not declared within 32 multiframes from the time BA or if the BNA is declared during the procedure.
- Once the BTA is declared, maintain the BTA condition unless the condition following holds. During each of the two consecutive multiframes, a multiframe marker is improper or the BNA is declared for the reference burst in question. If the condition holds, declare the BTNA condition and terminate the BTS.
- After the BTA declaration, UW's in the miss status do not violate the correct sequence. For the UW's in the miss status, a dummy UW detection timing is generated internally. The dummy UW detection timing is obtained by adding 32 msec to the UW detection timing in the first frame of the previous multiframe.

The outputs of this procedure are declaration of the BTNA or BTA condition and UW detection timing for the multiframe marker.

### 7.6.1.5   Reference Burst Identification (RBID)

For both reference bursts to declare the BQ and BNQ, the RBID subprocedure inspects the status code in the CDC for every control frame. Initially, the BNQ condition is assumed.

The input information for the RBID includes the BTA/BTNA condition declared by the BTS procedure and the reference burst data (reference terminal status code in the CDC).

The steps of RBID proceed as follows:

- Take one block of input data at a time for a control frame for both the reference bursts. Inspect the CDC message at address zero to check the status codes for each reference burst, for a series of control frames.
- For each of the reference bursts examine the status code PRB (primary reference burst), SRB (secondary reference burst), or inoperative—regardless of BTA or BTNA for all the control frames during which the CDC loss may occur.
- For each of the reference bursts use the decision table shown in Table 7.12 for every examined status code paired with the status code in the previous control frame to establish the preliminary status identification.
- After the preliminary status identification for both reference bursts, the final status identification is established at each control frame for both reference bursts by using the decision table shown in Table 7.13. Declare the BQ condition for the reference burst in the BTA condition whose final status identification is determined to be PRB or SRB. Otherwise, keep the reference burst in the BNQ condition.
- This status (BQ or BNQ) becomes effective at time $t_r + 2T_M$ where
  $t_r$ = the time of SORMF of the multiframe containing the status code
  $T_M$ = the duration of the receiving multiframe.

**Table 7.12**
Decision table: Preliminary status identification.

| Status Code Decoded in the Current Control Frame<br>· · · · · · · · · · · · · · · · · · · · · · · · · ·<br>Status Code Decoded in the<br>Previous Control Frame | PRB | SRB | Inoperative | CDC Lost at Address 0 |
|---|---|---|---|---|
| PRB | PRB | X | Inoperative | PRB |
| SRB | PRB | SRB | Inoperative | SRB |
| Inoperative | X | SRB | Inoperative | Inoperative |
| CDC lost at | PRB | SRB | Inoperative | X |

X:    Unidentified status
PRB:  Primary reference burst
SRB:  Secondary reference burst

**Table 7.13**

Final status identification

| RB1 Preliminary Status \ RB2 Preliminary Status | PRB | SRB | Inoperative | CDC |
|---|---|---|---|---|
| PRB | Retain Previous Status | RB1 = PRB<br>RB2 = SRB | RB1 = PRB<br>RB2 = I | RB1 = SRB<br>RB2 = X |
| SRB | RB1 = SRB<br>RB2 = PRB | Retain Previous Status | RB1 = SRB<br>RB2 = I | RB1 = SRB<br>RB2 = X |
| Inoperative | RB1 = I<br>RB2 = PRB | RB1 = I<br>RB2 = SRB | RB1 = I<br>RB2 = I | RB1 = I<br>RB2 = X |
| X | RB1 = X<br>RB2 = PRB | RB1 = X<br>RB2 = SRB | RB1 = X<br>RB2 = I | RB1 = X<br>RB2 = X |

I : Inoperative
X: Unidentified status

The procedure outputs are the final status identification for both the reference bursts and declaration of the BNQ or BQ.

### 7.6.2 Receive Frame Synchronization (RFS) Procedure

The RFS procedure entails selecting one of the received reference burst RB1 or RB2 as the source of timing in accordance with Table 7.14. From this timing the SORMF is determined (see Table 7.15).

The RFS takes the status of reference bursts, BQ/BNQ, and UW detection timing as inputs, and selects a reference burst for timing source and establishes

**Table 7.14**

Receive frame timing source selection.

| RB1 \ RB2 | BQ | BNQ |
|---|---|---|
| BQ | Timing taken from PRB<br>or<br>internal timing locked to it | Timing taken from RB1<br>or<br>internal timing locked to it |
| BNQ | Timing taken from RB2<br>or<br>internal timing locked to it | No Timing Available |

BQ : Burst qualified
BNQ: Burst not qualified

**Table 7.15**

Start of receive multiframe (SORMF).

| Source of Timing | SORMF |
|:---:|:---:|
| RB1 | $t_{r_1} - T1_i$ |
| RB2 | $t_{r_2} - T2_i$ |

$t_r$ = instant of reception of the last symbol of the unique word of the reference burst from which timing is derived.

$T1_i$ = Time displacement of RB1 from the SOF.

$T2_i$ = Time displacement of RB2 from the SOF.

Note: $T1_i$ and $T2_i$ are less than 120,832 symbols (one frame period).

receive timing. The RFS procedure limits the rate of change of SORMF to a maximum of 4 symbols per second for a reference station. If neither RB1 nor RB2 is identified as a timing source (i.e., NO TIMING AVAILABLE), then SORMF and SORF are not generated.

### 7.6.3  Transmit Frame Acquisition and Synchronization (TFAS)

The TFAS procedure consists of the TFA procedure and the TFS procedure. The TFAS procedure establishes SOTMF and maintains it. This procedure is performed only by the controlled reference station.

#### 7.6.3.1  Transmit Frame Acquisition (TFA)

The TFA procedure establishes SOTMF by transmitting short bursts (burst without data). This procedure is performed only by the inoperative reference station. The procedure inputs are: status of reference bursts, BQ/BNQ, BTP, CDC message (control code, $D_n$), and SORMF.

The steps of the procedure are:

- The TFA procedure operates only when PRB in BQ is being received. Take the control code and delay $D_n$ only from the PRB in this procedure.
- The TFA procedure starts if the BTP number at address 0 in the CDC of the PRB is identical to the BTP number in the current BTP number of the reference station, unless manually inhibited. (However, in the operational network either PRB or SRB is zero.)
- If IAP1 code is received, start transmission of the principal reference burst without data.
- IF an IAP2 code is received, continue transmission of the principal burst using the current $D_n$.
- Upon receiving a SYNC code, terminate the TFA procedure. Start the transmission of all bursts with data and then start the TFS procedure.
- If an IAP2 code is not received within 128 multiframes after the first transmission of the principal reference burst, or if a DNTX code is received during

the transmission of the principal reference burst, promptly sound the maintenance alarm and stop transmission within 100 ms. Restart the TFA procedure, if the above conditions have occurred once.

- Terminate the TFA procedure and restart, if any one of the following conditions has occurred:
  (a) The BNQ condition has been declared for the RB which was PRB at the start of the TFA procedure.
  (b) The CDC message has been lost.
  (c) The control code sequence has been violated.

The procedure outputs are SOTMF and transmission of principal burst.

### 7.6.3.2 Transmit Frame Synchronization (TFS)

The TFS procedure maintains the SOTMF. The inputs to the procedure include the status of reference bursts, CDC messages, BTP, and SORMF.

The steps of the procedure are:

- Generate SOTMF as SORMF + $D_n \cdot D_n$ is determined in accordance with the CDC message logic matrix (Table 7.16).
- Terminate the TFS procedure, stop transmission of all reference bursts within 100 ms, and restart TFA procedure:

**Table 7.16**
CDC message logic matrix.

| | | CDC Message Content From PRB | | | |
|---|---|---|---|---|---|
| | | SYNC received | SYNC missed* once | SYNC missed* more than once | DNTX |
| CDC message content from SRB | SYNC received | SYNC declared and delay from PRB | SYNC and last delay from PRB | SRB SYNC and delay from SRB | |
| | SYNC missed* once | | | SYNC and last delay from SRB | |
| | SYNC missed* more than once | | | SYNC lost | DNTX |
| | DNTX | | | | |

\* Note: SYNC code is missed when one of the following conditions occurs:
— Loss of the CDC message
— IAP1 or IAP2 code is received.

— if SYNC Loss is declared by CDC message logic matrix, or

— if RFS procedure terminates.

● Terminate the TFS procedure, stop transmission of all bursts. Promptly sound a maintenance alarm, if DNTX code is declared by CDC message logic matrix.

The outputs of the procedure are SOTMF and transmission of reference bursts.

## 7.7    TNS COMPUTER PROGRAM STRUCTURE

Based on the network protocols described in Sections 7.3 to 7.6 a computer program is developed [12,17,18]. The overall TDMA Network Simulation (TNS) program consists of three controllers, six models, and five basic modules. The three controllers are simulation, display, and event. The six network models are reference station, traffic station, burst reception, burst transmission, satellite, and external events. Except for display, the event controller controls all the simulation activities of the program by calling the models based on the program input data, which include the station and satellite parameters and transmit and receive reference condensed time plans.

The satellite model receives the bursts transmitted from the burst transmission model and sends them back to the burst reception model of the terminal as determined in accordance with the beam connection matrix and beam coverage.

The external event model generates the stochastic phenomena (such as UW misdetection) and the terminal and transponder function failure effect. The terminal switch operation, the terminal operator command generation, and the setting of the simulation execution mode of each terminal are also performed by this model. Within each station model there are a number of other functional modules, which are shown in Figure 7.55.

### 7.7.1    Event Controller

The TNS program includes six models to represent the TDMA network as described in Section 7.7. Each subprogram is called upon by the event controller when a network event needs to be simulated. There are two kinds of events, namely, external events and internal events. The external events are the ones planned in the simulation case. The internal events are the ones generated in the simulation. The internal events include the following:

● Burst reception model event
● Burst transmission model event
● Satellite model event
● Reference station (RS) model event
● Traffic station (TS) model event

The event controller is called by the simulation controller. The functions of the event controller are:

```
┌─────────────────────────────────┐
│           Controllers           │
├─────────────────────────────────┤
│                                 │
│         • Simulation            │
│         • Event                 │
│         • Display               │
│                                 │
└─────────────────────────────────┘
```

```
┌─────────────────────────────────┐
│             Models              │
├─────────────────────────────────┤
│         • Reference Station     │
│         • Traffic Station       │
│         • Satellite             │
│         • External Event        │
│         • Burst Transmission    │
│         • Burst Reception       │
└─────────────────────────────────┘
```

```
┌─────────────────────────────────┐
│             Modules             │
├─────────────────────────────────┤
│    • Acquisition And Synchronization │
│    • Network Coordination       │
│    • Network Control            │
│    • Housekeeping               │
│    • Data Management            │
└─────────────────────────────────┘
```

**Figure 7.55** The structure of TNS program.

- Initialize the models and set the model control status flag to RUN after the simulation starts. The model control status flag is used to indicate whether or not the simulation is in progress.
- Search for the first event in the event stack.
- Call the adequate subprogram corresponding to the event.
- Take in the operator command when the event is at PAUSE or when an alarm sounds as a result of simulation. Stop the process of event control holding the model control status flag of RUN, if the operator command is at RETURN. Continue the process if the operator command is at CONTINUE.
- Stop the process, set the model control status flag to INITIAL, if the searched event is at STOP.
- When the processing by the subprogram is completed, send the result of the subprogram to the output device specified in the simulation case.
- Before the onset of simulation calculate the Chebyshev coefficients used for satellite position prediction.

The flow chart of the processes performed by the event controller is shown in Figures 7.56(a) to 7.56(c). In the program the subprograms are actually stacked instead of being parallel as shown.

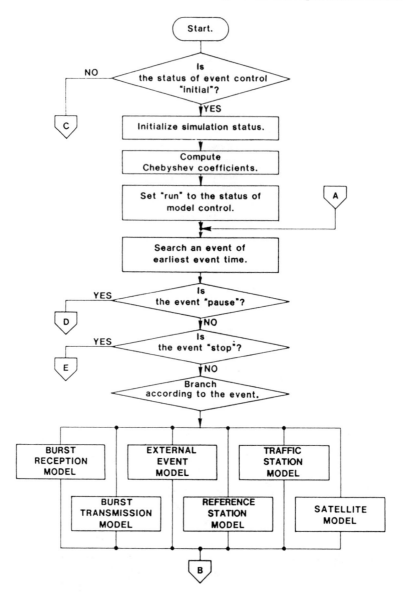

**Figure 7.56(a)** Flow chart of event controller program.

### 7.7.2  Satellite Model

The satellite model simulates the function of the satellite and has the following functions.

● Check burst overlapping for each transponder and activate burst overlapping alarm on detecting burst overlapping.

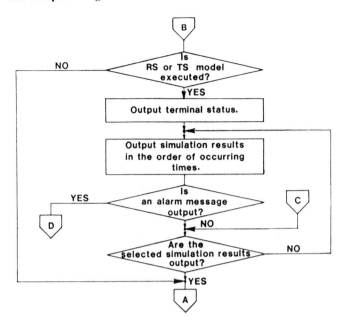

**Figure 7.56(b)** Flow chart of event controller program (continued).

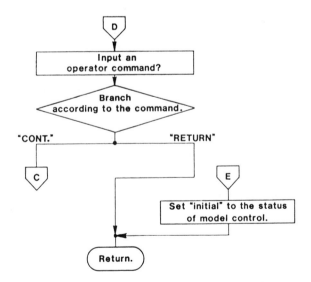

**Figure 7.56(c)** Flow chart of event controller program (continued).

- Simulate transponder failure.
- Obtain a downlink transponder number, making use of the uplink transponder number and the beam connection matrices.
- Obtain terminal numbers in the downlink beam coverage in accordance with transmit side transponder number and the beam coverage matrix.

- Calculate downlink propagation delay.
- Register the event to activate the burst reception model at the terminal on arrival of burst.

A block diagram and a flow chart of the satellite model are shown in Figures 7.57 and 7.58(a) and 7.58(b), respectively.

### 7.7.3  The Station Models

In the program, the network consists of up to four reference station models and up to twelve traffic station models. Each reference station is capable of performing the protocol functions as described in Section 7.3.1, that is, network coordination, network control, acquisition, and synchronization, which are referred to in the program as separate modules. During simulation the reference station model includes the functions of transmission and reception of reference bursts as well as traffic bursts. These functions are transmit data management and receive data management, as shown in Figure 7.59. During the performance of the most complex function in the network, the reference station model is controlled either by the reference station or (RS) model control module.

Without including network control and network coordination functions the traffic station (TS) model is shown in Figure 7.60. The housekeeping module function is detailed in Figure 7.61.

### 7.7.4  Burst Reception Model

The burst reception model simulates the reception of bursts transmitted from the satellite model. Functions of this model are as follows:

- Transfer the designated burst data from the transmission buffer in the satellite model to the reception buffer in the terminal model.
- Register the event initially to initiate the terminal model to receive the bursts. If the terminal is a reference terminal, the RS model starting event is registered, otherwise if it is a traffic terminal, the TS model starting event is registered.

A block diagram and the flow chart of the burst reception model are shown in Figures 7.62 and 7.63, respectively.

### 7.7.5  Burst Transmission Model

The burst transmission model simulates the transmission of bursts from the RS model or the TS model to the satellite model. Functions of this model are as follows:

- Modify transmit burst data in accordance with the following function failures if they occur to the terminal to be simulated: stop of transmission function, transmission synchronization function failure, UW miss detection, SC Loss, CDC Loss, and HBER. When the stop of transmission function occurs, the processing of this model is terminated.

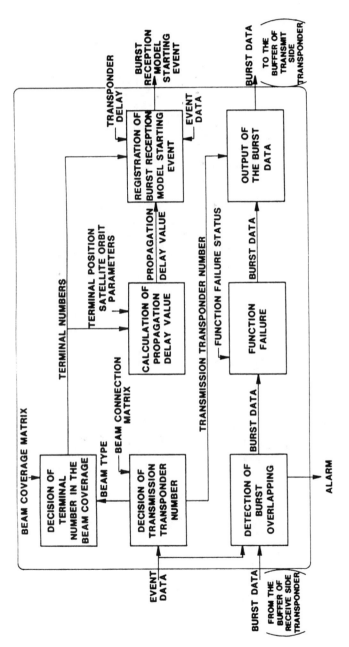

Figure 7.57 Block diagram of satellite model.

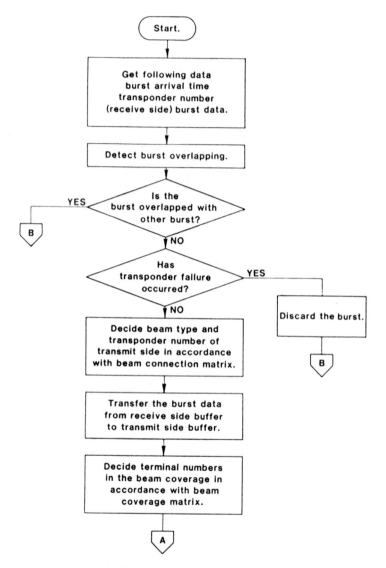

**Figure 7.58(a)** Flow chart of satellite model.

- Calculate the burst propagation delay from the terminal to the satellite.
- Register the satellite model starting event to start up the satellite model at the time of the arrival of the burst at the satellite.
- Transfer the designated burst data from the transmission buffer in the terminal model to the reception buffer in the satellite model. Figure 7.64 shows the block diagram of the burst transmission model and Figures 7.65(a) and 7.65(b) give the flow chart of this model.

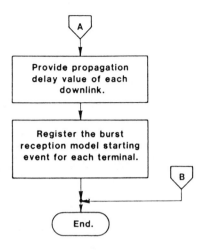

**Figure 7.58(b)** Flow chart of satellite model (continued).

## 7.7.6   External Event Model

The external event model as shown in Figures 7.66(a) and 7.66(b) provides the following five types of events:

- Terminal Switch Operation
  This event causes the switch operation effect on the terminal model. The following items are included in this event group: system start-up, initiate, inhibit, or cancel BTP changes; initiate and inhibit automatic traffic rearrangement; inoperative to secondary promotion; inhibit TFA switch; and select satellite position determination switch, i.e., external/prediction/ranging.
- Terminal Operator Command
  This event causes the operator command effect on the terminal model. The commands are: authorize acquisition, hold IAP2, cancel/hold IAP2, restart TFA, restart SC loss counter, restart burst transmission.
- Stochastic Phenomena
  This event causes the stochastic phenomena effect on transmitted bursts, which include UW miss detection, UW false detection, SC loss, and CDC loss.
- Terminal Function Failure
  This event causes the function failure effect on the terminal model. In addition to the events listed under stochastic phenomena, the group also performs stops reception and transmission functions, false decoding for SC and CDC, and transponder failure.
- Simulation Execution Mode Control
  This event controls the execution mode of simulation and determines the output format of the simulation results. This event consists of the following: simulation start or multiframe mode execution, terminal event output, received

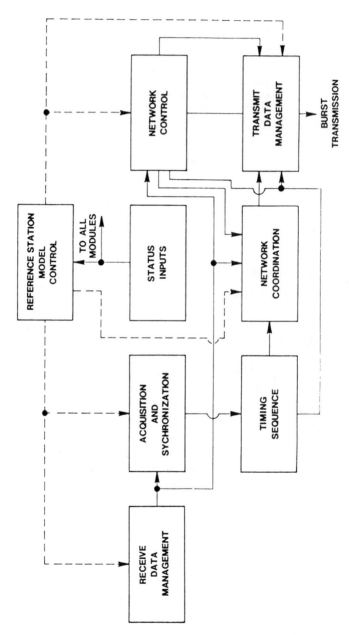

**Figure 7.59** Reference station model.

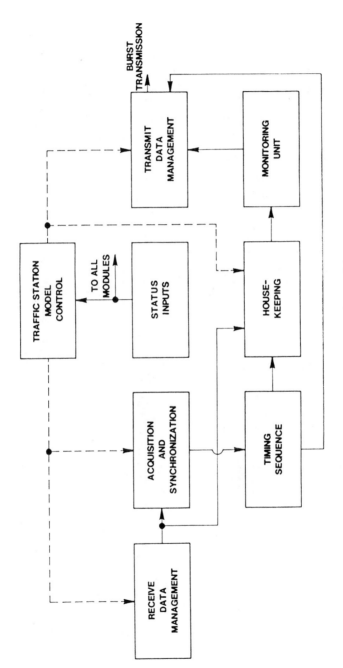

**Figure 7.60** Traffic station model.

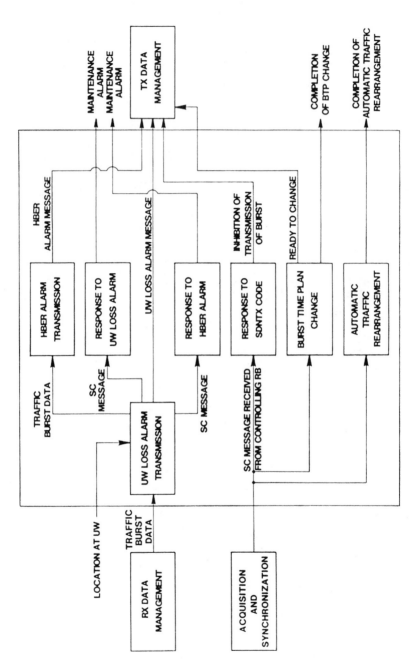

**Figure 7.61** Block diagram of housekeeping module HBER: High Bit Error Rate.

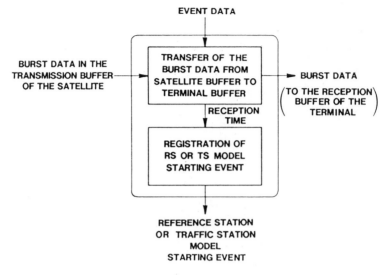

**Figure 7.62** Block diagram of burst reception model.

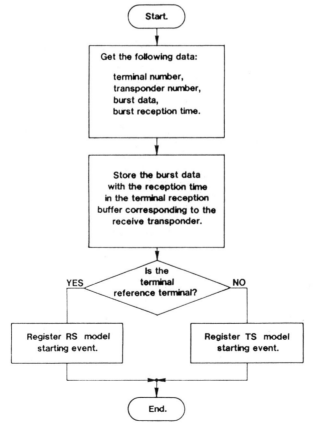

**Figure 7.63** Flow chart of burst reception model.

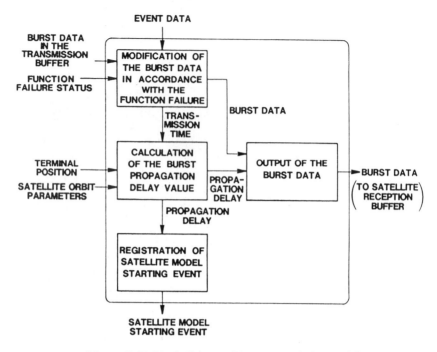

**Figure 7.64** Block diagram of burst transmission model.

and transmitted data at terminal output, data arrived at satellite output, and received delay output.

## 7.8    MODULAR OPERATIONS

The third-level protocol operation of the network is concerned with the set of modules as shown in Figure 7.55. The functions of network coordination and network control modules were described in Sections 7.3.4 and 7.5. The housekeeping module of a traffic terminal has already been shown in Figure 7.61. Described next will be the acquisition and synchronization and the data management modules.

### 7.8.1    Acquisition and Synchronization Module (ASM)

The protocol functions for acquisition and synchronization in both the reference station and the traffic station were described in Section 7.6. This section will give in detail relations in terms of logic flow diagrams. The block diagrams of the modules for the reference and traffic stations are shown in Figures 7.67 and 7.68, respectively. Each figure indicates the protocol input and output. Both modules receive burst data from their corresponding RX data management module. The ASM provides SORMF and timing source as its output. Superframe and TFA-SYNC information are provided (for the reference station ASM control frame). TFAS-SYNC is provided at the output of its ASM for the traffic terminal.

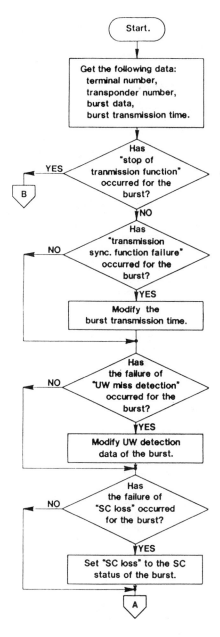

**Figure 7.65(a)** Flow chart of burst transmission model.

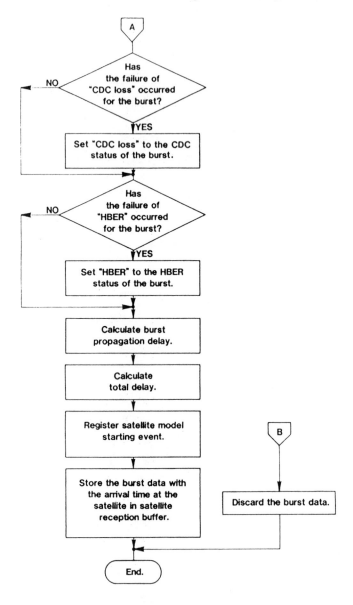

**Figure 7.65(b)** Flow chart of burst transmission model (continued).

### 7.8.1.1  ASSR

For both frame mode and multiframe mode operations the basic ASSR unit flow chart is given in Figure 7.69, with the two reference bursts designated as RB1

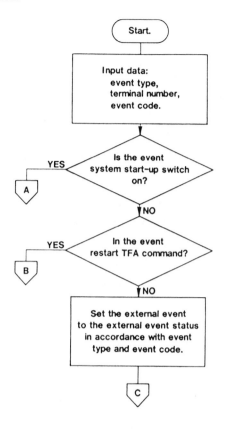

**Figure 7.66(a)** Flow chart of external event model.

and RB2 (recall that 16 frames makes a multiframe). The protocol steps of the ASSR for the multiframe case are described as follows:

Step 1: Recognize the controlling reference burst data.

Step 2: Choose the simulation execution mode (frame mode or multiframe mode) for the SMA, GMA, and SSR procedures. Processing is executed from Step 3 if it is in the multiframe mode or from Step 4 if it is in frame mode.

Step 3: For executing the SMA, GMA, or SSR subunit in the multiframe mode, the status and processing flow diagram is shown in Figures 7.70(a) and 7.70(b). The BTS subunit is executed in the multiframe mode (if BA is declared). Figure 7.71 shows the status and processing flow chart of the BTS subunit (multiframe mode).

Step 4: For executing the SMA, GMA, or SSR subunit in frame mode, the status and processing flow diagram of SMA, GMA, and SSR (frame mode) is shown

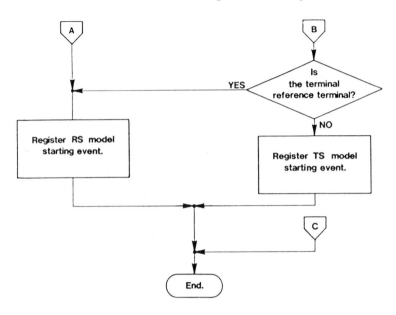

**Figure 7.66(b)** Flow chart of external event model (continued).

in Figures 7.72(a) and 7.72(b). The BTS subunit is executed in the frame
mode (if BA is declared). Figures 7.73(a) and 7.73(b) show the protocols of
BTS in the frame mode.

Step 5: When BTA is declared by the BTS subunit and all UWs in a multiframe
are correctly decoded, SC and CDC messages are decoded in addresses 0 and
1, respectively. Figures 7.74(a) and 7.74(b) show the protocol functions of
CDC decoding.

Step 6: After the CDC message has been decoded in address 0, the RRID subunit
is executed. The preliminary status identification of RB1 and RBS is executed
first before their final status identification.

### 7.8.1.2  Receive Frame Synchronization (RFS)

The RFS unit simulates the RFS procedure. Functions of this unit are as follows:

- Select timing source reference burst.
- Determine SORMF.
- Limit displacement of SORMF (RS only).

The protocol for the RFS function is shown in Figure 7.75.

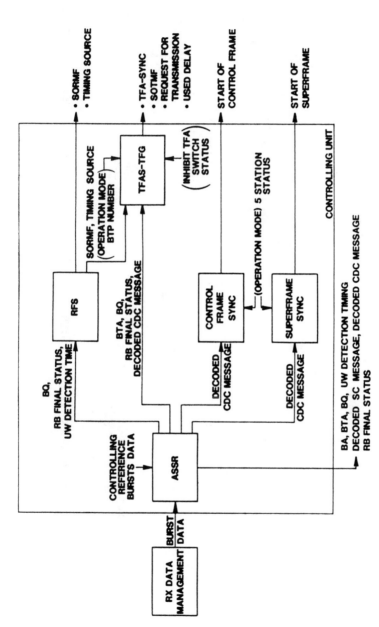

**Figure 7.67** Block diagram of reference station acquisition and synchronization module.

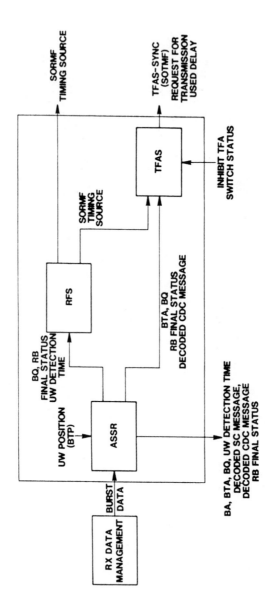

Figure 7.68 Block diagram of traffic station acquisition and synchronization module.

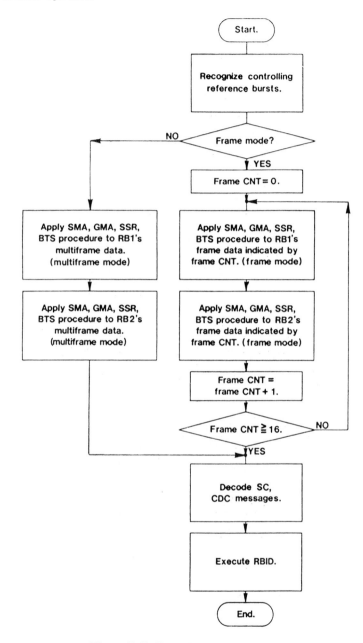

**Figure 7.69** Flow chart of ASSR unit.

### 7.8.1.3 Control Frame Synchronization

Identified by an address corresponding to the controlled terminal's short numbers, each reference station transmits delay and control information in a multiframe

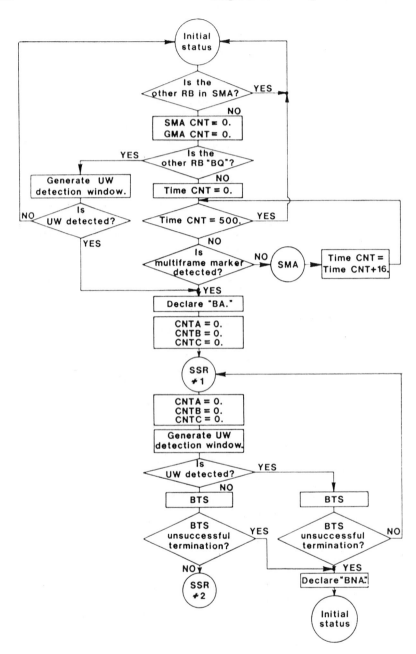

**Figure 7.70(a)** Flow chart of SMA, GMA, and SSR subunit (multiframe mode).

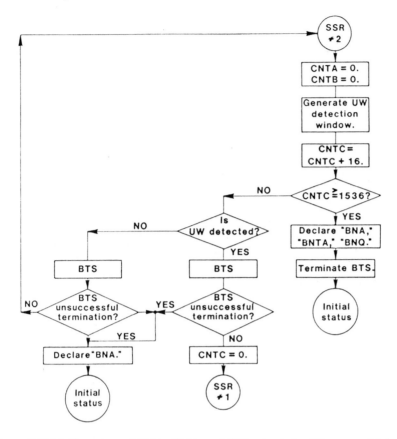

**Figure 7.70(b)** Flow chart of SMA, GMA, and SSR subunit (multiframe mode) (continued).

to each controlled terminal. Succesive multiframes contain up to 32 cyclic addresses. A synchronization procedure is established to ensure that both the primary and the secondary reference stations contained in one coverage area address the same controlled station in the same multiframe. The master primary reference station initiates the station address cycle at multiframe zero, which is phase locked to the superframe.

Each of the other reference stations will address terminal No. 0 in the transmit multiframe, whose SOTMF is at time

$$T_K + 23\,T_M + D_n$$

where $T_K$ is the SORMF of the multiframe addressing Station No. 0 (see Figure 7.76). The control frame synchronization unit inside the ASM simulates the control frame generation procedure of MP and the control frame synchronization procedure of the other reference stations on a multiframe basis. Figure

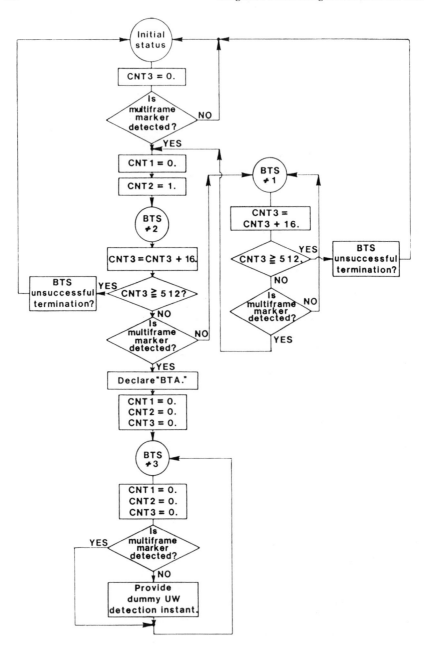

**Figure 7.71** Flow chart of BTS subunit (multiframe mode).

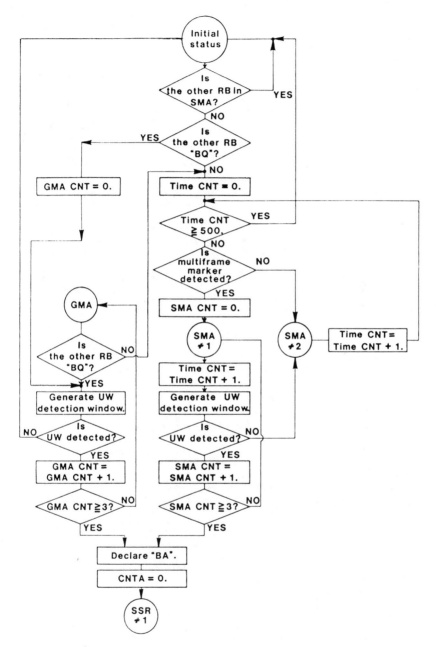

**Figure 7.72(a)** Flow chart of SMA, GMA, and SSR subunit (frame mode).

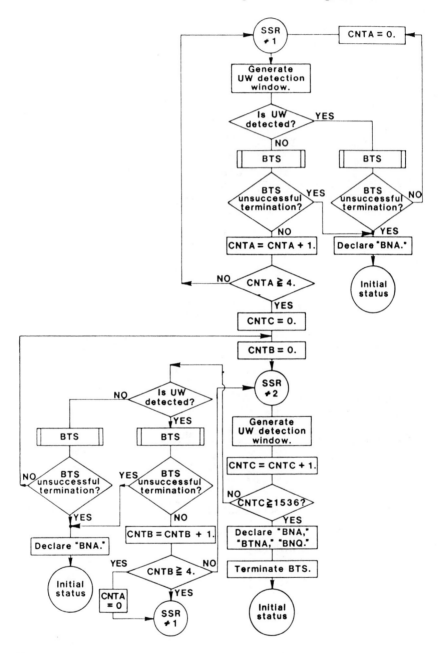

**Figure 7.72(b)** Flow chart of SMA, GMA, and SSR subunit (frame mode) (continued).

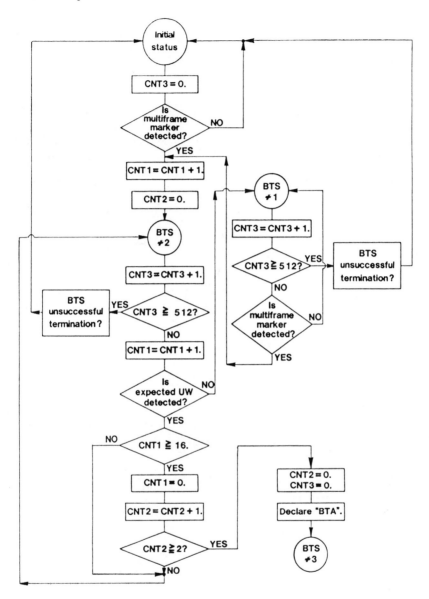

**Figure 7.73(a)** Flow chart of BTS subunit (frame mode).

7.77 shows the status and processing flow charts of the control frame synchronization unit.

### 7.8.1.4 Superframe Synchronization

The superframe is used to coordinate sequential acquisition and burst time plan changes. A synchronization procedure is established to ensure that both the primary and secondary station superframes start at the same time.

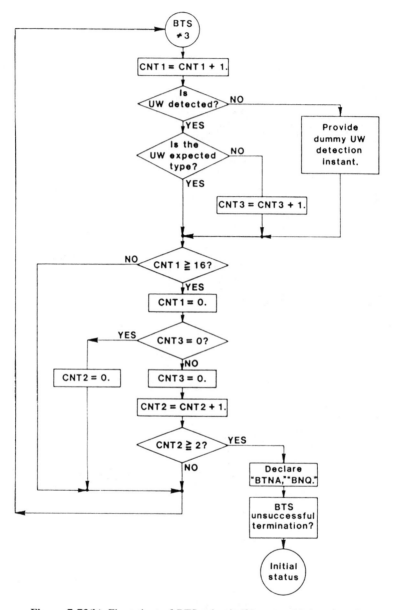

**Figure 7.73(b)** Flow chart of BTS subunit (frame mode) (continued).

The MP reference station initiates the synchronization procedure of transmitting a Start of Superframe (SS) code in the CDC of the first multiframe of the superframe. As each of the other stations decodes the SS code, it generates an SS instant at SOTMF at time

$$T_J + 503 \, T_M + D_n$$

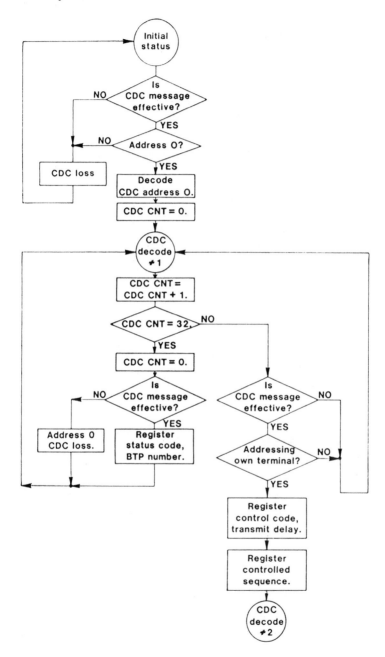

**Figure 7.74(a)** Flow chart of CDC decoding subunit.

where $T_J$ is the SORMF of the multiframe containing the SS code. The value of $D_n$ is the last value of $D_n$ received prior to implementing the SS instant. If the reference station fails to receive the SS code in either the PRB or SRB, it

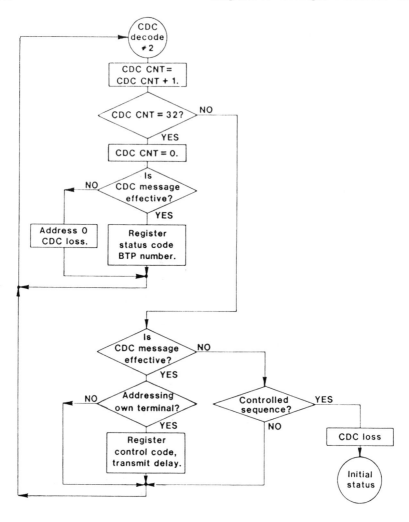

**Figure 7.74(b)** Flow chart of CDC decoding subunit (continued).

generates an SS instant at $512\ T_M$ from the last SS instant. Also if the reference station fails to receive the SS code for two consecutive superframes, it sounds an alarm and continues to generate SS instants at intervals of $512\ T_M$ from the last SS instant. If during the synchronization phase UW detects the SS code in the PRB and SRB in different multiframes it sounds an alarm and generates SS instants at intervals of $512\ T_M$ from the last SS instant. The timings for achieving superframe synchronization are shown in Figure 7.78.

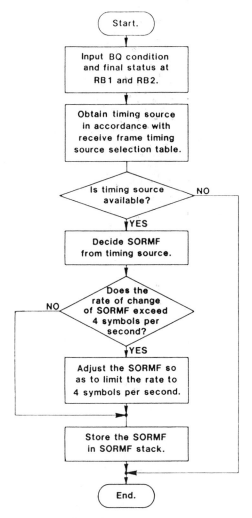

**Figure 7.75** Flow chart of RFS unit.

The superframe synchronization unit simulates the superframe generation procedure of the MP and the superframe synchronization procedure of the other reference stations on a multiframe basis. The status and processing flow charts of the superframe synchronization unit are shown in Figures 7.79(a) and 7.79(b).

### 7.8.1.5 The TFAS—TFG

The Transmit Frame Acquisition and Synchronization-Transmit Frame Generation (TFAS-TFG) unit simulates TFAS and TFG procedures only after the superframe and control frame synchronization of the station is established. One of the following three subunits that compose the TFAS-TFG unit is selected on the basis of the role of the station at that time.

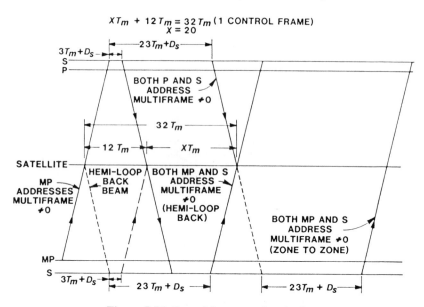

**Figure 7.76** Control frame synchronization.

- TFAS (TFA, TFS)

  The TFAS subunit is executed by the RS model only when the role of the station is I or S. Figure 7.80 shows the block diagram of this subunit. In the I station, the TFA subunit and the TFS subunit are executed. On the other hand, in an S station only the TFS subunit is executed. The status and processing flow charts for the TFAS subunit as a whole are shown in Figures 7.81(a) to 7.81(d).

- TFS in the primary role

  The TFS in the primary role subunit simulates the TFS procedure when the reference station is started up as P. The functional block diagram of this subunit is the same as Figure 7.80 without the TFA input. In this subunit, at first the coincidence of the BTP number in the CDC message of the received controlling reference burst and the BTP number in the current BTP of the station are checked. Next the reception of the SYNC code from the controlling reference station is confirmed, and finally the TFS subunit which is the same as that for the I or S station is executed. The TFS protocol in the primary role is shown in Figures 7.82 and 7.83.

- TFG

  The TFG subunit (block diagram shown in Figure 7.84) simulates the TFG procedure performed by the master primary reference station. In this subunit, synchronization of all transmit bursts from the MPR is achieved and also all transmit bursts are monitored by means of the information from the controlling reference station. The TFG subunit protocol is shown in Figure 7.85.

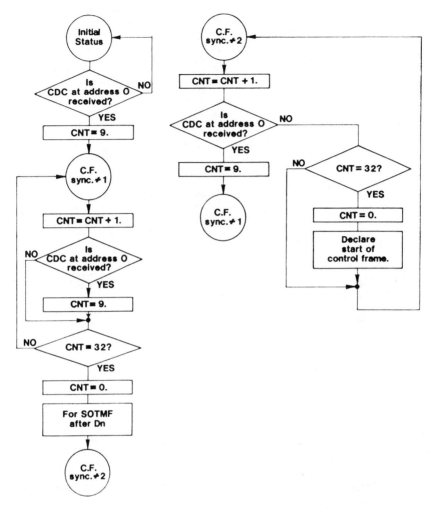

**Figure 7.77** Flow charts of control frame synchronization unit.

### 7.8.2 Data Management Modules

Both the reference station and traffic station models use data management modules. Each station model has the received (RX) and the transmitted (TX) data management modules. The functions of the data management modules are the same for reference stations and traffic stations, except for the difference in the CDC message. The protocol aspect of the data management modules is discussed separately in the following sections.

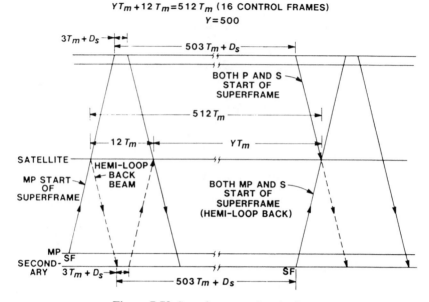

**Figure 7.78** Superframe synchronization.

### 7.8.2.1  RX (Receive) Data Management

The burst data received at a station is stored in the reception buffer. The RX data management module extracts from the reception buffer the burst data received by the station for one multiframe period. The RX data management module also produces the effect of stochastic phenomena and station failure on the received burst data. The functions of the RX data management module are as follows:

- Determine the 32 msec interval to receive the burst data.
- Extract the burst data from the reception buffer received in the time interval.
- Produce the stochastic phenomena effect on the burst data. Table 7.17 lists the stochastic phenomena, burst data to be modified in accordance with the stochastic phenomena, and the burst data modification processes.
- Produce the terminal function failure effect on the burst data. The terminal function failures include stop of reception function, UW miss detection, UW false detection, SC loss, CDC loss, SC false decoding, CDC false decoding, HBER.

The protocol of the process performed by the RX data management is shown in Figure 7.86.

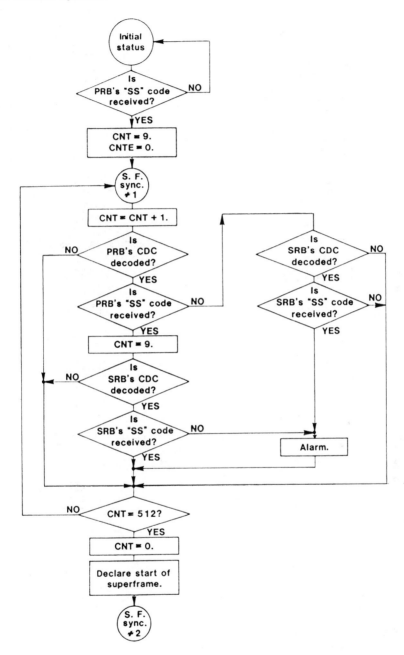

**Figure 7.79(a)** Flow chart of superframe synchronization unit.

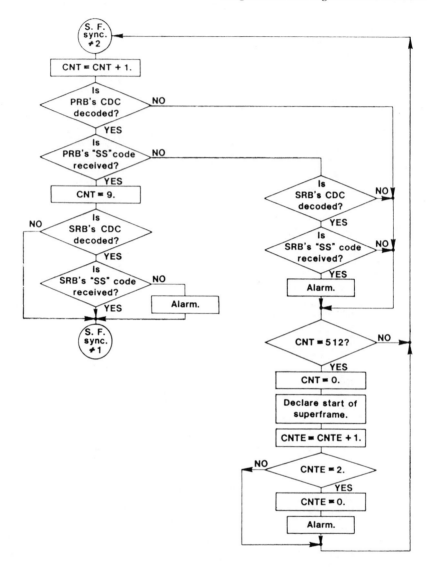

**Figure 7.79(b)** Flow chart of superframe synchronization unit (continued).

### 7.8.2.2  TX (Transmit) Data Management

The TX data management module generates burst data to be transmitted in accordance with the data provided by other modules in the station model. The functions of the TX data management module are as follows:

- Determine the burst type (RB1, RB2, TB) to be transmitted.
- Select the SC message to be transmitted. The SC message with a higher priority is selected.

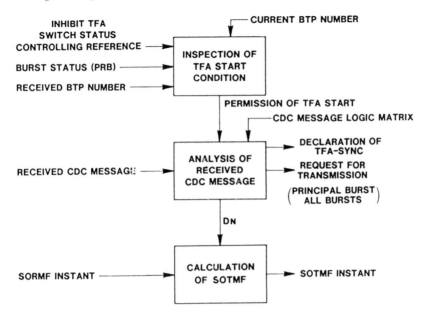

INHIBIT TFA
SWITCH STATUS
CONTROLLING REFERENCE

BURST STATUS (PRB)

RECEIVED BTP NUMBER

CURRENT BTP NUMBER

INSPECTION OF
TFA START
CONDITION

PERMISSION OF TFA START

CDC MESSAGE LOGIC MATRIX

RECEIVED CDC MESSAGE

ANALYSIS OF
RECEIVED
CDC MESSAGE

DECLARATION OF
TFA-SYNC

REQUEST FOR
TRANSMISSION

(PRINCIPAL BURST
  ALL BURSTS)

DN

SORMF INSTANT

CALCULATION
OF SOTMF

SOTMF INSTANT

**Figure 7.80** Block diagram of TFAS subunit.

- Generate the CDC message to be transmitted in accordance with the TAS-TSS control data (RS model only). The change of control code SYNC to IAP2 in the Hold IAP2 condition is also made by this module.
- Register the burst transmission model starting event.

The block diagram of the TX data management module is shown in Figure 7.87. The protocol of the TX data management process is shown in Figure 7.88.

## 7.9   PROGRAM DESIGNATIONS

The MAIN Simulation Controller of the TNS programs controls the Event Controller (EVTCNT) and the Display Controller (DSPC) has the capabilities of editing, storing, printing, and displaying.

There are six main subprograms controlled by the event controller (EVTCNT), as shown in Figure 7.89. From these six, there are two subprograms: the reference station model control (RTESMC) and the traffic station model control (TMODSC). The RTESMC is also shown in the figure, where the first letter R in each subprogram denotes a reference station. The main RTESMC subprograms include the control mechanisms of acquisition and synchronization (RASYNC), network coordination (RNCCOO), and network control (RNNCOS). The subprogram RKTRKP, which is activated by the SDNTX code, is included here only to complete the RTESMC calling series. There are five functions that are tied to RASYNC. These functions, all related to acquisition and synchronization, are RAASSR, RARFS, RACFSY, RASFSY, and RATASG, as shown in Figure

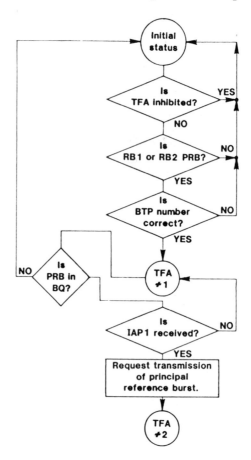

**Figure 7.81(a)** Flow chart of TFAS subunit.

7.89. The detail functions of RCNCOO and RNNCOS are shown in Figures 7.90 and 7.91, respectively.

Although in the INTELSAT TDMA specifications, APD (acquisition phase detection), SPD (synchronization phase detection), SGT (selective do not transmit code generation for terminal control), TBM (traffic burst monitoring), and SUA procedures are detailed separately for network operation and protocols, separate discussions are not warranted for our purpose here. The APD and SPD are merely unique word detection procedures. SUA is merely a procedure with a different window size for unique word detection in the network start-up. However, the TNS programs faithfully contain all the detailed procedures.

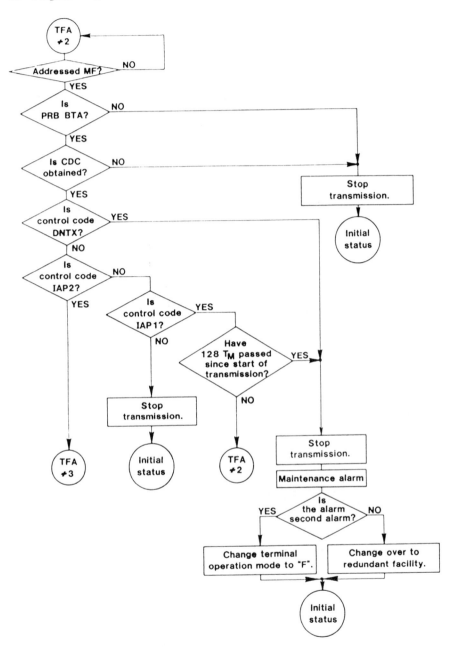

**Figure 7.81(b)** Flow chart of TFAS subunit (continued).

The computer program designations for the traffic station model control (TMODSC) are shown in Figure 7.92, in which the first letter T of each program denotes the traffic station. The acquisition and synchronization control (TACQSY)

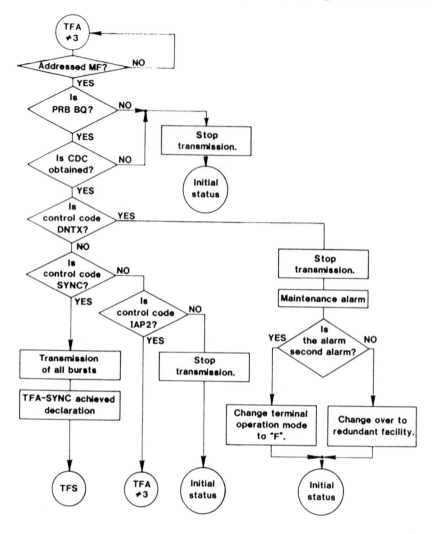

**Figure 7.81(c)** Flow chart of TFAS subunit (continued).

and housekeeping control (THCONT) subprograms of the TMODSC are also shown in Figure 7.92. This figure also shows the unique word loss and high bit error rate alarm subprograms.

## 7.9.1 Program Parameters

Network configuration is determined in accordance with program input data, namely, the locations of stations, the burst time plan of the stations, the beam coverage of the satellite, and the beam connection of satellites. These parameters

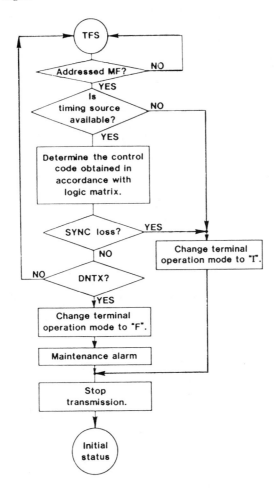

**Figure 7.81(d)** Flow chart of TFAS subunit (continued).

and data ranges are listed in Table 7.18. During the simulation, events such as UW miss detection, SC message loss, and station failure can be generated at designated burst stations and times in accordance with the program input data. During the execution of simulation, the designated items of network operation data, such as burst transmission time, reception time, CDC message content, SC message content, events, time of events, and status of procedures, are stored for the analysis of network operation.

The simulation program calls for a large quantity of data. In order for the data to be reused with modifications, the simulation program is so designed that the amount of initial input data is as small as possible and file management is as simple as possible.

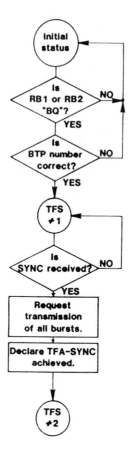

**Figure 7.82** Flow chart of TFS in the primary role.

The input data are categorized into three groups, namely, network parameters, event parameters, and control parameters. Network parameters characterize the network model to be simulated. Even parameters specify the events to occur in the simulation. Control parameters control the execution and selection of items to be simulated. The following is a summary of the main items.

1. Network parameters
   - Network configuration
   - Burst Time Plan—both CTP and RCTP
   - Reference terminal model
   - Traffic terminal model
   - Satellite model and position prediction parameters
2. Event parameters
   - Switch operation of the network stations
   - Operator command at the network stations

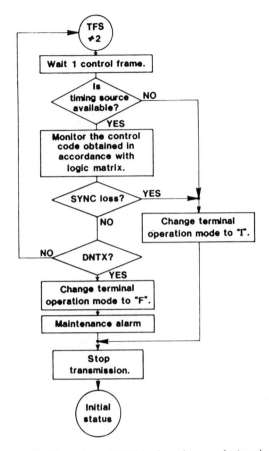

**Figure 7.83** Flow chart of TFS in the primary role (continued).

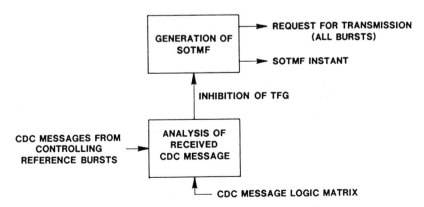

**Figure 7.84** Block diagram of TFG subunit.

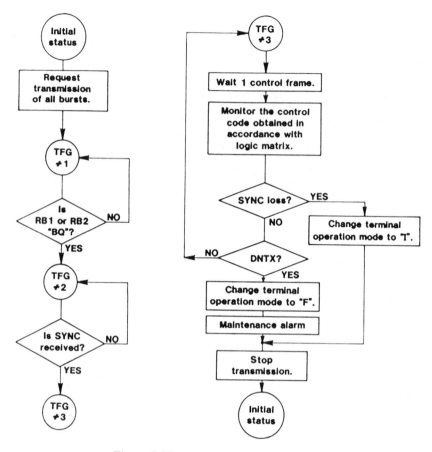

**Figure 7.85** Flow chart of TFG subunit.

- Function failure
- Stochastic phenomena
3. Control parameters
    - Execution control; start, stop, and pause; selection of execution mode
    - Output control: items of output data, output device selection

Since it is inefficient to create a set of input data every time the simulation program is executed, a special database, or milestone, is used to facilitate systematic management of input data. The database, a data set in secondary storage, contains all the input data, namely, network parameters, event parameters, and control parameters. To execute a simulation, a suitable milestone is first prepared, with modification if necessary. Modification of data in the milestone is made by interactive operation and the updated data set can be stored as another milestone.

Output data are classified into the following six groups:

**Table 7.17**

Stochastic phenomena and modification of burst data.

| Stochastic Phenomena | Burst Data To Be Modified | Burst Data Modification Process |
|---|---|---|
| UW miss detection | Unique Word Status<br><br>0: miss detection<br>1: correct detection | 1) Get UW miss detection rate: $r$<br>2) Generate uniform random number: $p$ ($0 \leqq p \leqq 1$)<br>3) Compare $r$ with $p$<br>$p > r$; retain previous status<br>$p \leqq r$; miss detection status |
| UW false detection | Unique Word False Status<br><br>0: true detection<br>1: false detection | 1) Get UW false detection rates: $r$<br>2) Generate uniform random number: $p$ ($0 \leqq P \leqq 1$)<br>3) Compare $r$ with $p$<br>$p > r$; true detection status<br>$P \leqq r$; false detection status |
| SC loss | SC Status<br><br>0: SC lost<br>1: Correct decoding | 1) Get Sc loss rate: $r$<br>2) Generate uniform random number: $P$ ($0 \leqq P \leqq 1$)<br>3) Compare $r$ with $p$<br>$p > r$; retain previous status<br>$p \leqq r$; SC lost |
| CDC loss | CDC Status<br><br>0: CDC lost<br>1: Correct decoding | 1) Get CDC loss rate: $r$<br>2) Generate uniform random number: $P$ ($0 \leqq p \leqq 1$)<br>3) Compare $r$ with $p$<br>$P > r$; retain previous status<br>$P \leqq r$; CDC lost |

- Events that occurred at a terminal and the satellite
- Status of terminals and the satellite
- Received data at a terminal (CDC, SC)
- Transmitted data at a terminal (CDC, SC)
- Data arrived at the satellite
- Transmit delay ($D_n$)

Some or all of these items can be selected for output. While the simulation program is running, data can be displayed at the interactive work station to monitor simulation progress.

## 7.9.2   Network Operations

Prior to the network operation test the protocol functions of the network elements, individual procedures, various controllers, models, and modules are pretested with a variety of input data. Next, the protocol functions of the network are

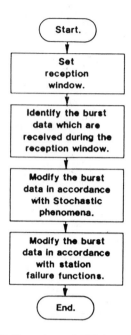

**Figure 7.86** Flow chart of RX data management.

evaluated as a whole. This evaluation is referred to as the network operation test. As network operation depends on many parameters, such as the network configuration, beam coverage, number of transponders, and number of stations, the results of network operation are presented in three examples. The first example illustrates testing the fundamental protocol functions of the network with the simplest network configuration and minimum number of stations and transponders. For this case, we use two reference stations, one traffic station with a single transponder in the simple loop-back coverage, and two BTPs (BTP1 and BTP2) used for the BTP change procedure. The second example makes use of four reference stations and six traffic terminals, with five transponders and twenty-one bursts.

The second example contains both hemispheric and zonal beam coverages. The third example is the actual first phase of the INTELSAT TDMA network introduced in the Atlantic Ocean Region. The network consists of four reference stations, eight traffic stations, and two transponders. All three examples are illustrated separately in the following.

*Example 7.10:* In the simple beam loop-back configuration shown in Figure 7.93, T6 and T7 are designated, respectively, as the master primary and secondary reference stations and T5 is designated as the traffic station. The transponder number is 51 with a transponder delay of 50 symbols. The beam connection is from East Zone to East Zone. Assuming the nominal satellite position located

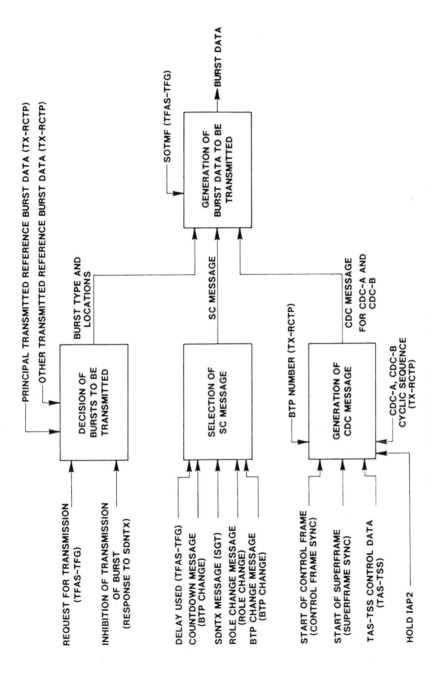

**Figure 7.87** Block diagram of TX data management module.

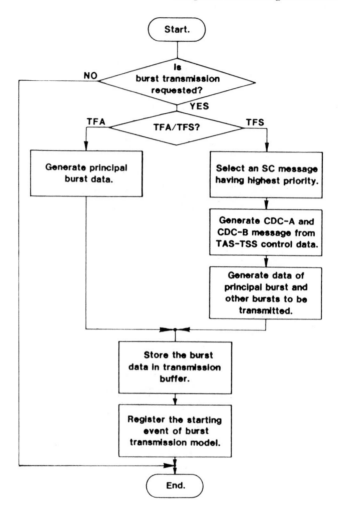

**Figure 7.88** Flow chart of TX data management.

at X = 298,2678, Y = −29,826,730, Z = −52,058 (all in meters), the satellite orbiting parameters are as follows:

| | |
|---|---|
| Semimajor axis of the orbit | = 42185650 (meters) |
| Orbital eccentricity | = 0.0001 |
| Orbital inclination to the equator | = 0.1 (degree) |
| Longitude at the ascending mode | = 0 (degree) |
| Argument of perigee | = 45 (degrees) |
| Latest time at perifocal passage | = 0 (msec) |

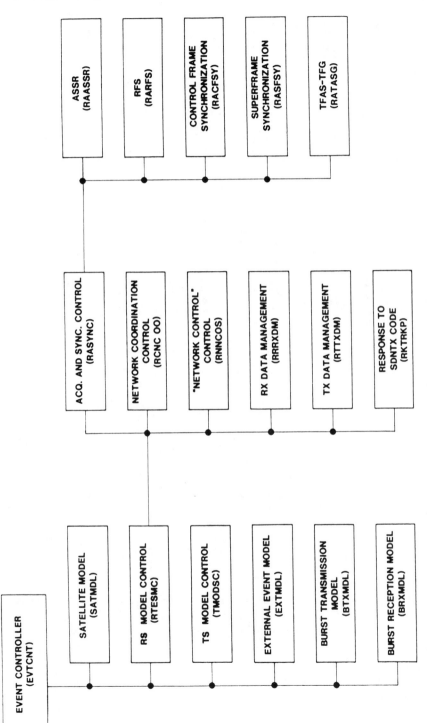

**Figure 7.89** The function of controller, model, and module.

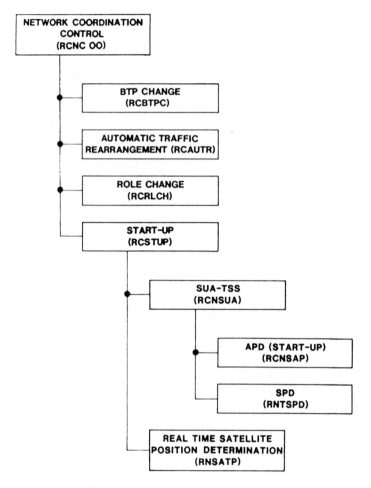

**Figure 7.90** Reference station network coordination module.

The locations of the stations are shown in Table 7.19.

The two BTPs with burst occurrences are shown in Figure A.6.1 of Appendix 6. Timing begins with the start of frame (SOF), and the numbers on top are in symbols. The burst numbers from T5, T6, and T7 are denoted as B10, B101, and B102, respectively. The destinations from T5 are T6 and T7; for the reference stations from T6 and T7 they are T6, T7, and T5.

All the program timing is equal to the number of symbols in a frame multiplied by 16, plus the number of symbol duration. This is done only to facilitate counting. The events of this simple network start-up protocol simulation are illustrated in Figures 7.94(a) and 7.94(b). Initially, the secondary station is in

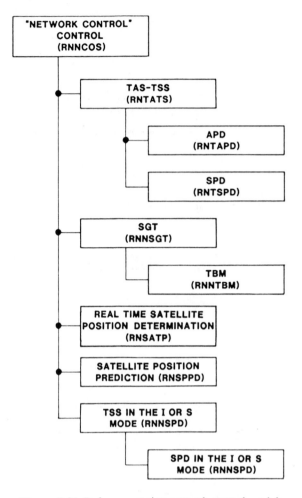

**Figure 7.91** Reference station network control module.

the inoperative mode, the required protocol to bring this station into operative mode is shown. The establishment of SOTMF and synchronization achieved status for the bursts are also indicated. Figure 7.94 is provided by K. Koga.

In this example, for the burst time plan change from BTP1 to BTP2 the program input milestone is essentailly the same as that for the network start-up and station promotion, except the event parameters and the control parameters are different. These parameters are listed as follows:

```
                        CONTROL PARAMETERS
0000 T=      0    OUT= 2    T*=   6    DEV=3
  ** EVENTS OF T NO.6 ARE PUT OUT
0010 T=      0    OUT= 1    T*=   6    DEV=1    CYC=3
  ** SUPERFRAME CYCLE STATUS OUTPUT
0020 T=      0    OUT= 1    T*=   5    DEV=1    CYC=3
  ** SUPERFRAME CYCLE STATUS OUTPUT
0030 T=      0    OUT= 1    T*=   7    DEV=1    CYC=3
  ** SUPERFRAME CYCLE STATUS OUTPUT


  ** T:TIME   EXEC:EXECUTION CONTROL CODE   OUT:OUTPUT CONTROL CODE
     NUMBER   DEV:DEVICE   CYC:OUTPUT CYCLE


0040 T=      0    OUT= 2    T*=   5    DEV=3
  ** EVENTS OF T NO.5 ARE PUT OUT
0050 T=      0    OUT= 2    T*=   7    DEV=2
  ** EVENTS OF T NO.7 ARE PUT OUT
0060 T=      0    EXEC= 1    DATA=   1
  ** SIMULATION START
0070 T= 25000    EXEC= 5
  ** PAUSE


  ** T:TIME   EXEC:EXECUTION CONTROL CODE   OUT:OUTPUT CONTROL CODE
     NUMBER   DEV:DEVICE   CYC:OUTPUT CYCLE
```

```
                        EVENT PARAMETERS
0000 T=      16    SW= 1    T*=   6    STS=1
  ** SYSTEM START-UP SWITCH ON
0010 T=      16    SW= 8    T*=   6    STS=1
  ** SAT. POSITION PREDICTION SW ON
0020 T=      16    SW= 8    T*=   7    STS=2
  ** SAT. POSITION RANGING SW ON
0030 T= 18400    OPE= 1    T*=   7    DATA=   5,   6,   7,   0
  ** TERMINALS 5,6,7 ARE AUTHORIZED


  ** T:TIME   SW:SWITCH OPERATION CODE   OPE:OPERATOR COMMAND CODE
     T*:TERMINAL NUMBER   STS:STATUS   FAIL:FAILURE CODE   B*:BURST NUMBER


0040 T= 18600    OPE= 1    T*=   6    DATA=   5,   0,   0,   0
  ** TERMINAL 5 IS AUTHORIZED
0050 T= 26000    SW= 3    T*=   6    STS=1
  ** INITIATE BTP CHANGE
```

Where the output control parameter OUT = 2 denotes station or satellite events, output events $T^* = 6$ means the terminal number is 6; DEV = 1, 2, 3 represent, respectively, line printout, CRT, and both line printout and CRT display; SW = 1, 3, 8 represent, respectively, the network start-up switch for MP and P, the BTP change switch for MP, and the satellite position determination method selection switch. STS = 1, 2 (when SW = 8), respectively, is for the prediction and ranging method; OPE = 1 denotes authorization of acquisition for the reference stations; EXEC = 1, 5 represent, respectively, simulation start and pause; CYC = 3 signifies the superframe output cycle. The example sequence of the BTP change protocol is shown in Figures 7.95(a) and 7.95(b), which are provided by K. Koga.

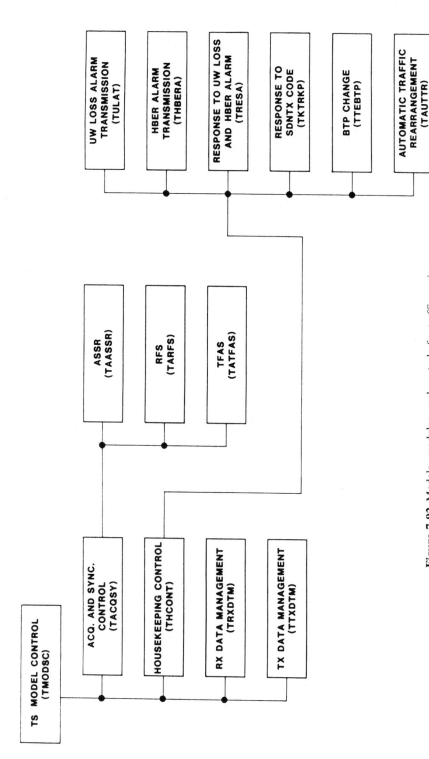

**Figure 7.92** Models, modules, and controls for traffic stations.

**Table 7.18**

Network configuration parameters.

| | Description | Accuracy or Data Range | |
|---|---|---|---|
| **Beam Connection** | Channel* | 0–4 | |
| | Up Link Beam Coverage** | 0–5 | |
| | Down Link Beam Coverage*** | 0–5 | |
| | Transponder Delay | 0000–9999 | SYMBOL |
| | Note: Repeat until all transponders are defined. | | |
| **Earth Station** | Terminal No. | 0000–1999 | |
| | West Hemisphere Coverage | 0–1 | |
| | East Hemisphere Coverage | 0–1 | |
| | West Zone Coverage | 0–1 | |
| | East Zone Coverage | 0–1 | |
| | Terminal Position X | 1 m | |
| | Terminal Position Y | 1 m | |
| | Terminal Position Z | 1 m | |
| | Internal Propagation Delay | 0000–9999 | SYMBOL |
| | Note: Report until all terminals are defined in ascending order of the terminal number. | | |
| **Nominal Satellite Position** | X | 1 m | |
| | Y | 1 m | |
| | Z | 1 m | |
| **Satellite Orbit Parameters** | a: seminar axis of the orbit | 35000–50000 km | |
| | e: orbital eccentricity | $0-10^{-3}$ | |
| | i: orbital inclination to the equator | 0–0.5 deg. | |
| | r: longitude of the ascending node | 0–360 deg. | |
| | w: argument of perigee | 0–360 deg. | |
| | T: latest time of perifocal passage | $0-(2^{31}-1)$ mS. | |

```
*Channel              0 : not used
                      1 : Satellite Channel Number 1, 2
                      2 : Satellite Channel Number 3, 4
                      3 : Satellite Channel Number 5, 6
                      4 : Satellite Channel Number 7, 8
**,***Beam Coverage   0 : not used
                      1 : West Hemisphere Beam
                      2 : East Hemisphere Beam
                      3 : not used
                      4 : West Zone Beam
                      5 : East Zone Beam
```

**Table 7.19**

Station locations.

| Station | Location in Meters | | |
| | X | Y | Z |
|---|---|---|---|
| T6 (MP) | −747,010 | −5,315,256 | 3,485,690 |
| T7 (S) | 834,229 | −4,691,240 | 4,272,743 |
| T5 (T) | 274,375 | −5,235,388 | 3,670,889 |

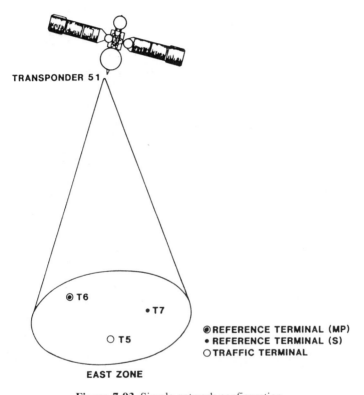

**Figure 7.93** Simple network configuration.

In addition to the network start-up and BTP change, other evaluations can be performed on the TNS programs, such as traffic rearrangement, stochastic phenomena generation, satellite function failure, station function failure, and other detailed protocol functions.

*Example 7.11:* A more complete detailed network configuration is shown in Figure 7.96 comprising a total of ten stations in the network. Four are reference stations and the rest are traffic stations. Designated are the stations, beam coverages, and transponder numbers. The locations of the stations with their corresponding delays are listed in Table A.6.1 of Appendix 6. The satellite location

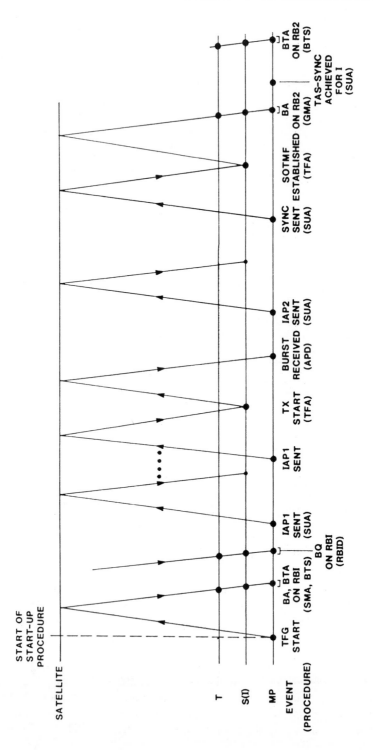

**Figure 7.94(a)** Network start-up protocol of Example 7.10.

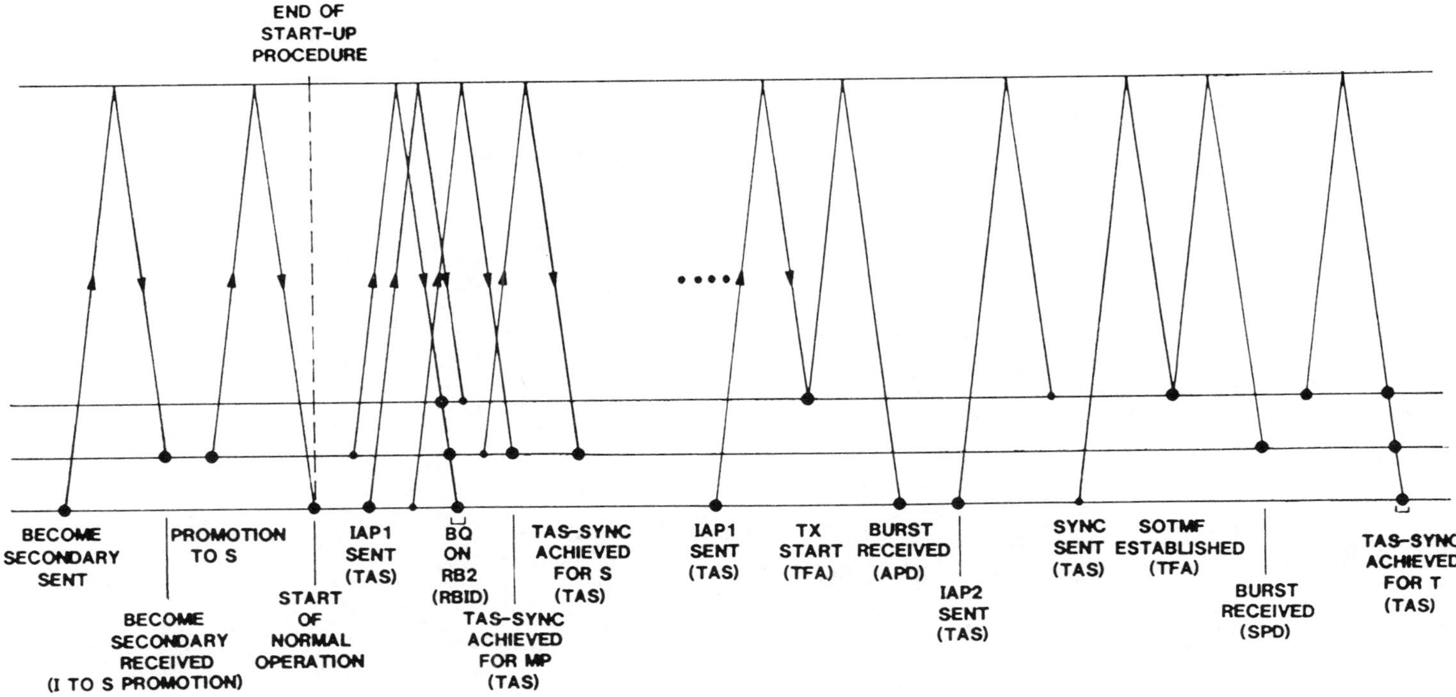

**Figure 7.94(b)** Network start-up protocol of Example 7.10 (continued).

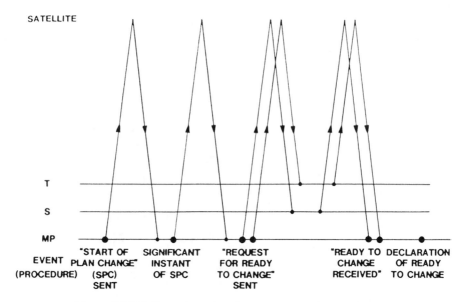

**Figure 7.95(a)**  Burst time plan change protocol of Example 7.10.

is at 330.5°E. The transponder delay variations are less than three symbols. The TDMA burst numbers for the reference stations and for the traffic stations are listed in Tables A.6.2 and A.6.3, respectively. For simulation purposes, the principle bursts as well as the transponder numbers are also indicated in these tables.

The burst timing formats corresponding to BTP3 and BTP4 are shown, respectively, in Figures A.6.2 and A.6.3 of Appendix 6, and are used for the preparation of BTP change. In both BTP3 and BTP4, each of the reference stations in WZ transmits two reference bursts and each of the reference stations in EZ transmits three reference bursts. The traffic terminal in WH (excluding WZ) transmits one traffic burst and the traffic terminals in WZ transmit two traffic bursts. Each traffic terminal in EH transmits one traffic burst and each traffic terminal in EZ transmits three traffic bursts. Samples of BTP3 and BTP4 contents, respectively, are listed in Tables A.6.4 and A.6.5 of Appendix 6, where the beginning and end bursts are marked in terms of symbol timing. The burst position refers either to the end of unique word or the start of frame position transmit in the burst.

*Example 7.12:*   The actual INTELSAT TDMA network, first introduced through the INTELSAT V satellite in the Atlantic Ocean region, initially consists of eight countries, as shown in Table 7.20. In this table the station number, station location, station representation, and station operational status are also listed.

There are four reference stations (1, 2, 33, 34) and eight traffic stations (3, 4, 5, 6, 8, 10, 11, 12). With the beam coverage and the transponder connection the stations are shown in Figure 7.97, where the functions of the reference stations are designated. The initial burst time plan for this network is shown in

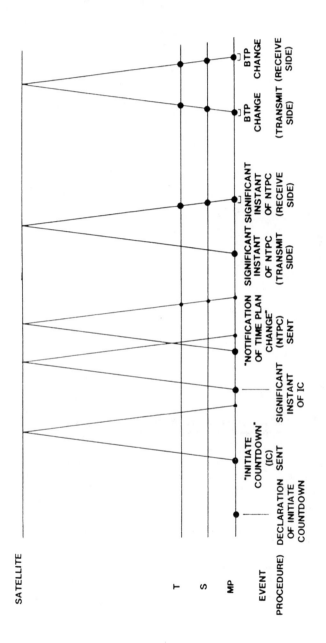

**Figure 7.95(b)**  Burst time plan change protocol of Example 7.10 (continued).

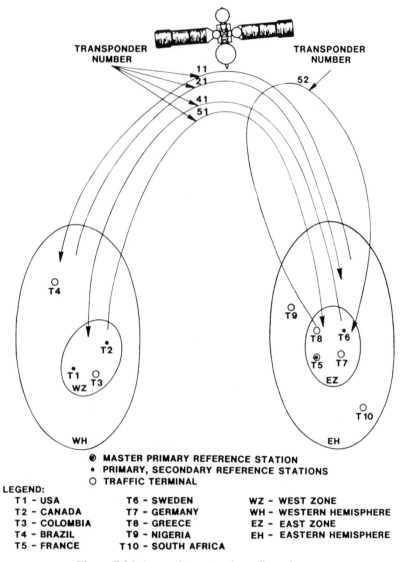

**TRANSPONDER NUMBER**

**TRANSPONDER NUMBER**

11
21
41
51

52

O
T4

O
T9

O
T8   T6
T5   O
EZ   T7

T2

T1   O
WZ   T3

O
T10

WH

EH

◉  MASTER PRIMARY REFERENCE STATION
•  PRIMARY, SECONDARY REFERENCE STATIONS
O  TRAFFIC TERMINAL

LEGEND:

| | | |
|---|---|---|
| T1 - USA | T6 - SWEDEN | WZ - WEST ZONE |
| T2 - CANADA | T7 - GERMANY | WH - WESTERN HEMISPHERE |
| T3 - COLOMBIA | T8 - GREECE | EZ - EAST ZONE |
| T4 - BRAZIL | T9 - NIGERIA | EH - EASTERN HEMISPHERE |
| T5 - FRANCE | T10 - SOUTH AFRICA | |

**Figure 7.96** A complete network configuration.

Figure 7.98, in which the reference bursts (RB1, RB2) and the duration of the traffic bursts are clearly indicated in terms of thousands of symbols.

Each symbol is 1/60 μsec. The East to West hemispheric transponder handles the upper three rows of the time plan, i.e., rows 1, 2, and 3. The West to East hemispheric transponder handles the lower three rows of the traffic burst time plan. The countries where the reference stations are located are also shown in scale with their corresponding reference burst.

Both the reference burst format and the traffic burst format of the INTELSAT TDMA network are shown in Figure 3.31 of Volume I. In terms of burst for-

**Table 7.20**

Network stations of Example 7.12

| Country | Station Designation | Operation Status | Sta. Numb. | Station Location | | |
|---|---|---|---|---|---|---|
| | | | | X | Y | Z |
| Canada (CAN) | MVL-4A | RS | 1 | 1956698 | −4133202 | 4446196 |
| | | TS | 3 | " | " | " |
| U.S.A. (USA) | ETM-1A | RS | 2 | 879620 | −4858041 | 4038117 |
| | | TS | 8 | " | " | " |
| Germany (D) | RAI-5A | RS | 33 | 4195813 | 824368 | 4732468 |
| | | TS | 11 | " | " | " |
| France (F) | PBD-6A | RS | 34 | 4239701 | 287877 | 4756338 |
| | | TS | 4 | " | " | " |
| Italy (I) | LRO-1A | TS | 5 | 4358520 | 722250 | 4600254 |
| Great Britain (UK) | MDY-2A | TS | 6 | 3919331 | −194243 | 5028102 |
| Belgium (BEL) | LSV-2A | TS | 10 | 4072047 | 366644 | 4895375 |
| South Africa (AFS) | PRE-3A | TS | 12 | 5079617 | 2666744 | −2786871 |

MVL: Mill Village  
ETM: Etam  
RAI : Raisting  
PBD : Pleumeur-Bodour  
LRO : Lario  
MDY: Madley  

LSV : Lessive  
PRE : Pretoria  
RS  : reference station  
TS  : traffic station  
Station locations: in meters

mating, the only difference between the two types of bursts is that reference bursts contain CDC (control delay channel) segments for network protocol and such bursts do not carry traffic. The actual burst formation procedure for the burst time plan, as shown in Figure 7.99, is described as follows: After starting, the procedure is to first check the availability of the transponder. Depending on the location of the station, the East to West or West to East hemisphere transponder is determined. If the transponder is available and can be allocated, the next step is to check the availability of the communicating station or stations for which a link or links are to be established. If both the transponder and the desired station are not available, then burst formation is ceased; otherwise, the procedure goes to the next step, selecting burst type in terms of either single destination or multiple destination transmission. In either case, as shown in Figure 7.99, burst assignment in the transponder and the required preamble are provided through control logic (CL) circuits to form a burst with a string of binary digits.

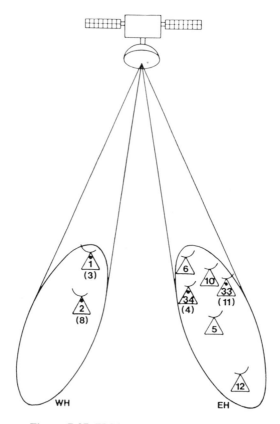

**Figure 7.97** TDMA network of Example 7.12.

## 7.10 THEORETICAL BASIS OF PROTOCOL INFORMATION

Gallager considers the basic limitations on the amount of protocol information that must be transmitted in a very large data communication network [19,20]. His network model may be interpreted to include intersatellite links as shown in Figure 7.100, where $S_0$, $S_1$, $S_2$ are the satellites. In network terminology, as described in Chapter 7 of Volume I, they are called nodes. $l_{0,1}$, $l_{1,2}$, and $l_{2,0}$ are the intersatellite links. $T_0$, $T_1$ . . . and $R_0$, $R_1$ . . . denote transmit and receive stations, respectively. Messages are transmitted from each transmitter to its designated receiver. The protocol information in such a network is essentially the control information, such as destination, number, and type of messages, message origin, routing and message distribution. In some cases network synchronization and timing are also considered as part of the protocol information.

Gallager observed that a protocol is a source code in information theory for representing such network control information. Through the rate-distortion theory of a transmitter source (see Section 7.10.1) a lower bound to the average protocol information about message arrival times, which must be transmitted in order to limit the averaged delay, can be derived. Because of its fundamental significance

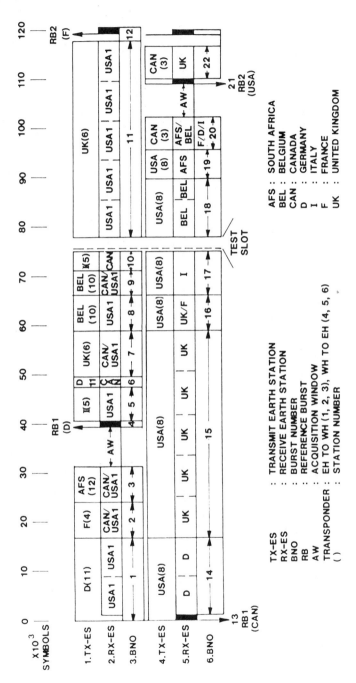

**Figure 7.98** TDMA traffic burst time plan for Example 7.12.

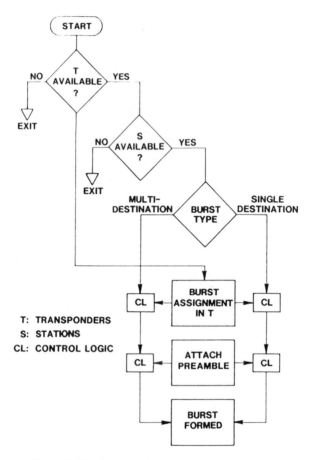

**Figure 7.99** Procedure for TDMA burst formation.

in protocol information evaluation, we proceed to study the results obtained by Gallager on data communication networks in the context of satellite applications.

### 7.10.1 The Rate Distortion Theory

The rate distortion theory is a part of the study of source coding with a fidelity criterion in information theory. A simple example is the analog-to-digital and digital-to-analog conversion process in which a difference exists in the reproduction of the original analog source due either to quantization or overload noise. The difference between the original source and the binary representation of the source is considered as distortion. The fidelity criterion is the maximum tolerable value of this average distortion. One of the objectives in source encoding is to find the minimum number of binary digits required to encode a source so that it can be reproduced to meet a given fidelity criterion.

Gallager has shown that if a transmission channel has a capacity that is too

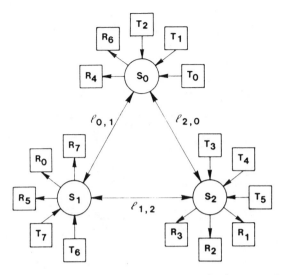

**Figure 7.100** Inter-satellite linked data communication network model.

small to reliably transmit the minimum rate binary sequence required to represent the source with a given fidelity criterion, then the fidelity criterion cannot be met, no matter what kind of processing is used between the source and the channel. This is, indeed, a powerful conclusion regarding the transmission rate and the maximum value of distortion.

Consider a discrete memoryless source with possible $0, 1, \ldots, K - 1$ letter occurrence for each source letter, $0 \leq k \leq K - 1$. The probability of such letter occurrence is $Q(k)$. When the set of source letters terminates at the destination they are represented by the letters $0, 1, \ldots, J - 1$. The probability of destination letter $j$ conditional on the source letter $k$ is $P(j/k)$. In terms of $Q(k)$ and $P(j/k)$ the average distortion is

$$\bar{d} = \sum_k \sum_j Q(k) P(j/k) d(k/j) \tag{7.5}$$

where $d(k/j)$ is the distortion measure, which is a numerical value in practice if the source letter $k$ is represented by the letter $j$ at the destination. The rate distortion function of a source relative to the given distortion measure is defined as

$$R(d^*) = \min \bar{I}(Q;P)$$
$$P : \bar{d} \leq d^* \tag{7.6}$$

where $\bar{I}(Q;P)$ is the average mutual information, which is identical in expression but different in meaning, as shown in Section 2.6.1 of Chapter 2 in Volume I for transmission channels

$$\bar{I}(Q;P) = \sum_{k=0}^{K-1} \sum_{j=0}^{J-1} Q(k) P(j/k) ln \frac{P(j/k)}{\sum_i Q(i) P(j/i)} . \tag{7.7}$$

The minimization of $\bar{I}(Q;P)$ for calculating $R(d^*)$ is the overall conditional probabilities $P(j/k)$ subject to the constraint that the average distortion is less than or equal to $d^*$. $R(d^*)$ is considered as the rate of the source, in nats per symbol relative to the fidelty criterion $d^*$.

### 7.10.2   A Transmitter Model

We consider a simplified two-satellite, single intersatellite link case first. Assume satellite A antenna covers a set of $K$ transmit earth stations (sources) and through intersatellite link $\ell_{A,B}$ satellite B covers $K$ number of receiving stations. For a nonbroadcasting (such as telephone conversation) case, transmit earth station $T_k$, $1 \leq K \leq K$ transmits messages only to the receiving station $R_k$. The timing of the transmitting stations is synchronous, as in the TDMA operation. If messages from the set of transmitting stations are assumed to be transmitted at random times and with random message duration, then we may model such a transmitter as an independent Markov source, which is characterized, as shown in Figure 7.101, by the busy and idle states and the corresponding transition probabilities. The transitions from idle to busy and from busy to busy state are indicated by the two possible binary symbols 0,1; and the idle symbol $i$ is shown for the idle to idle state and the transition from busy state toward the idle state. From idle state to the transmission state each symbol has a transition probability of $\delta/2$; the transition probability from the busy state to the idle state is $\varepsilon$. For simplicity we assume $\delta$ and $\varepsilon$ are the same for all the transmitters in the network. For two symbols to remain in the busy state there is a probability of $(1-\varepsilon)/2$ for each symbol. For an idle symbol to remain in the idle state this probability is $1-\delta$; because of the single idle symbol we do not divide $1 - \delta$ by 2.

The steady state probability of the busy state is the ratio of the sum of the transition probabilities for busy symbols to the total transition probabilities, i.e.,

$$P_s = \frac{\dfrac{\delta}{2} + \dfrac{\delta}{2}}{\dfrac{\delta}{2} + \dfrac{\delta}{2} + \varepsilon} = \frac{\delta}{\delta + \varepsilon} . \tag{7.8}$$

For geometrically distributed message length and idle length the average message length is $1/\varepsilon$ and the average idle length is $1/\delta$. The average recurrence time

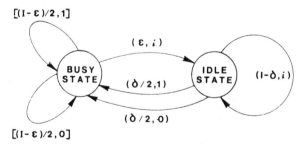

**Figure 7.101** A Markov transmitter.

between messages is $1/\varepsilon + 1/\delta$. The amount of information or entropy of these Markov sources is given as [21]

$$H(\varepsilon,\delta) = \frac{\delta}{\delta+\varepsilon} + \frac{\delta}{\delta+\varepsilon} H(\varepsilon) + \frac{\varepsilon}{\delta+\varepsilon} H(\delta)$$

$$H(x) = \log_2 \frac{1}{x} + (1-x) \log_2 \frac{1}{1-x} \tag{7.9}$$

The first term on the right-hand side of (7.9) is considered the entropy of the messages; the second term is considered the entropy in the length of the message. Together the two terms are the message entropy. The last term in (7.9) is considered as protocol because $H(\delta)$ is the amount of information from idle state to busy state, as shown in Figure 7.101. If delay is experienced in the network, then the protocol information will be distorted. This we shall consider next.

### 7.10.3   Network with Delays

Almost all practical communication networks encounter delays, particularly the ones with satellites. We have used queuing analyses for delays arising from multiple earth station access in a satellite system (Chapter 3, Volume I) and from onboard satellite processing (Chapter 6, Volume I). We are now interested in how network delays affect protocol information.

For a given transmitter-receiver pair in the network let $X_i$ and $Y_i$ be the message arrival times to the network and let $i = 1, 2, \ldots$ be the message arrival timer to the receiver. Assuming the arrival process to be Poisson with rate $\lambda$, the interarrival times $T_i = X_i - X_{i-1}$ with $X_0 = 0$ are independent, each with a probability density function $\lambda e^{-\lambda t}$. For any integer $N$ a joint probability measure $P(N)$ can be defined over the ensembles $X^N = (X_1, \ldots X_N)$ and $Y = (Y_1, \ldots, Y_N)$. As in the case of study transmission channel characterization (Chapter 2, Volume I), an average mutual information $I(X^N;Y^N)$ can be evaluated based on $P(N)$ and the two ensembles $X^N$ and $Y^N$. This mutual information is the amount of information provided to the receiver about message arrival times at the transmitter. For a given $P(N)$ the corresponding expected delay per message is

$$\overline{D}(N) = \frac{1}{N} \sum_{i=1}^{N} E(Y_i - X_i) \tag{7.10}$$

where $Y_i - X_i$ is the network delay and $E(Y_i - X_i)$ denotes the expected delay over the ensemble $i$. If $D(N)$ is bound by a maximum value $d$, then we want to obtain a lower bound on the amount of transmitted protocol information about message arrival times, subject to the constraint of the expected delay. By minimizing the mutual information, this problem, as it turns out, is analogous to the $N$th order rate-distortion function defined as

$$R(d) = \lim_{N \to \infty} \inf R_N(d) \tag{7.11}$$

$$R_N(d) = \inf_{P(N,d)} \frac{1}{N} I(X^N;Y^N) \qquad (7.12)$$

where $P(N,d)$ denotes the set of probability $P(N)$ with $D(N) \leq d$ and inf denotes the limit inferior where the sequence is bounded from below.

Thus the problem of the average protocol information about message arrival times which need to be transmitted for limiting the expected delay per message to $d$ can be lower bound by $R(d)$. This relationship of network delay to the distortion measure was given by Gallager. He also proved that when $N = 1$ (single message arriving and delivering), $R_1(d)$ from (7.12) is a lower bound to $R_N(d)$ for all $N > 1$, and to $R(d)$ of (7.11). As proved in [21]

$$R_1(d) = \frac{1}{\log_2(1 - e^{-\lambda d})} \qquad (7.13)$$

in bits/message. The probability $P(N = 1)$ which achieves $R_1(d)$ is defined implicitly in (7.14)

$$Y_1 = \max(X_1,d) + Z \qquad (7.14)$$

where $Z$ is a non-negative random variable, independent of $X_1$, with probability density

$$p(z) = (\lambda + p)e^{-(\lambda+p)z} \qquad (7.15)$$

$$p = \frac{\lambda e^{-\lambda d}}{1 - e^{-\lambda d}}. \qquad (7.16)$$

It can be observed that not only $p(z)$, but $R_1(d)$, $R_N(D)$, as well as $R(d)$, are also a function of $\lambda d$, the product of the message arrival rate and the maximum value of delay. This is a very interesting simple result, because the amount of transmitted protocol information can be evaluated in terms of the traffic parameter $\lambda$ and the delay parameter $d$.

Gallager has provided algorithmic strategies for minimizing protocol information with delay constraints. He also evaluated the amount of protocol bits needed per message in terms of $\lambda d$. For $\lambda d \leq 1$ the difference between the strategies and the lower bound is negligible. For $\lambda d \gg 1$ the differences exist and at $\lambda d = 1$ it takes 1.0 bit protocol per message.

The discussion of transmitter-receiver pair connection in the above analysis can be extended to broadcasting type satellite application by introducing virtual stations. Variable length source codes can be made of equal length protocols by using a large number of stations. The use of rate distortion theory and the idle to busy information to describe protocols shall provide the groundwork for future analysis of flow control and routing from an information theoretical viewpoint.

## 7.11 QUEUEING NETWORKS

Queueing principles have been used in Volume I to analyze and to model multiple access systems and message processing phenomena with satellite onboard pro-

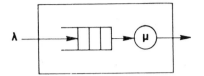

**Figure 7.102** Single-stage open network.

cessing capability. In practical congested satellite communication networks the phenomenon of waiting is as real as waiting to cash a check in a local bank. In this section, however, we single out only a basic issue, which can be related to the loops inside the protocol functions of any satellite network. This basic issue is the equivalence of open and closed networks with queues. The reason we are interested in such equivalence is for the purposes of understanding and simplifying more complicated analysis.

### 7.11.1    Open and Closed Networks

It is both interesting and important to note that the equivalence of open and closed networks exists. Consider a simple open network of single-stage queue as shown in Figure 7.102, assuming a Poisson message arrival rate $\lambda$ and a single processor with exponential processing time and processing rate $\mu$. This queue has a message storage capacity $N > n$, where $n$ is the total number of messages. If there are more than $N$ messages arriving at the stage, they will be blocked without being processed.

Now we consider the closed network as shown in Figure 7.103, which is comprised of two queueing stages with feedback. When a network is closed it means that messages neither arrive nor depart from the network. The single-stage queue A at the top of Figure 7.103 is identical to the single stage of Figure 7.102, with the same queueing parameters. The single-stage queue B has an arrival rate $\mu$ and a processing rate $\lambda$. If there are a total of $\mu$ messages in the network and there are $n_A$ number of messages in processor A, then the number of messages at processor B is $n - n_A$. As long as $n - n_A > 0$, processor B is working with arrival rate $\mu$ and departure rate $\lambda$. When $n = n_A$, processor B becomes idle and it fails to produce a message to processor A. This is equivalent to the blocking situation described previously for the open network. As a consequence the single-stage simple open network is equivalent to the two-stage closed network. The equivalence refers to the queueing conditions described above. It is almost intuitively obvious that by induction an $m$-stage open network is equivalent to an $(m + 1)$-stage closed network.

Assume in general that there are $m$ queueing stages in a satellite link and let the $i$th stage have $r$ parallel servers (processors, or transponders, for example), as described in Chapter 3, Volume I. We assume at stage $i$ the message arrival and service rates are $\lambda_i$ and $\mu_i$, respectively, with an exponentially distributed service time with mean $1/\mu_i$. A message, or TDMA burst, completes its process at the $i$th stage and proceeds directly to the $j$th stage with probability $Pij$. Let

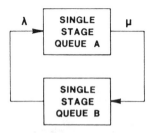

**Figure 7.103** Two-stage closed network.

$\mu_i$ denote the number of messages in process plus the number of messages in queue at stage $i$. The state of the network is determined by $\{\mu_i : i = 1, 2, \ldots, m\}$, where $\mu_1 + \mu_2 + \ldots + \mu_m = \mu$. We denote $\lambda_0$ as the link message arrival rate and consider $\lambda_i$ the multiple accessed message rate at stage $i$. Then in terms of these arrival rates a general cascaded queueing link of $m$ stages is shown in Figure 7.104. The function of each stage is shown in Figure 7.105. The reason that at each stage the output rate $\lambda_i$, or the input arrival rate to the $i + 1$ stage, is the same as the multiple accessed input arrival rate output message is that at each stage the distribution is the same as its input message distribution. Analogous to the two-stage cyclic queue example described previously, a closed queueing system of $m + 1$ stages is shown in Figure 7.106, where the function of each stage is the same as in Figure 7.105. The number of messages in operation at the first stage of the closed system is the same as that for the open network, and this number is

$$n_1 = n - \sum_{i=2}^{m} n_i .$$

The equivalence of open and closed networks has been proved in general by Gordon and Newell [22]. The technique was first applied to congestion control by Pennotti and Schwartz [23].

The effect of a multiple access message in a queueing network may be represented by replacing the processing rate by $\mu_i - \lambda_i$ for the $i$th stage. The difference of $\mu_i - \lambda_i$ means a reduction of the processing rate $\lambda_i$. Network throughput depends not only on the number of stages, but it is also a factor in the total number of messages. For simplicity, let $\lambda_i = \lambda$ and $\mu_i = \mu$ for all $i$. Then the closed network of Figure 7.106 can be reduced to Figure 7.107 with identical $\mu$ stages. In this case it can be shown that the network throughput is [24,25].

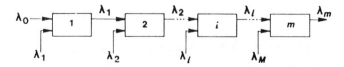

**Figure 7.104** Multiple stage queues—open system.

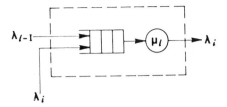

**Figure 7.105**  A queueing stage.

$$\lambda = \frac{n\mu}{n + m - 1}.$$   (7.17)

It is of interest to note that the processing rate of the $m$ identical stage network becomes $m\mu$.

## 7.12  MESSAGE CONGESTION

Congestion refers to the condition of message transmission in a satellite network when a user cannot get its message to the intended destination through the network immediately. This message-blocking phenomenon can be caused either by an insufficient arrangement at the earth station or resource limitations at the satellite. When blocking occurs either message loss or/and message delay can take place. Since message occurrence is a statistical phenomenon, the condition of congestion can be analyzed as a stochastic process with queues. For telephone calls and telephone switching exchanges the theory of congestion has been developed, and theoretical treatment has been provided by Syski [26]. Although the mechanism and channel characteristics are different in a terrestrial network and a satellite network, some of the fundamentals in regard to congestion are similar. For example, calls are turned away at the earth station if all the transponders are fully occupied. The statistical description of message occurrence is the same for both cases, as will be discussed next. The congestion processes for both terrestrial and satellite networks can be modeled as an information source emitting messages probabilistically.

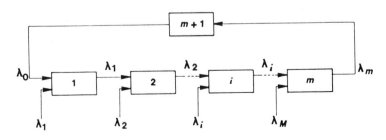

**Figure 7.106**  Multiple stage queues—closed system.

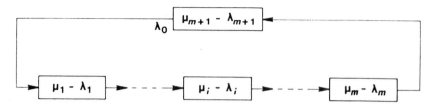

**Figure 7.107** Simplified closed system queues from Figure 7.106.

## 7.12.1   Congestion Process as Information Source

It was shown by Syski that any congestion process can be regarded as an information source. The ensemble of all the messages that are capable of being produced by the system is considered as the information source. The reason that a congestion process can be regarded as an information source may be explained as follows: given a satellite network with a fixed number of transponders, if the utilization of the transponders is not prearranged, then the time of transponder occupancy or waiting conditions determine the completion of messages. The process of such message completion in the network is probabilistic and thus the process can be modeled as an information source. The only unrealistic aspect of this model is the possibility of message blocking. If the blocking probability is small, such analysis then is useful.

Similar to the discussion of the rate-distortion theory for protocol information in Section 7.10.1, consider a discrete memoryless source with possible 0, 1, . . ., $K - 1$ message occurrence. For each source message $k, 0 \le k \le K - 1$, the probability of such message occurrence is $Q(k)$. When the set of source messages terminates at the destination, they are represented by the messages 0, 1, . . . , $J - 1$. The probability of the destination message $j$ conditional on the source message $k$ is $p(j/k)$.

The congestion process takes place during the transition between the messages $j$ and $k$, and the conditional probability $p(j/k)$ forms the quantitive basis for the process. $Q(k)$ then can be considered as the probability of equilibrium for a queueing state. The conditional entropy of the message information source $j$ given $k$ with time dependency is defined as the average value of the entropy of $p(j/k)$, that is

$$H(j/k,t) = - \sum_k Q(k) \sum_j p(j/k,t) \log p(j/k, t). \qquad (7.18)$$

$H(j/K, t)$ gives the amount of selective information generated by the congestion process regarded as information source. For $t = 0$, $H(0) = 0$; for $t \mapsto \infty$ the conditional entropy becomes the marginal entropy

$$H = - \sum_j p(j) \log p(j). \qquad (7.19)$$

For geometrically distributed messages with simple $M/M/1$ queue (see Volume I, Appendix 4 and Appendix 5),

$$H = - \frac{\rho \log \rho + (1 - \rho) \log (1 - \rho)}{1 - \rho} \tag{7.20}$$

with $\rho = \lambda/\mu$. For negative exponential distribution of the form $\lambda e^{-\lambda t}$

$$H = - \log \frac{\lambda}{e}.$$

Thus entropy of an information source $H(j/k,t)$, in this case the congestion process, is dependent not only on time but also message distribution.

### 7.12.2  Congestion Control

Congestion control refers to the area of studies concerned with the prediction, design, and management of traffic routing distribution in a communication network. The objective of such study is to provide strategies and to prevent congestion from occurring for a given network configuration and traffic condition. This is a very important issue for networks that involve satellites because of the limitations of satellite resources and the limited flexibilities of a satellite system. This applies to both circuit-switched or packet-(message) switched satellite networks.

Techniques of congestion control have developed through flow control for computer communication networks such as ARPA, Cyclade, and TYMNET. A comparative survey of flow control schemes is provided by Gerla and Kleinrock [27]. Congestion controls by input limitation have been treated by Lam and Reiser [28] and Saad and Schwartz [29]; optimum end-to-end flow control in networks has been analyzed by Bharath-Kumas [30]; control techniques and performance analyses of flow control mechanisms are given by Pennotti and Schwartz [23], Reiser [31], Davis [32], Chou and Gerla [33], Wong and Unsoy [34], and Hsieh and Kraimeche [35]; an excellent summary of routing and flow control in data networks is provided by M. Schwartz [24].

The first question is, from a user-to-user viewpoint, what is the most appropriate model for congestion control in satellite communication. To answer this question we refer the reader to Volume I, Chapter 3, Section 3.6. In Section 3.6.4 in particular we modeled the satellite link as a series of cascades. Further attention has been given in the previous section for such queue networks. We assumed that in both sections the transmit and receive earth stations as well as the satellite have limited message buffer capabilities. We also mentioned the fact that difficulty exists in analyzing such problems. However, for some special cases restrictions may be placed on the flow of messages of each source-destination (user-to-user) pair in order to avoid or minimize congestion. When the restrictions are directly applied at the source user (outside of the network), these are called end-to-end controls. When the control is done on the messages inside the network, they are referred to as local controls.

From a user-to-user viewpoint in a satellite queueing multiple stage link, long queues or even blockage will occur at one or more stages if the link arrival rate increases significantly. As a consequence the delay of message through the link increases, or the throughput of the message decreases. Assuming that the processing capabilities of the stages or the link capacity can no longer be improved, an obvious solution is to limit the number of messages per unit time at the input to the link, that is to say, maintaining the message flow by controlling the arrival rate as result of the link queueing capability.

Local control may be accomplished through the buffer and processing control of each stage in the link. Depending on the particular type of processors, processing techniques may vary and elastic buffers or buffer sharing may be used. Analysis for both end-to-end and local control may be obtained for specific links with specific parameters, which is beyond the scope of this discussion.

The effectiveness and tradeoffs of the control mechanism for both end-to-end and local controls have not been specifically addressed with regard to satellite networks. It is reasonable to expect that new results and solutions will be forthcoming in this area.

## 7.13   CONCLUDING REMARKS

In this chapter we first address the issue of integrated international digital satellite network services in Section 7.1 and 7.2. In summary, this envisioned network structure consists of two basic network elements. The first one is the integrated services digital network, which has been evolved from the terrestrial telephone networks and which is to be standardized by member international organizations; the second network element has been developed primarily from the viewpoint of satellite communication.

The issues concerning ISDN with satellites have been recently addressed by many, among them notably CCITT Study Group XVIII, CCIR Study Group IV, Guenin, Lucas, and Montaudoin [36], and Lee [37]. Lee observed the possible integration problem between satellite networks and terrestrial networks, and he offered some preliminary suggestions as to how the integration can best be handled with minimum impact on both networks. Nevertheless, a truly integrated global satellite network remains a challenge. In this chapter some criteria have been discussed with a view toward the establishment of such a network.

From Section 7.3 to Section 7.6 the INTELSAT TDMA network elements are treated in detail with emphasis on network protocols. Workable protocol arrangements including computer test results are demonstrated in Section 7.7 to 7.9. Unfortunately, we have been able to say little about their optimal use or theoretical background. The objective in describing the INTELSAT TDMA network protocols is not only to present the inner workings of the network but also to speculate on how some of its merit may be adapted as an international standard in the development of the integrated services digital satellite network.

In the context of digital communication for satellite applications, Section 7.10 highlights the connection between network protocol information and rate-distortion theory. Such connection was observed and theoretically analyzed by R. Gallager [19]. It is quite illuminating to note that all the rate-distortion functions (hence the protocol information) can be expressed in terms of the simple product of message arrival rate and the maximum value of delay. Not included in

Section 7.10 are the algorithmic strategies for minimizing protocol information with delay constraints.

Section 7.11 discusses the equivalence between an open network of series queues and a closed network of feedback queues. The analyses and applications to protocol of such queueing networks have been thoroughly investigated by M. Schwartz [24,38]. As in previous discussions, only those points specifically relevant to satellite communication have been brought forth.

Many ideas and practical results of network protocol have been advanced with computer communications. A sample of activities in this area can be found in the references [39, 40, 41]. As for international contributions, reference [41] includes a comparative flow control survey paper by L. Kleinrock and M. Gerla, discussions of bit and character oriented data link control protocols by D. Carlson and J. Conrad, respectively, high-level protocol by I. Ioda, network interconnection protocol approaches by J. Postel, protocol design methods by G. Bochmann and C. Sunshine, and protocol representation modeled with finite state automation and Petri nets by A. Danthine for both interface and end-to-end applications. Protocol modeling for specification and validation can be effectively represented and analyzed by Petri nets. The theory of Petri nets can be found in [42,43,44]. In the same issue a tracking algorithm is presented and analyzed by P. Zafiropulo, C. West, H. Rudin, D. Cowan, and D. Brand for the purpose of error-free protocols, an OSI reference model is outlined by H. Zimmermann, F. Tobagi presents the performance of a number of multiple access protocols in terms of bandwidth utilization and message delay, and P. Green, Jr., not only introduces recent trends on network architectures and protocols but explains why modern protocol structures are layered.

As a message-carrying mechanism, every satellite network has the potential problem of traffic congestion. Thus congestion control is a very important part of network design and operation. Section 7.12 provides a general discussion suggesting that a congestion process can be modeled as an information source and the simple types of congestion control techniques can be rapidly applied to satellite networks.

In the absence of unique theories on protocol and congestion specifically for satellite communication, the well-developed principles from telephone traffic and computer communication disciplines will prove to be valuable assets for satellite applications.

## 7.14 REFERENCES

1. Woolfe, R. *Videotex: The New Television/Telephone Information Services*. Hayden, 1980.
2. Hass, T. "International Standardization and ISDN." *Journal of Telecommunication Networks* 1, 4 (Winter 1982).
3. International Telegraph and Telephone Consultative Committee (CCITT) Study Group XVIII—Report No. R3 to R15 for 1981–1984, March 1983.
4. Mahoney, J., et al. "Users View of the Network." *Bell Systems Technical Journal*, 54, 5 (May–June 1975).

5. Zimmerman, H. "OSI Reference Model—The ISO Model of Architecture for Open System Interconnection." *IEE Transactions on Communications Technology* COM-28 (April 1980): 425–432.

6. CCITT Provisional Recommendations X.3, X.25, X.28,and X.29 on Packet-Switched Data Transmission Services, 1978.

7. Meijer, A., and Peeters, P. *Computer Network Architectures.* Rockville, Md.: Computer Science Press, 1982.

8. Rutledge, J. "OSI and SNA: A Perspective." *Journal of Telecommunication Networks* 1, 1 (Spring 1982).

9. INTELSAT TDMA Reference Station Equipment Specifications, BG Temp, 47–108, September 1981.

10. INTELSAT TDMA/DSI Traffic Terminal Specifications, BG-42-65 (Rev.1), June 1981.

11. Pontano, B., Dicks, J. L., Colby, R., Forcina, G., and Phiel, J. "The INTELSAT TDMA/DSI Systems." *IEEE Journal on Selected Areas in Communications* vol. SAC-1, no. 1 (January 1983).

12. Wu, W., Forcina, G., Koga, K., Shinonaga, H., Mauro, H., and Kondo, G. "INTELSAT TDMA Network Simulation Program." *Proceedings of the First International and Canadian Satellite Communication Conference,* June 1983.

13. Campanella, S., and Colby, R. "Network Control for Multibeam TDMA and SS/TDMA." *IEEE Journal on Selected Areas in Communications* vol. SAC-1, no. 1 (January 1983).

14. "TDMA CTP Transmission Protocols," INTELSAT Documentation, BG/T-45-15, 26 January 1983.

15. CCITT Yellow Book, Volume VIII—FASCICLE VIII.2, Data Communication Networks Services and Facilities, Terminal Equipment and Interfaces, Recommendations X.1–X.29, November 1980.

16. CCITT Orange Book, Volume VIII.2. *Public Data Networks,* 1977.

17. Wu, W. Statement of Work For INTELSAT Contract "Development of TDMA Network Simulation Programs." April 1982.

18. Progress Reports For INTELSAT Contract INTEL-233, "Development of TDMA Network Simulation Programs," by KDD. September 1982, January 1983.

19. Gallager, R. "Basic Limit on Protocol Information in Data Communications Networks." *IEEE Transactions on Information Theory* IT-22, 24 (July 1976): 385–398.

20. Gallager, R. "Applications of Information Theory to Data Communication Networks." *New Concepts in Multi-User Communication,* edited by J. Skwirzynski. Alphen aan den Kijn, The Netherlands: Sijthoff & Noordhoff, 1981.

21. Gallager, R. *Information Theory and Reliable Communication.* New York: John Wiley & Sons, 1968.

22. Gordon, W., and Newell, G. "Closed Queueing Networks with Exponential Servers." *Operation Research* (April 1967): 254–265.

23. Pennotti, M., and Schwartz, M. "Congestion Control in Store and Forward Tandem Links." *IEEE Transactions on Communications Technology* COM-23, 12 (December 1975): 1434–1443.

24. Schwartz, M. "Routing and Flow Control in Data Networks." *New Concepts in Multi-User Communication,* edited by J. Skwirznski. Alphen aan den Rijn, The Netherlands: Sijthoff & Noordhoff, 1981.

25. Kobayashi, H. *Modeling and Analysis: An Introduction to System Performance Evaluation Methodology.* Reading, Mass.: Addison-Wesley, 1978.

26. Syski, R. *Introduction to Congestion Theory in Telephone Systems*, Published for Automatic Telephone and Electric Company Limited by Oliver and Boyd, 1960.
27. Gerla, M., and Kleinrock, L. "Flow Control: A Comparative Survey." *IEEE Transactions on Communication Technology* COM-28, 4 (April 1980): 553–574.
28. Lam, S., and Reiser, M. "Congestion Control of Store-and-Forward Networks by Input Buffer Limits." *IEEE Transactions on Communications Technology* COM-27, 1 (January 1979): 127–134.
29. Saad, S., and Schwartz, M. "Input Buffer Limiting Mechanisms for Congestion Control." *Proceedings of the ICC* (June 1980).
30. Bharath-Kumar, K. "Optimum End-to-End Flow Control in Networks." *Proceedings of the ICC* (1980): 23.3.1–23.3.6.
31. Reiser, M. "Performance Evaluation of Data Communications Systems." *Proceedings of the IEEE*, 70, 2 (February 1982): 171–196.
32. Davis, D. "The Control of Congestion in Packet Switching Networks." *IEEE Transaction on Communications Technology* COM-20 (June 1972).
33. Chou, W., and Gerla, M. "A Unified Flow and Congestion Control Model for Packet Networks." *Proceedings of Third International Conference on Computer Communications*, Toronto, Canada, August 1976.
34. Wong, J., and Unsoy, M. "Analysis of Flow Control in Switched Data Networks." IFIP Congress, August 1977.
35. Hsieh, W., and Kraimeche, B. "Performance Analysis of an End-to-End Flow Control Mechanism in a Packet-Switched Network." *Journal of Telecommunication Networks* 3,1 (Spring 1983): 103-116.
36. Guenin, J., Lucas, F., and Montaudoin, P. "The Role of Satellite in Achieving ISDN." *ICC* (1981): 19.5-1 to 19.5-5.
37. Lee, J. "Symbiosis Between a Terrestrial-Based Integrated Services Digital Network and a Digital Satellite Network." *IEEE Journal on Selected Areas in Communication* vol. SAC-1, no. 1 (January 1983): 103–109.
38. Schwartz, M. "Performance Analysis of the SNA Virtual Route Pacing Control." *IEEE Transactions on Communications Technology* COM-30, 1 (January 1982): 172–184.
39. Special Issue on Computer Communications, edited by R. L. Pickholtz, *IEEE Transactions on Communications Technology* COM-25, 1 (January 1977).
40. Chu, W. *Advances in Computer Communications and Networking*. Artech House, 1979.
41. Special Issue on Computer Network Architecture and Protocols, edited by P. Green, Jr. *IEEE Transactions on Communications Technology* COM-28, 4 (April 1980).
42. Danthine, A. "Petri Nets for Protocol Modelling and Verification." *Proceedings on Computer Networks Teleprocessing Symposium*, vol. Ii, Budapest, Hungary (October 1977).
43. Brauer, W. Ed., Proc. Advanced Course General Net. Theory Processes Systems, Lecture Notes in Computer Science, vol. 84. New York: Springer-Verlag, 1980.
44. Berthelot, G., and Terrat, R. "Petri Nets Theory for the Correctness of Protocols." *IEEE Transactions on Communications Technology* COM-30, 12 (December 1982): 2497–2505.

## 7.15 PROBLEMS

1. Verify that all the transmission rates, or channel types, suggested by the

CCITT can be expressed as the products of distinctive prime powers.
2. Write a short essay and express your concept on the issue and architecture of an ideal integrated services digital satellite network.
3. Block error probability $P_{bk}$ is defined as the probability of at least one error in a block of $n$ digits.

    (a)   Assuming error occurrence is independent with bit error rate $p$, express $P_{bk}$ in terms of $n$ and $p$.

    (b)   For statistical not independent errors the error event can be described by a two-state Markov chain. Let 1 denote the error state and 0 denote the errorless state. The state transition probabilities are shown in Figure 7.108.

Let $X_i = 1$, or $0$ denote error, or no error in the present bit, and $X_{i-1} = 1$, or $0$ denote error, or no error in the previous bit. Then

$$\alpha = P(x_i = 1/x_{i-1} = 1)$$

$$\beta = P(x_i = 0/x_{i-1} = 0)$$

$$\delta = P(x_i = 0/x_{i-1} = 1)$$

$$\xi = P(x_i = 1/x_{i-1} = 0)$$

show that

$$\beta = 1 - \frac{(1 - \alpha)\,p}{1 - p}.$$

    (c)   Express $P_{bk}$ in terms of $\beta$, $p$ and $n$.

    (d)   Evaluate $P_{bk}$ in (c) if $\alpha = p$, $\alpha > p$, and $\alpha < p$. What is the condition for $P_{bk}$ to be valid?

4. As a possible performance criterion, what are the disadvantages of using error free second and/or block error rate in network evaluation? Why is the concept of hypothetical distance, or hypothetical reference connections, inadequate when it applies to satellite communication?
5. Why does the seven-layer basic reference model for open system interconnection (OSI) become less efficient when it applies to satellite communication?
6. In terms of the layered protocols closely examine the differences between ISO's OSI and the Systems Network Architecture (SNA) developed earlier by IBM.

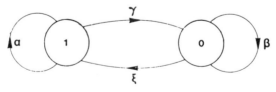

**Figure 7.108** A Two-state Markov chain for dependent error events.

7. What are the basic differences between CCITT Recommendations X.25 and X.75?

8. Design a simple digital satellite transmission network using the first three-layer protocol logic building blocks of the open system interconnections.

9. For the traffic terminal acquisition support (TAS) procedure in the INTEL-SAT TDMA network what are the tradeoffs between the common window and the dedicated window acquisition arrangements? Are the options necessary?

10. The information theoretical basis of protocols involves a distortion measure, $R(d^*)$, which can be obtained by minimizing the corresponding average mutual information. On the other hand, the capacity of a transmission channel can be calculated by maximizing its corresponding average mutual information.

    (a) Investigate the differences between the two types of mutual information.

    (b) Examine the ranges and techniques for the minimization and maximization of the mutual information.

11. For an intersatellite-linked network without broadcasting (a) show that the amount of information from a transmitting station is

$$H(\varepsilon,\delta) = \frac{\delta}{\delta + \varepsilon} [1 + H(\varepsilon)] + \frac{\varepsilon}{\delta + \varepsilon} H(\delta)$$

where

$$H(x) = x\log_2 \frac{1}{x} + (1-x)\log_2 \frac{1}{1-x}$$

and $\delta/2$ are the transition probabilities from busy to idle and from idle to busy (transmitting) states.

    (b) Evaluate the asymptotic protocol information

$$\varepsilon H(\delta) / (\delta + \varepsilon).$$

12. For an $M/M/1$ quencing network with geometrically distributed messages of arrival and departure rate $\lambda$ and $\mu$ respectively, show that for a large amount of time $t$ the marginal entropy is

$$H = - \frac{\rho\log\rho + (1-\rho)\log(1-\rho)}{1 - \rho}$$

where

$$\rho = \lambda / \mu.$$

# AFTERWORD

Satellite communication is indeed a multifaceted subject that encompasses many diversified and specialized areas. This two-volume work touches upon only a few areas that may be considered as essential elements. The choice of subjects and the depth of treatment reflect the author's judgment and background.

Chapter 3 of this second volume addresses the fundamental issues concerning various applications in satellite communication that have deep roots in combinatorial set theory. If a reader has followed the discussions faithfully, then the conclusion may be reached that this second volume has shown the origins of many identified applications in satellite communication. The objective here is to bring about an understanding of the basic processes involved; through these processes we will be able to know the limitations, to see the alternatives, and to develop better solutions.

We advocated cryptology and channel coding techniques as useful elements for digital satellite communication, and we spent Chapters 4, 5, and 6 working our way from principles, through performance, to practical designs. As the last part of Volume II, Chapter 7 should not be considered as an element of satellite communication. Rather, satellite-related issues become elements in a larger-scale network context. In this chapter we proposed criteria and provided guidelines for the establishment of an integrated service digital satellite network for effective worldwide communication.

Throughout these two volumes a number of unsolved problems have been brought forth. The reader is encouraged to provide solutions, some of which are so fundamental that useful results may be considered as breakthroughs. Please send solutions to Dr. William Wu, c/o Computer Science Press, 1803 Research Blvd., Rockville, Maryland 20850. Most solutions to the problems for all the chapters are available. They have been organized with the help of Patricio Northland and may be obtained by professors who have adopted *Elements of Digital Satellite Communication* for a class. Requests for the solutions should include the name of the course, where the course is being taught, as well as class enrollment.

For any comments or suggestions, please contact the publisher.

# Appendix 1

# TABLES OF PRIMITIVE POLYNOMIALS OVER GF (*p*) FOR *p* ≠ 2 AND EVEN POWERS OF *p* [A.1]

**Table A.1.1**

$p = 3, m = 1, n^2 = 9, 8 = 2^3, \varphi(8)/2 = 2.$

| $\alpha_1$ | $\alpha_2$ |
|---|---|
| 1 | 2 |
| 2 | 2 |

**Table A.1.2**

$p = 3, m = 2, n^2 = 81, 80 = 2^4 \cdot 5, \varphi(80)/4 = 8.$

| $\alpha_1$ | $\alpha_2$ | $\alpha_3$ | $\alpha_4$ |
|---|---|---|---|
| 1 | 0 | 0 | 2 |
| 2 | 0 | 0 | 2 |
| 0 | 0 | 1 | 2 |
| 2 | 1 | 1 | 2 |
| 2 | 2 | 1 | 2 |
| 0 | 0 | 2 | 2 |
| 1 | 1 | 2 | 2 |
| 1 | 2 | 2 | 2 |

### Table A.1.3
$p = 3$, $m = 3$, $n^2 = 729$, $728 = 2^3 \cdot 7 \cdot 13$, $\varphi(728)/6 = 48$.

| $\alpha_1$ | $\alpha_2$ | $\alpha_3$ | $\alpha_4$ | $\alpha_5$ | $\alpha_6$ | $\alpha_1$ | $\alpha_2$ | $\alpha_3$ | $\alpha_4$ | $\alpha_5$ | $\alpha_6$ |
|---|---|---|---|---|---|---|---|---|---|---|---|
| 1 | 0 | 0 | 0 | 0 | 2 | 0 | 1 | 2 | 1 | 1 | 2 |
| 2 | 0 | 0 | 0 | 0 | 2 | 0 | 1 | 0 | 2 | 1 | 2 |
| 1 | 0 | 1 | 0 | 0 | 2 | 2 | 2 | 0 | 2 | 1 | 2 |
| 2 | 0 | 2 | 0 | 0 | 2 | 1 | 0 | 1 | 2 | 1 | 2 |
| 1 | 2 | 0 | 1 | 0 | 2 | 2 | 1 | 1 | 2 | 1 | 2 |
| 2 | 2 | 0 | 1 | 0 | 2 | 1 | 2 | 1 | 2 | 1 | 2 |
| 1 | 2 | 1 | 1 | 0 | 2 | 2 | 1 | 2 | 2 | 1 | 2 |
| 2 | 2 | 2 | 1 | 0 | 2 | 0 | 0 | 0 | 0 | 2 | 2 |
| 1 | 1 | 0 | 2 | 0 | 2 | 0 | 1 | 1 | 0 | 2 | 2 |
| 2 | 1 | 0 | 2 | 0 | 2 | 0 | 0 | 2 | 0 | 2 | 2 |
| 2 | 0 | 1 | 2 | 0 | 2 | 2 | 1 | 2 | 0 | 2 | 2 |
| 2 | 2 | 1 | 2 | 0 | 2 | 2 | 2 | 2 | 0 | 2 | 2 |
| 1 | 0 | 2 | 2 | 0 | 2 | 1 | 1 | 0 | 1 | 2 | 2 |
| 1 | 2 | 2 | 2 | 0 | 2 | 0 | 2 | 0 | 1 | 2 | 2 |
| 0 | 0 | 0 | 0 | 1 | 2 | 0 | 1 | 1 | 1 | 2 | 2 |
| 0 | 0 | 1 | 0 | 1 | 2 | 2 | 0 | 2 | 1 | 2 | 2 |
| 1 | 1 | 1 | 0 | 1 | 2 | 2 | 1 | 2 | 1 | 2 | 2 |
| 1 | 2 | 1 | 0 | 1 | 2 | 0 | 2 | 2 | 1 | 2 | 2 |
| 0 | 1 | 2 | 0 | 1 | 2 | 0 | 1 | 0 | 2 | 2 | 2 |
| 2 | 1 | 0 | 1 | 1 | 2 | 1 | 2 | 0 | 2 | 2 | 2 |
| 0 | 2 | 0 | 1 | 1 | 2 | 1 | 1 | 1 | 2 | 2 | 2 |
| 1 | 0 | 1 | 1 | 1 | 2 | 2 | 0 | 2 | 2 | 2 | 2 |
| 1 | 1 | 1 | 1 | 1 | 2 | 1 | 1 | 2 | 2 | 2 | 2 |
| 0 | 2 | 1 | 1 | 1 | 2 | 2 | 2 | 2 | 2 | 2 | 2 |

### Table A.1.4
$p = 5$, $m = 1$, $h^2 = 25$, $24 = 2^3 \cdot 3$, $\varphi(24)/2 = 4$.

| $\alpha_1$ | $\alpha_2$ |
|---|---|
| 1 | −2 |
| 4 | 2 |
| 2 | 3 |
| 3 | 3 |

### Table A.1.5
$p = 5, m = 2, n^2 = 625, 624 = 2^4 \cdot 3 \cdot 13, \varphi(624)/4 = 48.$

| $\alpha_1$ | $\alpha_2$ | $\alpha_3$ | $\alpha_4$ | $\alpha_1$ | $\alpha_2$ | $\alpha_3$ | $\alpha_4$ | $\alpha_1$ | $\alpha_2$ | $\alpha_3$ | $\alpha_4$ |
|---|---|---|---|---|---|---|---|---|---|---|---|
| 1 | 2 | 0 | 2 | 0 | 1 | 3 | 2 | 4 | 4 | 1 | 3 |
| 4 | 2 | 0 | 2 | 3 | 2 | 3 | 2 | 1 | 0 | 2 | 3 |
| 2 | 3 | 0 | 2 | 2 | 3 | 3 | 2 | 3 | 0 | 2 | 3 |
| 3 | 3 | 0 | 2 | 1 | 0 | 4 | 2 | 0 | 1 | 2 | 3 |
| 3 | 0 | 1 | 2 | 2 | 0 | 4 | 2 | 2 | 1 | 2 | 3 |
| 4 | 0 | 1 | 2 | 4 | 2 | 4 | 2 | 3 | 4 | 2 | 3 |
| 1 | 2 | 1 | 2 | 1 | 3 | 4 | 2 | 2 | 0 | 3 | 3 |
| 4 | 3 | 1 | 2 | 0 | 4 | 4 | 2 | 4 | 0 | 3 | 3 |
| 0 | 4 | 1 | 2 | 2 | 2 | 0 | 3 | 0 | 1 | 3 | 3 |
| 2 | 0 | 2 | 2 | 3 | 2 | 0 | 3 | 3 | 1 | 3 | 3 |
| 4 | 0 | 2 | 2 | 1 | 3 | 0 | 3 | 2 | 4 | 3 | 3 |
| 0 | 1 | 2 | 2 | 4 | 3 | 0 | 3 | 3 | 0 | 4 | 3 |
| 2 | 2 | 2 | 2 | 1 | 0 | 1 | 3 | 4 | 0 | 4 | 3 |
| 3 | 3 | 2 | 2 | 2 | 0 | 1 | 3 | 4 | 1 | 4 | 3 |
| 1 | 0 | 3 | 2 | 1 | 1 | 1 | 3 | 0 | 4 | 4 | 3 |
| 3 | 0 | 3 | 2 | 0 | 4 | 1 | 3 | 1 | 4 | 4 | 3 |

### Table A.1.6
$p = 7, m = 1, n^2 = 49, 48 = 2^4 \cdot 3, \varphi(48)/2 = 8.$

| $\alpha_1$ | $\alpha_2$ |
|---|---|
| 1 | 3 |
| 2 | 3 |
| 5 | 3 |
| 6 | 3 |
| 2 | 5 |
| 3 | 5 |
| 4 | 5 |
| 5 | 5 |

### Table A.1.7
$p = 11, m = 1, n^2 = 121, 120 = 2^3 \cdot 3 \cdot 5, \varphi(120)/2 = 16.$

| $\alpha_1$ | $\alpha_2$ | $\alpha_1$ | $\alpha_2$ | $\alpha_1$ | $\alpha_2$ | $\alpha_1$ | $\alpha_2$ |
|---|---|---|---|---|---|---|---|
| 4 | 2 | 2 | 6 | 1 | 7 | 1 | 8 |
| 5 | 2 | 3 | 6 | 4 | 7 | 3 | 8 |
| 6 | 2 | 8 | 6 | 7 | 7 | 8 | 8 |
| 7 | 2 | 9 | 6 | 10 | 7 | 10 | 8 |

## Table A.1.8
$p = 13, m = 1, n^2 = 169, 168 = 2^3 \cdot 3 \cdot 7, \varphi(168)/2 = 24.$

| $\alpha_1$ | $\alpha_2$ | $\alpha_1$ | $\alpha_2$ | $\alpha_1$ | $\alpha_2$ | $\alpha_1$ | $\alpha_2$ |
|---|---|---|---|---|---|---|---|
| 1 | 2 | 2 | 6 | 2 | 7 | 4 | 11 |
| 4 | 2 | 3 | 6 | 3 | 7 | 5 | 11 |
| 6 | 2 | 4 | 6 | 6 | 7 | 6 | 11 |
| 7 | 2 | 9 | 6 | 7 | 7 | 7 | 11 |
| 9 | 2 | 10 | 6 | 10 | 7 | 8 | 11 |
| 12 | 2 | 11 | 6 | 11 | 7 | 9 | 11 |

## Table A.1.9
$p = 17, m = 1, n^2 = 289, 288 = 2^5 \cdot 3^2, \varphi(288)/2 = 48.$

| $\alpha_1$ | $\alpha_2$ | $\alpha_1$ | $\alpha_2$ | $\alpha_1$ | $\alpha_2$ | $\alpha_1$ | $\alpha_2$ |
|---|---|---|---|---|---|---|---|
| 1 | 3 | 2 | 6 | 1 | 10 | 2 | 12 |
| 6 | 3 | 6 | 6 | 3 | 10 | 3 | 12 |
| 7 | 3 | 8 | 6 | 4 | 10 | 5 | 12 |
| 10 | 3 | 9 | 6 | 13 | 10 | 12 | 12 |
| 11 | 3 | 11 | 6 | 14 | 10 | 14 | 12 |
| 16 | 3 | 15 | 6 | 16 | 10 | 15 | 12 |
| 3 | 5 | 1 | 7 | 2 | 11 | 4 | 14 |
| 5 | 5 | 4 | 7 | 7 | 11 | 6 | 14 |
| 8 | 5 | 5 | 7 | 8 | 11 | 7 | 14 |
| 9 | 5 | 12 | 7 | 9 | 11 | 10 | 14 |
| 12 | 5 | 13 | 7 | 10 | 11 | 11 | 14 |
| 14 | 5 | 16 | 7 | 15 | 11 | 13 | 14 |

## Table A.1.10
$p = 19, m = 1, n^2 = 361, 360 = 2^3 \cdot 3^2 \cdot 5, \varphi(360)/2 = 48.$

| $\alpha_1$ | $\alpha_2$ | $\alpha_1$ | $\alpha_2$ | $\alpha_1$ | $\alpha_2$ | $\alpha_1$ | $\alpha_2$ |
|---|---|---|---|---|---|---|---|
| 1 | 2 | 10 | 3 | 3 | 13 | 11 | 14 |
| 4 | 2 | 11 | 3 | 4 | 13 | 12 | 14 |
| 7 | 2 | 12 | 3 | 6 | 13 | 13 | 14 |
| 8 | 2 | 18 | 3 | 9 | 13 | 18 | 14 |
| 11 | 2 | 2 | 10 | 10 | 13 | 4 | 15 |
| 12 | 2 | 4 | 10 | 13 | 13 | 5 | 15 |
| 15 | 2 | 6 | 10 | 15 | 13 | 6 | 15 |
| 18 | 2 | 9 | 10 | 16 | 13 | 9 | 15 |
| 1 | 3 | 10 | 10 | 1 | 14 | 10 | 15 |
| 7 | 3 | 13 | 10 | 6 | 14 | 13 | 15 |
| 8 | 3 | 15 | 10 | 7 | 14 | 14 | 15 |
| 9 | 3 | 17 | 10 | 8 | 14 | 15 | 15 |

### Table A.1.11
$p = 23$, $m = 1$, $n^2 = 529$, $528 = 2^4 \cdot 3 \cdot 11$, $\varphi(528)/2 = 80$.

| $\alpha_1$ | $\alpha_2$ | $\alpha_1$ | $\alpha_2$ | $\alpha_1$ | $\alpha_2$ | $\alpha_1$ | $\alpha_2$ | $\alpha_1$ | $\alpha_2$ |
|---|---|---|---|---|---|---|---|---|---|
| 2 | 5 | 2 | 10 | 1 | 14 | 3 | 17 | 4 | 20 |
| 4 | 5 | 3 | 10 | 3 | 14 | 4 | 17 | 7 | 20 |
| 5 | 5 | 6 | 10 | 5 | 14 | 6 | 17 | 8 | 20 |
| 8 | 5 | 10 | 10 | 10 | 14 | 11 | 17 | 10 | 20 |
| 15 | 5 | 13 | 10 | 13 | 14 | 12 | 17 | 13 | 20 |
| 18 | 5 | 17 | 10 | 18 | 14 | 17 | 17 | 15 | 20 |
| 19 | 5 | 20 | 10 | 20 | 14 | 19 | 17 | 16 | 20 |
| 21 | 5 | 21 | 10 | 22 | 14 | 20 | 17 | 19 | 20 |
| 1 | 7 | 3 | 11 | 5 | 15 | 1 | 19 | 5 | 21 |
| 2 | 7 | 7 | 11 | 9 | 15 | 2 | 19 | 6 | 21 |
| 4 | 7 | 8 | 11 | 10 | 15 | 7 | 19 | 7 | 21 |
| 9 | 7 | 9 | 11 | 11 | 15 | 11 | 19 | 9 | 21 |
| 14 | 7 | 14 | 11 | 12 | 15 | 12 | 19 | 14 | 21 |
| 19 | 7 | 15 | 11 | 13 | 15 | 16 | 19 | 16 | 21 |
| 21 | 7 | 16 | 11 | 14 | 15 | 21 | 19 | 17 | 21 |
| 22 | 7 | 20 | 11 | 18 | 15 | 22 | 19 | 18 | 21 |

### Table A.1.12
$p = 29$, $m = 1$, $n^2 = 841$, $840 = 2^3 \cdot 3 \cdot 5 \cdot 7$, $\varphi(840)/2 = 96$.

| $\alpha_1$ | $\alpha_2$ | $\alpha_1$ | $\alpha_2$ | $\alpha_1$ | $\alpha_2$ | $\alpha_1$ | $\alpha_2$ | $\alpha_1$ | $\alpha_2$ | $\alpha_1$ | $\alpha_2$ |
|---|---|---|---|---|---|---|---|---|---|---|---|
| 5 | 2 | 1 | 8 | 6 | 11 | 7 | 15 | 2 | 19 | 5 | 26 |
| 7 | 2 | 7 | 8 | 9 | 11 | 9 | 15 | 4 | 19 | 6 | 26 |
| 11 | 2 | 10 | 8 | 10 | 11 | 11 | 15 | 7 | 19 | 8 | 26 |
| 14 | 2 | 14 | 8 | 11 | 11 | 12 | 15 | 8 | 19 | 12 | 26 |
| 15 | 2 | 15 | 8 | 18 | 11 | 17 | 15 | 21 | 19 | 17 | 26 |
| 18 | 2 | 19 | 8 | 19 | 11 | 18 | 15 | 22 | 19 | 21 | 26 |
| 22 | 2 | 22 | 8 | 20 | 11 | 20 | 15 | 25 | 19 | 23 | 26 |
| 24 | 2 | 28 | 8 | 23 | 11 | 22 | 15 | 27 | 19 | 24 | 26 |
| 1 | 3 | 3 | 10 | 1 | 14 | 4 | 18 | 3 | 21 | 2 | 27 |
| 2 | 3 | 5 | 10 | 3 | 14 | 8 | 18 | 4 | 21 | 3 | 27 |
| 9 | 3 | 9 | 10 | 8 | 14 | 13 | 18 | 6 | 21 | 6 | 27 |
| 14 | 3 | 10 | 10 | 13 | 14 | 14 | 18 | 12 | 21 | 13 | 27 |
| 15 | 3 | 19 | 10 | 16 | 14 | 15 | 18 | 17 | 21 | 16 | 27 |
| 20 | 3 | 20 | 10 | 21 | 14 | 16 | 18 | 23 | 21 | 23 | 27 |
| 27 | 3 | 24 | 10 | 26 | 14 | 21 | 18 | 25 | 21 | 26 | 27 |
| 28 | 3 | 26 | 10 | 28 | 14 | 25 | 18 | 26 | 21 | 27 | 27 |

**Table A.1.13**

$p = 31,\ m = 1,\ n^2 = 961,\ 960 = 2^6 \cdot 3 \cdot 5,\ \varphi(960)/2 = 128.$

| $\alpha_1$ | $\alpha_2$ | $\alpha_1$ | $\alpha_2$ | $\alpha_1$ | $\alpha_2$ | $\alpha_1$ | $\alpha_2$ | $\alpha_1$ | $\alpha_2$ | $\alpha_1$ | $\alpha_2$ | $\alpha_1$ | $\alpha_2$ | $\alpha_1$ | $\alpha_2$ |
|---|---|---|---|---|---|---|---|---|---|---|---|---|---|---|---|
| 2 | 3 | 2 | 11 | 1 | 12 | 1 | 13 | 1 | 17 | 2 | 21 | 1 | 22 | 1 | 24 |
| 5 | 3 | 3 | 11 | 3 | 12 | 4 | 13 | 2 | 17 | 5 | 21 | 4 | 22 | 3 | 24 |
| 6 | 3 | 4 | 11 | 4 | 12 | 6 | 13 | 3 | 17 | 7 | 21 | 5 | 22 | 4 | 24 |
| 7 | 3 | 5 | 11 | 10 | 12 | 8 | 13 | 6 | 17 | 8 | 21 | 7 | 22 | 5 | 24 |
| 8 | 3 | 6 | 11 | 11 | 12 | 9 | 13 | 7 | 17 | 11 | 21 | 9 | 22 | 7 | 24 |
| 10 | 3 | 9 | 11 | 12 | 12 | 10 | 13 | 8 | 17 | 12 | 21 | 10 | 22 | 8 | 24 |
| 14 | 3 | 11 | 11 | 14 | 12 | 12 | 13 | 9 | 17 | 13 | 21 | 14 | 22 | 12 | 24 |
| 15 | 3 | 15 | 11 | 15 | 12 | 13 | 13 | 11 | 17 | 15 | 21 | 15 | 22 | 13 | 24 |
| 16 | 3 | 16 | 11 | 16 | 12 | 18 | 13 | 20 | 17 | 16 | 21 | 16 | 22 | 18 | 24 |
| 17 | 3 | 20 | 11 | 17 | 12 | 19 | 13 | 22 | 17 | 18 | 21 | 17 | 22 | 19 | 24 |
| 21 | 3 | 22 | 11 | 19 | 12 | 21 | 13 | 23 | 17 | 19 | 21 | 21 | 22 | 23 | 24 |
| 23 | 3 | 25 | 11 | 20 | 12 | 22 | 13 | 24 | 17 | 20 | 21 | 22 | 22 | 24 | 24 |
| 24 | 3 | 26 | 11 | 21 | 12 | 23 | 13 | 25 | 17 | 23 | 21 | 24 | 22 | 26 | 24 |
| 25 | 3 | 27 | 11 | 27 | 12 | 25 | 13 | 28 | 17 | 24 | 21 | 26 | 22 | 27 | 24 |
| 26 | 3 | 28 | 11 | 28 | 12 | 27 | 13 | 29 | 17 | 26 | 21 | 27 | 22 | 28 | 24 |
| 29 | 3 | 29 | 11 | 30 | 12 | 30 | 13 | 30 | 17 | 29 | 21 | 30 | 22 | 30 | 24 |

# Appendix 2

# TABLE OF DIFFERENCE SETS [A.2]

Set Elements

| $v$ | $k$ | $\lambda$ | Set Elements |
|---|---|---|---|
| 7 | 3 | 1 | 1  2  4 |
| 13 | 4 | 1 | 0  1  3  9 |
| 21 | 5 | 1 | 3  6  7  12  14 |
| 31 | 6 | 1 | 1  5  11  24  25  27 |
| 57 | 8 | 1 | 1  6  7  9  19  38  42  49 |
| 73 | 9 | 1 | 1  2  4  8  16  32  37  55  64 |
| 91 | 10 | 1 | 0  1  3  9  27  49  56  61  77  81 |
| 133 | 12 | 1 | 1  11  16  40  41  43  52  56  60  74  78  121  128 |
| 183 | 14 | 1 | 0  2  3  10  26  39  43  56  61  77  78  109  121  130  136  141  155 |
| 273 | 17 | 1 | 1  2  4  8  16  32  64  91  117  128  137  182  195  205  234  239  256 |
| 307 | 18 | 1 | 0  1  3  9  27  50  55  76  98  117  129  133  157  189  199  222  293  299 |
| 381 | 20 | 1 | 0  1  19  28  96  118  151  153  176  202  240  254  290  296  300  307  337  359  361  366 |
| 553 | 24 | 1 | 1  23  52  90  108  120  152  151  153  176  202  240  223  232  272  355  359  407  411  431  438 ; 513  515  529  548 |
| 651 | 26 | 1 | 1  5  25  42  71  107  81  125  129  173  209  210  217  243  310  354  355  357  387  399  412  434  462  468 ; 483  521  535  561  625  633  673  729 |
| 757 | 28 | 1 | 0  1  3  9  27  43  43  173  220  243  310  404  409  445  455  466  470  505  543  582  599  645  659 ; 519  578  608  641  653  660 |
| 871 | 30 | 1 | 1  24  29  69  151  167  216  234  259  263  295  321  329  414  488  543  582  599  645  659 ; 683  689  696  716  731  819  820  822  831  841 |

Set Elements

| $v$ | $k$ | $\lambda$ | Set Elements |
|---|---|---|---|
| 993 | 32 | 1 | 0 1 23 31 60 66 77 84 87 195 253 257 291 331 401 416 468 473 515 590 606 618 711 713 752 761 841 867 892 912 961 980 |
| 1057 | 33 | 1 | 1 2 4 8 16 32 55 64 110 128 139 220 256 278 299 339 349 440 453 512 529 556 598 678 698 703 755 793 880 906 925 991 1024 |
| 1407 | 38 | 1 | 0 1 37 63 205 274 289 302 314 316 321 362 414 420 436 465 469 486 550 621 644 652 655 711 731 844 854 924 981 1098 1122 1152 1187 1230 1248 1316 1325 1369 |
| 1723 | 42 | 1 | 1 31 41 99 152 187 232 271 313 345 361 421 508 594 613 618 638 672 679 710 761 772 773 775 879 897 988 996 1011 1017 1063 1067 1194 1207 1216 1243 1271 1542 1579 1612 1681 1707 |
| 1893 | 44 | 1 | 0 5 43 147 164 215 284 308 356 375 439 457 459 537 627 642 655 715 721 744 755 777 780 807 854 981 1104 1230 1262 1338 1359 1373 1468 1507 1592 1647 1663 1673 1704 1779 1840 1849 1886 |
| 2257 | 48 | 1 | 1 47 102 136 149 163 232 243 263 280 353 572 622 792 848 890 918 994 999 1037 1076 1112 1136 1204 1289 1324 1342 1375 1429 1481 1487 1578 1702 1710 1742 1813 1875 1876 1878 1885 1897 1901 1942 2057 2135 2150 2179 2209 |
| 2451 | 50 | 1 | 0 7 49 170 223 267 343 348 382 491 508 688 750 828 886 894 977 986 989 1105 1123 1177 1190 1300 1356 1561 1634 1716 1745 1747 1775 1849 1869 1892 1937 2000 2021 2101 2139 2171 2208 2269 2346 2365 2411 2425 2436 |
| 2863 | 54 | 1 | 1 21 53 72 189 225 292 345 386 417 443 451 473 561 566 575 644 904 929 953 966 999 1103 1107 1113 1161 1199 1234 1246 1347 1368 1410 1411 1413 1428 1487 1510 1700 1729 1838 1845 2060 2076 2104 2165 2233 2416 2443 2527 2639 2679 2718 2729 2809 |
| 3541 | 60 | 1 | 1 59 63 79 102 135 176 266 348 366 373 374 376 462 595 608 629 761 820 883 938 962 1009 1120 1211 1379 1421 1530 1557 1607 1701 1713 1745 1850 1919 2187 2227 2244 2312 2342 2347 2396 2407 2471 2477 2523 2728 2747 2827 2875 2920 3198 3236 3251 3265 3302 3338 3450 3459 3481 |
| 3783 | 62 | 1 | 0 1 61 220 239 246 268 314 400 570 572 613 641 650 720 723 756 845 950 1042 1107 1205 1216 1271 1313 1347 1404 1488 1492 1628 1681 1702 1820 1836 1871 2071 2161 2206 2289 2299 2307 2319 2366 2418 2490 2522 2724 3034 3199 3216 3230 3243 3346 3441 3490 3495 3607 3652 3663 3721 3744 3759 |

## Set Elements

| v | k | λ | Set Elements |
|---|---|---|---|

**v = 4161, k = 65, λ = 1**

```
   1    2    4    8   16   32   64  128  256  285  307  357  399  425  512  570  614  714  751  798
 850 1024 1117 1140 1228 1387 1428 1502 1551 1596 1605 1700 1761 1847 2043 2048 2081 2223 2234 2259
2280 2293 2456 2639 2774 2856 2883 2961 3004 3102 3121 3192 3210 3227 3400 3522 3561 3641 3694 3861
3901 4011 4031 4086 4096
```

**v = 4557, k = 68, λ = 1**

```
   0    1   25   67  192  213  256  305  328  350  363  481  491  544  600  615  665  820  858  897
 998 1150 1318 1359 1516 1519 1536 1675 1723 1760 2045 2102 2107 2153 2207 2548 2629 2658 2762 2774
2802 2857 2888 2904 2977 2984 3036 3068 3174 3285 3359 3481 3508 3542 3578 3744 3748 3750 3826 3977
3995 4021 4124 4138 4459 4470 4489 4549
```

**v = 5113, k = 72, λ = 1**

```
   1   71  222  260  419  423  499  510  819  829  877  897  911 1179 1361 1502 1668 1707 1732 1885
1901 1906 2003 2004 2006 2033 2143 2331 2346 2388 2435 2580 2616 2704 2803 2856 2869 2931 2950 3065
3121 3325 3369 3421 3581 3598 3635 3714 3845 3876 3989 4001 4061 4156 4162 4184 4207 4225 4233 4268
4292 4342 4375 4382 4468 4597 4719 4751 4921 4930 4976 5041
```

**v = 5403, k = 74, λ = 1**

```
   0   23   73  275  352  470  546  644  686  732  734  864  885  900  967  971  979  985 1004
1148 1228 1262 1291 1451 1470 1496 1535 1666 1679 1720 1801 1892 2037 2392 2427 2703 2728 2752 2759
2811 2820 3041 3164 3196 3266 3306 3442 3596 3606 3639 3701 3788 3866 3894 3995 4046 4084 4104 4275
4636 4653 4683 4749 4752 4809 4950 4955 5117 5172 5265 5276 5292 5329
```

**v = 6321, k = 80, λ = 1**

```
   0    1   14   17   43   79  255  490  578  717  784  822  993 1037 1106 1143 1182 1324 1343 1415
1479 1537 1728 1779 1803 1836 1906 1989 2375 2384 2540 2595 2712 2733 2783 2785 2881 3007 3014 3063
3246 3375 3397 3460 3594 3676 3771 4214 4229 4275 4316 4328 4709 4756 4824 4884 4916 4943 4961 5027
5047 5101 5191 5201 5226 5231 5393 5399 5427 5534 5545 5655 5802 5850 5951 5959 5982 6071 6075 6241
```

**v = 6643, k = 82, λ = 1**

```
   0    3    9   27   32   81   96  151  199  243  288  453  457  509  584  597  614  729  803
 864  904  932 1003 1133 1359 1371 1493 1527 1745 1752 1791 1842 1856 2044 2187 2225 2384 2409 2419
2482 2525 2592 2712 2796 2833 2956 3009 3056 3233 3292 3399 3418 3478 3554 3611 3791 3802 4019 4077
4113 4190 4429 4479 4495 4581 4730 4763 5110 5235 5256 5373 5414 5526 5568 5588 5696 5905 5927 6132
6397 6561
```

**v = 6973, k = 84, λ = 1**

```
   1   83  219  242  274  275  277  293  472  591  598  691  699  823  900 1043 1056 1103 1129 1201
1544 1569 1635 1823 1906 1945 2044 2061 2120 2190 2233 2300 2508 2523 2629 2638 2786 2791 2893
3037 3058 3218 3280 3400 3429 3504 3599 3677 3687 3691 3711 3741 3927 3972 4041 4077 4174 4311
4496 4624 4713 4765 4792 4830 4836 4876 4917 4939 4970 5007 5183 5352 5491 5503 5552 5687 5851 5947
6140 6182 6514 6561 6889
```

Set Elements

| v | k | λ | Set Elements |
|---|---|---|---|
| 8011 | 90 | 1 | 1 50 82 89 341 342 344 403 411 631 647 855 927 931 1024 1043 1084 1264 1300 1354 |
| | | | 1382 1407 1506 1546 1572 1789 1914 2115 2232 2262 2373 2393 2598 2718 2727 2749 2833 2911 3015 3065 |
| | | | 3148 3160 3165 3511 3546 3721 3785 3796 3823 3917 3959 3972 3982 3996 4030 4140 4185 4331 4387 4450 |
| | | | 4535 4564 4691 4706 4828 5058 5076 5109 5646 5720 5806 5812 5858 5915 6085 6186 6316 6384 6405 6510 |
| | | | 6583 6914 7012 7221 7298 7406 7798 7878 7921 7965 |
| 9507 | 98 | 1 | 1 13 68 97 137 360 568 611 657 670 696 717 833 889 963 1070 1071 1073 1107 1122 |
| | | | 1261 1378 1402 1503 1984 1989 2054 2163 2225 2301 2308 2670 2748 2793 2802 2825 2843 2896 3000 3008 |
| | | | 3169 3186 3211 3527 3782 3929 4128 4257 4536 4594 4725 4745 4818 5209 5215 5253 5367 5371 5588 5598 |
| | | | 5670 5790 5847 6034 6113 6124 6246 6338 6399 6426 6566 6596 6671 6687 6921 7221 7243 7561 7609 7829 |
| | | | 7848 7862 7948 8091 8233 8296 8360 8560 8720 8817 8930 9011 9098 9126 9224 9374 9409 9440 |

# Appendix 3

# $n = 4$ RANDOM MULTIPLE ACCESS SEQUENCES

W0:

|   |   |    |    |
|---|---|----|----|
| 1 | 3 | 4  | 12 |
| 2 | 4 | 5  | 13 |
| 3 | 5 | 6  | 14 |
| 4 | 6 | 7  | 0  |
| 5 | 7 | 8  | 1  |
| 6 | 8 | 9  | 2  |
| 7 | 9 | 10 | 3  |
| 8 | 10| 11 | 4  |
| 9 | 11| 12 | 5  |
| 10| 12| 13 | 6  |
| 11| 13| 14 | 7  |
| 12| 14| 0  | 8  |
| 13| 0 | 1  | 9  |
| 14| 1 | 2  | 10 |
| 0 | 2 | 3  | 11 |

W1:

|   |   |    |    |
|---|---|----|----|
| 0 | 5 | 10 | 15 |
| 1 | 6 | 11 | 15 |
| 2 | 7 | 12 | 15 |
| 3 | 8 | 13 | 15 |
| 4 | 9 | 14 | 15 |

M (FREQUENCY X TIME)

|    |    |    |    |
|----|----|----|----|
| 0  | 1  | 2  | 3  |
| 4  | 5  | 6  | 7  |
| 8  | 9  | 10 | 11 |
| 12 | 13 | 14 | 15 |
| 16 | 17 | 18 | 19 |
| 20 | 21 | 22 | 23 |
| 24 | 25 | 26 | 27 |
| 28 | 29 | 30 | 31 |
| 32 | 33 | 34 | 35 |
| 36 | 37 | 38 | 39 |
| 40 | 41 | 42 | 43 |
| 44 | 45 | 46 | 47 |
| 48 | 49 | 50 | 51 |
| 52 | 53 | 54 | 55 |
| 56 | 57 | 58 | 59 |
| 60 | 61 | 62 | 63 |

B: 0

|   |    |    |    |
|---|----|----|----|
| 4 | 12 | 15 | 48 |
| 5 | 13 | 17 | 49 |
| 6 | 14 | 18 | 50 |
| 7 | 15 | 19 | 51 |

A 1:

|   |   |   |   |
|---|---|---|---|
| 0 | 1 | 2 | 3 |
| 3 | 0 | 1 | 2 |
| 2 | 3 | 0 | 1 |
| 1 | 2 | 3 | 0 |

U 1:

|   |    |    |    |
|---|----|----|----|
| 4 | 13 | 18 | 51 |
| 7 | 12 | 17 | 50 |
| 6 | 15 | 16 | 49 |
| 5 | 14 | 19 | 48 |

A 2:

|   |   |   |   |
|---|---|---|---|
| 0 | 3 | 1 | 2 |
| 2 | 0 | 3 | 1 |
| 1 | 2 | 0 | 3 |
| 3 | 1 | 2 | 0 |

U 2:

|   |    |    |    |
|---|----|----|----|
| 4 | 15 | 17 | 50 |
| 6 | 12 | 19 | 49 |
| 5 | 14 | 16 | 51 |
| 7 | 13 | 18 | 48 |

A 3:

|   |   |   |   |
|---|---|---|---|
| 0 | 2 | 3 | 1 |
| 1 | 0 | 2 | 3 |
| 3 | 1 | 0 | 2 |
| 2 | 3 | 1 | 0 |

U 3:

|   |    |    |    |
|---|----|----|----|
| 4 | 14 | 19 | 49 |
| 5 | 12 | 18 | 51 |
| 7 | 13 | 16 | 50 |
| 6 | 15 | 17 | 48 |

B: 1

|    |    |    |    |
|----|----|----|----|
| 8  | 16 | 20 | 52 |
| 9  | 17 | 21 | 53 |
| 10 | 18 | 22 | 54 |
| 11 | 19 | 23 | 55 |

A 1:

| 0 | 1 | 2 | 3 |
|---|---|---|---|
| 3 | 0 | 1 | 2 |
| 2 | 3 | 0 | 1 |
| 1 | 2 | 3 | 0 |

U 1:

| 8 | 17 | 22 | 55 |
|---|----|----|----|
| 11 | 16 | 21 | 54 |
| 10 | 19 | 20 | 53 |
| 9 | 18 | 23 | 52 |

A 2:

| 0 | 3 | 1 | 2 |
|---|---|---|---|
| 2 | 0 | 3 | 1 |
| 1 | 2 | 0 | 3 |
| 3 | 1 | 2 | 0 |

U 2:

| 8 | 19 | 21 | 54 |
|---|----|----|----|
| 10 | 16 | 23 | 53 |
| 9 | 18 | 20 | 55 |
| 11 | 17 | 22 | 52 |

A 3:

| 0 | 2 | 3 | 1 |
|---|---|---|---|
| 1 | 0 | 2 | 3 |
| 3 | 1 | 0 | 2 |
| 2 | 3 | 1 | 0 |

U 3:

| 8 | 19 | 23 | 53 |
|---|----|----|----|
| 9 | 16 | 22 | 55 |
| 11 | 17 | 20 | 54 |
| 10 | 19 | 21 | 52 |

B: 2

| 12 | 20 | 24 | 56 |
|----|----|----|----|
| 13 | 21 | 25 | 57 |
| 14 | 22 | 26 | 58 |
| 15 | 23 | 27 | 59 |

A 1:

| 0 | 1 | 2 | 3 |
|---|---|---|---|
| 3 | 0 | 1 | 2 |
| 2 | 3 | 0 | 1 |
| 1 | 2 | 3 | 0 |

U 1:

| 12 | 21 | 26 | 59 |
|----|----|----|----|
| 15 | 20 | 25 | 58 |
| 14 | 23 | 24 | 57 |
| 13 | 22 | 27 | 56 |

A 2:

| 0 | 3 | 1 | 2 |
|---|---|---|---|
| 2 | 0 | 3 | 1 |
| 1 | 2 | 0 | 3 |
| 3 | 1 | 2 | 0 |

U 2:

| 12 | 23 | 25 | 58 |
|----|----|----|----|
| 14 | 20 | 27 | 57 |
| 13 | 22 | 24 | 59 |
| 15 | 21 | 26 | 56 |

A 3:

| 0 | 2 | 3 | 1 |
|---|---|---|---|
| 1 | 0 | 2 | 3 |
| 3 | 1 | 0 | 2 |
| 2 | 3 | 1 | 0 |

U 3:

| 12 | 22 | 27 | 57 |
|----|----|----|----|
| 13 | 20 | 26 | 59 |
| 15 | 21 | 24 | 58 |
| 14 | 23 | 25 | 56 |

B: 3

| 16 | 24 | 28 | 0 |
|----|----|----|---|
| 17 | 25 | 29 | 1 |
| 18 | 26 | 30 | 2 |
| 19 | 27 | 31 | 3 |

A 1:

| 0 | 1 | 2 | 3 |
|---|---|---|---|
| 3 | 0 | 1 | 2 |
| 2 | 3 | 0 | 1 |
| 1 | 2 | 3 | 0 |

U 1:

| 16 | 25 | 30 | 3 |
|----|----|----|---|
| 19 | 24 | 22 | 2 |
| 18 | 27 | 23 | 1 |
| 17 | 26 | 31 | 0 |

∆ 2:

| | | | |
|---|---|---|---|
| 0 | 3 | 1 | 2 |
| 2 | 0 | 3 | 1 |
| 1 | 2 | 0 | 3 |
| 3 | 1 | 2 | 0 |

U 2:

| | | | |
|---|---|---|---|
| 16 | 27 | 29 | 2 |
| 18 | 24 | 31 | 1 |
| 17 | 26 | 28 | 3 |
| 19 | 25 | 30 | 0 |

∆ 3:

| | | | |
|---|---|---|---|
| 0 | 2 | 3 | 1 |
| 1 | 0 | 2 | 3 |
| 3 | 1 | 0 | 2 |
| 2 | 3 | 1 | 0 |

U 3:

| | | | |
|---|---|---|---|
| 16 | 26 | 31 | 1 |
| 17 | 24 | 30 | 3 |
| 19 | 25 | 28 | 2 |
| 18 | 27 | 29 | 0 |

B: 4

| | | | |
|---|---|---|---|
| 20 | 28 | 32 | 4 |
| 21 | 29 | 33 | 5 |
| 22 | 30 | 34 | 6 |
| 23 | 31 | 35 | 7 |

∆ 1:

| | | | |
|---|---|---|---|
| 0 | 1 | 2 | 3 |
| 3 | 0 | 1 | 2 |
| 2 | 3 | 0 | 1 |
| 1 | 2 | 3 | 0 |

U 1:

| | | | |
|---|---|---|---|
| 20 | 29 | 34 | 7 |
| 23 | 28 | 33 | 6 |
| 22 | 31 | 32 | 5 |
| 21 | 30 | 35 | 4 |

∆ 2:

| | | | |
|---|---|---|---|
| 0 | 3 | 1 | 2 |
| 2 | 0 | 3 | 1 |
| 1 | 2 | 0 | 3 |
| 3 | 1 | 2 | 0 |

U 2:

| | | | |
|---|---|---|---|
| 20 | 31 | 33 | 6 |
| 22 | 28 | 35 | 5 |
| 21 | 30 | 32 | 7 |
| 23 | 29 | 34 | 4 |

∆ 3:

| | | | |
|---|---|---|---|
| 0 | 2 | 3 | 1 |
| 1 | 0 | 2 | 3 |
| 3 | 1 | 0 | 2 |
| 2 | 3 | 1 | 0 |

U 3:

| | | | |
|---|---|---|---|
| 20 | 30 | 35 | 5 |
| 21 | 28 | 34 | 7 |
| 23 | 29 | 32 | 6 |
| 22 | 31 | 33 | 4 |

B: 5

| | | | |
|---|---|---|---|
| 24 | 32 | 36 | 8 |
| 25 | 33 | 37 | 9 |
| 26 | 34 | 38 | 10 |
| 27 | 35 | 39 | 11 |

∆ 1:

| | | | |
|---|---|---|---|
| 0 | 1 | 2 | 3 |
| 3 | 0 | 1 | 2 |
| 2 | 3 | 0 | 1 |
| 1 | 2 | 3 | 0 |

U 1:

| | | | |
|---|---|---|---|
| 24 | 33 | 38 | 11 |
| 27 | 32 | 37 | 10 |
| 26 | 35 | 36 | 9 |
| 25 | 34 | 39 | 8 |

∆ 2:

| | | | |
|---|---|---|---|
| 0 | 3 | 1 | 2 |
| 2 | 0 | 3 | 1 |
| 1 | 2 | 0 | 3 |
| 3 | 1 | 2 | 0 |

U 2:

| | | | |
|---|---|---|---|
| 24 | 35 | 37 | 10 |
| 26 | 32 | 39 | 9 |
| 25 | 34 | 36 | 11 |
| 27 | 33 | 38 | 8 |

Δ 3:

| | | | |
|---|---|---|---|
| 0 | 2 | 3 | 1 |
| 1 | 0 | 2 | 3 |
| 3 | 1 | 0 | 2 |
| 2 | 3 | 1 | 0 |

U 3:

| | | | |
|---|---|---|---|
| 24 | 34 | 39 | 9 |
| 25 | 32 | 38 | 11 |
| 27 | 33 | 36 | 10 |
| 26 | 35 | 37 | 8 |

θ: 6

| | | | |
|---|---|---|---|
| 28 | 36 | 40 | 12 |
| 29 | 37 | 41 | 13 |
| 30 | 38 | 42 | 14 |
| 31 | 39 | 43 | 15 |

Δ 1:

| | | | |
|---|---|---|---|
| 0 | 1 | 2 | 3 |
| 3 | 0 | 1 | 2 |
| 2 | 3 | 0 | 1 |
| 1 | 2 | 3 | 0 |

U 1:

| | | | |
|---|---|---|---|
| 28 | 37 | 42 | 15 |
| 31 | 36 | 41 | 14 |
| 30 | 39 | 40 | 13 |
| 29 | 38 | 43 | 12 |

Δ 2:

| | | | |
|---|---|---|---|
| 0 | 3 | 1 | 2 |
| 2 | 0 | 3 | 1 |
| 1 | 2 | 0 | 3 |
| 3 | 1 | 2 | 0 |

U 2:

| | | | |
|---|---|---|---|
| 28 | 39 | 41 | 14 |
| 30 | 36 | 43 | 13 |
| 29 | 38 | 40 | 15 |
| 31 | 37 | 42 | 12 |

Δ 3:

| | | | |
|---|---|---|---|
| 0 | 2 | 3 | 1 |
| 1 | 0 | 2 | 3 |
| 3 | 1 | 0 | 2 |
| 2 | 3 | 1 | 0 |

U 3:

| | | | |
|---|---|---|---|
| 28 | 38 | 43 | 13 |
| 29 | 36 | 42 | 15 |
| 31 | 37 | 40 | 14 |
| 30 | 39 | 41 | 12 |

B: 7

| | | | |
|---|---|---|---|
| 32 | 40 | 44 | 15 |
| 33 | 41 | 45 | 17 |
| 34 | 42 | 46 | 18 |
| 35 | 43 | 47 | 19 |

Δ 1:

| | | | |
|---|---|---|---|
| 0 | 1 | 2 | 3 |
| 3 | 0 | 1 | 2 |
| 2 | 3 | 0 | 1 |
| 1 | 2 | 3 | 0 |

U 1:

| | | | |
|---|---|---|---|
| 32 | 41 | 46 | 19 |
| 35 | 40 | 45 | 18 |
| 34 | 43 | 44 | 17 |
| 33 | 42 | 47 | 16 |

Δ 2:

| | | | |
|---|---|---|---|
| 0 | 3 | 1 | 2 |
| 2 | 0 | 3 | 1 |
| 1 | 2 | 0 | 3 |
| 3 | 1 | 2 | 0 |

U 2:

| | | | |
|---|---|---|---|
| 32 | 43 | 45 | 18 |
| 34 | 40 | 47 | 17 |
| 33 | 42 | 44 | 19 |
| 35 | 41 | 46 | 16 |

Δ 3:

| | | | |
|---|---|---|---|
| 0 | 2 | 3 | 1 |
| 1 | 0 | 2 | 3 |
| 3 | 1 | 0 | 2 |
| 2 | 3 | 1 | 0 |

U 3:

| | | | |
|---|---|---|---|
| 32 | 42 | 47 | 17 |
| 33 | 40 | 46 | 19 |
| 35 | 41 | 44 | 18 |
| 34 | 43 | 45 | 16 |

B: 8

| 36 | 44 | 48 | 20 |
| 37 | 45 | 49 | 21 |
| 38 | 46 | 50 | 22 |
| 39 | 47 | 51 | 23 |

A 1:

| 0 | 1 | 2 | 3 |
| 3 | 0 | 1 | 2 |
| 2 | 3 | 0 | 1 |
| 1 | 2 | 3 | 0 |

U 1:

| 36 | 45 | 50 | 23 |
| 39 | 44 | 49 | 22 |
| 38 | 47 | 48 | 21 |
| 37 | 45 | 51 | 20 |

A 2:

| 0 | 3 | 1 | 2 |
| 2 | 0 | 3 | 1 |
| 1 | 2 | 0 | 3 |
| 3 | 1 | 2 | 0 |

U 2:

| 36 | 47 | 49 | 22 |
| 38 | 44 | 51 | 21 |
| 37 | 46 | 48 | 23 |
| 39 | 45 | 50 | 20 |

A 3:

| 0 | 2 | 3 | 1 |
| 1 | 0 | 2 | 3 |
| 3 | 1 | 0 | 2 |
| 2 | 3 | 1 | 0 |

U 3:

| 36 | 46 | 51 | 21 |
| 37 | 44 | 50 | 23 |
| 39 | 45 | 48 | 22 |
| 38 | 47 | 49 | 20 |

B: 9

| 40 | 48 | 52 | 24 |
| 41 | 49 | 53 | 25 |
| 42 | 50 | 54 | 26 |
| 43 | 51 | 55 | 27 |

A 1:

| 0 | 1 | 2 | 3 |
| 3 | 0 | 1 | 2 |
| 2 | 3 | 0 | 1 |
| 1 | 2 | 3 | 0 |

U 1:

| 40 | 49 | 54 | 27 |
| 43 | 48 | 53 | 26 |
| 42 | 51 | 52 | 25 |
| 41 | 50 | 55 | 24 |

A 2:

| 0 | 3 | 1 | 2 |
| 2 | 0 | 3 | 1 |
| 1 | 2 | 0 | 3 |
| 3 | 1 | 2 | 0 |

U 2:

| 40 | 51 | 53 | 26 |
| 42 | 48 | 55 | 25 |
| 41 | 50 | 52 | 27 |
| 43 | 49 | 54 | 24 |

A 3:

| 0 | 2 | 3 | 1 |
| 1 | 0 | 2 | 3 |
| 3 | 1 | 0 | 2 |
| 2 | 3 | 1 | 0 |

U 3:

| 40 | 50 | 55 | 25 |
| 41 | 48 | 54 | 27 |
| 43 | 49 | 52 | 26 |
| 42 | 51 | 53 | 24 |

B: 10

| 44 | 52 | 56 | 28 |
| 45 | 53 | 57 | 29 |
| 46 | 54 | 58 | 30 |
| 47 | 55 | 59 | 31 |

A 1:

| 0 | 1 | 2 | 3 |
| 3 | 0 | 1 | 2 |
| 2 | 3 | 0 | 1 |
| 1 | 2 | 3 | 0 |

U 1:

```
44  53  58  31
47  52  57  30
46  55  56  29
45  54  59  28
```

Δ 2:

```
0   3   1   2
2   0   3   1
1   2   0   3
3   1   2   0
```

U 2:

```
44  55  57  30
46  52  59  29
45  54  56  31
47  53  58  28
```

Δ 3:

```
0   2   3   1
1   0   2   3
3   1   2   2
2   3   1   0
```

U 3:

```
44  54  59  29
45  52  58  31
47  53  56  30
46  55  57  29
```

B:11

```
49  56  0   32
49  57  1   33
50  58  2   34
51  59  3   35
```

Δ 1:

```
0   1   2   3
3   0   1   2
2   3   0   1
1   2   3   0
```

U 1:

```
48  57  2   35
51  56  1   34
50  59  0   33
49  58  3   32
```

Δ 2:

```
0   3   1   2
2   0   3   1
1   2   0   3
3   1   2   0
```

U 2:

```
48  59  1   34
50  56  3   33
49  58  0   35
51  57  2   32
```

Δ 3:

```
0   2   3   1
1   0   2   3
3   1   0   2
2   3   1   0
```

U 3:

```
49  58  3   33
49  55  2   35
51  57  0   34
50  59  1   32
```

B:12

```
52  0   4   36
53  1   5   37
54  2   6   38
55  3   7   39
```

Δ 1:

```
0   1   2   3
3   0   1   2
2   3   0   1
1   2   3   0
```

U 1:

```
52  1   6   39
55  0   5   38
54  3   4   37
53  2   7   36
```

Δ 2:

```
0   3   1   2
2   0   3   1
1   2   0   3
3   1   2   0
```

| U 2: | | | |
|---|---|---|---|
| 52 | 3 | 5 | 38 |
| 54 | 0 | 7 | 37 |
| 53 | 2 | 4 | 39 |
| 55 | 1 | 6 | 36 |

| Δ 3: | | | |
|---|---|---|---|
| 0 | 2 | 3 | 1 |
| 1 | 0 | 2 | 3 |
| 3 | 1 | 0 | 2 |
| 2 | 3 | 1 | 0 |

| U 3: | | | |
|---|---|---|---|
| 52 | 2 | 7 | 37 |
| 53 | 0 | 6 | 39 |
| 55 | 1 | 4 | 38 |
| 54 | 3 | 5 | 36 |

| B:13 | | | |
|---|---|---|---|
| 56 | 4 | 8 | 40 |
| 57 | 5 | 9 | 41 |
| 58 | 6 | 10 | 42 |
| 59 | 7 | 11 | 43 |

| Δ 1: | | | |
|---|---|---|---|
| 0 | 1 | 2 | 3 |
| 3 | 0 | 1 | 2 |
| 2 | 3 | 0 | 1 |
| 1 | 2 | 3 | 0 |

| U 1: | | | |
|---|---|---|---|
| 55 | 5 | 10 | 43 |
| 59 | 4 | 9 | 42 |
| 58 | 7 | 8 | 41 |
| 57 | 6 | 11 | 40 |

| Δ 2: | | | |
|---|---|---|---|
| 0 | 3 | 1 | 2 |
| 2 | 0 | 3 | 1 |
| 1 | 2 | 0 | 3 |
| 3 | 1 | 2 | 0 |

| U 2: | | | |
|---|---|---|---|
| 56 | 7 | 9 | 42 |
| 58 | 4 | 11 | 41 |
| 57 | 6 | 8 | 43 |
| 59 | 5 | 10 | 40 |

| Δ 3: | | | |
|---|---|---|---|
| 0 | 2 | 3 | 1 |
| 1 | 0 | 2 | 3 |
| 3 | 1 | 0 | 2 |
| 2 | 3 | 1 | 0 |

| U 3: | | | |
|---|---|---|---|
| 56 | 6 | 11 | 41 |
| 57 | 4 | 10 | 43 |
| 59 | 5 | 8 | 42 |
| 58 | 7 | 9 | 40 |

| B:14 | | | |
|---|---|---|---|
| 0 | 8 | 12 | 44 |
| 1 | 9 | 13 | 45 |
| 2 | 10 | 14 | 46 |
| 3 | 11 | 15 | 47 |

| Δ 1: | | | |
|---|---|---|---|
| 0 | 1 | 2 | 3 |
| 3 | 0 | 1 | 2 |
| 2 | 3 | 0 | 1 |
| 1 | 2 | 3 | 0 |

| U 1: | | | |
|---|---|---|---|
| 0 | 9 | 14 | 47 |
| 3 | 8 | 13 | 46 |
| 2 | 11 | 12 | 45 |
| 1 | 10 | 15 | 44 |

| Δ 2: | | | |
|---|---|---|---|
| 0 | 3 | 1 | 2 |
| 2 | 0 | 3 | 1 |
| 1 | 2 | 0 | 3 |
| 3 | 1 | 2 | 2 |

| U 2: | | | |
|---|---|---|---|
| 0 | 11 | 13 | 46 |
| 2 | 8 | 15 | 45 |
| 1 | 10 | 12 | 47 |
| 3 | 9 | 14 | 44 |

| Δ 3: | | | |
|---|---|---|---|
| 0 | 2 | 3 | 1 |
| 1 | 0 | 2 | 3 |
| 3 | 1 | 0 | 2 |
| 2 | 3 | 1 | 0 |

U 3:

```
0  10  15  45
1   8  14  47
3   9  12  46
2  11  13  44
```

B:15

```
0  20  40  60
1  21  41  61
2  22  42  52
3  23  43  63
```

A 1:

```
0  1  2  3
3  0  1  2
2  3  0  1
1  2  3  0
```

U 1:

```
0  21  42  63
3  20  41  62
2  23  40  61
1  22  43  60
```

A 2:

```
0  3  1  2
2  0  3  1
1  2  0  3
3  1  2  0
```

U 2:

```
0  23  41  62
2  20  43  61
1  22  40  53
3  21  42  60
```

A 3:

```
0  2  3  1
1  0  2  3
3  1  0  2
2  3  1  0
```

U 3:

```
0  22  43  61
1  20  42  63
3  21  40  62
2  23  41  60
```

B:16

```
4  24  44  60
5  25  45  61
6  26  46  62
7  27  47  63
```

A 1:

```
0  1  2  3
3  0  1  2
2  3  0  1
1  2  3  0
```

U 1:

```
4  25  46  63
7  24  45  62
6  27  44  61
5  26  47  60
```

A 2:

```
0  3  1  2
2  0  3  1
1  2  0  3
3  1  2  0
```

U 2:

```
4  27  45  62
6  24  47  61
5  26  44  63
7  25  46  60
```

A 3:

```
0  2  3  1
1  0  2  3
3  1  0  2
2  3  1  0
```

U 3:

```
4  26  47  61
5  24  46  63
7  25  44  62
6  27  45  60
```

B:17

```
 8  28  48  60
 9  29  49  61
10  30  50  62
11  31  51  63
```

U 1:

| | | | |
|---|---|---|---|
| 12 | 33 | 54 | 63 |
| 15 | 32 | 53 | 62 |
| 14 | 35 | 52 | 61 |
| 13 | 34 | 55 | 60 |

Δ 1:

| | | | |
|---|---|---|---|
| 0 | 1 | 2 | 3 |
| 3 | 0 | 1 | 2 |
| 2 | 3 | 0 | 1 |
| 1 | 2 | 3 | 0 |

Δ 2:

| | | | |
|---|---|---|---|
| 0 | 3 | 1 | 2 |
| 2 | 0 | 3 | 1 |
| 1 | 2 | 0 | 3 |
| 3 | 1 | 2 | 0 |

U 1:

| | | | |
|---|---|---|---|
| 8 | 29 | 50 | 63 |
| 11 | 28 | 49 | 62 |
| 10 | 31 | 48 | 61 |
| 9 | 30 | 51 | 60 |

U 2:

| | | | |
|---|---|---|---|
| 12 | 35 | 53 | 62 |
| 14 | 32 | 55 | 61 |
| 13 | 34 | 52 | 63 |
| 15 | 33 | 54 | 60 |

Δ 2:

| | | | |
|---|---|---|---|
| 0 | 3 | 1 | 2 |
| 2 | 0 | 3 | 1 |
| 1 | 2 | 0 | 3 |
| 3 | 1 | 2 | 0 |

Δ 3:

| | | | |
|---|---|---|---|
| 0 | 2 | 3 | 1 |
| 1 | 0 | 2 | 3 |
| 3 | 1 | 0 | 2 |
| 2 | 3 | 1 | 0 |

U 2:

| | | | |
|---|---|---|---|
| 8 | 31 | 49 | 62 |
| 10 | 29 | 51 | 61 |
| 9 | 30 | 48 | 63 |
| 11 | 29 | 50 | 60 |

U 3:

| | | | |
|---|---|---|---|
| 12 | 34 | 55 | 51 |
| 13 | 32 | 54 | 63 |
| 15 | 33 | 52 | 62 |
| 14 | 35 | 53 | 60 |

Δ 3:

| | | | |
|---|---|---|---|
| 0 | 2 | 3 | 1 |
| 1 | 0 | 2 | 3 |
| 3 | 1 | 0 | 2 |
| 2 | 3 | 1 | 0 |

U 3:

| | | | |
|---|---|---|---|
| 8 | 30 | 51 | 61 |
| 9 | 28 | 50 | 63 |
| 11 | 29 | 48 | 62 |
| 10 | 31 | 49 | 60 |

B:19

| | | | |
|---|---|---|---|
| 16 | 36 | 55 | 60 |
| 17 | 37 | 57 | 61 |
| 18 | 38 | 58 | 62 |
| 19 | 39 | 59 | 63 |

Δ 1:

| | | | |
|---|---|---|---|
| 0 | 1 | 2 | 3 |
| 3 | 0 | 1 | 2 |
| 2 | 3 | 0 | 1 |
| 1 | 2 | 3 | 0 |

B:18

| | | | |
|---|---|---|---|
| 12 | 32 | 52 | 60 |
| 13 | 33 | 53 | 61 |
| 14 | 34 | 54 | 62 |
| 15 | 35 | 55 | 63 |

Δ 1:

| | | | |
|---|---|---|---|
| 0 | 1 | 2 | 3 |
| 3 | 0 | 1 | 2 |
| 2 | 3 | 0 | 1 |
| 1 | 2 | 3 | 0 |

U 1:

| | | | |
|---|---|---|---|
| 16 | 37 | 58 | 63 |
| 19 | 36 | 57 | 62 |
| 18 | 39 | 56 | 61 |
| 17 | 38 | 59 | 60 |

A 2:

| 0 | 3 | 1 | 2 |
|---|---|---|---|
| 2 | 0 | 3 | 1 |
| 1 | 2 | 0 | 3 |
| 3 | 1 | 2 | 0 |

A 3:

| 0 | 2 | 3 | 1 |
|---|---|---|---|
| 1 | 0 | 2 | 3 |
| 3 | 1 | 0 | 2 |
| 2 | 3 | 1 | 0 |

U 2:

| 16 | 39 | 57 | 62 |
|----|----|----|----|
| 18 | 36 | 59 | 61 |
| 17 | 38 | 55 | 53 |
| 19 | 37 | 58 | 60 |

U 3:

| 16 | 38 | 59 | 61 |
|----|----|----|----|
| 17 | 36 | 58 | 63 |
| 19 | 37 | 55 | 52 |
| 18 | 39 | 57 | 60 |

# Appendix 4

# CODES FOR ERROR CONTROLS

This appendix lists 1,10 0 useful codes belonging to the 20 categories listed in Table A.4.1. In the tables most block codes are represented by three numbers: $n$ = code length, $k$ = number of information digits, and $d_{\min}$ = minimum distance of the code. Primitive BCH codes are represented by BCH(P). Nonprimitive BCH codes are represented by BCH(NP). One-step threshold decodable tree codes are characterized in order by decoding constraint length $n_A$, effective constraint length $n_e$, number of encoder outputs in bit time $n_o$, number of encoder inputs in bit time $k_o$, and the code minimum distance $d_{\min} = J + 1$, where $J$ is the number of parity check equations.

Nonbinary codes are indicated by their corresponding GF($p^m$) and Lee distance measures. Viterbi and sequential decodable codes are listed by code rate $R$, code constraint length $K$ or $n_A$, free-distance measure $D(F)$, or minimum distance $M$. In the case of tree codes for Viterbi and sequential decoding, generator polynomials are given. For random multiple access codes the integer residues for each code generation are also tabulated. The generation and decoding algorithm for a code or a set of codes can be obtained from the corresponding numbered references.

Codes that can be modified or derived from other codes such as product codes, concatenated codes, quasi-cyclic codes, and other useful codes are not listed here.

Codes can be stored in a computer data bank and be updated as the need increases. A simple code search program can be used to search for a code or a set of codes from a set of design constraints given as input data.

The flow diagram of the search program is shown in Figure A.4.1. The input data are: $k$ = block message length, $PB$ = decoding bit error rate desired, $SNR$ = $E_b/N_o$ in dB required, $TOL$ = percentage accuracy in $P_{BB}$ calculation for block code, and $ACC$ = percentage accuracy in the error rate $P_{BC}$ calculation for tree codes. After the program searches from the code list, the program outputs either block codes in terms of $n$, $k$, $d_{\min}$ and code rate $k/n$ or tree codes in terms of $n_A$, $n_e$, $n_o$, $k_o$, $d_{\min}$ and code rate $k_o/n_o$. This program also can be linked to a codec design program where a codec design configuration becomes readily available for a specific chosen code.

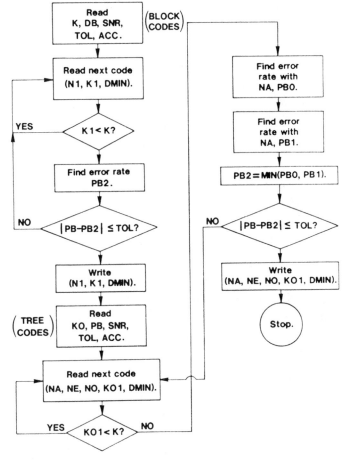

**Figure A.4.1** Flow diagram of code search.

**Table A.4.1**
Code designations for Table A.4.2.

| | |
|---|---|
| **1** Random error correction binary block (cyclic, BCH, linear group) | **11** Quadratic residue |
| | **12** Random multiple access |
| **2** Burst error correction | **13** Cyclic compound channel |
| **3** Majority logic decodable block | **14** Shortened projective geometry |
| **4** Tree—Threshold | **15** Shortened Euclidean geometry |
| **5** —Viterbi | **16** Nonlinear codes (not listed) |
| **6** —Sequential | **17** Polynomial |
| **7** Nonbinary, GF($p$) | **18** Primitive binary RM |
| **8** Reed-Muller | **19** Nonprimitive binary RM |
| **9** Goppa | **20** Binary Srivastava |
| **10** Equidistance, Maximum-distance | |

**Table A.4.2**
Code list for error correction.

| Code Type | Code Length $n$ | Inf. Dig. $k$ | Min. Dist. $d_{min}$ | Code Type | Code Length $n$ | Inf. Dig. $k$ | Min. Dist. $d_{min}$ |
|---|---|---|---|---|---|---|---|
| | | | | 1 | 31 | 21 | 5 |
| | | | | 1 | 31 | 20 | 6 |
| 1 | 15 | 9 | 3 | 1 | 31 | 16 | 6 |
| 1 | 15 | 8 | 4 | 1 | 31 | 16 | 7 |
| 1 | 15 | 7 | 3 | 1 | 31 | 16 | 5 |
| 1 | 15 | 7 | 5 | 1 | 31 | 15 | 6 |
| 1 | 15 | 6 | 6 | 1 | 31 | 15 | 8 |
| 1 | 15 | 5 | 3 | 1 | 31 | 11 | 11 |
| 1 | 15 | 5 | 7 | 1 | 31 | 10 | 12 |
| 1 | 15 | 4 | 8 | 1 | 31 | 6 | 15 |
| 1 | 15 | 4 | 6 | 1 | 31 | 5 | 16 |
| 1 | 15 | 3 | 5 | 1 | 31 | 1 | 31 |
| 1 | 15 | 2 | 10 | 1 | 33 | 23 | 3 |
| 1 | 15 | 1 | 15 | 1 | 33 | 22 | 6 |
| 1 | 17 | 9 | 5 | 1 | 33 | 21 | 3 |
| 1 | 17 | 8 | 6 | 1 | 33 | 21 | 4 |
| 1 | 17 | 1 | 17 | 1 | 33 | 20 | 6 |
| 1 | 21 | 16 | 3 | 1 | 33 | 20 | 4 |
| 1 | 21 | 15 | 3 | 1 | 33 | 13 | 3 |
| 1 | 21 | 15 | 4 | 1 | 33 | 13 | 10 |
| 1 | 21 | 14 | 4 | 1 | 33 | 12 | 6 |
| 1 | 21 | 13 | 4 | 1 | 33 | 12 | 10 |
| 1 | 21 | 13 | 3 | 1 | 33 | 11 | 3 |
| 1 | 21 | 12 | 5 | 1 | 33 | 11 | 11 |
| 1 | 21 | 12 | 4 | 1 | 33 | 10 | 6 |
| 1 | 21 | 12 | 3 | 1 | 33 | 10 | 12 |
| 1 | 21 | 11 | 4 | 1 | 33 | 3 | 11 |
| 1 | 21 | 11 | 6 | 1 | 33 | 2 | 22 |
| 1 | 21 | 10 | 5 | 1 | 33 | 1 | 33 |
| 1 | 21 | 10 | 4 | 1 | 35 | 29 | 4 |
| 1 | 21 | 9 | 3 | 1 | 35 | 27 | 4 |
| 1 | 21 | 9 | 4 | 1 | 35 | 25 | 4 |
| 1 | 21 | 9 | 6 | 1 | 35 | 24 | 4 |
| 1 | 21 | 9 | 8 | 1 | 35 | 23 | 3 |
| 1 | 21 | 8 | 6 | 1 | 35 | 22 | 4 |
| 1 | 21 | 7 | 3 | 1 | 35 | 20 | 3 |
| 1 | 21 | 7 | 8 | 1 | 35 | 20 | 6 |
| 1 | 21 | 6 | 8 | 1 | 35 | 19 | 4 |
| 1 | 21 | 6 | 7 | 1 | 35 | 19 | 6 |
| 1 | 21 | 6 | 6 | 1 | 35 | 18 | 4 |
| 1 | 21 | 5 | 10 | 1 | 35 | 17 | 6 |
| 1 | 21 | 4 | 9 | 1 | 35 | 16 | 6 |
| 1 | 21 | 3 | 12 | 1 | 35 | 16 | 4 |
| 1 | 21 | 3 | 7 | 1 | 35 | 16 | 7 |
| 1 | 21 | 2 | 14 | 1 | 35 | 15 | 8 |
| 1 | 21 | 1 | 21 | 1 | 35 | 15 | 4 |
| 1 | 23 | 12 | 7 | 1 | 35 | 13 | 8 |
| 1 | 23 | 11 | 8 | 1 | 35 | 12 | 8 |
| 1 | 23 | 1 | 23 | 1 | 35 | 11 | 5 |
| 1 | 25 | 5 | 5 | 1 | 35 | 10 | 10 |
| 1 | 25 | 4 | 10 | 1 | 35 | 8 | 8 |
| 1 | 25 | 1 | 25 | 1 | 35 | 7 | 14 |
| 1 | 27 | 9 | 3 | 1 | 35 | 7 | 5 |
| 1 | 27 | 8 | 6 | 1 | 35 | 6 | 10 |
| 1 | 27 | 7 | 6 | 1 | 35 | 5 | 7 |
| 1 | 27 | 6 | 6 | 1 | 35 | 4 | 15 |
| 1 | 27 | 3 | 9 | 1 | 35 | 4 | 14 |
| 1 | 27 | 2 | 18 | 1 | 35 | 3 | 20 |
| 1 | 27 | 1 | 27 | | | | |
| 1 | 31 | 26 | 3 | | | | |
| 1 | 31 | 25 | 4 | 1 | 39 | 26 | 6 |
| 1 | 31 | 21 | 5 | 1 | 39 | 25 | 4 |

| Code Type | Code Length $n$ | Inf. Dig. $k$ | Min. Dist. $d_{min}$ | Code Type | Code Length $n$ | Inf. Dig. $k$ | Min. Dist. $d_{min}$ |
|---|---|---|---|---|---|---|---|
| 1 | 39 | 25 | 3  | 1 | 63 | 57 | 3 |
| 1 | 39 | 24 | 4  | 1 | 63 | 54 | 3 |
| 1 | 39 | 24 | 6  | 1 | 63 | 51 | 3 |
| 1 | 39 | 15 | 3  | 1 | 63 | 50 | 6 |
| 1 | 39 | 15 | 10 | 1 | 63 | 49 | 5 |
| 1 | 39 | 14 | 6  | 1 | 63 | 49 | 4 |
| 1 | 39 | 14 | 10 | 1 | 63 | 48 | 5 |
| 1 | 39 | 13 | 3  | 1 | 63 | 47 | 6 |
| 1 | 39 | 13 | 12 | 1 | 63 | 46 | 6 |
| 1 | 39 | 12 | 6  | 1 | 63 | 45 | 3 |
| 1 | 39 | 12 | 12 | 1 | 63 | 45 | 5 |
| 1 | 39 | 3  | 13 | 1 | 63 | 45 | 7 |
| 1 | 39 | 2  | 26 | 1 | 63 | 44 | 6 |
| 1 | 39 | 1  | 39 | 1 | 63 | 44 | 8 |
| 1 | 41 | 21 | 9  | 1 | 63 | 43 | 4 |
| 1 | 41 | 20 | 10 | 1 | 63 | 43 | 5 |
| 1 | 41 | 1  | 41 | 1 | 63 | 43 | 7 |
| 1 | 43 | 29 | 6  | 1 | 63 | 43 | 4 |
| 1 | 43 | 28 | 6  | 1 | 63 | 42 | 6 |
| 1 | 43 | 15 | 13 | 1 | 63 | 42 | 5 |
| 1 | 43 | 14 | 14 | 1 | 63 | 40 | 7 |
| 1 | 43 | 1  | 43 | 1 | 63 | 40 | 8 |
| 1 | 45 | 35 | 4  | 1 | 63 | 39 | 3 |
| 1 | 45 | 34 | 4  | 1 | 63 | 39 | 4 |
| 1 | 45 | 33 | 3  | 1 | 63 | 39 | 5 |
| 1 | 45 | 33 | 4  | 1 | 63 | 37 | 7 |
| 1 | 45 | 32 | 4  | 1 | 63 | 37 | 9 |
| 1 | 45 | 31 | 4  | 1 | 63 | 36 | 3 |
| 1 | 45 | 30 | 4  | 1 | 63 | 36 | 5 |
| 1 | 45 | 29 | 4  | 1 | 63 | 36 | 6 |
| 1 | 45 | 29 | 5  | 1 | 63 | 36 | 11 |
| 1 | 45 | 28 | 4  | 1 | 63 | 35 | 8 |
| 1 | 45 | 28 | 6  | 1 | 63 | 35 | 12 |
| 1 | 45 | 27 | 3  | 1 | 63 | 34 | 8 |
| 1 | 45 | 27 | 4  | 1 | 63 | 34 | 9 |
| 1 | 45 | 27 | 6  | 1 | 63 | 34 | 11 |
| 1 | 45 | 26 | 4  | 1 | 63 | 33 | 7 |
| 1 | 45 | 26 | 6  | 1 | 63 | 33 | 8 |
| 1 | 45 | 25 | 4  | 1 | 63 | 33 | 3 |
| 1 | 45 | 25 | 5  | 1 | 63 | 33 | 5 |
| 1 | 45 | 24 | 4  | 1 | 63 | 33 | 6 |
| 1 | 45 | 23 | 7  | 1 | 63 | 32 | 10 |
| 1 | 45 | 21 | 8  | 1 | 63 | 32 | 6 |
| 1 | 45 | 19 | 7  | 1 | 63 | 32 | 8 |
| 1 | 45 | 18 | 8  | 1 | 63 | 32 | 12 |
| 1 | 45 | 15 | 7  | 1 | 63 | 31 | 6 |
| 1 | 45 | 13 | 5  | 1 | 63 | 31 | 7 |
| 1 | 45 | 11 | 9  | 1 | 63 | 31 | 9 |
| 1 | 45 | 9  | 12 | 1 | 63 | 31 | 11 |
| 1 | 45 | 7  | 15 | 1 | 63 | 31 | 12 |
| 1 | 47 | 24 | 11 | 1 | 63 | 30 | 5 |
| 1 | 51 | 35 | 5  | 1 | 63 | 29 | 11 |
| 1 | 51 | 27 | 5  | 1 | 63 | 27 | 3 |
| 1 | 51 | 27 | 9  | 1 | 63 | 27 | 6 |
| 1 | 51 | 24 | 10 | 1 | 63 | 27 | 7 |
| 1 | 51 | 19 | 9  | 1 | 63 | 27 | 8 |
| 1 | 51 | 11 | 15 | 1 | 63 | 25 | 15 |
| 1 | 51 | 9  | 19 | 1 | 63 | 25 | 9 |
| 1 | 55 | 25 | 11 | 1 | 63 | 25 | 12 |
| 1 | 55 | 21 | 15 | 1 | 63 | 25 | 13 |
| 1 | 55 | 5  | 11 | 1 | 63 | 25 | 15 |
| 1 | 57 | 37 | 3  | 1 | 63 | 24 | 15 |

| Code Type | Code Length $n$ | Inf. Dig. $k$ | Min. Dist. $d_{min}$ | |
|---|---|---|---|---|
| 1 | 63 | 24 | 13 | |
| 1 | 63 | 24 | 11 | |
| 1 | 63 | 24 | 9 | |
| 1 | 63 | 24 | 8 | |
| 1 | 63 | 15 | 24 | |
| 1 | 63 | 13 | 9 | |
| 1 | 63 | 12 | 9 | |
| 1 | 63 | 12 | 15 | |
| 1 | 63 | 10 | 15 | |
| 1 | 65 | 53 | 5 | |
| 1 | 65 | 41 | 5 | |
| 1 | 65 | 37 | 5 | |
| 1 | 65 | 17 | 13 | |
| 1 | 127 | 57 | 23 | |
| 1 | 127 | 50 | 27 | |
| 1 | 127 | 43 | 29 | |
| 1 | 127 | 36 | 31 | |
| 1 | 127 | 22 | 47 | |
| 1 | 127 | 9 | 63 | |
| 1 | 255 | 247 | 3 | |
| 1 | 2863 | 54 | 1 | |
| 1 | 7 | 4 | 3 | BCH(P) |
| 1 | 15 | 11 | 3 | BCH(P) |
| 1 | 15 | 7 | 5 | BCH(P) |
| 1 | 15 | 5 | 5 | BCH(P) |
| 1 | 31 | 26 | 3 | BCH(P) |
| 1 | 31 | 21 | 5 | BCH(P) |
| 1 | 31 | 16 | 7 | BCH(P) |
| 1 | 31 | 11 | 11 | BCH(P) |
| 1 | 31 | 6 | 15 | BCH(P) |
| 1 | 63 | 57 | 3 | BCH(P) |
| 1 | 63 | 51 | 5 | BCH(P) |
| 1 | 63 | 45 | 7 | BCH(P) |
| 1 | 63 | 39 | 9 | BCH(P) |
| 1 | 63 | 36 | 11 | BCH(P) |
| 1 | 63 | 30 | 13 | BCH(P) |
| 1 | 63 | 24 | 15 | BCH(P) |
| 1 | 63 | 18 | 21 | BCH(P) |
| 1 | 63 | 15 | 23 | BCH(P) |
| 1 | 63 | 10 | 27 | BCH(P) |
| 1 | 63 | 7 | 31 | BCH(P) |
| 1 | 127 | 120 | 3 | BCH(P) |
| 1 | 127 | 113 | 5 | BCH(P) |
| 1 | 127 | 105 | 7 | BCH(P) |
| 1 | 127 | 99 | 9 | BCH(P) |
| 1 | 127 | 92 | 11 | BCH(P) |
| 1 | 127 | 85 | 13 | BCH(P) |
| 1 | 127 | 78 | 15 | BCH(P) |
| 1 | 127 | 71 | 19 | BCH(P) |
| 1 | 127 | 64 | 21 | BCH(P) |
| 1 | 127 | 57 | 23 | BCH(P) |
| 1 | 127 | 50 | 27 | BCH(P) |
| 1 | 127 | 43 | 29 | BCH(P) |
| 1 | 127 | 36 | 31 | BCH(P) |
| 1 | 127 | 29 | 43 | BCH(P) |
| 1 | 127 | 22 | 47 | BCH(P) |
| 1 | 127 | 15 | 55 | BCH(P) |
| 1 | 127 | 8 | 63 | BCH(P) |
| 1 | 255 | 247 | 3 | BCH(P) |
| 1 | 255 | 239 | 5 | BCH(P) |
| 1 | 255 | 231 | 7 | BCH(P) |
| 1 | 255 | 223 | 9 | BCH(P) |

| Code Type | Code Length $n$ | Inf. Dig. $k$ | Min. Dist. $d_{min}$ | |
|---|---|---|---|---|
| 1 | 255 | 215 | 11 | BCH(P) |
| 1 | 255 | 207 | 13 | BCH(P) |
| 1 | 255 | 199 | 15 | BCH(P) |
| 1 | 255 | 191 | 17 | BCH(P) |
| 1 | 255 | 187 | 19 | BCH(P) |
| 1 | 255 | 179 | 21 | BCH(P) |
| 1 | 255 | 171 | 23 | BCH(P) |
| 1 | 255 | 163 | 25 | BCH(P) |
| 1 | 255 | 155 | 27 | BCH(P) |
| 1 | 255 | 147 | 29 | BCH(P) |
| 1 | 255 | 139 | 31 | BCH(P) |
| 1 | 255 | 131 | 37 | BCH(P) |
| 1 | 255 | 123 | 39 | BCH(P) |
| 1 | 255 | 115 | 43 | BCH(P) |
| 1 | 255 | 107 | 45 | BCH(P) |
| 1 | 255 | 93 | 47 | BCH(P) |
| 1 | 255 | 91 | 51 | BCH(P) |
| 1 | 255 | 87 | 53 | BCH(P) |
| 1 | 255 | 79 | 55 | BCH(P) |
| 1 | 255 | 71 | 59 | BCH(P) |
| 1 | 255 | 63 | 61 | BCH(P) |
| 1 | 255 | 55 | 63 | BCH(P) |
| 1 | 255 | 47 | 85 | BCH(P) |
| 1 | 255 | 45 | 87 | BCH(P) |
| 1 | 255 | 73 | 91 | BCH(P) |
| 1 | 255 | 29 | 95 | BCH(P) |
| 1 | 255 | 21 | 111 | BCH(P) |
| 1 | 255 | 13 | 119 | BCH(P) |
| 1 | 255 | 9 | 127 | BCH(P) |
| 1 | 511 | 502 | 3 | BCH(P) |
| 1 | 511 | 493 | 5 | BCH(P) |
| 1 | 511 | 484 | 7 | BCH(P) |
| 1 | 511 | 475 | 9 | BCH(P) |
| 1 | 511 | 466 | 11 | BCH(P) |
| 1 | 511 | 457 | 13 | BCH(P) |
| 1 | 511 | 448 | 15 | BCH(P) |
| 1 | 511 | 439 | 17 | BCH(P) |
| 1 | 511 | 430 | 19 | BCH(P) |
| 1 | 511 | 421 | 21 | BCH(P) |
| 1 | 511 | 412 | 23 | BCH(P) |
| 1 | 511 | 403 | 25 | BCH(P) |
| 1 | 511 | 394 | 27 | BCH(P) |
| 1 | 511 | 385 | 29 | BCH(P) |
| 1 | 511 | 376 | 31 | BCH(P) |
| 1 | 511 | 367 | 33 | BCH(P) |
| 1 | 511 | 358 | 37 | BCH(P) |
| 1 | 511 | 349 | 39 | BCH(P) |
| 1 | 511 | 340 | 41 | BCH(P) |
| 1 | 511 | 331 | 43 | BCH(P) |
| 1 | 511 | 322 | 45 | BCH(P) |
| 1 | 511 | 313 | 47 | BCH(P) |
| 1 | 511 | 304 | 51 | BCH(P) |
| 1 | 511 | 295 | 53 | BCH(P) |
| 1 | 511 | 286 | 55 | BCH(P) |
| 1 | 511 | 277 | 57 | BCH(P) |
| 1 | 511 | 268 | 59 | BCH(P) |
| 1 | 511 | 259 | 61 | BCH(P) |
| 1 | 511 | 250 | 63 | BCH(P) |
| 1 | 511 | 241 | 73 | BCH(P) |
| 1 | 511 | 239 | 75 | BCH(P) |
| 1 | 511 | 229 | 77 | BCH(P) |
| 1 | 511 | 220 | 79 | BCH(P) |

| Code Type | Code Length $n$ | Inf. Dig. $k$ | Min. Dist. $d_{min}$ | | Code Type | Code Length $n$ | Inf. Dig. $k$ | Min. Dist. $d_{min}$ | |
|---|---|---|---|---|---|---|---|---|---|
| 1 | 511 | 211 | 83 | BCH(P) | 1 | 1023 | 638 | 85 | BCH(P) |
| 1 | 511 | 202 | 85 | BCH(P) | 1 | 1023 | 628 | 87 | BCH(P) |
| 1 | 511 | 193 | 87 | BCH(P) | 1 | 1023 | 618 | 89 | BCH(P) |
| 1 | 511 | 184 | 91 | BCH(P) | 1 | 1023 | 608 | 91 | BCH(P) |
| 1 | 511 | 175 | 93 | BCH(P) | 1 | 1023 | 598 | 93 | BCH(P) |
| 1 | 511 | 166 | 95 | BCH(F) | 1 | 1023 | 588 | 95 | BCH(P) |
| 1 | 511 | 157 | 103 | BCH(P) | 1 | 1023 | 578 | 99 | BCH(P) |
| 1 | 511 | 148 | 107 | BCH(P) | 1 | 1023 | 573 | 101 | BCH(P) |
| 1 | 511 | 139 | 109 | BCH(P) | 1 | 1023 | 563 | 103 | BCH(P) |
| 1 | 511 | 130 | 111 | BCH(P) | 1 | 1023 | 553 | 105 | BCH(P) |
| 1 | 511 | 121 | 117 | BCH(P) | 1 | 1023 | 543 | 107 | BCH(P) |
| 1 | 511 | 112 | 119 | BCH(P) | 1 | 1023 | 533 | 109 | BCH(P) |
| 1 | 511 | 103 | 123 | BCH(P) | 1 | 1023 | 523 | 111 | BCH(P) |
| 1 | 511 | 94 | 125 | BCH(P) | 1 | 1023 | 513 | 115 | BCH(P) |
| 1 | 511 | 85 | 127 | BCH(P) | 1 | 1023 | 503 | 117 | BCH(P) |
| 1 | 511 | 76 | 171 | BCH(P) | 1 | 1023 | 493 | 119 | BCH(P) |
| 1 | 511 | 67 | 175 | BCH(P) | 1 | 1023 | 483 | 121 | BCH(P) |
| 1 | 511 | 58 | 183 | BCH(P) | 1 | 1023 | 473 | 123 | BCH(P) |
| 1 | 511 | 49 | 187 | BCH(P) | 1 | 1023 | 463 | 125 | BCH(P) |
| 1 | 511 | 40 | 191 | BCH(P) | 1 | 1023 | 453 | 127 | BCH(P) |
| 1 | 511 | 31 | 219 | BCH(P) | 1 | 1023 | 443 | 147 | BCH(P) |
| 1 | 511 | 29 | 223 | BCH(F) | 1 | 1023 | 433 | 149 | BCH(P) |
| 1 | 511 | 19 | 239 | BCH(P) | 1 | 1023 | 423 | 151 | BCH(P) |
| 1 | 511 | 10 | 243 | BCH(P) | 1 | 1023 | 413 | 155 | BCH(P) |
| 1 | 1023 | 1013 | 3 | BCH(P) | 1 | 1023 | 403 | 157 | BCH(P) |
| 1 | 1023 | 1003 | 5 | BCH(P) | 1 | 1023 | 393 | 159 | BCH(P) |
| 1 | 1023 | 993 | 7 | BCH(P) | 1 | 1023 | 383 | 165 | BCH(P) |
| 1 | 1023 | 983 | 9 | BCH(P) | 1 | 1023 | 378 | 167 | BCH(P) |
| 1 | 1023 | 973 | 11 | BCH(P) | 1 | 1023 | 368 | 171 | BCH(P) |
| 1 | 1023 | 963 | 13 | BCH(P) | 1 | 1023 | 358 | 173 | BCH(F) |
| 1 | 1023 | 953 | 15 | BCH(P) | 1 | 1023 | 348 | 175 | BCH(P) |
| 1 | 1023 | 943 | 17 | BCH(P) | 1 | 1023 | 338 | 179 | BCH(P) |
| 1 | 1023 | 933 | 19 | BCH(P) | 1 | 1023 | 328 | 181 | BCH(P) |
| 1 | 1023 | 923 | 21 | BCH(F) | 1 | 1023 | 318 | 183 | BCH(P) |
| 1 | 1023 | 913 | 23 | BCH(P) | 1 | 1023 | 308 | 187 | BCH(P) |
| 1 | 1023 | 903 | 25 | BCH(P) | 1 | 1023 | 298 | 189 | BCH(P) |
| 1 | 1023 | 893 | 27 | BCH(P) | 1 | 1023 | 288 | 191 | BCH(P) |
| 1 | 1023 | 883 | 29 | BCH(F) | 1 | 1023 | 278 | 205 | BCH(P) |
| 1 | 1023 | 873 | 31 | BCH(P) | 1 | 1023 | 268 | 207 | BCH(P) |
| 1 | 1023 | 863 | 33 | BCH(P) | 1 | 1023 | 258 | 213 | BCH(P) |
| 1 | 1023 | 853 | 35 | BCH(P) | 1 | 1023 | 248 | 215 | BCH(P) |
| 1 | 1023 | 848 | 37 | BCH(P) | 1 | 1023 | 238 | 219 | BCH(P) |
| 1 | 1023 | 838 | 39 | BCH(P) | 1 | 1023 | 228 | 221 | BCH(P) |
| 1 | 1023 | 828 | 41 | BCH(P) | 1 | 1023 | 218 | 223 | BCH(P) |
| 1 | 1023 | 818 | 43 | BCH(P) | 1 | 1023 | 208 | 231 | BCH(P) |
| 1 | 1023 | 808 | 45 | BCH(P) | 1 | 1023 | 203 | 235 | BCH(P) |
| 1 | 1023 | 798 | 47 | BCH(P) | 1 | 1023 | 193 | 237 | BCH(P) |
| 1 | 1023 | 788 | 49 | BCH(P) | 1 | 1023 | 183 | 239 | BCH(P) |
| 1 | 1023 | 778 | 51 | BCH(P) | 1 | 1023 | 173 | 245 | BCH(P) |
| 1 | 1023 | 768 | 53 | BCH(P) | 1 | 1023 | 163 | 247 | BCH(P) |
| 1 | 1023 | 758 | 55 | BCH(P) | 1 | 1023 | 153 | 251 | BCH(P) |
| 1 | 1023 | 748 | 57 | BCH(P) | 1 | 1023 | 143 | 253 | BCH(P) |
| 1 | 1023 | 738 | 59 | BCH(P) | 1 | 1023 | 133 | 255 | BCH(P) |
| 1 | 1023 | 728 | 61 | BCH(P) | 1 | 1023 | 123 | 341 | BCH(P) |
| 1 | 1023 | 718 | 63 | BCH(P) | 1 | 1023 | 121 | 343 | BCH(P) |
| 1 | 1023 | 708 | 69 | BCH(P) | 1 | 1023 | 111 | 347 | BCH(P) |
| 1 | 1023 | 698 | 71 | BCH(P) | 1 | 1023 | 101 | 351 | BCH(P) |
| 1 | 1023 | 688 | 73 | BCH(P) | 1 | 1023 | 91 | 363 | BCH(P) |
| 1 | 1023 | 678 | 75 | BCH(P) | 1 | 1033 | 86 | 367 | BCH(P) |
| 1 | 1023 | 668 | 77 | BCH(P) | 1 | 1023 | 76 | 375 | BCH(P) |
| 1 | 1023 | 658 | 79 | BCH(P) | 1 | 1023 | 66 | 379 | BCH(F) |
| 1 | 1023 | 648 | 83 | BCH(P) | 1 | 1023 | 56 | 383 | BCH(P) |

Left table:

| Code Type | Code Length $n$ | Inf. Dig. $k$ | Min. Dist. $d_{min}$ | Code |
|---|---|---|---|---|
| 1 | 1023 | 45 | 439 | BCH(P) |
| 1 | 1023 | 36 | 447 | BCH(P) |
| 1 | 1023 | 26 | 479 | BCH(P) |
| 1 | 1023 | 16 | 295 | BCH(P) |
| 1 | 1023 | 11 | 511 | BCH(P) |
| 1 | 21 | 12 | 5 | BCH(NP) |
| 1 | 21 | 6 | 7 | BCH(NP) |
| 1 | 21 | 4 | 9 | BCH(NP) |
| 1 | 23 | 12 | 5 | BCH(NP) |
| 1 | 25 | 5 | 5 | BCH(NP) |
| 1 | 27 | 3 | 9 | BCH(NP) |
| 1 | 33 | 13 | 5 | BCH(NP) |
| 1 | 35 | 11 | 5 | BCH(NP) |
| 1 | 35 | 9 | 7 | BCH(NP) |
| 1 | 35 | 4 | 15 | BCH(NP) |
| 1 | 39 | 15 | 7 | BCH(NP) |
| 1 | 39 | 3 | 13 | BCH(NP) |
| 1 | 43 | 15 | 7 | BCH(NP) |
| 1 | 45 | 29 | 5 | BCH(NP) |
| 1 | 45 | 23 | 7 | BCH(NP) |
| 1 | 45 | 11 | 9 | BCH(NP) |
| 1 | 45 | 7 | 15 | BCH(NP) |
| 1 | 45 | 5 | 21 | BCH(NP) |
| 1 | 47 | 24 | 5 | BCH(NP) |
| 1 | 49 | 7 | 7 | BCH(NP) |
| 1 | 49 | 4 | 21 | BCH(NP) |
| 1 | 51 | 35 | 5 | BCH(NP) |
| 1 | 51 | 27 | 9 | BCH(NP) |
| 1 | 51 | 19 | 11 | BCH(NP) |
| 1 | 51 | 11 | 17 | BCH(NP) |
| 1 | 51 | 9 | 19 | BCH(NP) |
| 1 | 55 | 15 | 5 | BCH(NP) |
| 1 | 55 | 5 | 11 | BCH(NP) |
| 1 | 57 | 21 | 5 | BCH(NP) |
| 1 | 57 | 3 | 19 | BCH(NP) |
| 1 | 65 | 41 | 5 | BCH(NP) |
| 1 | 65 | 29 | 7 | BCH(NP) |
| 1 | 65 | 17 | 11 | BCH(NP) |
| 1 | 65 | 5 | 13 | BCH(NP) |
| 1 | 69 | 36 | 5 | BCH(NP) |
| 1 | 69 | 14 | 15 | BCH(NP) |
| 1 | 69 | 3 | 23 | BCH(NP) |
| 1 | 71 | 36 | 7 | BCH(NP) |
| 1 | 73 | 55 | 5 | BCH(NP) |
| 1 | 73 | 46 | 9 | BCH(NP) |
| 1 | 73 | 37 | 11 | BCH(NP) |
| 1 | 73 | 28 | 13 | BCH(NP) |
| 1 | 73 | 13 | 17 | BCH(NP) |
| 1 | 73 | 10 | 25 | BCH(NP) |
|  |  |  | B |  |
| 2 | 7 | 3 | 2 |  |
| 2 | 15 | 9 | 3 |  |
| 2 | 27 | 17 | 5 |  |
| 2 | 34 | 22 | 6 |  |
| 2 | 50 | 34 | 8 |  |
| 2 | 67 | 54 | 6 |  |
| 2 | 103 | 88 | 7 |  |
| 2 | 63 | 55 | 3 |  |
| 2 | 85 | 75 | 4 |  |
| 2 | 131 | 119 | 5 |  |
| 2 | 169 | 155 | 6 |  |
| 2 | 121 | 112 | 3 |  |

Right table:

| Code Type | Code Length $n$ | Inf. Dig. $k$ | Min. Dist. $d_{min}$ |
|---|---|---|---|
| 2 | 290 | 277 | 5 |
| 2 | 511 | 499 | 4 |
| 2 | 1023 | 1011 | 4 |
| 3 | 7 | 4 | 3 |
| 3 | 7 | 1 | 7 |
| 3 | 15 | 11 | 3 |
| 3 | 15 | 7 | 5 |
| 3 | 15 | 5 | 7 |
| 3 | 15 | 1 | 15 |
| 3 | 21 | 11 | 5 |
| 3 | 31 | 26 | 3 |
| 3 | 31 | 16 | 7 |
| 3 | 31 | 6 | 15 |
| 3 | 31 | 1 | 31 |
| 3 | 51 | 16 | 3 |
| 3 | 63 | 57 | 5 |
| 3 | 63 | 49 | 5 |
| 3 | 63 | 45 | 7 |
| 3 | 63 | 42 | 7 |
| 3 | 63 | 36 | 9 |
| 3 | 63 | 24 | 15 |
| 3 | 63 | 22 | 15 |
| 3 | 63 | 13 | 21 |
| 3 | 63 | 7 | 31 |
| 3 | 63 | 1 | 63 |
| 3 | 73 | 45 | 9 |
| 3 | 95 | 69 | 5 |
| 3 | 85 | 68 | 1 |
| 3 | 85 | 24 | 21 |
| 3 | 89 | 44 | 1 |
| 3 | 93 | 27 | 1 |
| 3 | 105 | 61 | 1 |
| 3 | 117 | 50 | 1 |
| 3 | 127 | 120 | 3 |
| 3 | 127 | 99 | 7 |
| 3 | 127 | 64 | 15 |
| 3 | 127 | 29 | 31 |
| 3 | 127 | 8 | 63 |
| 3 | 127 | 1 | 127 |
| 3 | 195 | 76 | 1 |
| 3 | 255 | 247 | 3 |
| 3 | 255 | 211 | 5 |
| 3 | 255 | 219 | 7 |
| 3 | 255 | 163 | 15 |
| 3 | 255 | 91 | 31 |
| 3 | 255 | 21 | 85 |
| 3 | 255 | 1 | 255 |
| 3 | 315 | 58 | 3 |
| 3 | 381 | 65 | 1 |
| 3 | 455 | 49 | 9 |
| 3 | 511 | 502 | 3 |
| 3 | 511 | 475 | 7 |
| 3 | 511 | 466 | 7 |
| 3 | 511 | 448 | 9 |
| 3 | 511 | 392 | 15 |
| 3 | 511 | 274 | 31 |
| 3 | 511 | 184 | 63 |
| 3 | 511 | 133 | 73 |
| 3 | 511 | 46 | 127 |
| 3 | 511 | 10 | 255 |
| 3 | 511 | 1 | 511 |

| Code Type | Code Length n | Inf. Dig. k | Min. Dist. $d_{min}$ | | Code Type | Code Length n | Inf. Dig. k | Min. Dist. $d_{min}$ |
|---|---|---|---|---|---|---|---|---|
| 3 | 819 | 65 | 51 | | 3 | 4096 | 299 | 512* |
| 3 | 1023 | 1013 | 3 | | 3 | 4359 | 4112 | 17 |
| 3 | 1023 | 988 | 5 | | 3 | 4369 | 4112 | 1 |
| 3 | 1023 | 973 | 7 | | 3 | 4369 | 1545 | 273 |
| 3 | 1023 | 968 | 7 | | 3 | 4681 | 4555 | 9 |
| 3 | 1023 | 893 | 15 | | 3 | 4681 | 3105 | 73 |
| 3 | 1023 | 868 | 15 | | 3 | 4681 | 590 | 585 |
| 3 | 1023 | 848 | 15 | | 3 | 5461 | 5411 | 5 |
| 3 | 1023 | 813 | 31 | | 3 | 5461 | 4900 | 21 |
| 3 | 1023 | 781 | 33 | | 3 | 5461 | 3195 | 85 |
| 3 | 1023 | 748 | 21 | | 3 | 5461 | 1064 | 341 |
| 3 | 1023 | 648 | 31 | | 3 | 5461 | 119 | 1365 |
| 3 | 1023 | 638 | 31 | | 3 | 16383 | 16369 | 3 |
| 3 | 1023 | 438 | 63 | | 3 | 16383 | 16320 | 5 |
| 3 | 1023 | 388 | 63 | | 3 | 16383 | 16278 | 7 |
| 3 | 1023 | 386 | 63 | | 3 | 16383 | 15914 | 15 |
| 3 | 1023 | 288 | 85 | | 3 | 16383 | 14913 | 31 |
| 3 | 1023 | 186 | 127 | | 3 | 16383 | 12711 | 63 |
| 3 | 1023 | 176 | 127 | | 3 | 16383 | 9908 | 127 |
| 3 | 1023 | 76 | 255 | | 3 | 16383 | 6476 | 255 |
| 3 | 1023 | 55 | 255 | | 3 | 21845 | 21780 | 1 |
| 3 | 1023 | 31 | 341 | | 3 | 21845 | 21790 | 1 |
| 3 | 1023 | 11 | 511 | | 3 | 21845 | 21140 | 1 |
| 3 | 1023 | 1 | 1023 | | 3 | 21845 | 20884 | 17 |
| 3 | 1197 | 402 | 1 | | 3 | 21845 | 19860 | 1 |
| 3 | 1365 | 595 | 17 | | 3 | 21845 | 17749 | 17 |
| 3 | 1365 | 76 | 341 | | 3 | 21845 | 16628 | 85 |
| 3 | 2047 | 2036 | 3 | | 3 | 21845 | 8908 | 273 |
| 3 | 2047 | 1981 | 7 | | 3 | 21845 | 8536 | 341 |
| 3 | 2047 | 1816 | 15 | | 3 | 21845 | 3709 | 273 |
| 3 | 2047 | 1486 | 31 | | 3 | 21845 | 2136 | 1365 |
| 3 | 3855 | 128 | 25 | | 3 | 21845 | 176 | 5461 |
| 3 | 4076 | 3302 | 32 | | | | | |
| 3 | 4096 | 1586 | 128 | | | | | |

**Table A.4.3**
Code list for error correction.

| Code Type | $n_A$ | $n_e$ | $n_o$ | $k_o$ | $d_{min}$ | | Code Type | $n_A$ | $n_e$ | $n_o$ | $k_o$ | $d_{min}$ |
|---|---|---|---|---|---|---|---|---|---|---|---|---|
| 4 | 3 | 3 | 3 | 1 | 3 | | 4 | 24 | 13 | 4 | 1 | 8 |
| 4 | 4 | 4 | 2 | 1 | 3 | | 4 | 24 | 12 | 3 | 2 | 6 |
| 4 | 5 | 5 | 5 | 1 | 5 | | 4 | 25 | 13 | 5 | 4 | 3 |
| 4 | 9 | 7 | 3 | 1 | 5 | | 4 | 25 | 13 | 5 | 4 | 3 |
| 4 | 9 | 7 | 3 | 2 | 3 | | 4 | 30 | 29 | 10 | 1 | 19 |
| 4 | 10 | 10 | 10 | 1 | 10 | | 4 | 30 | 27 | 5 | 1 | 14 |
| 4 | 10 | 9 | 5 | 1 | 7 | | 4 | 33 | 30 | 3 | 1 | 11 |
| 4 | 12 | 11 | 2 | 1 | 5 | | 4 | 33 | 17 | 3 | 1 | 8 |
| 4 | 12 | 9 | 4 | 1 | 6 | | 4 | 35 | 30 | 5 | 1 | 15 |
| 4 | 14 | 11 | 2 | 1 | 5 | | 4 | 36 | 22 | 2 | 1 | 7 |
| 4 | 15 | 13 | 3 | 1 | 7 | | 4 | 36 | 16 | 6 | 5 | 3 |
| 4 | 15 | 13 | 5 | 1 | 9 | | 4 | 40 | 39 | 10 | 1 | 25 |
| 4 | 15 | 10 | 3 | 1 | 6 | | 4 | 42 | 21 | 3 | 2 | 5 |
| 4 | 16 | 10 | 4 | 1 | 3 | | 4 | 42 | 20 | 3 | 2 | 5 |
| 4 | 20 | 20 | 10 | 1 | 15 | | | | | | | |
| 4 | 20 | 18 | 5 | 1 | 11 | | | | | | | |
| 4 | 24 | 22 | 2 | 1 | 7 | | 4 | 44 | 19 | 4 | 1 | 10 |
| 4 | 24 | 22 | 3 | 1 | 9 | | 4 | 44 | 17 | 4 | 3 | 6 |
| 4 | 24 | 16 | 2 | 1 | 6 | | 4 | 45 | 37 | 5 | 1 | 17 |
| 4 | 24 | 13 | 3 | 2 | 6 | | 4 | 49 | 19 | 7 | 6 | 3 |

| Code Type | $n_A$ | $n_c$ | $n_o$ | $k_o$ | $d_{min}$ | Code Type | $n_A$ | $n_c$ | $n_o$ | $k_o$ | $d_{min}$ |
|---|---|---|---|---|---|---|---|---|---|---|---|
| 4 | 50 | 49 | 10 | 1 | 27 | 4 | 507 | 130 | 3 | 2 | 12 |
| 4 | 54 | 40 | 3 | 1 | 13 | 4 | 512 | 105 | 4 | 3 | 9 |
| 4 | 55 | 47 | 5 | 1 | 19 | 4 | 539 | 91 | 7 | 6 | 6 |
| 4 | 64 | 22 | 8 | 7 | 3 | 4 | 545 | 95 | 5 | 4 | 7 |
| 4 | 65 | 55 | 5 | 1 | 21 | 4 | 545 | 85 | 5 | 4 | 7 |
| 4 | 65 | 25 | 5 | 4 | 4 | 4 | 576 | 133 | 3 | 2 | 12 |
| 4 | 65 | 24 | 5 | 4 | 4 | 4 | 589 | 157 | 3 | 2 | 13 |
| 4 | 66 | 26 | 3 | 1 | 10 | 4 | 589 | 152 | 3 | 2 | 13 |
| 4 | 69 | 55 | 3 | 1 | 15 | 4 | 615 | 113 | 5 | 4 | 8 |
| 4 | 69 | 31 | 3 | 2 | 6 | 4 | 615 | 113 | 5 | 4 | 8 |
| 4 | 69 | 30 | 3 | 2 | 6 | 4 | 615 | 109 | 5 | 4 | 8 |
| 4 | 70 | 69 | 10 | 1 | 34 | 4 | 615 | 109 | 5 | 4 | 8 |
| 4 | 72 | 56 | 2 | 1 | 11 | 4 | 630 | 106 | 6 | 5 | 7 |
| 4 | 72 | 37 | 2 | 1 | 9 | 4 | 660 | 135 | 4 | 3 | 10 |
| 4 | 80 | 65 | 5 | 1 | 23 | 4 | 690 | 76 | 6 | 5 | 6 |
| 4 | 80 | 31 | 4 | 3 | 5 | 4 | 702 | 49 | 9 | 8 | 4 |
| 4 | 90 | 89 | 10 | 1 | 39 | 4 | 715 | 61 | 11 | 10 | 4 |
| 4 | 92 | 46 | 2 | 1 | 10 | 4 | 720 | 105 | 8 | 7 | 6 |
| 4 | 96 | 31 | 6 | 5 | 4 | 4 | 753 | 183 | 3 | 2 | 14 |
| 4 | 108 | 68 | 3 | 1 | 17 | 4 | 755 | 179 | 3 | 2 | 14 |
| 4 | 117 | 37 | 3 | 1 | 12 | 4 | 812 | 166 | 4 | 3 | 11 |
| 4 | 120 | 43 | 3 | 2 | 7 | 4 | 832 | 136 | 4 | 3 | 10 |
| 4 | 120 | 41 | 3 | 2 | 7 | 4 | 861 | 183 | 3 | 2 | 14 |
| 4 | 135 | 41 | 5 | 4 | 5 | 4 | 867 | 219 | 3 | 2 | 15 |
| 4 | 135 | 41 | 5 | 4 | 5 | 4 | 867 | 211 | 3 | 2 | 15 |
| 4 | 135 | 39 | 5 | 4 | 5 | 4 | 882 | 127 | 7 | 6 | 7 |
| 4 | 135 | 39 | 5 | 4 | 5 | 4 | 895 | 161 | 5 | 4 | 9 |
| 4 | 148 | 46 | 4 | 3 | 6 | 4 | 895 | 161 | 5 | 4 | 9 |
| 4 | 148 | 45 | 4 | 3 | 6 | 4 | 895 | 145 | 5 | 4 | 9 |
| 4 | 159 | 50 | 3 | 1 | 14 | 4 | 895 | 145 | 5 | 4 | 9 |
| 4 | 165 | 57 | 3 | 2 | 8 | 4 | 935 | 113 | 5 | 4 | 8 |
| 4 | 165 | 54 | 3 | 2 | 8 | 4 | 935 | 113 | 5 | 4 | 8 |
| 4 | 168 | 45 | 4 | 3 | 6 | 4 | 1002 | 211 | 3 | 2 | 15 |
| 4 | 194 | 43 | 8 | 7 | 4 | 4 | 1011 | 249 | 3 | 2 | 16 |
| 4 | 199 | 51 | 6 | 5 | 5 | 4 | 1011 | 241 | 3 | 2 | 16 |
| 4 | 237 | 73 | 3 | 2 | 9 | 4 | 1175 | 145 | 5 | 4 | 9 |
| 4 | 237 | 73 | 3 | 2 | 9 | 4 | 1175 | 145 | 5 | 4 | 9 |
| 4 | 248 | 66 | 4 | 3 | 7 | 4 | 1176 | 149 | 8 | 7 | 7 |
| 4 | 248 | 64 | 4 | 3 | 7 | 4 | 1224 | 241 | 3 | 2 | 16 |
| 4 | 251 | 59 | 5 | 4 | 6 | 4 | 1233 | 81 | 9 | 8 | 5 |
| 4 | 251 | 59 | 5 | 4 | 6 | 4 | 1336 | 106 | 8 | 7 | 6 |
| 4 | 255 | 61 | 5 | 4 | 6 | 4 | 1352 | 199 | 4 | 3 | 12 |
| 4 | 255 | 61 | 5 | 4 | 6 | 4 | 1365 | 273 | 3 | 2 | 17 |
| 4 | 270 | 61 | 5 | 4 | 6 | 4 | 1368 | 121 | 9 | 8 | 6 |
| 4 | 270 | 61 | 5 | 4 | 6 | 4 | 1396 | 141 | 6 | 5 | 8 |
| 4 | 270 | 59 | 5 | 4 | 6 | 4 | 1425 | 95 | 15 | 14 | 4 |
| 4 | 270 | 59 | 5 | 4 | 6 | 4 | 1445 | 181 | 5 | 4 | 10 |
| 4 | 287 | 61 | 7 | 6 | 5 | 4 | 1445 | 181 | 5 | 4 | 10 |
| 4 | 309 | 91 | 3 | 2 | 10 | 4 | 1506 | 307 | 3 | 2 | 18 |
| 4 | 309 | 89 | 3 | 2 | 10 | 4 | 1529 | 101 | 11 | 10 | 5 |
| 4 | 368 | 83 | 4 | 3 | 8 | 4 | 1664 | 73 | 13 | 12 | 4 |
| 4 | 368 | 64 | 4 | 1 | 22 | 4 | 1716 | 111 | 12 | 11 | 5 |
| 4 | 384 | 76 | 6 | 5 | 6 | | | | | | |
| 4 | 384 | 71 | 8 | 7 | 5 | | | | | | |
| 4 | 393 | 111 | 3 | 2 | 11 | 4 | 2090 | 91 | 16 | 15 | 4 |
| | | | | | | 4 | 2217 | 421 | 3 | 2 | 21 |
| | | | | | | 4 | 2400 | 507 | 3 | 2 | 23 |
| 4 | 415 | 83 | 5 | 4 | 7 | 4 | 2499 | 553 | 3 | 2 | 24 |
| 4 | 415 | 83 | 5 | 4 | 7 | 4 | 2720 | 361 | 4 | 3 | 16 |
| 4 | 449 | 85 | 4 | 3 | 8 | 4 | 2740 | 313 | 5 | 4 | 13 |
| 4 | 468 | 111 | 3 | 2 | 11 | 4 | 2740 | 313 | 5 | 4 | 13 |
| 4 | 507 | 133 | 3 | 2 | 12 | 4 | 2757 | 651 | 3 | 2 | 26 |

| Code Type | $n_A$ | $n_c$ | $n_u$ | $k_u$ | $d_{min}$ | | Code Type | $n_A$ | $n_c$ | $n_u$ | $k_u$ | $d_{min}$ |
|---|---|---|---|---|---|---|---|---|---|---|---|---|
| 4 | 2640 | 115 | 20 | 19 | 4 | | 4 | 8556 | 765 | 6 | 5 | 18 |
| 4 | 2934 | 524 | 6 | 5 | 15 | | 4 | 8659 | 256 | 18 | 17 | 6 |
| 4 | 3072 | 165 | 12 | 11 | 6 | | 4 | 8815 | 925 | 5 | 4 | 22 |
| 4 | 3098 | 197 | 8 | 7 | 8 | | 4 | 8815 | 925 | 5 | 4 | 22 |
| 4 | 3165 | 365 | 5 | 4 | 14 | | 4 | 9215 | 271 | 19 | 18 | 6 |
| 4 | 3165 | 365 | 5 | 4 | 14 | | 4 | 9455 | 1013 | 5 | 4 | 23 |
| 4 | 3216 | 450 | 4 | 3 | 18 | | 4 | 9455 | 1013 | 5 | 4 | 23 |
| 4 | 3510 | 141 | 15 | 14 | 5 | | 4 | 9507 | 1561 | 3 | 2 | 40 |
| 4 | 3696 | 703 | 3 | 2 | 27 | | | | | | | |
| 4 | 3750 | 331 | 6 | 5 | 12 | | | | | | | |
| 4 | 4116 | 571 | 4 | 3 | 20 | | 4 | 9925 | 241 | 25 | 24 | 5 |
| 4 | 4235 | 211 | 11 | 10 | 7 | | 4 | 9981 | 1641 | 3 | 2 | 41 |
| 4 | 4290 | 181 | 13 | 12 | 6 | | 4 | 10008 | 1306 | 4 | 3 | 30 |
| 4 | 4383 | 813 | 3 | 2 | 29 | | 4 | 10448 | 639 | 8 | 7 | 14 |
| 4 | 4860 | 253 | 10 | 9 | 8 | | 4 | 10491 | 1723 | 3 | 2 | 42 |
| 4 | 4990 | 694 | 4 | 3 | 22 | | 4 | 11109 | 1402 | 4 | 3 | 32 |
| 4 | 5015 | 545 | 5 | 4 | 17 | | 4 | 11358 | 951 | 6 | 5 | 20 |
| 4 | 5015 | 545 | 5 | 4 | 17 | | 4 | 11554 | 337 | 17 | 16 | 7 |
| 4 | 5270 | 161 | 17 | 16 | 5 | | 4 | 11690 | 595 | 10 | 9 | 12 |
| 4 | 5316 | 903 | 3 | 2 | 32 | | 4 | 11995 | 1991 | 3 | 2 | 45 |
| 4 | 5512 | 253 | 13 | 12 | 7 | | 4 | 11998 | 217 | 37 | 36 | 4 |
| 4 | 5700 | 1657 | 3 | 2 | 33 | | 4 | 12078 | 551 | 11 | 10 | 11 |
| 4 | 5750 | 133 | 23 | 22 | 4 | | 4 | 13156 | 433 | 13 | 12 | 9 |
| 4 | 5757 | 291 | 11 | 10 | 8 | | 4 | 13315 | 301 | 21 | 20 | 6 |
| 4 | 5947 | 181 | 19 | 18 | 5 | | 4 | 13376 | 841 | 8 | 7 | 16 |
| 4 | 6304 | 226 | 16 | 15 | 6 | | 4 | 13508 | 421 | 16 | 15 | 8 |
| 4 | 6310 | 685 | 5 | 4 | 19 | | 4 | 13860 | 205 | 35 | 34 | 4 |
| 4 | 6310 | 695 | 5 | 4 | 19 | | 4 | 14271 | 2353 | 3 | 2 | 49 |
| 4 | 6552 | 274 | 14 | 13 | 7 | | 4 | 14445 | 303 | 15 | 14 | 8 |
| 4 | 6578 | 127 | 22 | 21 | 4 | | 4 | 14822 | 359 | 18 | 17 | 7 |
| 4 | 6709 | 151 | 26 | 25 | 4 | | 4 | 14858 | 223 | 38 | 37 | 4 |
| 4 | 6750 | 145 | 25 | 24 | 4 | | 4 | 14938 | 316 | 22 | 21 | 6 |
| 4 | 6533 | 1191 | 3 | 2 | 35 | | 4 | 15212 | 197 | 32 | 31 | 4 |
| 4 | 7056 | 163 | 28 | 27 | 4 | | 4 | 16375 | 505 | 15 | 14 | 9 |
| 4 | 7192 | 975 | 4 | 3 | 26 | | 4 | 16462 | 201 | 29 | 28 | 5 |
| 4 | 7460 | 761 | 5 | 4 | 20 | | 4 | 19270 | 241 | 41 | 40 | 4 |
| 4 | 7460 | 761 | 5 | 4 | 20 | | 4 | 19782 | 247 | 42 | 41 | 4 |
| 4 | 7570 | 406 | 10 | 9 | 10 | | 4 | 15809 | 301 | 31 | 30 | 5 |
| 4 | 7791 | 201 | 21 | 20 | 5 | | 4 | 19823 | 253 | 43 | 42 | 4 |
| 4 | 7845 | 1333 | 3 | 2 | 37 | | 4 | 23046 | 271 | 46 | 45 | 4 |
| 4 | 8130 | 1407 | 3 | 2 | 38 | | 4 | 26750 | 275 | 50 | 49 | 4 |
| 4 | 8216 | 337 | 13 | 12 | 8 | | | | | | | |

**Table A.4.4**

| Code Type | R | K | D(F) |
|---|---|---|---|
| 5 | 1/4 | 7 | (1111001,1011011,1100101,1011011) |
| 5 | 1/5 | 5 | (11101,10011,11111,11011,10101) |
| 5 | 1/7 | 7 | (11111001,10100111,11110111, 11C110C1,10010101,10011111, 11100101) |
| 5 | 1/2 | 3 | 5 | (111,101) |
| 5 | 1/2 | 4 | 6 | (1111,1101) |
| 5 | 1/2 | 5 | 7 | (11101,10011) |
| 5 | 1/2 | 6 | 8 | (111101,101011) |
| 5 | 1/2 | 7 | 10 | (1111001,1011011) |
| 5 | 1/2 | 8 | 10 | (11111001,10100111) |

## Table A.4.4

| Code Type | $R$ | $K$ | $D(F)$ | |
|---|---|---|---|---|
| 5 | 1/2 | 9 | 12 | (111101011,1011100011) |
| 5 | 1/3 | 3 | 8 | (111,111,101) |
| 5 | 1/3 | 4 | 10 | (1111,1101,1011) |
| 5 | 1/3 | 5 | 12 | (11111,11011,10101) |
| 5 | 1/3 | 6 | 13 | (111101,101011,1001111) |
| 5 | 1/3 | 7 | 14 | (1111001,1100101,10110111) |
| 5 | 1/3 | 8 | 16 | (11110111,11011001,10010101) |
| 5 | 2/3 | 6 | 5 | (101111,011001,110010) |
| 5 | 2/3 | 9 | 7 | (10110110,01111001,11110111) |
| 5 | 3/4 | 6 | 4 | (100001,010011,001110,111101) |

## Table A.4.5

| Code Type | $R$ | $n_A$ | $M$ | |
|---|---|---|---|---|
| 6 | 1/2 | 96 | 47 | $G(1)=(533,533,676,737,355,3)$, |
| | | | | $G(2)=(733,533,676,737,355,3)$,OCTAL |
| 6 | 1/2 | 79 | 35 | $G(1)=(533,533,676,737)$, |
| | | | | $G(2)=(733,533,676,737)$,$\overline{0}$ |
| 6 | 1/2 | 79 | 35 | $G(1)=(400,000,000,000)$, |
| | | | | $G(2)=(715,473,701,317)$,OCTAL |

## Table A.4.6

| Code Type | $n$ | $k$ | $d_{min}$ | | |
|---|---|---|---|---|---|
| 7 | 2 | 1 | 3 | GF(5) | LEE DISTANCE |
| 7 | 3 | 2 | 3 | GF(7) | LEE DISTANCE |
| 7 | 3 | 1 | 5 | GF(7) | LEE DISTANCE |
| 7 | 5 | 4 | 3 | GF(11) | LEE DISTANCE |
| 7 | 5 | 3 | 5 | GF(11) | LEE DISTANCE |
| 7 | 5 | 2 | 7 | GF(11) | LEE DISTANCE |
| 7 | 5 | 1 | 9 | GF(11) | LEE DISTANCE |
| 7 | 6 | 4 | 3 | GF(5) | LEE DISTANCE |
| 7 | 6 | 3 | 5 | GF(5) | LEE DISTANCE |
| 7 | 8 | 7 | 3 | GF(17) | LEE DISTANCE |
| 7 | 8 | 6 | 5 | GF(17) | LEE DISTANCE |
| 7 | 8 | 5 | 7 | GF(17) | LEE DISTANCE |
| 7 | 8 | 4 | 9 | GF(17) | LEE DISTANCE |
| 7 | 8 | 3 | 11 | GF(17) | LEE DISTANCE |
| 7 | 12 | 10 | 3 | GF(5) | LEE DISTANCE |
| 7 | 12 | 8 | 5 | GF(5) | LEE DISTANCE |
| 7 | 15 | 13 | 3 | GF(11) | LEE DISTANCE |
| 7 | 15 | 12 | 5 | GF(11) | LEE DISTANCE |
| 7 | 15 | 10 | 7 | GF(11) | LEE DISTANCE |
| 7 | 15 | 8 | 9 | GF(11) | LEE DISTANCE |
| 7 | 15 | 7 | 11 | GF(11) | LEE DISTANCE |
| 7 | 24 | 22 | 3 | GF(17) | LEE DISTANCE |
| 7 | 24 | 22 | 3 | GF(7) | LEE DISTANCE |
| 7 | 24 | 21 | 5 | GF(17) | LEE DISTANCE |
| 7 | 24 | 20 | 5 | GF(7) | LEE DISTANCE |
| 7 | 24 | 19 | 7 | GF(17) | LEE DISTANCE |
| 7 | 24 | 18 | 7 | GF(7) | LEE DISTANCE |
| 7 | 24 | 17 | 9 | GF(17) | LEE DISTANCE |

| Code Type | $n$ | $k$ | $d_{min}$ | |
|---|---|---|---|---|
| 7 | 24 | 16 | 11 | GF(17) LEE DISTANCE |
| 7 | 60 | 58 | 3 | GF(11) LEE DISTANCE |
| 7 | 60 | 56 | 5 | GF(11) LEE DISTANCE |
| 7 | 60 | 54 | 7 | GF(11) LEE DISTANCE |
| 7 | 60 | 52 | 9 | GF(11) LEE DISTANCE |
| 7 | 60 | 50 | 11 | GF(17) LEE DISTANCE |
| 7 | 62 | 59 | 3 | GF(5) LEE DISTANCE |
| 7 | 62 | 55 | 5 | GF(5) LEE DISTANCE |
| 7 | 72 | 70 | 3 | GF(17) LEE DISTANCE |
| 7 | 72 | 68 | 5 | GF(17) LEE DISTANCE |
| 7 | 72 | 66 | 7 | GF(17) LEE DISTANCE |
| 7 | 72 | 64 | 9 | GF(17) LEE DISTANCE |
| 7 | 72 | 63 | 11 | GF(17) LEE DISTANCE |
| 7 | 144 | 142 | 3 | GF(17) LEE DISTANCE |
| 7 | 144 | 140 | 5 | GF(17) LEE DISTANCE |
| 7 | 144 | 139 | 7 | GF(17) LEE DISTANCE |
| 7 | 144 | 136 | 9 | GF(17) LEE DISTANCE |
| 7 | 171 | 168 | 3 | GF(7) LEE DISTANCE |
| 7 | 171 | 165 | 5 | GF(7) LEE DISTANCE |
| 7 | 171 | 162 | 7 | GF(7) LEE DISTANCE |
| 7 | 312 | 309 | 3 | GF(5) LEE DISTANCE |
| 7 | 312 | 304 | 5 | GF(5) LEE DISTANCE |
| 7 | 665 | 662 | 3 | GF(11) LEE DISTANCE |
| 7 | 665 | 659 | 5 | GF(11) LEE DISTANCE |
| 7 | 665 | 656 | 7 | GF(11) LEE DISTANCE |
| 7 | 665 | 653 | 9 | GF(11) LEE DISTANCE |
| 7 | 665 | 650 | 11 | GF(11) LEE DISTANCE |
| 8 | 15 | 11 | 3 | |
| 8 | 15 | 5 | 7 | |
| 8 | 31 | 6 | 15 | |
| 8 | 48 | 11 | 15 | |
| 8 | 56 | 3 | 31 | |
| 8 | 63 | 57 | 3 | |
| 8 | 63 | 42 | 7 | |
| 8 | 63 | 22 | 15 | |
| 8 | 63 | 7 | 31 | |
| 8 | 93 | 13 | 31 | |
| 8 | 96 | 39 | 15 | |
| 8 | 112 | 18 | 31 | |
| 8 | 112 | 3 | 63 | |
| 8 | 127 | 99 | 7 | |
| 8 | 127 | 64 | 15 | |
| 8 | 127 | 29 | 31 | |
| 8 | 127 | 8 | 63 | |
| 8 | 192 | 105 | 15 | |
| 8 | 192 | 51 | 31 | |
| 8 | 192 | 15 | 63 | |
| 8 | 224 | 67 | 31 | |
| 8 | 224 | 21 | 63 | |
| 8 | 224 | 3 | 127 | |
| 8 | 240 | 26 | 63 | |
| 8 | 240 | 4 | 127 | |
| 8 | 240 | 5 | 127 | |
| 8 | 255 | 247 | 3 | |
| 8 | 255 | 219 | 7 | |
| 8 | 255 | 93 | 31 | |
| 8 | 255 | 37 | 63 | |
| 8 | 255 | 9 | 127 | |
| 8 | 384 | 157 | 31 | |
| 8 | 384 | 66 | 63 | |
| 8 | 384 | 17 | 127 | |

| Code Type | $n$ | $k$ | $d_{\min}$ |
|---|---|---|---|
| 8 | 448 | 199 | 31 |
| 8 | 448 | 89 | 63 |
| 8 | 448 | 30 | 127 |
| 8 | 448 | 24 | 127 |
| 8 | 448 | 4 | 255 |
| 8 | 448 | 3 | 255 |
| 8 | 480 | 104 | 63 |
| 8 | 496 | 35 | 127 |
| 8 | 496 | 5 | 255 |
| 8 | 511 | 502 | 3 |
| 8 | 511 | 466 | 7 |
| 8 | 511 | 382 | 15 |
| 8 | 511 | 256 | 31 |
| 8 | 511 | 130 | 63 |
| 8 | 511 | 46 | 127 |
| 8 | 511 | 10 | 255 |
| 8 | 1023 | 1013 | 3 |
| 8 | 1023 | 963 | 7 |
| 8 | 1023 | 849 | 15 |
| 8 | 1023 | 639 | 31 |
| 9 | 1023 | 386 | 63 |
| 8 | 1023 | 176 | 127 |
| 8 | 1023 | 56 | 255 |
| 8 | 1023 | 11 | 510 |
| 9 | 31 | 16 | 7 |
| 9 | 84 | 63 | 7 |
| 9 | 95 | 60 | 11 |
| 9 | 101 | 52 | 15 |
| 9 | 141 | 85 | 15 |
| 9 | 157 | 117 | 11 |
| 9 | 160 | 136 | 7 |
| 9 | 270 | 207 | 15 |
| 9 | 273 | 192 | 19 |
| 9 | 279 | 172 | 23 |
| 9 | 291 | 164 | 27 |
| 9 | 312 | 295 | 7 |
| 9 | 312 | 267 | 11 |

| Code Type | $N$ | $K$ | $D$ |
|---|---|---|---|
| 9 | 280 | 217 | 11 |
| 9 | 288 | 236 | 13 |
| 9 | 296 | 235 | 15 |
| 9 | 304 | 234 | 17 |
| 9 | 312 | 233 | 19 |
| 9 | 320 | 232 | 21 |
| 9 | 328 | 231 | 23 |
| 9 | 336 | 230 | 25 |
| 9 | 352 | 236 | 27 |
| 9 | 360 | 235 | 29 |
| 10 | 32 | 19 | 6 |
| 10 | 32 | 9 | 10 |
| 10 | 128 | 105 | 6 |
| 10 | 128 | 102 | 8 |
| 10 | 128 | 82 | 10 |
| 10 | 128 | 70 | 16 |
| 10 | 128 | 55 | 18 |
| 10 | 128 | 37 | 22 |
| 10 | 128 | 13 | 34 |
| 10 | 192 | 6196 | 128 |
| 10 | 256 | 195 | 10 |
| 10 | 256 | 103 | 32 |
| 10 | 256 | 77 | 34 |

| Code Type | $n$ | $k$ | $d_{min}$ |
|---|---|---|---|
| 10 | 512 | 478 | 6 |
| 10 | 512 | 414 | 16 |
| 10 | 512 | 390 | 16 |
| 10 | 512 | 366 | 18 |
| 10 | 512 | 316 | 32 |
| 10 | 512 | 298 | 22 |
| 10 | 512 | 285 | 34 |
| 10 | 1024 | 977 | 8 |
| 10 | 1024 | 923 | 10 |
| 10 | 1024 | 323 | 74 |
| 10 | 2048 | 2001 | 6 |
| 10 | 2048 | 1990 | 8 |
| 10 | 2048 | 1927 | 10 |
| 10 | 2048 | 1836 | 16 |
| 10 | 2048 | 1615 | 22 |
| 10 | 2048 | 1414 | 64 |
| 10 | 2048 | 1351 | 66 |
| 10 | 2048 | 744 | 128 |
| 10 | 2048 | 645 | 130 |
| 10 | 4096 | 2349 | 128 |
| 10 | 4096 | 2229 | 130 |
| 10 | 8192 | 8130 | 6 |
| 10 | 8192 | 6069 | 130 |
| 11 | 17 | 6 | 7 |
| 11 | 16 | 13 | 11 |
| 11 | 42 | 21 | 9 |
| 11 | 72 | 33 | 15 |
| 11 | 72 | 28 | 17 |
| 11 | 73 | 35 | 13 |
| 11 | 80 | 40 | 13 |
| 11 | 82 | 40 | 15 |
| 11 | 84 | 33 | 19 |
| 11 | 92 | 45 | 15 |
| 11 | 94 | 45 | 17 |

## Table A.4.7

| Code Type | $N$ | MODULUS | $D$ | RESIDUES |
|---|---|---|---|---|
| 12 | 2 | 3 | 1 | 0,1 |
| 12 | 3 | 8 | 2 | 0,5,6 |
| 12 | 4 | 15 | 3 | 0,2,3,11 |
| 12 | 5 | 24 | 4 | 0,3,15,16,19 |
| 12 | 7 | 49 | 6 | 0,1,3,10,30,35,37 |
| 12 | 8 | 63 | 7 | 0,5,7,13,37,47,48,49 |
| 12 | 9 | 80 | 8 | 0,12,34,47,48,65,71,73 |

## Table A.4.8

| Code Type | $n$ | $k$ | $d_{min}$ |
|---|---|---|---|
| 13 | 73 | 55 | 5 |
| 13 | 73 | 46 | 9 |
| 13 | 73 | 37 | 11 |
| 13 | 85 | 61 | 7 |
| 13 | 85 | 53 | 9 |
| 13 | 85 | 45 | 13 |
| 13 | 95 | 37 | 15 |
| 13 | 99 | 56 | 9 |
| 13 | 89 | 45 | 11 |
| 13 | 89 | 34 | 13 |
| 13 | 89 | 33 | 9 |
| 13 | 89 | 24 | 11 |
| 13 | 127 | 106 | 7 |
| 13 | 127 | 99 | 9 |
| 13 | 127 | 92 | 11 |
| 13 | 127 | 85 | 13 |
| 13 | 127 | 79 | 15 |
| 13 | 127 | 71 | 19 |
| 13 | 127 | 71 | 11 |
| 13 | 127 | 64 | 21 |
| 13 | 127 | 53 | 15 |
| 13 | 127 | 47 | 15 |
| 13 | 127 | 39 | 19 |
| 13 | 151 | 121 | 5 |
| 13 | 151 | 106 | 9 |
| 13 | 151 | 91 | 11 |
| 13 | 151 | 76 | 15 |
| 13 | 151 | 70 | 11 |
| 13 | 255 | 223 | 9 |
| 13 | 255 | 215 | 11 |
| 13 | 255 | 207 | 13 |
| 13 | 255 | 199 | 15 |
| 13 | 255 | 191 | 17 |
| 13 | 255 | 187 | 19 |
| 13 | 255 | 184 | 11 |
| 13 | 255 | 173 | 21 |
| 13 | 255 | 172 | 13 |
| 13 | 255 | 170 | 15 |
| 13 | 255 | 156 | 17 |
| 13 | 255 | 156 | 15 |
| 13 | 255 | 148 | 17 |
| 13 | 255 | 144 | 19 |
| 13 | 255 | 135 | 21 |
| 13 | 257 | 225 | 5 |
| 13 | 257 | 209 | 7 |
| 13 | 257 | 193 | 9 |
| 13 | 257 | 177 | 11 |
| 13 | 257 | 161 | 13 |
| 13 | 257 | 145 | 15 |
| 13 | 257 | 120 | 13 |
| 13 | 341 | 301 | 9 |
| 13 | 341 | 291 | 11 |
| 13 | 341 | 286 | 13 |
| 13 | 341 | 276 | 15 |
| 13 | 341 | 262 | 11 |
| 13 | 391 | 332 | 9 |
| 13 | 391 | 325 | 11 |
| 13 | 381 | 311 | 13 |
| 13 | 391 | 297 | 15 |
| 13 | 391 | 268 | 13 |
| 13 | 391 | 246 | 15 |

| Code Type | $n$ | $k$ | $d_{min}$ |
|---|---|---|---|
| 13 | 511 | 431 | 11 |
| 13 | 511 | 429 | 13 |
| 13 | 511 | 407 | 15 |
| 13 | 511 | 402 | 17 |
| 13 | 511 | 389 | 19 |
| 13 | 511 | 376 | 21 |
| 13 | 1023 | 914 | 13 |
| 13 | 1023 | 902 | 15 |
| 13 | 1023 | 902 | 17 |
| 13 | 1023 | 889 | 19 |
| 13 | 1023 | 872 | 21 |
| 14 | 10 | 6 | 3 |
| 14 | 11 | 6 | 4 |
| 14 | 14 | 4 | 7 |
| 14 | 15 | 10 | 4 |
| 14 | 19 | 12 | 4 |
| 14 | 21 | 15 | 3 |
| 14 | 25 | 10 | 7 |
| 14 | 33 | 24 | 4 |
| 14 | 35 | 13 | 8 |
| 14 | 41 | 20 | 7 |
| 14 | 47 | 32 | 5 |
| 14 | 58 | 41 | 6 |
| 14 | 79 | 43 | 7 |
| 14 | 90 | 35 | 15 |
| 14 | 120 | 21 | 32 |
| 14 | 227 | 99 | 21 |
| 14 | 246 | 28 | 63 |
| 14 | 500 | 35 | 127 |
| 15 | 48 | 6 | 21 |
| 15 | 192 | 79 | 21 |
| 15 | 192 | 8 | 85 |
| 15 | 240 | 14 | 85 |
| 15 | 448 | 102 | 137 |
| 15 | 768 | 517 | 21 |
| 15 | 768 | 161 | 85 |
| 15 | 768 | 12 | 341 |
| 15 | 960 | 240 | 85 |
| 15 | 560 | 20 | 341 |
| 15 | 1008 | 26 | 341 |
| 15 | 3584 | 2137 | 73 |
| 15 | 3584 | 267 | 385 |
| 15 | 2840 | 120 | 273 |
| 15 | 4028 | 414 | 385 |
| 17 | 21 | 13 | 5 |
| 17 | 63 | 51 | 3 |
| 17 | 63 | 27 | 7 |
| 17 | 63 | 16 | 15 |
| 17 | 73 | 29 | 9 |
| 17 | 85 | 61 | 5 |
| 17 | 255 | 236 | 3 |
| 17 | 255 | 126 | 15 |
| 17 | 255 | 16 | 95 |
| 18 | 7 | 4 | 3 |
| 18 | 15 | 11 | 3 |
| 18 | 15 | 5 | 7 |
| 18 | 31 | 25 | 3 |
| 18 | 31 | 16 | 7 |
| 18 | 31 | 6 | 15 |
| 18 | 63 | 57 | 3 |
| 18 | 63 | 42 | 7 |
| 18 | 63 | 22 | 15 |

| Code Type | n | k | $d_{min}$ | Code Type | n | k | $d_{min}$ |
|---|---|---|---|---|---|---|---|
| 18 | 63 | 7 | 31 | 19 | 585 | 520 | 10 |
| 19 | 127 | 120 | 3 | 19 | 585 | 184 | 74 |
| 18 | 127 | 90 | 7 | | | | |
| 19 | 127 | 64 | 15 | | | | |
| 18 | 127 | 29 | 31 | 19 | 1365 | 1043 | 22 |
| 18 | 127 | 8 | 63 | 19 | 1365 | 483 | 86 |
| 19 | 255 | 247 | 3 | 19 | 1365 | 78 | 342 |
| 19 | 255 | 219 | 7 | 20 | 13 | 5 | 5 |
| 18 | 255 | 163 | 15 | 20 | 28 | 130 | 7 |
| 18 | 255 | 93 | 31 | 20 | 29 | 19 | 5 |
| 18 | 255 | 37 | 63 | 20 | 58 | 28 | 11 |
| 18 | 255 | 9 | 127 | 20 | 59 | 35 | 9 |
| 18 | 511 | 502 | 3 | 20 | 60 | 42 | 7 |
| 18 | 511 | 466 | 7 | 20 | 61 | 49 | 5 |
| 18 | 511 | 382 | 15 | 20 | 121 | 79 | 13 |
| 18 | 511 | 256 | 31 | 20 | 122 | 87 | 11 |
| 18 | 511 | 130 | 63 | 20 | 123 | 95 | 9 |
| 18 | 511 | 46 | 127 | 20 | 124 | 103 | 7 |
| 18 | 511 | 13 | 255 | 20 | 125 | 111 | 5 |
| 19 | 1023 | 1013 | 3 | 20 | 249 | 201 | 13 |
| 18 | 1023 | 968 | 7 | 20 | 250 | 210 | 11 |
| 18 | 1023 | 849 | 15 | 20 | 251 | 219 | 9 |
| 19 | 1023 | 639 | 31 | 20 | 252 | 228 | 7 |
| 18 | 1023 | 386 | 63 | 20 | 253 | 237 | 5 |
| 18 | 1023 | 176 | 127 | 20 | 505 | 451 | 13 |
| 18 | 1023 | 56 | 255 | 20 | 506 | 461 | 11 |
| 19 | 21 | 11 | 6 | 20 | 507 | 471 | 9 |
| 19 | 73 | 45 | 10 | 20 | 508 | 481 | 7 |
| 19 | 85 | 68 | 6 | 20 | 509 | 491 | 5 |
| 19 | 85 | 24 | 22 | 20 | 1020 | 990 | 7 |
| 19 | 273 | 191 | 18 | 20 | 1021 | 1001 | 5 |
| 19 | 341 | 315 | 6 | | | | |
| 19 | 341 | 195 | 22 | | | | |
| 19 | 341 | 45 | 86 | | | | |

# Appendix 5

# CONSULTATIVE COMMITTEE FOR SPACE DATA SYSTEMS RECOMMENDATION FOR TELEMETRY CHANNEL CODING [A.3]

This appendix contains an abridged version of the Recommendation developed jointly by the following organizations: Centre National D'Etudes Spatiales (CNES)/ France, Deutsche Forschungs-und Versuchsanstalt für Luft und Raumfahrt e.V (DFVLR)/West Germany, European Space Agency (ESA)/Europe, Instituto de Pesquisas Espaciais (INPE)/Brazil, National Aeronautics and Space Administration (NASA)/USA, and National Space Development Agency (NASDA)/Japan.

Actually, our interest in this Recommendation goes beyond telemetry applications. The Recommendation can be applied to message transmission of many digital satellite communication systems.

## 1 INTRODUCTION

### 1.1 Purpose

The purpose of this document is to establish a common Recommendation for space telemetry channel coding systems to provide cross-support among missions and facilities of member Agencies of the Consultative Committee for Space Data Systems (CCSDS). In addition, it provides focusing for the development of multi-mission support capabilities within the respective agencies to eliminate the need for arbitrary, unique capabilities for each mission.

Telemetry channel coding is a method by which data can be sent from a source to a destination by processing data so that distinct messages are created which are easily distinguishable from one another. This allows reconstruction of the data with low error probability, thus improving the performance of the channel.

This document was prepared by the CCSDS primarily for the purpose of facilitating the cross-support concept through standardizing key items of data systems compatibility. While the CCSDS has no power of enforcement, it is expected that this recommendation will be incorporated into each respective

Agency's data systems standards, and through them, will apply to all missions that wish to utilize telemetry channel coding for cross-support.

## 1.2  Scope

Several space telemetry channel coding schemes are described in this document. The characteristics of the codes are specified only to the extent necessary to ensure interoperability and cross-support. The specification does not attempt to quantify the relative coding gain or the merits of each approach discussed, nor the design requirements for encoders or decoders.

This recommendation does not require that coding be used on all cross-supported missions. However, for those planning to use coding, the recommended codes to be used are those described in this document.

The rate ½ convolutional code recommended for cross-support is described in Section 2. Depending on performance requirements, this code alone may be satisfactory.

Users of the NASA Tracking and Data Relay Satellite System (TDRSS) may be required to use periodic convolutional interleaving in addition to the convolutional code above. This approach is described in Section 3.

Where a greater coding gain is needed than can be provided by the convolutional code alone, a standard Reed-Solomon outer code may be concatenated for improved performance. The specification of the Reed-Solomon code selected for cross-support is given in Section 4. It should be noted that if a spacecraft, utilizing the services of TDRSS, incorporates Reed-Solomon coding, it is the responsibility of the user project to provide the required Reed-Solomon decoding.

## 1.3  Applicability

This Recommendation applies to telemetry channel coding applications of space missions anticipating cross-support among CCSDS member Agencies at the coding layer. In addition, it serves as a guideline for the development of compatible internal Agency Standards in this field, based on good engineering practice.

## 1.4  Bit Numbering Convention and Nomenclature

The following ''Caution'' should be observed when interpreting the bit numbering convention which is used throughout this CCSDS Recommedation:

## CAUTION

In this document, the following convention is used to identify each bit in an N-bit field.

The first bit in the field to be transmitted (i.e., the most left justified when drawing a figure) is defined to be ''Bit 0''; the following bit is defined to be "Bit 1" and so on up to "Bit N − 1". When the field is used to express a binary value (such as a counter), the Most Significant Bit (MSB) shall be the first transmitted bit of the field, i.e., "Bit 0".

```
Bit 0                                                                    Bit N − 1
|                                                                               |
v                                                                               v
 _____
|                                                                               |
|                           N-BIT DATA FIELD                                   |
|_____|
^
|_____ First bit transmitted = MSB
```

In accordance with modern data communications practice, spacecraft data fields are often grouped into 8-bit "words" as an octet. This term will be used throughout this Recommendation.

## 2   CONVOLUTIONAL CODING

The basic code selected for cross-support is a rate of ½, constraint-length 7 convolutional code. It may be used alone, as described in this section, or in conjunction with enhancements described in the following sections. While slightly different conventions of this code, currently in use by some member Agencies, may continue to be supported for an interim period, it is the recommendation of the CCSDS to universally adopt the single convention described herein.

This recommendation is a non-systematic code and a specific decoding procedure, with the following characteristics:[1,2]

| | |
|---|---|
| a. Nomenclature: | Convolutional code with maximum-likelihood (Viterbi) decoding. |
| b. Code rate: | ½ |
| c. Constraint length: | 7 bits |
| d. Connection vectors: | G1 = 1111001;    G2 = 1011011 |
| e. Phase relationship: | G1 is associated with first symbol |
| f. Symbol inversion: | On output path of G2 |

An encoder block diagram with the recommended convention is shown in Fig. A.5.1.

It is recommended that soft bit decisions with 3-bit quantization be used whenever constraints (such as location of decoder) permit.

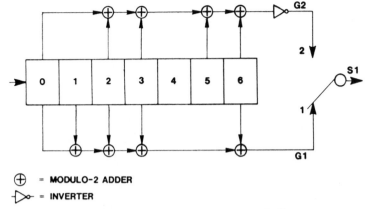

⊕ = MODULO-2 ADDER

▷- = INVERTER

**Figure A.5.1** Convolutional encoder block diagram.

# 3 CONVOLUTIONAL CODING WITH INTERLEAVING FOR TRACKING AND DATA RELAY SATELLITE OPERATIONS

## 3.1 Introduction

Users of the TDRSS S-band Single Access (SSA) Channel, where the channel symbol rate exceeds 300 ks/s, will be required to employ interleaving in conjunction with the convolutional code which has been described in Section 2. Users are cautioned that if such interleaving is not used under these conditions, the Goddard Space Flight Center Networks Directorate does not guarantee the specified performance and will not be obligated to troubleshoot the system in case of problems.

It should be noted that this interleaving is totally separate and distinct from the interleaving used in conjunction with the Reed-Solomon code described in Section 4.

---

Footnote 1: The following upper bounds to the data rates for spacecraft telemetry reception may exist because of symbol synchronizer or decoder limitations:

(a) ESA: Maximum symbol rate is obtained at a data rate of 1 Mb/s NRZ-L, or 500 kb/s split-phase.

(b) NASA-GSFC: Maximum symbol rate is obtained at a data rate of 3 Mb/s NRZ.

(c) NASA-JPL: Current maximum symbol rate is 250 ksymbols/s. A planned upgrade will be decoder-limited at 250 kb/s.

Footnote 2: When suppressed-carrier modulation systems are used, NRZ-M or NRZ-L may be used as an encoding waveform. If the user contemplates conversion of his encoding waveform from NRZ-L to NRZ-M, such conversion should be performed on-board at the input to the convolutional encoder. Correspondingly, the conversion on the ground from NRZ-M to NRZ-L should be performed at the output of the convolutional decoder. This is shown in Figure A.5.2.

CAUTION: When a fixed pattern in the symbol stream is used to provide node synchronizatin for the Viterbi decoder, care must be taken to account for any translation of the pattern due to the encoding waveform conversion.

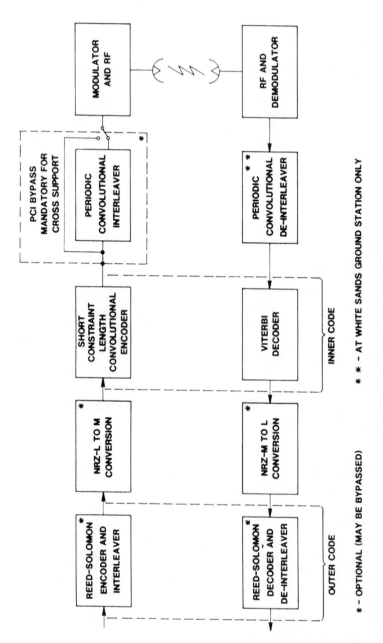

**Figure A.5.2** Coding system block diagram.

## 3.2  Description

The type of interleaving required is called "Periodic Convolutional Interleaving" and is specified in the TDRSS User's Guide.

## 3.3  Bypass Capability

A TDRSS-compatible spacecraft using the Periodic Convolutional Interleaving specified in this section must be capable of bypassing its Periodic Convolutional Interleaver in the event direct support from a non-TDRSS ground tracking station is desired. This is because the interference that this interleaving is designed to protect against is not harmful in this configuration, and moreover, the necessary de-interleavers do not exist at these ground stations.

## 4  REED-SOLOMON CODING

## 4.1  Introduction

While a convolutional code provides good forward error correction capability in a Gaussian noise channel, significant additional improvement (particularly to correct bursts of errors from the Viterbi decoder) can be obtained by concatenating a Reed-Solomon (RS) code with the convolutional code. The Reed-Solomon code forms the *outer* code, while the convolutional code is the *inner* code. The overhead associated with the Reed-Solomon code is comparatively low, and the improvement in the error performance can often provide the nearly-error-free channel required to support efficient automated ground handling of space mission telemetry.

The user is cautioned that the RS outer code described in this section is not intended for use except when concatenated with the inner convolutional code described in Section 2.

The TDRS System does not furnish any Reed-Solomon decoding services.

## 4.2  Specification

The parameters of the selected Reed-Solomon code are as follows:

(a)   $J = 8$ bits per RS symbol

(b)   $E = 16$ RS symbol error correction capability within a Reed-Solomon codeword.

(c)   General characteristics of Reed-Solomon codes

    1.  $J$, $E$, and $I$, the depth of interleaving, are independent parameters.

    2.  $n = 2^J - 1 = 255$ symbols per RS codeword.

    3.  $2E$ is the number of RS symbols among $n$ symbols of an RS codeword representing checks.

4. $k = n - 2E$ is the number of RS symbols among $n$ RS symbols of an RS codeword representing information.

(d)    Field generator polynomial:

$$F(x) = x^8 + x^7 + x^2 + x + 1$$

over GF(2)

(e)    Code generator polynomial:

$$g(x) = \prod_{j=112}^{143} (x - a^{11j}) = \sum_{i=0}^{32} G_i x^i$$

over GF($2^8$),
where $F(a) = 0$.

It should be recognized that $a^{11}$ is a primitive element in GF($2^8$) and that $F(x)$ and $g(x)$ characterize a (255,223) Reed-Solomon code.

(f)    The selected code is a systematic code. This results in a systematic code-block.

(g)    Symbol Interleaving

Symbol interleaving is accomplished in a manner functionally described with the aid of Figure A.5.3. (It should be noted that this functional description does not necessarily correspond to the physical implementation of an encoder.)

Data bits to be encoded into a single Reed-Solomon codeblock enter at the port labeled "IN." Switches S1 and S2 are synchronized together and advance from encoder to encoder in the sequence 1, 2, ... $I$, 1 ... , spending one RS symbol time (8 bits) in each position.

One codeblock will be formed from 223$I$ RS symbols entering "IN." In this functional representation, a space of 32$I$ RS symbols in duration is required between each entering set of 223$I$ RS information symbols.

Due to the action of S1, each encoder accepts 223 of these symbols, each symbol spaced $I$ symbols apart (in the original stream). These 223 symbols

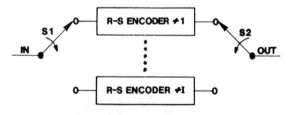

**Figure A.5.3** Functional representation of RS interleaving.

are passed directly to the output of each encoder. The synchronized action of S2 reassembles the symbols at the port labeled "OUT" in the same way as they entered at "IN."

Following this, each encoder outputs its 32 check symbols, one symbol at a time, as it is sampled in sequence by S2.

If, for $I = 5$, the original symbol stream is

$$d_1^1 \ldots d_1^5 \; d_2^1 \ldots d_2^5 \ldots \ldots d_{223}^1 \ldots d_{223}^5 \; [\; 32 \times 5 \text{ spaces}]$$

then the output is the same sequence followed by the $[32 \times 5]$ parity symbols as shown below:

$$p_1^1 \ldots p_1^5 \ldots p_{32}^1 \ldots p_{32}^5$$

where $\qquad d_1^i \; d_2^i \ldots d_{223}^i \; p_1^i \ldots p_{32}^i$

is the RS codeword produced by the $i^{th}$ encoder. If $q$ virtual fill symbols are used in each codeword, then replace 223 by $(223 - q)$ in the above discussion.

With this method of interleaving, the original $kI$ consecutive information symbols that entered the encoder appear unchanged at the output of the encoder with $2EI$ RS check symbols appended.

The recommended value of interleaving depth is $I = 5$, but $I = 1$ is permitted. (See Footnote 3.)

(h)    Maximum Code Block Length

The maximum code block length, in RS symbols, is given by:

$$L_{max} = nI = (2^J - 1)I = 255I$$

(i)    Shortened Code Block Length[4]

A shortened code block length may be used to accommodate frame lengths smaller than the maximum. However, since the Reed-Solomon code is a block code, the decoder must always operate on a full block basis. To achieve a full code block, "virtual fill" must be added to make up the difference between the shortened block and the maximum code block length. The characteristics and limitations on virtual fill are covered in paragraph (j). Since the virtual fill is not transmitted, both encoder and decoder must be set to insert it with the proper length for the encoding and decoding processes to be carried out properly.

---

Footnote 3: Users of TDRSS are cautioned that, under some special RFI circumstances, additional measures may have to be employed to obtain a required performance. One such approach may be to utilize a Reed-Solomon code with a different depth of interleaving. In addition to $I = 1$ and $I = 5$, ESA will support $I = 8$.

Footnote 4: It should be noted that shortening the block length in this way changes the overall performance to a degree dependent on the amount of virtual fill used. Since it incorporates no virtual fill, the Packet Telemetry transfer frame length recommended allows full performance.

When an encoder (initially cleared at the start of a block) receives $kJ - Q$ symbols representing information (where $Q$, representing fill, is a multiple of $I$, and is less than $kI$), $2EI$ check symbols are computed over $kI$ symbols, of which the leading $Q$ symbols are treated as all-zero symbols. A $(nI - Q,$ $kI - Q)$ shortened codeblock results where the leading $Q$ symbols (all zeros) are neither entered into the encoder nor transmitted.

(j)   Partitioning and Virtual Fill

The codeblock is partitioned as shown in Figure A.5.4.

The Reed-Solomon Check Symbols consist of the trailing $2EI$ symbols ($2EIJ$ bits) of the codeblock.

The Transfer Frame is defined by the CCSDS Recommedation for Packet Telemetry. For $I = 5$, it has a length of 8920 bits, which includes the 32-bit RS codeblock marker used to synchronize the Reed-Solomon codeblock.

The Transmitted Codeblock consists of what is actually transmitted on the space telemetry channel (i.e., it consists of all bits transmitting from the beginning of one RS codeblock marker to the beginning of the next RS codeblock marker). For $I = 5$ and with no virtual fill, this is 10,200 bits.

If used, virtual fill shall:

a)  consist of all zeros.

b)  not be transmitted.

c)  not change in length during a tracking pass.

d)  be inserted, prior to decoding, only at the beginning of the codeblock (i.e., before the transmitted frame sync word).

e)  be inserted only in integer multiples of $8I$ bits.

(k)   Symbol representation and ordering for transmission

Each 8-bit Reed-Solomon symbol is an element of the finite field GF(256). Since GF(256) is a vector space of dimension 8 over the binary field GF(2), the actual 8-bit representation of a symbol is a function of the particular basis that is chosen. One basis for GF(256) over GF(2) is the set $\{1, a^1,$

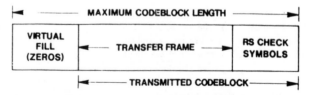

**Figure A.5.4** Codeblock partitioning.

$a^2, \ldots a^7\}$. This means that any element $b$ of GF(256) has a representation of the form

$$b = u_7 a^7 + u_6 a^6 + \ldots + u_1 a^1 + u_0 1$$

where each $u_i$ is either a zero or a one.

There is another basis $\{1_0, 1_1, \ldots 1_7\}$ called the "dual basis" to the above basis. It has the property that

$$Tr(1_i a^j) = \begin{cases} 1 \ if \ i = j \\ 0 \ \text{otherwise} \end{cases}$$

for each $j = 0, 1, \ldots, 7$. The function $Tr(z)$, called the "trace," is defined by

$$Tr(z) = \sum_{k=0}^{7} z^{2^k}$$

for each element $z$ of GF(256). Each Reed-Solomon symbol can also be represented as

$$b = z_0 1_0 + z_1 1_1 + \ldots + z_7 1_7$$

where each $z_i$ is either a zero or a one. The representation recommended in this document is the dual basis eight bit string $z_0, z_1, \ldots, z_7$, transmitted in that order (with $z_0$ first). The relationship between the two representations is given by the two equations

$$[z_0, \ldots, z_7] = [u_7, \ldots, u_0] \begin{bmatrix} 1 & 0 & 0 & 0 & 1 & 1 & 0 & 1 \\ 1 & 1 & 1 & 0 & 1 & 1 & 1 & 1 \\ 1 & 1 & 1 & 0 & 1 & 1 & 0 & 0 \\ 1 & 0 & 0 & 0 & 0 & 1 & 1 & 0 \\ 1 & 1 & 1 & 1 & 1 & 0 & 1 & 0 \\ 1 & 0 & 0 & 1 & 1 & 0 & 0 & 1 \\ 1 & 0 & 1 & 0 & 1 & 1 & 1 & 1 \\ 0 & 1 & 1 & 1 & 1 & 0 & 1 & 1 \end{bmatrix}$$

and

$$[u_7, \ldots, u_0] = [z_0, \ldots, z_7] \begin{bmatrix} 1 & 1 & 0 & 0 & 0 & 1 & 0 & 1 \\ 0 & 1 & 0 & 0 & 0 & 0 & 1 & 0 \\ 0 & 0 & 1 & 0 & 1 & 1 & 1 & 0 \\ 1 & 1 & 1 & 1 & 1 & 1 & 0 & 1 \\ 1 & 1 & 1 & 1 & 0 & 0 & 0 & 0 \\ 0 & 1 & 1 & 1 & 1 & 0 & 0 & 1 \\ 1 & 0 & 1 & 0 & 1 & 1 & 0 & 0 \\ 1 & 1 & 0 & 0 & 1 & 1 & 0 & 0 \end{bmatrix}$$

(l)   Synchronization

Codeblock synchronization of the Reed-Solomon decoder is achieved by synchronization of the 32-bit RS codeblock marker. This marker may be part of the Packet Telemetry Transfer Frame or it may be part of the RS encoding layer.

(m)   Ambiguity Resolution

The ambiguity between true and complemented data must be resolved so that only true data is provided to the Reed-Solomon decoder. Data in NRZ-L form is normally resolved using the 32-bit RS codeblock marker, while NRZ-M data is self-resolving.

# Appendix 6

# NETWORK OPERATION PARAMETERS FOR EXAMPLES 7.10, 7.11 [A.4]

**Table A.6.1**
Station locations and corresponding delays for Example 7.11

| Terminal | Latitude , Longitude | Geocentric Coordinates (meters) | Terminal Delay (symbols) |
|----------|----------------------|--------------------------------|--------------------------|
| 1 (USA) | 39° 16′ 50″ N<br>280° 15′ 47″ E | X =   879,620<br>Y = 4,858,041<br>Z = 4,038,117 | 620 |
| 2 (CAN.) | 44° 11′ 41″ N<br>295° 20′ 00″ E | X = 1,956,698<br>Y = 4,133,202<br>Z = 4,446,196 | 346 |
| 3 (COL.) | 5° 09′ 38″ N<br>286° 19′ 20″ E | X = 1,785,240<br>Y = 6,096,268<br>Z =   573,694 | 830 |
| 4 (BRAZ.) | 22° 44′ 43″ S<br>317° 12′ 55″ E | X = 4,316,961<br>Y = 3,995,412<br>Z = 2,466,012 | 600 |
| 5 (FR.) | 48° 13′ 17″ N<br>3° 53′ 04″ E | X = 4,239,701<br>Y =   287,877<br>Z = 4,756,338 | 590 |
| 6 (SWED.) | 58° 42′ 17″ N<br>11° 22′ 46″ E | X = 3,247,988<br>Y =   653,696<br>Z = 5,450,132 | 776 |
| 7 (GERM.) | 47° 54′ 02″ N<br>11° 06′ 56″ E | X = 4,195,813<br>Y =   824,368<br>Z = 4,732,468 | 800 |
| 8 (GREEK) | 38° 49′ 25″ N<br>22° 41′ 13″ E | X = 4,584,603<br>Y = 1,916,553<br>Z = 3,998,616 | 490 |
| 9 (NIGERIA) | 7° 36′ 12″ N<br>3° 27′ 17″ E | X = 6,310,575<br>Y =   380,966<br>Z =   843,917 | 660 |
| 10 (S.AFRICA) | 25° 54′ 31″ S<br>27° 41′ 57″ E | X = 5,079,620<br>Y = 2,666,766<br>Z = 2,786,846 | 720 |

**Table A.6.2**

Burst number for reference stations.

| Terminal Number | Transmitted RB's | | | Controlling RB's | | Other RB's To Be Received | | Received Burst Number | | | | | |
|---|---|---|---|---|---|---|---|---|---|---|---|---|---|
| 1 | *8 (21) | 15 (51) | | 1 (11) | 4 (11) | 11 (41) | 12 (41) | *2 (11) | *3 (11) | 5 (11) | 6 (11) | 13 (41) | 14 (41) |
| 2 | *9 | 17 | | | | | | | | | | | |
| 5 | 1 (11) | 11 (41) | *18 (52) | *18 | *20 | 15 | 17 | *7 (21) | *8 (21) | *9 (21) | *10 (21) | 16 (51) | *19 (52) |
| 6 | 4 (11) | 12 (41) | *20 (52) | (52) | (52) | (51) | (51) | *21 (52) | | | | | |

* Principal Burst

(No.) Transponder Number

## Table A.6.3
Burst number for traffic stations.

| Terminal Number | Transmitted RB's | | | Controlling RB's | | Other RB's To Be Received | | Received Burst Number | | | |
|---|---|---|---|---|---|---|---|---|---|---|---|
| 3 | *7 (21) | 16 (51) | | 11 (41) | 12 (41) | | | *2 (11) | *3 (11) | 14 (11) | |
| 4 | *10 (21) | | | 1 (11) | 4 (11) | | | *2 (11) | *3 (11) | 5 (11) | 6 (11) |
| 7 | 6 (11) | 13 (41) | *21 (52) | *18 (52) | *20 (52) | 15 (51) | 17 (51) | 16 (51) | *19 (52) | *21 (52) | |
| 8 | 5 (11) | 14 (41) | *19 (52) | *18 (52) | *20 (52) | 15 (51) | 17 (51) | 16 (51) | 21 (52) | | |
| 9 | *2 (11) | | | *8 (21) | *9 (21) | | | *7 (21) | *10 (21) | | |
| 10 | *3 (11) | | | *8 | *9 (21) | | (21) | *7 | *10 (21) | (21) | |

* Principal Burst

(No.) Transponder Number

**Table A.6.4**

BTP3 contents.

| Burst No. | Terminal No. | Destination Terminal No. | Transponder No. | Burst Beginning | Burst Position | Burst End |
|---|---|---|---|---|---|---|
| 1 | 5 | | 11 | 10041 | 10240 | 10328 |
| 2 | 9 | 3,4 | 11 | 15001 | 15200 | 33421 |
| 3 | 10 | 3,4 | 11 | 35801 | 36000 | 60052 |
| 4 | 6 | | 11 | 69801 | 70000 | 70088 |
| 5 | 8 | 4 | 11 | 75401 | 75600 | 83860 |
| 6 | 7 | 4 | 11 | 89945 | 90144 | 101500 |
| 7 | 3 | 9,10 | 21 | 7001 | 7200 | 23060 |
| 8 | 1 | | 21 | 29801 | 30000 | 30088 |
| 9 | 2 | | 21 | 49801 | 50000 | 50088 |
| 10 | 4 | 9,10 | 21 | 61817 | 62016 | 85000 |
| 11 | 5 | | 41 | 19801 | 20000 | 20088 |
| 12 | 6 | | 41 | 64793 | 64992 | 65080 |
| 13 | 7 | 3 | 41 | 73401 | 73600 | 87640 |
| 14 | 8 | 3 | 41 | 89801 | 90000 | 90088 |
| 15 | 1 | | 51 | 41801 | 42000 | 42088 |
| 16 | 3 | 7,8 | 51 | 45289 | 45488 | 71080 |
| 17 | 2 | | 51 | 79801 | 80000 | 80088 |
| 18 | 5 | | 52 | −199 | 0 | 88 |
| 19 | 8 | 7 | 52 | 4601 | 4800 | 19608 |
| 20 | 6 | | 52 | 35801 | 36000 | 36088 |
| 21 | 7 | 8 | 52 | 60793 | 60992 | 73868 |

## Table A.6.5
BTP4 contents.

| Burst No. | Terminal No. | Destination Terminal No. | Transponder No. | Burst Beginning | Burst Position | Burst End |
|-----------|--------------|--------------------------|-----------------|-----------------|----------------|-----------|
| 1 | 5 | | 11 | 19801 | 20000 | 20088 |
| 2 | 9 | 3,4 | 11 | 29801 | 30000 | 40000 |
| 3 | 10 | 3,4 | 11 | 42809 | 43008 | 52500 |
| 4 | 6 | | 11 | 79801 | 80000 | 80088 |
| 5 | 8 | 4 | 11 | 54809 | 55008 | 64060 |
| 6 | 7 | 4 | 11 | 66809 | 67008 | 76560 |
| 7 | 3 | 9,10 | 21 | 12361 | 12560 | 25010 |
| 8 | 1 | | 21 | 4841 | 5040 | 5128 |
| 9 | 2 | | 21 | 48297 | 48496 | 48584 |
| 10 | 4 | 9,10 | 21 | 29801 | 30000 | 42160 |
| 11 | 5 | | 41 | 4841 | 5040 | 5128 |
| 12 | 6 | | 41 | 16281 | 16480 | 16568 |
| 13 | 7 | 3 | 41 | 55801 | 56000 | 64120 |
| 14 | 8 | 3 | 41 | 68793 | 68992 | 78000 |
| 15 | 1 | | 51 | 8409 | 8608 | 8696 |
| 16 | 3 | 7,8 | 51 | 30809 | 31008 | 48500 |
| 17 | 2 | | 51 | 50809 | 51008 | 51096 |
| 18 | 5 | | 52 | −199 | 0 | 88 |
| 19 | 8 | 7 | 52 | 14809 | 15008 | 29060 |
| 20 | 6 | | 52 | 9801 | 10000 | 10088 |
| 21 | 7 | 8 | 52 | 82809 | 83008 | 95320 |

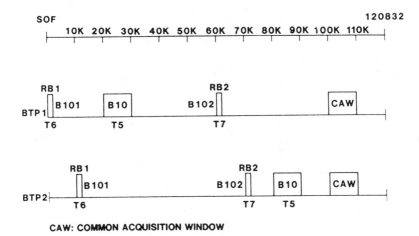

**Figure A.6.1** Burst time planes for Example 7.10 (Section 7.9.2—Network operations).

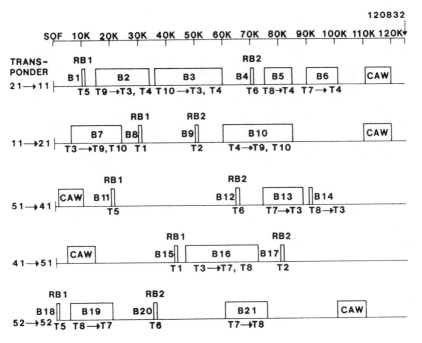

**Figure A.6.2** Burst timing format (BTP3) for Example 7.11.

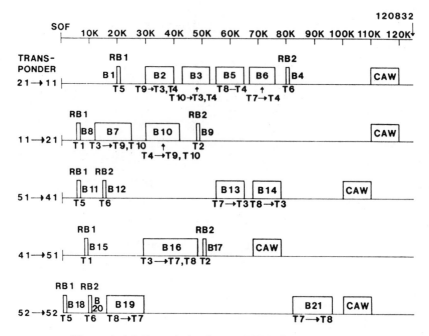

**Figure A.6.3** Burst timing format (BTP4) for Example 7.11.

## APPENDIX REFERENCES

**A.1.** Alanen, J. and Knuth, D. "Table of Finite Fields." *Sankhyā*, Series A, 26, 4 (December 1964): 305–328.

**A.2.** Baumert, L.D. *Cyclic Difference Sets*, Lecture Notes in Mathematics No. 182, California Institute of Technology, Pasadena, California, 1971. New York: Springer-Verlag.

**A.3.** Recommendation for Space Data System Standards: Telemetry Channel Coding, Draft "Blue Book," Issue O, by the Consultative Committee for Space Data Systems, February 1984.

**A.4.** *TNS Program—Example Executions*, Tokyo, Japan: Kokusai Denshin Denwa Co., Ltd., May 1984.

# Glossary of Acronyms

ACI—acquisition cycle intervals

ACK—positive acknowledgment

$\{A_i\}$—parity check

$A_k$—permutation matrix

(ALIB)—additive-like instantaneous block

APD—acquisition phase UW detection

APP—a posteriori probability (decoding)

ARPA—Advanced Research Projects Agency

ARQ—automatic repeat request (scheme)

ASM—acquisition and synchronization module

ASSR—acquisition and steady state reception

AWGN—additive white Gaussian noise

BA—burst acquired

BCH—Bose-Chaudhuri-Hocquenghem

BEE—bit-error evaluation

BIB—balanced incomplete block

$B_j$—RMA sequence generation matrix

BL—back-search limit

BLAP—balanced link access protocol

BNA—burst not acquired

BQ—burst qualified

BSC—binary symmetric channel

BSS—binary symmetric source

BTP—burst time plan

BTS—burst time synchronization

CCIF—Technical Consultative Committee for Telegraphy

CCIR—International Radio Consultative Committee

CCIT—Technical Consultative Committee for Telephone

CCITT—International Telegraph and Telephone Consultative Committee

CCSDS—Consultative Committee for Space Data Systems

CDC—control delay channel

CF—control frame

C/I—carrier-to-interference ratio

$C_{lim}$—computational limit

CMF—control multiframe

CML-IC—current mode logic-integrated circuit

$C/N_{AVA}$—available signal-to-noise ratio

$C/N_{REQ}$—required signal-to-noise ratio

CNES—Centre National D'Etudes Spatiales (France)

COMSAT—Communications Satellite Corporation

CRC—cyclic redundancy checks

CSOC—convolutional self-orthogonal codes

$C_T$—capacity of AWGN channel

CTP—condensed time plans

$d$—minimum distance

$D$—weight or distance in a branch of the state diagram

$\{D\}$—difference set

$D_A(M)$—acquisition delay value for Station M

DCE—data circuit terminating equipment

$\{D_E\}$—Euclidean difference set

DFVLR—Deutsche Forschungs-und Versuchsanstalt für Luft und Raumfahrte.V

$D(k)$—degree of $\sigma(k)$ of error-locating polynomial $\sigma(x)$

DLE—data link escape

$d_m$—minimum distance

$d_{min}$—minimum distance of RMA sequence

$D_n$—time delay for Station $n$

DNTX—Do Not Transmit
$\{D_p\}$—projective geometry set
DSA—difference set array
DTE—data station (terminal) equipment
DTN—Digital Telephone Network
(Germany)

$E$—encryption algorithm
$\bar{E}$—error sequence
$(E_b/N_0)$—information bit-energy-to-
noise-density ratio
EBD—error burst probability
distribution function
$E(D)$—error sequence polynomial
EFS—error free second
$EG(t,n)$—Euclidean geometry
$E_j(D)$—error sequence (sub-generator $j$)
$\text{erf}(\alpha)$—error function
$\overline{\text{erf}}(\alpha) = 1 - \text{erf}(\alpha)$—complementary
error function
$E(\rho,Q)$—channel computational cutoff
rate
$E_s$—symbol signaling energy
ESA—European Space Agency
ETX—end (of test)
EUTELSAT—European
Telecommunication Satellite

$f$—number of frequency divisions in an
RMA sequence
$F_A$—auxiliary framing
$F(A_k)[B_j]$—RMA mapping
FEC—forward error correction
FEP—finite Euclidean plane
FFT—fast Fourier transform
$f(m_i,c_{k,i})$—combiner function
$f(m,n,\epsilon)$—false detection probability
$f[R(i),K(i+1)]$—transformation
function
FSR—feedback shift register

$G$—general linear group of linear
nondegenerative transformations
$G$—code generator matrix
gcd—greatest common divisor
$g(\text{CRC-CCITT})$—generator polynomial
$g(\text{CRC-12})$—code generator polynomial
with 6-bit characters and 2 characters
per block

$g(\text{CRC-16})$—code generator polynomial
with 8-bit characters and 2 characters
per codeword
$G(D)$—code generator polynomial
GMA—gated mode acquisition
$g(\text{MARISAT})$—code generator
polynomial of MARISAT
$g(\text{SCPC})$—generator polynomial of the
SCPC ARQ code
GSFC—Goddard Space Flight Center
$g(x)$—code generator polynomial

$H$—code parity check matrix
HBER—high bit error rate
$H_D$—hard decision decoding
HDLC—high-level data link control
HRX—hypothetical reference
connections
$h(x)$—code parity check polynomial

$\bar{I}$—encoded information sequence
$I(A)$—mutual information
IAP1—Initial Acquisition Phase 1
IAP2—Initial Acquisition Phase 2
IDSN—Integrated Digital Satellite
Network
IF—intermediate frequency
INPE—Instituto de Pesquisas Espaciais
(Brazil)
INTELSAT—International
Telecommunications Satellite
Organization
IOCTF—INTELSAT Operation Center
TDMA Facility
$\bar{I}(Q;P)$—the average mutual information
IR—image rejection
ISDN—Integrated Service Digital
Network
ISDSN—Integrated Services Digital
Satellite Network
ISO—International Organization for
Standardization
ITU—International Telecommunication
Union
IUE—International Ultraviolet Explorer

$J(n,m)$—Jacobi symbol

$k$—information digits for block code
$k_0$—digits for convolutional code

$K(i+1)$—set of iteration-dependent keys for encryption

$K_D$—decryption key

KDD—Kokusai Denshin Denwa

$K_E$—cryptographic key

$L$—length of a given path

LAP—link access procedure

LCM—least common multiple

LFSR—linear feedback shift register

$L(i)$—$i$th iteration of left half of message

$M$—time/frequency matrix

MF—multiframe

$M_i(x)$—minimum polynomials

MMF—measurement multiframe

MPR or $M$—master primary reference station

MRD—maximum radius decoding

$M_s$—shift register stages

MSA—multiple stack algorithm

MTP—master time plan (p. 452)

MTP—message transfer part (p. 403)

$M_{xj}$—metric associated with state $x_j$

$n_A$—constraint length

NASDA—National Space Development Agency (Japan)

NBCEC—nonbinary character error-correcting

NCK—negative acknowledgment

$n_e$—effective constraint length

NEC—Nippon Electric Corporation

NRZ—no return to zero

OSI—open system interconnection

$P(A)$—probability of system becoming insecure

PABX—private automatic branch exchange

PC-1—Permuted Choice 1

PC-2—Permuted Choice 2

$P_E$—first-event error probability

P(EFS)—percentage of EFS

$P_e(n_0)$—probability that an arbitrary sub-block of $n_0$ bits is decoded in error

$P_{est}$—erasure-estimate error probability

$P_{fd}$—false detected error probability

PG$(t,n)$—projective geometry

$P_i$—probability of an overlap at the $i$th digit

$P(i,j)$—geometric point

$P\ell(I/Q)$—symbol transition probability

$P_{i,j,\ldots,z}$—probability of $z$ overlaps

$P_k$—the pairwise error probability for an incorrect path that differs in $k$ symbols from the correct path

$P_m(\lambda_Q)$—miss-detection error probability due to $\lambda_Q$

$P_m(\lambda_0,\lambda_1,\ldots,\lambda_{m-1})$—multinomial distribution

PR—primary reference station

PRB—primary reference burst

$P_{ud}$—probability that the received word differs from the transmitted word

$P'_w$—coded word error rate

$p_w(i)$—word error rate of exactly $i$ errors in a codeword that cannot be corrected

$P_w$—word error rate without coding

$P(z)$—probability that there are exactly $z$ overlaps in $n$ symbols

$Q$—quadrature (p. 23)

$Q$—quantization levels (pp. 237, 257)

$Q(k)$—channel input probability

QR—quadratic residue code

$q(x)$—quotient polynomial

$\overline{R}$—total average information rate of RMA

$R$—right half of message

RAM—random access memory

$R_b$—bit rate of transmission

RBID—reference burst identification

RCTP—retransmitted CTP from the reference station

$R(d^*)$—rate of the source in nats per symbol relative to the fidelity criterion $d^*$

R(D)—received message sequence

$R_e$—bit rate after coding

RF—radio frequency

RFA—RF amplifiers

RFS—receive frame synchronization

$R(i)$—$i$th iteration of right-hand message

$R_{ij}$—branch metric

$r^{(i)}(x)$—remainder polynomials

$R_j$—row matrix

RMA—random multiple access

$RN(i)$—random numbers generated by $S(i)$

$RN(j)$—random numbers generated by $S(j)$

ROM—read-only-memory

$R_p$—number of codewords

$R'_p$—repeated codewords in ARQ

$R_s$—symbol signaling rate

RS—Reed-Solomon

RS or R—reference station

$r(x)$—remainder polynomial

$R_\sigma$—reliability indicator

$\bar{S}$—Syndrome sequence

SABM—set asynchronous balanced mode

SC—service channel

SCPC—single channel per carrier

$S(D)$—syndrome sequence in D domain

$S_D$—$(E_b/N_o)$ with soft decision

s.d.s—simple difference set

SGT—selective Do Not Transmit code generation for terminal control

$S(i)$— station $i$

SI—significant instant

SID—ship identification

$S(j)$—station $j$

$S_j$—power sums

SMA—search mode acquisition

SNA—systems network architecture

SNR—required signal-to-noise ratios

$snr\ (c)$—coded case of required SNR

$snr\ (i)$—uncoded case of required signal-to-noise ratios

$snr\ (0)$—required $E_b/N_o$ to achieve $P_o$ if coding is not used

SOF—start of frame

SORF—start of receive frame

SORMF—start of receive multiframe

SOTF—start of transmit frame

SOTMF—start of transmit multiframe

SPD—synchronization phase detection

SR or S—secondary reference station

SRB—secondary reference burst

s/s—symbols per second

SS—signaling system or start of superframe

SSA—S-band single access or single stack algorithm

SSR—steady state reception

STDN—Spaceflight Tracking and Data Network

STX—start of test or start transmit

SUA—SPD procedure with a different window size for unique word acqusition

TAS—terminal acquisition support

TASS—terminal acquisition and synchronization support

TBM—traffic burst monitoring

$T(D,L,N)$—end-to-end transfer function

TDRSS—Tracking Delay Relay Satellite System

TE—Tong-Ebert code

TF—time frequency

TFAS—transmit frame acquisition and synchronization

TFG—transmit frame generation

$\begin{bmatrix} t \\ k \end{bmatrix}$—Gaussian coefficients

TNS—TDMA network simulation

TS (or T)—traffic station

TSS—terminal synchronization support

TTL—transistor-transistor logic

$U(j,k)$—ordered $B_j$ matrices

UW—unique word

VA—Viterbi algorithm

VLSI—very large-scale integrated

$[W]_0$—$(n^2 - 1) \times n$ matrix

$[W]_1$—$(n + 1) \times n$ matrix

$W_B$—system bandwidth

WED—weighted erasure decoding

$W^i(D)$—composite parity check

$W(j)$—number of codewords of weight $j$

$W(Z)$—weight enumerator

$W_\alpha(\vec{Z})$—analog weight of a given error pattern $\vec{Z}$

$X_{ij}$—transmitted code symbols
$X_{ij}^{(m)}$—code symbol of the $m$th path

$Y_{ij}$—corresponding received (demodulated) symbol

$\vec{Z}$—error sequence of minimum Hamming weight

$\vec{Z}_m$—number of possible error sequences

$\bigoplus\limits_{i}^{j}$—modulo $-2$ sum of the $m_i$'s in the term of $\alpha^j$

$\alpha(E)$—Pareto exponent
$\{\alpha_i\}$—measure of the reliability of the received digits
$\beta_j^{(\tau)}$—transformed $\beta_j$
$\Delta$—running threshold spacing
$\Delta_G$—coding gain
$\eta_A$—throughput efficiency
$\eta_{AF}$—hybrid throughput efficiency
$\eta_c$—coding efficiency
$\eta_r$—repetition efficiency
$\eta_w$—delay factor
$\mu_t$—channel utilization factor
$\rho$—cross-correlation coefficient
$\sigma_1, \sigma_2, \ldots \sigma_E$—elementary symmetric functions
$\sigma(x)$—error-locator polynomial
$\tau(k)$—step in BCH decoding
$\tau_w$—round-trip time delay through satellites
$\tau_w'$—delay in ARQ
$\varphi(m)$—Euler function

# Author Index

# Subject Index

# VOLUME I
## Author Index

# VOLUME I
# Subject Index